METHODS IN MOLECULAR BIOLOGY

Series Editor
John M. Walker
School of Life and Medical Sciences
University of Hertfordshire
Hatfield, Hertfordshire, AL10 9AB, UK

For further volumes:
http://www.springer.com/series/7651

Computational Methods for GPCR Drug Discovery

Edited by

Alexander Heifetz

Evotec (UK) Ltd., Abingdon, Oxfordshire, UK

Humana Press

Editor
Alexander Heifetz
Evotec (UK) Ltd.
Abingdon, Oxfordshire, UK

ISSN 1064-3745 ISSN 1940-6029 (electronic)
Methods in Molecular Biology
ISBN 978-1-4939-7464-1 ISBN 978-1-4939-7465-8 (eBook)
https://doi.org/10.1007/978-1-4939-7465-8

Library of Congress Control Number: 2017957977

Printed on acid-free paper

This Humana Press imprint is published by Springer Nature
The registered company is Springer Science+Business Media, LLC
The registered company address is: 233 Spring Street, New York, NY 10013, U.S.A.

Preface

G protein-coupled receptors (GPCRs) have enormous physiological and biomedical importance, being the primary site of action of approximately 40% of prescribed drugs. Although the human genome encodes more than 850 different GPCR proteins, to date drugs have only been developed against 50 of these. Thus, there is a unique opportunity to design new therapies for a huge number of unexploited but potentially tractable targets. Recent advances in GPCR pharmacology and structural biology together with developments in computational modeling have resulted in a resurgence in the number of GPCR drug discovery campaigns.

This book provides a unique overview of modern computational strategies and techniques employed in the field of GPCR drug discovery, including structure- and ligand-based approaches and cheminformatics. It is demonstrated how these computational approaches can be used to address key issues in drug discovery such as receptor structure modeling, protein-ligand interactions, GPCR function, flexibility and dynamics, ligand binding kinetics, positions of water molecules and their role in ligand binding, calculation of the free energy of binding (affinity), prediction of the effects of mutations on ligand binding, interconversion between agonists and antagonists, deorphanization of GPCRs, and discovery of biased and allosteric modulators. A review of these techniques will allow a diverse audience, including structural and molecular biologists, computational and medicinal chemists, pharmacologists and drug designers, to navigate through and effectively deploy these advances.

Abingdon, Oxfordshire, UK ***Alexander Heifetz***

Contents

Contributors

MATTEO ALDEGHI • *Evotec (UK) Ltd., Abingdon, Oxfordshire, UK*
PATRICIA ANGELLOZ-NICOUD • *Research Informatics, Evotec (France) SAS, Toulouse, France*
JOHAN ÅQVIST • *Department of Cell and Molecular Biology, Uppsala University, Biomedical Center, Uppsala, Sweden*
DAMIAN BARTUZI • *Department of Synthesis and Chemical Technology of Pharmaceutical Substances with Computer Modelling Lab, Medical University of Lublin, Lublin, Poland*
PHILIP C. BIGGIN • *Department of Biochemistry, Structural Bioinformatics and Computational Biochemistry, University of Oxford, Oxford, UK*
MIKE J. BODKIN • *Evotec (UK) Ltd., Abingdon, Oxfordshire, UK*
ANDREA BORTOLATO • *Heptares Therapeutics Ltd., Welwyn Garden City, Hertfordshire, UK*
EWA I. CHUDYK • *Evotec (UK) Ltd., Abingdon, Oxfordshire, UK*
ANTONELLA CIANCETTA • *Molecular Recognition Section, Laboratory of Bioorganic Chemistry, National Institute of Diabetes and Digestive and Kidney Diseases, National Institutes of Health, Bethesda, MD, USA*
JASON B. CROSS • *University of Texas MD Anderson Cancer Center, Houston, TX, USA*
CONSTANTINO DIAZ • *Research Informatics, Evotec (France) SAS, Toulouse, France*
IWAN J.P. DE ESCH • *Department of Medicinal Chemistry, Amsterdam Institute for Molecules Medicines and Systems, Vrije Universiteit Amsterdam, Amsterdam, The Netherlands*
MAURICIO ESGUERRA • *Department of Cell and Molecular Biology, Uppsala University, Biomedical Center, Uppsala, Sweden*
DMITRI G. FEDOROV • *CD-FMat, National Institute of Advanced Industrial Science and Technology (AIST), Ibaraki, Japan*
SŁAWOMIR FILIPEK • *Biological and Chemical Research Centre, Faculty of Chemistry, University of Warsaw, Warsaw, Poland*
MARTA FILIZOLA • *Department of Pharmacological Sciences, Icahn School of Medicine at Mount Sinai, New York, NY, USA*
DAVID GLORIAM • *Department of Drug Design and Pharmacology, University of Copenhagen, Copenhagen, Denmark*
CHRIS DE GRAAF • *Department of Medicinal Chemistry, Amsterdam Institute for Molecules Medicines and Systems, Vrije Universiteit Amsterdam, Amsterdam, The Netherlands*
DONG GUO • *Jiangsu Key Laboratory of New Drug Research and Clinical Pharmacy, Xuzhou Medical University, Xuzhou, Jiangsu, China*
HUGO GUTIÉRREZ-DE-TERÁN • *Department of Cell and Molecular Biology, Uppsala University, Biomedical Center, Uppsala, Sweden*
PAUL C.D. HAWKINS • *OpenEye Scientific Software, Santa Fe, NM, USA*
GEORGE HEDGER • *Department of Biochemistry, Structural Bioinformatics and Computational Biochemistry, University of Oxford, Oxford, UK*
ALEXANDER HEIFETZ • *Evotec (UK) Ltd., Oxfordshire, UK; Division of Biosciences, Research Department of Structural and Molecular Biology, Institute of Structural and Molecular Biology, University College London, London, UK*
ADRIAAN P. IJZERMAN • *Division of Medicinal Chemistry, Leiden Academic Centre for Drug Research (LACDR), Leiden University, Leiden, The Netherlands*

Kenneth A. Jacobson • *Molecular Recognition Section, Laboratory of Bioorganic Chemistry, National Institute of Diabetes and Digestive and Kidney Diseases, National Institutes of Health, Bethesda, MD, USA*

Jakub Jakowiecki • *Biological and Chemical Research Centre, Faculty of Chemistry, University of Warsaw, Warsaw, Poland*

Tim James • *Evotec (UK) Ltd., Abingdon, Oxfordshire, UK*

Willem Jespers • *Department of Cell and Molecular Biology, Uppsala University, Biomedical Center, Uppsala, Sweden*

Agnieszka A. Kaczor • *Department of Synthesis and Chemical Technology of Pharmaceutical Substances with Computer Modelling Lab, Faculty of Pharmacy with Division of Medical Analytics, Medical University of Lublin, Lublin, Poland; Department of Pharmaceutical Chemistry, School of Pharmacy, University of Eastern Finland, Kuopio, Finland*

Albert J. Kooistra • *Department of Medicinal Chemistry, Amsterdam Institute for Molecules Medicines and Systems, Vrije Universiteit Amsterdam, Amsterdam, The Netherlands; Centre for Molecular and Biomolecular Informatics (CMBI), Radboud University Medical Center, Nijmegen, The Netherlands*

Paul Labute • *Chemical Computing Group Inc., Montreal, QC, Canada*

Dorota Latek • *Biological and Chemical Research Centre, Faculty of Chemistry, University of Warsaw, Warsaw, Poland*

Kristen A. Marino • *Department of Pharmacological Sciences, Icahn School of Medicine at Mount Sinai, New York, NY, USA*

Jonathan S. Mason • *Heptares Therapeutics Ltd., Welwyn Garden City, Hertfordshire, UK*

Dariusz Matosiuk • *Department of Synthesis and Chemical Technology of Pharmaceutical Substances with Computer Modelling Lab, Medical University of Lublin, Lublin, Poland*

Przemysław Miszta • *Biological and Chemical Research Centre, Faculty of Chemistry, University of Warsaw, Warsaw, Poland*

Inaki Morao • *Evotec (UK) Ltd., Abingdon, Oxfordshire, UK*

Ana Oliveira • *Department of Cell and Molecular Biology, Uppsala University, Biomedical Center, Uppsala, Sweden*

Daniel F. Ortwine • *Discovery Chemistry, Genentech, Inc., South San Francisco, CA, USA*

Emilie Pihan • *Research Informatics, Evotec (France) SAS, Toulouse, France*

Andrew Potterton • *Structural and Molecular Biology, University College London, London, UK; Evotec (UK) Ltd., Oxfordshire, UK*

Roger Robinson • *Evotec (UK) Ltd., Abingdon, Oxfordshire, UK*

Ismael Rodríguez-Espigares • *Department of Experimental and Health Sciences, Research Programme on Biomedical Informatics (GRIB), Hospital del Mar Medical Research Institute (IMIM), Pompeu Fabra University (UPF), Barcelona, Spain*

Ewelina Rutkowska • *Biological and Chemical Research Centre, Faculty of Chemistry, University of Warsaw, Warsaw, Poland*

Jessica Sallander • *Department of Cell and Molecular Biology, Uppsala University, Biomedical Center, Uppsala, Sweden*

Laurie Sarrat • *Evotec (UK) Ltd., Abingdon, Oxfordshire, UK*

Jana Selent • *Department of Experimental and Health Sciences, Research Programme on Biomedical Informatics (GRIB), Hospital del Mar Medical Research Institute (IMIM), Pompeu Fabra University (UPF), Barcelona, Spain*

Robert T. Smith • *Heptares Therapeutics Ltd., Welwyn Garden City, Hertfordshire, UK*

Michelle Southey • *Evotec (UK) Ltd., Abingdon, Oxfordshire, UK*

GUNTHER STAHL • *OpenEye Scientific Software, Köln, Germany*
TOMASZ MACIEJ STEPNIEWSKI • *Department of Experimental and Health Sciences, Research Programme on Biomedical Informatics (GRIB), Hospital del Mar Medical Research Institute (IMIM), Pompeu Fabra University (UPF), Barcelona, Spain*
CHRISTOFER S. TAUTERMANN • *Department for Medicinal Chemistry, Boehringer Ingelheim Pharma, GmbH & Co KG, Biberach an der Riss, Germany*
BENJAMIN G. TEHAN • *Heptares Therapeutics Ltd., Welwyn Garden City, Hertfordshire, UK*
SID TOPIOL • *3D-2drug, LLC, Fair Lawn, NJ, USA; Center for Healthcare Innovation, Stevens Institute of Technology, Hoboken, NJ, USA*
ANDREA TOWNSEND-NICHOLSON • *Division of Biosciences, Research Department of Structural and Molecular Biology, University College London, London, UK*
MARIA TURANT • *Biological and Chemical Research Centre, Faculty of Chemistry, University of Warsaw, Warsaw, Poland*
SILVANA VASILE • *Department of Cell and Molecular Biology, Uppsala University, Biomedical Center, Uppsala, Sweden*
MÁRTON VASS • *Department of Medicinal Chemistry, Amsterdam Institute for Molecules Medicines and Systems, Vrije Universiteit Amsterdam, Amsterdam, The Netherlands*
NAUSHAD VELGY • *Department of Biochemistry, Structural Bioinformatics and Computational Biochemistry, University of Oxford, Oxford, UK*
STEFAN VERHOEVEN • *Netherlands eScience Center, Amsterdam, The Netherlands*

Chapter 1

Current and Future Challenges in GPCR Drug Discovery

Sid Topiol

Abstract

GPCRs play a pervasive physiological role and, in turn, are the leading target class for pharmaceuticals. Beginning with the determination of the structure of rhodopsin, and dramatically accelerating since the reporting of the first ligand-mediated GPCR X-ray structures, our understanding of the structural and functional characteristics of these proteins has grown dramatically. Deploying this now rapidly emerging information for drug discovery has already been extensively demonstrated through a watershed of studies appearing in numerous scientific reports. Included in these expositions are areas such as sites and characteristics of ligand to GPCR binding, protein activation, effector bias, allosteric mechanisms, dimerization, polypharmacology and others. Computational chemistry studies are demonstrating an increasing role in capitalizing on the structural studies to further advance our understanding of these proteins as well as to drive drug discovery. Such drug discovery activities range from the design of orthosteric site inhibitors through, for example, allosteric modulators, biased ligands, partial agonists and bitopic ligands. Herein, these topics are outlined through specific examples in the hopes of providing a glimpse of the state of the field.

Key words GPCR, Structure-based drug discovery, X-ray structure, Allosteric modulators, Receptor bias

1 Introduction

As early as 50 years ago, when the first computer software programs were being written, computational chemistry tools were being developed to understand and guide drug properties and discovery. Ligands for G-protein coupled receptors [GPCRs], such as endogenous amines, and psychotropic drugs such as LSD acting on these receptors, were a common focus of these research efforts. Electrostatic point charge representations of ligands were often used, followed by molecular mechanics, low level quantum chemical (e.g., semi-empirical), and later ab initio methods. Calculations of static properties such as atomic point charges, electrostatic fields, and electron densities of small molecules pushed the envelope of computational hardware and software wherein LSD was a "large" molecule and the structure of its GPCR target receptors, such as

Alexander Heifetz (ed.), *Computational Methods for GPCR Drug Discovery*, Methods in Molecular Biology, vol. 1705,
https://doi.org/10.1007/978-1-4939-7465-8_1,

the 5-HT receptors, could only be imagined. Computational studies involving the target GPCR proteins were beyond reach both because of the computational limitations presented by such large systems and the lack of useful structural information about these proteins. For decades following this, computational chemistry software and the available hardware capabilities grew dramatically allowing for far more sophisticated, accurate, and rapidly generated information. On the experimental side, significant advances in areas such as molecular biology, protein crystallography and structure determination, and NMR provided the means for detailed investigations of proteins at the atomic level. These experimental approaches have been successfully applied to soluble proteins for many years so that both small molecule and protein computational methods were deployed and advanced extensively for these targets. For membrane bound proteins such as GPCRs, the powerful approach of X-ray structure determination remained elusive, thus limiting computational drug discovery to ligand-based methods such as pharmacophore studies. While ligand-based methods have indeed been very successful, the much sought atomic level structural information, with its more powerful and far-reaching capabilities, remained a much sought after goal. The first glimpse of the architecture of these proteins came from electron microscopy studies of the related 7 transmembrane protein, bacterio-rhodopsin [1, 2]. While efforts were made to use this structure as a template for homology models of GPCRs of interest, the distal relationship between them did not allow for the suitable accuracy of models needed for drug discovery. It was not until 2000 that the first X-ray structure of a GPCR, the class A GPCR rhodopsin, was reported [3]. Although this was not a ligand-mediated GPCR, it provided a significant advance in the information needed for understanding the structure and function of GPCRs, especially for class A GPCRs. Extensive use was made of the structure of the transmembrane region of rhodopsin as a template for homology models for ligand-mediated GPCRs, but the greater structural accuracy needed for the most efficient drug design was still not achieved. This was exacerbated by the far more varied structure of the extracellular loops of these proteins which contributed to the differential involvement of this extracellular loop region which generally interacted directly with bound ligands at the orthosteric sites. The biggest informational breakthrough for ligand-mediated GPCRs came with the X-ray structure determinations of the first ligand-mediated (class A) GPCR, those of the β2-adrenergic receptor (β_2AR) [4, 5]. These first detailed atomic level structure reports of a ligand-mediated GPCR heralded the beginning of a new era of computer-aided drug discovery for GPCRs. The details of the orthosteric ligand's binding, including the involvement of residues from the extracellular loops, were seen in these first examples as well as in various functional features of GPCRs such as the so-called "ionic lock" and "tryptophan switch" which had previously been characterized and/or hypothesized by other experimental studies. Computational methods that had matured over years as applied to mostly soluble proteins were directly and instantly deployed toward these targets. The determination of these GPCR X-ray structures, and others solved since then, has involved a number of methods to overcome the challenges of crystallizing membrane bound proteins. Prominent among these for facilitating crystallization were the use of companion proteins which were either covalently bound (e.g., T4L, BRIL) at either side of the 7-TM (e.g., spliced into IC3 or attached at N-terminus) or non-covalently bound (e.g., antibodies and nanobodies), selective stabilizing mutations, bound high-affinity ligands, and the lipidic cubic phase methods. The power of structure-based drug-discovery (SBDD) when applied to these first structures was quickly demonstrated. In the now 10 years since those structures were reported, there have been numerous other X-ray structures reported covering examples of a number of class A subclasses, various activation states, other classes of GPCRs (B, C, and F) and yielding a watershed of tools for drug discovery and understanding of the detailed molecular mechanisms and parameters governing a host of physiological roles. Among the many class A X-ray structures, that of LSD bound to 5-HT_{2b} is now added to the arsenal [6] and begins to satisfy the imagination that has stirred over many years.

2 GPCR Structure: A Bottoms-Up Guided Tour

The term GPCRs refers to a broad range of proteins with a common architectural feature, i.e., a domain consisting of seven alpha helices which traverse the cell membrane alternatively from the extracellular (EC) side to the intracellular (IC) side (helix 1 or H1 or TM1) and back again (helix 2) etc. Helices 1 and 2 are connected on their intracellular side by an intracellular loop (IC1) while helices 2 and 3 are connected on their extracellular side by extracellular loop 2 (EC2) with corresponding connecting loops and nomenclature for all the helices. This description generally defines class A GPCRs, whereas non-class A GPCRs (classes B, C, and F) contain an additional extracellular domain. The most well-established effector proteins to which these proteins couple are G-proteins, which is the source of the name "GPCR." As it is now well established that an important role of these proteins is to couple to other effectors besides G-proteins a more universal name seems called for. Often, the name "7TM" is used for all of these proteins, owing to their common architectural feature. Nevertheless, as the name GPCR remains widely used and recognized for all of these proteins, we will use it herein.

As the primary role of GPCRs is by definition to couple with G-proteins, or more generally with effectors, on the intracellular side of the membrane bound GPCRs, we begin our tour of the architecture of GPCRs there. Various regions are highlighted which have gained recognition for their structural, functional, or ligand binding roles.

2.1 The Intracellular Rim

With the role of 7TM proteins to induce signal propagation to the intracellular region via interaction with their various effectors, this region serves as the initial conduit for this information transmission mechanism. The structures of the 7TM proteins and their changes in this region determine whether fruitful interactions with the effectors will take place (activation), to what extent these effective interactions will occur (intrinsic activity), and with which effectors these will occur (biased agonism). The various X-ray structures now available, together with a wealth of molecular biological, biophysical, and biochemical studies, include examples spanning these various possibilities. At the fully active protein extreme is the X-ray structure of the fully activated β_2AR receptor [7] in complex with a high-affinity agonist BI-167107 and its effector, the heterotrimeric GTP binding protein Gs. As with many of the 7TM X-ray structures, companion proteins used to aid in the crystallization are included in the structure. Here, there are two such proteins, the camelid nanobody Nb35 and T4L (replacing the N-terminus of the 7TM). In this case, the role of the camelid nanobody in helping to stabilize the active form of the 7TM protein was demonstrated through molecular dynamics simulations [8] an approach playing an increasing role in complementing X-ray structural information. In comparison with structures of the inactive state, this structure reveals a more extended conformation for helix 5 and an outward shift of helix 6 from the central helical transmembrane axis while helices 3 and 7 move slightly inward. Similar structural information for the transmembrane region is available for the fully inactive protein extreme which has generally been more accessible due to the greater availability of high affinity antagonists (versus agonists) to facilitate protein crystallization as illustrated by an X-ray structure of the adenosine 2a receptor (A_{2a}AR) [9]. Rhodopsin X-ray structures of the inactive state provide other, earlier examples. In addition to the X-ray structures of the active and inactive extremes, there are now a number of examples of various intermediate states including complexes with partial agonists, and demonstrating intermediate structural features. It is noteworthy that the first X-ray structures of GPCR proteins were those of rhodopsin, due in large part to the availability of large quantities of the protein for crystallization. Thus, while ligand-mediated GPCR (non-rhodopsin) proteins hold a central focus for much of the interest in this area because of their pharmacological role as drug targets, rhodopsin has played an early and continued role in unraveling structure/function information about 7TM proteins. A recent example of this is the reporting of an X-ray structure of rhodopsin bound to the β-arrestin effector [10] which starts to shed light on the GPCR structural differences corresponding to differential effector interaction, i.e., biased signaling. Compared to the β_2AR /Gs active structure, H6 in the Rhodopsin/arrestin active structure exhibits a **4A** lesser outward shift (10**A** vs. 14**A**). Supporting the relevance of this difference not being an artifact of the difference in GPCRs (rhodopsin versus β_2AR) smaller differences are also seen in TM1, TM4, TM5, and TM7 when comparing the rhodopsin/arrestin active structure to the active state structure of rhodopsin bound to a C-terminal peptide of Gα.

2.2 The Most Intracellular Ligand Site—To Date

While the role of these proteins is to communicate information from the extracellular region, generally via ligand (or light, in the case of rhodopsin) mediated signaling, to the intracellular region via interaction with various effectors, the location for the signal modulating ligand has traditionally been understood to be in the upper region of the protein for orthosteric as well as sites bordering these orthosteric sites (acting as selectivity sources or allosteric sites). In striking contrast to this, chemokine receptor X-ray structures for CCR2 and CCR9 demonstrate that inhibitor binding in the extreme IC region of the protein and immediately proximal to the effector binding region occurs [11, 12]. The CCR9 X-ray structure has only one bound ligand, the inhibitor vercirnon, which is bound at this site and juts out at the IC domain. Two simultaneous inhibitor ligands are bound in the CCR2 X-ray structure. The first inhibitor, CCR2-RA-[R], is bound in the same location as vercirnon in the CCR9 X-ray structure (*see* Fig. 1), whereas the second ligand, BMS-681, is bound in the assumed orthosteric site. The inhibitory role of the ligands at this extreme IC location is reflected in the protein structure wherein the outward movement of H6, required for the effector binding to the 7TM protein, is prevented by the inhibitor. Additionally, the inhibitor's position directly precludes effector binding. This IC region allosteric site uncovered in these studies is not known to have any endogenous role, and can be considered an illustration of a *man-made* site [15].

2.3 The B Site

The X-ray crystal structure of corticotropin-releasing factor receptor 1 (CRF_1R), a member of the secretin like class B GPCRs, in complex with the antagonist CP-376395 [16] revealed yet another *man-made* 7TM ligand binding site (B site) which is also much deeper than the classical orthosteric site. The location of CP-376395, a compound identified through screening studies, is further away from the IC region of the protein toward the EC

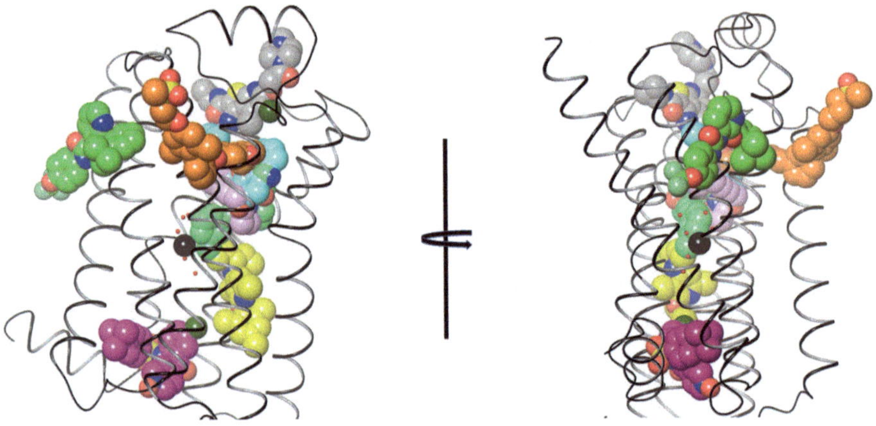

Fig. 1 Illustration of the range of GPCR ligand binding sites. Examples of binding sites of selected ligands in the 7TM domain, as seen from the side of the α-helical barrel. The ligands are superimposed with a ribbon representation of the 7TM domain using the X-ray structure of β_2AR. The identity of each ligand, the protein to which it is bound, and the Protein Data Bank (PDB) [13] accession number of the complex are as follows: vercirnon (purple carbon atoms) in CCR9 (PDB:5LWE); CP-376395 (*yellow* carbon atoms) in CRF_1R (PDB:4K5Y); sodium/water cluster (*black* sodium atom, *red* water molecules) in the A_{2a}AR (PDB:4EIY); mavoglurant (*spring green* carbon atoms) in mGluR5 (PDB:4009); iperoxo (*plum* carbon atoms) in the M2 receptor (PDB:4MQT); carazolol (*aqua* carbon atoms) in the β_2AR (PDB:2RH1); LY2119620 (*gray* carbon atoms) in the M2 receptor (PDB:4MQT). Two views are shown at 90° rotation as indicated. The molecular graphics were generated in Maestro [14]

region than that of the ligands in the CCR2 and CCR9 X-ray structures described above (*see* Fig. 1). Nevertheless, it is still far removed from the orthosteric binding sites expected for this class of proteins wherein peptide-like ligands are expected to bind in, e.g., an open, v-shaped EC cavity as found in the X-ray structure of the related class B receptor whose X-ray structure has also been solved [17]. CP-376395 is selective for CRF_1R over CRF_2R. Two characteristics of this site would suggest conflicting predictions regarding its potential as a source of selectivity. Pervasive dogma argues that allosteric sites offer greater opportunities for selectivity than orthosteric sites, a principle based on the expected conservation of residues among related proteins for common ligands at their orthosteric sites. In contrast, there is generally expected to be less variation in structure and sequence of 7TM proteins in the IC direction than the EC direction. In this case, differences in just two residues at this binding site between CRF_1R and CRF_2R could provide the explanation for the greater preference observed with CP-376395 for CRF_1R. Analysis of this structure also suggests that CP-376395 prevents the activating outward motion of TM6, thereby explaining its inhibitory effect and offering clues for design of ligands with desired intrinsic activity [16, 17].

2.4 The Ionic Lock ("D(E)RY")

As with a much of the early understanding of 7TM structure/function relationships, evidence for this feature as a characterization of the inactive state has its origins in rhodopsin X-ray structures. Using the Ballesteros-Weinstein numbering scheme [18], the ionic lock describes the structure of the cluster of residues D3.49, R3.50, Y3.51 ("DRY"), and E6.30 as a means for establishing the inactive state of the protein. In the structure of the inactive state of rhodopsin, R3.50 interacts with D3.49, E6.30, and T6.34. The ionic interaction of R3.50 with E6.30 forms the lock between helices 3 and 6 which is associated with the inactive state. X-ray structures of the inactive state of ligand-mediated class A GPCRs, e.g., the D3 dopamine receptor [19] have been found to include this ionic lock. Interestingly, similar ionic and/or polar hydrogen bonding networks are found in X-ray structures of inactive forms of class B [16, 17], class C [20, 21], and class F [22, 23] GPCRs. In active state structures of rhodopsin and the β_2AR [7, 24] the interaction of R3.50 with E6.30 is no longer present, but R3.50 interacts with Y5.58 instead. A number of X-ray structures with common ligands but varying ionic cluster interactions, along with X-ray structures with ligands of varying intrinsic activity, and molecular dynamics simulations, lead to an emerging picture that these active/inactive state ionic lock indicators are not guarantees of the activation state but serve as indicators of their propensities for the given state [15]. Moreover, they seem to contribute to the induction of the structural changes more proximal to the effector.

2.5 Internal Water Network and Its Sodium Site

Sodium has been shown to act as an allosteric modulator of 7TM proteins. A 1.8-Å high-resolution X-ray structure of the A_{2a}AR with the inhibitor ZM241385 bound [25] shows the position of a sodium atom at the center of a network of water molecules which traverse much of the transmembrane region and has three clusters whose central cluster contains the sodium atom (Fig. 1). This cluster is situated between the ionic lock and the so-called toggle switch (*see* below). This site can potentially serve as a ligand binding site as supported by a crystal structure of the 7TM region of a class C GPCR, the mGluR5 receptor [21] containing the bound negative allosteric modulator (NAM) mavoglurant whose lower portion overlaps spatially with this sodium/water cluster (Fig. 1). In the case of the A_{2a}AR, the orthosteric site is located in the more common upper region of the transmembrane as is the inhibitor also seen in the A_{2a}AR X-ray structure. For mGluR5 however, the orthosteric site resides in an extracellular domain separated from the 7TM domain by a "cysteine-rich" protein linker. The mavoglurant site in mGluR5 is thus another example of a *man-made* site [26]. The role of this sodium/water-cluster region to serve as an allosteric site to two very differently located orthosteric sites is more uniformly understood when viewed as serving a common function to modulate the same local transmembrane region.

In the case of the $A_{2a}AR$, active state structures are available for comparison [27, 28] and show that the hydrated sodium-ion induces a kinking in helices VI and VII. Ameloride is known to compete with this site for the $A_{2a}AR$ and it has been used for structure-based design [29–33].

2.6 The CWxP "Toggle Switch"

Analogous to the ionic lock, a highly conserved CWxP motif contains tryptophan W6.48 whose orientation had been hypothesized as a marker for the activation state of 7TM proteins. This has now been verified extensively in numerous GPCR X-ray structures where there is a shift in the position of the indole of W6.48 between the active and inactive state structures, albeit not a flipping of the indole ring as originally hypothesized. Interestingly, the driving forces for this indole positioning are varied. In the inactive state structures of rhodopsin [3], the histamine H1 receptor [34], and the muscarinic M2 receptors [35], the inhibitors (retinal in the case of rhodopsin) hold the corresponding indole of W6.48 in the same position by directly interacting with it. In other instances, such as the inhibitor bound inactive state X-ray structures of the β_2AR [4] or the dopamine D3 receptor [19], ligand interaction is with an aromatic ring of an intervening residue. Whereas this region is proximal to the endogenous ligand's binding sites in class A 7TM proteins, that is not the case for class C 7TM proteins such as mGluRs. It is thus interesting that X-ray structures of the 7TM domains of mGluRs show that allosteric inhibitor bound proteins with ligands at this man-made site (for mGluRs) [20, 21] have their corresponding tryptophan rings displace outward from the 7TM core by the bound ligand. The role of the differences in the structural features in this region in protein activation is becoming clearer. Comparing the active and inactive states of the $A_{2a}AR$ [27, 36] shows that the agonist sits much deeper in the pocket forming a series of hydrogen bonds with the protein as well as a steric clash with W6.48 which collectively induce more active like orientations and positions of H5 and H6.

2.7 The "Orthosteric" Pocket—The HUB

The approximately upper third region of the 7TM core generally serves as the binding site for endogenous ligands, particularly for class A 7TM proteins and, in turn, for most synthetic ligands; herein referring to this as the "HUB" region. In considering all GPCRs, many more ligands are accommodated at the HUB than effectors at the IC region. It is therefore intuitive that there is considerable diversity at this HUB site as described above. This diversity is rooted in multiple sources including significant amino acid variability between 7TM proteins in the EC direction, still greater variability in the 7TM connecting extracellular loops and greater structural variation (such as degree of openness) in this region.

The architecture of this HUB region makes it the region of first choice for ligands in general and for endogenous ligands for similar reasons. In addition to its relative diversity, it is for the most part an enclosed cavity, yet not as narrow as the more internal section of the 7TM which is therefore more limiting in the scope of ligands it can accommodate. It is also generally accessible from the extracellular side—the source of most modulators.

Relative to the 7TM helical axis, ligands are found to span a considerable vertical depth and lateral breadth of this HUB region, which may reasonably be considered as comprised of sub-regions. In some cases, such as inhibitors of the histamine H1 receptor [34] and the muscarinic M2 [35] and M3 [37] receptors, ligands reside at the extreme depth of this region (*see* Fig. 1), having no direct interaction with the extracellular loops that are often a source of selectivity as well as affinity. Nevertheless, single-residue differences in this region can account for selectivity of inhibitors between closely related proteins such as M2 and M3 [15]. More generally, ligands binding in the HUB region interact with the extracellular loops as well (*see*, e.g., the many structures for the β_2AR or the A_{2a}AR) and interactions with portions of the N-terminal regions are also seen (*see*, e.g., Ref. 38). In the lateral direction, small molecules can bind in nonoverlapping positions at the same depth within the 7TM domain (*see* Fig. 2). For example, the inhibitor AZD1283 in a P2Y12 X-ray structure [39] where AZD1283 laterally spans the TM region interacting with helices III-VII, and the inhibitor in the CB1 X-ray structure [38] occupy nonoverlapping locations as shown in Fig. 2.

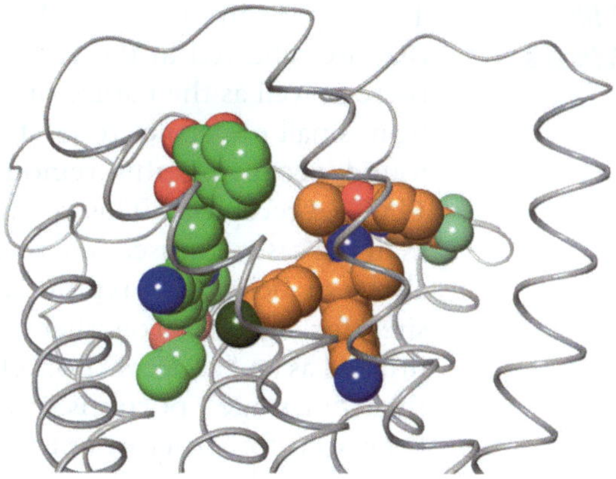

Fig. 2 Illustration of laterally parallel sub-sites at the orthosteric HUB site. Locations of two ligands which bind in the HUB site of the 7TM domain, as seen from the side of the α-helical barrel. The ligands are superimposed with a ribbon representation of the 7TM domain using the X-ray structure of β_2AR. The identity of each ligand, the protein to which it is bound, and the Protein Data Bank (PDB) [13] accession number of the complex are as follows: taranabant (*orange* carbon atoms) in CB1 (PDB:5 U09); AZD1283 (*green* carbon atoms) in the P2Y12 receptor (PDB:4NTJ). The molecular graphics were generated in Maestro [14]

While the structural characteristics of the orthosteric site are clearly related to the affinities of ligands at these sites, the more subtle sources of the varying intrinsic activities of these ligands are becoming clearer. Based on reported protein X-ray structures in different states for the same proteins, e.g., the adenosine and adrenergic GPCR proteins, the more inactive states are associated with more open pockets in this region containing slightly larger ligands, with increasing activating influence and decreasing openness of the site as one proceeds from inverse agonists through partial agonists to full agonists [40]. The relative openness here plays a role in structural changes propagating to the IC region related to the relevant active/inactive state. This trend is not universal as the ultimate criteria lie in the induced changes in the IC direction. Yet more subtle than understanding of intrinsic activity is the understanding of the mechanism of ligand bias for ligands binding here and inducing different effector interaction profiles. The structural bases for underlying biased agonism are emerging. For example, a comparison of the X-ray structure complexes of 5-HT_{1b} and 5-HT_{2b} [41, 42], each bound with ergotamine, a compound having bias for β-arrestin over G-protein at 5-HT_{1b} shows differences at the binding site in conformations of residues at helix 6 which, in turn, correspond to a less active like conformation of helix 6 in the IC region. The crystal structure of LSD bound to 5-HT_{2b} [6] adds to this picture and, coupled with molecular dynamics studies, indicates interaction with extracellular loop 2 (ECL2) modulates the LSD off rate thereby providing a kinetic component to enhanced β-arrestin interaction.

2.8 The EC Rim (Vestibule, Address Site, Etc.)

The upper rim of the 7TM domain has the most diverse features that are reflected in the variability in the types of ligands residing there as well as the nature of its usage. Ligands binding here range from small molecules to relatively large peptides. Small ligands are found occupying this region as seen, e.g., for the A_{2a}AR and *CXCR4* receptors. This region serves as an allosteric "vestibule" as in the case of muscarinic receptors [35, 37, 43] (Fig. 1). Regions above or below this region can combine with it to form binding sites for ligands. Together with the region below it, it is thus utilized as an "address" pocket in conjunction with ligands binding their "message" portions more deeply into the 7TM such as has been seen for structures of the opioid receptors [44–47]. Alternatively, ligands bound here can be found to also interact with portions of the protein in the N-terminus direction for class A (e.g., CB1 [38]) and, e.g., class B proteins where structural and mutational evidence indicates that endogenous ligand binding straddles the 7TM and EC domains [17].

2.9 Binding To and Through the External 7TM Wall

The overall barrel-like shape of the transmembrane region of the 7TM proteins, whose most commonly accepted orthosteric binding site lies inside the barrel-like structure, is consistent with a simple model for the trajectory of ligands engaging in interactions with these proteins. An alternate model for this trajectory has been considered for some time, wherein a ligand approaches and enters a 7TM protein from the outer side of the barrel [48–50]. With examples of X-ray structures showing ligands completely buried within the 7TM and covered by EC loops now available for, e.g., rhodopsin [3], S1p1 [51], and PAR1 [52], an external trajectory becomes a more plausible explanation for ligand entry. Indeed, there is now proof of ligands actually binding partially external to, or even completely external to the 7TM region (*see* GPR40 [53] and $P2Y_1$ [54] respectively). External interactions as modulators of GPCR activity are supported by other types of data. Both homo- and hetero-dimerization is known to play a role in the functioning of various GPCRs [55–58]. Binding of cholesterol to the external transmembrane region has been shown by X-ray structures (*see*, e.g., [4, 59]) as well as electron microscopy [60] and is believed to play a role in dimerization. Long time frame molecular dynamics investigations are helpful in examining these potential interactions [61] and new mass spectrometry-based tools are emerging to measure the dependence and degree of protein oligomerization due to membrane lipid binding mediation [62]. As GPCR signaling is dependent on changes in their helical positions and conformation in the IC regions to prepare for effector interaction, it is not surprising to find evidence that modulating such changes from the external side of their transmembrane region is possible. Taken together, the collection of structural information that is now available for 7TM proteins indicates that *all regions of 7TM proteins, inside and out, appear to provide potential sites for ligand modulation.*

2.10 EC Domains for GPCR Classes B, C, and F

The class B, C, and F 7TM receptor subgroups are differentiated in part from class A 7TM receptors by an additional domain at their N-terminus, the EC domain. X-ray structures for the 7TM domains have been determined for examples of all three of these protein subgroups and confirm their generally similar architecture to the class A GPCR proteins. While peptidic ligands for the class B 7TM proteins bind in between the 7TM and EC domains, the EC domains of the class C and F 7TM proteins contain structurally separated binding sites. Class C receptors contain a cysteine-rich linker region that connects the 7TM domain to the so-called Venus "flytrap" (VFT) domain to which the endogenous ligands bind, such as glutamate in the case of mGluR receptors. X-ray structures of the EC domains of class C and class F receptors with ligands bound have been determined. For the EC domain, the structural changes associated with the active versus inactive state of the VFT, as modulated by ligands, are understood. While X-ray structures of the VFT domains of mGluRs were first reported in 2000 [63], they have not been used extensively for structure-based design, albeit SBDD studies including additional X-ray structures are still being reported [64]. The belief that the transmembrane region offers an advantageous environment for drug discovery is responsible for this, although the rationales given, such as greater sequence variation in the 7TM domains, have been challenged [65]. There is no full-length structure of a Class C GPCR protein available at this time to establish the interplay between these domains. For the class F 7TM proteins, full-length structures have now been determined for the example of the smoothened receptor [66]. The smoothened proteins have a cysteine-rich C terminal domain which is connected through a linker to its 7TM domain. Structural insights into ligand-mediated communication from the 7TM to the EC domains have emerged and the stage is set for understanding the EC to 7TM communication mechanism as well.

3 Computational Drug-Discovery Approaches/Capabilities

The exploitation of X-ray structures of ligand-mediated 7TM proteins through computer-aided drug discovery followed rapidly and with striking successes after these structures became available and have been reviewed extensively. Three major factors have contributed to the precipitation and rapid growth of 7TM SBDD work. First, experimental tools and strategies, such as the use of companion proteins, lipidic cubic phase methods, and targeted stabilizing mutations, enabled the crystallization and X-ray structure determination of 7TM (and other classes of membrane bound) proteins. Second, computational chemistry and modeling methods, notably for protein structure studies, had been honed over many years with the available protein targets. While these targets were primarily soluble proteins, many of the computational approaches were independent of the differences. Third, the inherent architecture and most commonly targeted HUB ligand binding site are highly suitable for small molecule binding and modulation. The enclosed nature, probably not coincidentally, is essentially ideal for occupancy by a small molecule and in striking contrast to other extremes such as surface binding sites, inter-domain sites, etc. The "druggability" of this HUB site, the broad range of endogenous ligand modulators utilizing this HUB, the prominence of GPCR targeted drugs, and the pervasive role of GPCR proteins can all be argued to stem from these GPCR architectural characteristics.

3.1 Virtual Screening, High-Throughput Docking

As was demonstrated soon after the first X-ray structures for GPCR proteins were reported, high-throughput docking (HTD) campaigns using X-ray structure models for compounds acting at those targets, and in the same fashion (site, active/inactive state, etc.) are extremely effective for GPCR proteins. From a drug discovery perspective, there is little doubt that this is an extremely efficient and striking approach to begin studies of a target protein. Screening of large databases of preexisting compounds, from commercial or proprietary sources, can rapidly jump start a drug discovery program with identification of potent compounds and structure activity information. In silico HTD screening of large databases containing millions of compounds, using a number of different software systems, is already routinely conducted and used to rank and select as few as tens or hundreds of compounds for in vitro testing. Hit rates above 30% and yielding compounds with activities in the single-digit nanomolar range are common [15]. Integration of HTD methods with ligand-based methods or protein-based pharmacophore methods often further improves these successes. As expected, the success of these approaches depends on how close to this optimal paradigm one operates. Within a subgroup of closely related targets for proteins with common endogenous ligands (e.g., adrenergic or dopaminergic receptors) homology models based on X-ray structures of other members of the subgroup yield comparable results to those where the X-ray structure of the target of interest is used. However, in silico screening for an agonist using an antagonist bound structure of the same protein as a template is often more challenging than for an antagonist using a homology model based on an X-ray structure template of another protein in an inactive state within the same subgroup. This is because differing states of a protein have greater deviation in their binding site structures from their templates than common states within a subgroup where there are few amino acid changes. As one progresses to create and deploy homology models based on templates of X-ray structures outside a target sub-group the reliability of the homology model decreases. In part, this is due to the reduction in sequence identity, and consequentially reduction in structural similarity in the transmembrane helices. More elaborate protocols such as the use of multiple templates help improve the accuracy of the homology models. More significantly, the extracellular loops, and in some instances sections of the N-terminal, vary much more significantly in shape, length, fold etc., while contributing significantly to ligand binding at the HUB.

3.2 Structure-Based Drug Design

The strengths and weaknesses described for HTD pertain more generally to computational drug discovery for GPCR proteins. It has become commonplace to employ models of 7TM proteins in drug discovery activities as evidenced by the extensive reporting of these approaches in the medicinal chemistry literature. Unlike

HTD studies where the sheer number of ligand structures being investigated necessitates invoking considerable restrictions and approximations on the methodologies employed, studies of individual complexes allow for more sophisticated approaches to be employed leading to more accurate results. Thus, while most HTD studies use frozen, or nearly frozen protein structures in calculations with highly approximate models for the interaction energetics, this is not the case otherwise. Considerable protein flexibility ranging from the active site to the entire protein complex structure is common using molecular mechanics based approaches. More accurate quantum chemical ab initio methods are possible for more accurate evaluations of, e.g., ligand protein interactions, albeit this has not been as extensively reported. To explore processes such as conformational changes associated with activation, molecular mechanics-based molecular dynamics methods for extended time frame simulations are being reported more regularly. Additionally, complementing the increasing structural information on the involvement of explicit water molecules in various regions of GPCRs such as orthosteric sites or the sodium binding site seen in $A_{2a}AR$ crystal structures [25], computational methods are now more reliably examining the various roles of these waters. Finally, the increased capabilities of GPCR modeling are opening up new types of opportunities such as the use of homology modeling of orphan receptors to identify ligands for use in de-orphanizing these receptors [67].

4 Scope of Information: Lessons and Emerging Opportunities

The evermore rapidly emerging structural and functional information for GPCRs, especially from X-ray structure determinations, translates into various properties and processes of these systems of relevance to drug action and is now becoming more amenable to investigation and exploitation by computational approaches. The various aspects of drug discovery dominate the general interests here, with protein structure/function insights sought as well and these two goals are often inseparable. It is interesting to consider the range in scope, and potential already realized as well as anticipated.

4.1 Allosteric Sites: Another Look at the Other Site

The use of X-ray structures of GPCRs for discovery and design of ligands in the simplest approach, i.e., for the *same site and same activity* as the X-ray structure, is now well established and remarkably effective. As noted, at the most common (for class A) 7TM orthosteric HUB site, structural differences observed between inactive and active state structures provide a clear explanation for the deterioration of results when inactive structures are used (without other moderations or considerations) to identify activating compounds. It is reasonable to assume that similar differences would occur at other ligand binding sites, i.e., allosteric sites, such as those described herein, but there is as yet not much data available to test this. The HUB site serving as the orthosteric site for class A proteins, however, serves as an *allosteric* site for non-class A proteins. X-ray structures for the smoothened class F GPCR, with activators or inhibitors bound reveal differences between their binding sites [22, 23]. For the mGluRs, while X-ray structures in the 7TM domain are only available with inhibitors bound, SAR data finds that very small ligand changes result in the switch between activating and inactivating ligands [21, 26, 68]. Whether this apparent sensitivity is inherent in the role of this site as an allosteric site or simply a consequence of the still limited information is unclear. Relatedly, whereas by definition an allosteric site is a site other than that where the endogenous ligand binds (the orthosteric site), other implications for the allosteric nomenclature, e.g., the modulation of the orthosteric binding site events, may point to a different perspective. GPCRs are a category of proteins whose architecture and prominent functioning can be described as proteins whose communication with intracellular effectors is generally modulated by ligand interaction at the control-center/HUB site. It therefore seems logical to re-consider the EC domain sites of non-class A GPCRs as *operationally allosteric* sites as compared to their more unifying (with respect to class A GPCRs) HUB. By analogy to class A GPCRs, these 7TM sites would be directly involved in signal transmission in the IC direction as opposed to the usual indirect model wherein these 7TM sites modulate signaling in the EC direction (at the EC domain) which must then propagate back through the same 7TM domain—where they began. Evidence for both perspectives exists vis. the X-ray structure of the complete smoothened protein [66] shows evidence for ligand binding in the 7TM domain resulting in interactions from the 7TM with the EC domain which influence EC ligand binding whereas evidence for the direct ligand control at the IC region of non-class A GPCR proteins is provided by reports of a truncated mGluR5 protein without its EC domain which can be activated by a TM binding ligand [69]. The inherent machinery of GPCRs thus questions whether the usual roles, experimental analyses, and ligand design of allosteric modulators should be treated differently for non-class A GPCRs.

4.2 Multi-Target Tuning: Within and Between Subgroups, to Other Classes, Tuning In Vs. Out, Polypharmacology

A critical factor in the action of drugs is the profile of their activity at varying targets. This target profile is important even when only simple inhibition is considered at multiple targets and extends to considerations of varying activities at different sites of different proteins such as activators with respect to one site and inhibitors with respect to another. Indeed, poly-pharmacology has grown as a medicinal approach. The growing structural information that has become available now introduces a broad selection of opportunities to aid in the tuning (in or out) of various targets' activities ranging from closely related to distally related targets. Even at the extreme of very closely related targets for which selectivity design is particularly challenging, for example target proteins with common endogenous ligands, powerful structural data is now available. Thus, the available X-ray structures of four opioid receptors were used in SBDD studies to optimize selectivity [70, 71], as well as to explore effector bias. Similarly, the X-ray structure of one target among such closely related GPCRs can be used to optimize activity profiles [72]. Similar structure-based selectivity design is possible at the EC domains of class C mGluRs [64]. The use of X-ray structures for protein targets having a common endogenous ligand can now take on a broader scope as illustrated by the recent reports of X-ray structures for endogenous amine transporters such as the 5-HT transporter [73] which can also be used for drug discovery [74, 75]. Thus, the search for incorporating various activity profiles at 5-HT receptors along with SERT activity [76] lends itself to SBDD at both the GPCRs and the transporters of the corresponding endogenous ligands. More traditional considerations of drug profiles among different GPCRs is now rapidly becoming a more fruitful area for SBDD involvement. Such pairing of SBDD approaches is also being invoked when considering actions on a GPCR together with unrelated proteins such as the design of dual inhibitors of HIV Integrase and CCR5 [77]. Moreover, these types of pairing of structural information of GPCRs with one or more other proteins are an emerging area in SBDD studies of ligands with disparate activities (e.g., activators versus inhibitors) at their respective targets.

4.3 Site and Roles: Alternative Pressure Points

Considering the breath of non-HUB sites for which there is now evidence for ligand binding and protein modulation, it appears that almost any site is a possible site for effective ligand binding and protein action modulation. Driven by ligand binding at these sites, or *pressure* points, GPCRs seem to operate through a model somewhat akin to a collection of 7 chop sticks. Relatively rigid transmembrane helices adjust their positions, with additional localized (e.g., ionic lock) structural changes in specific residue conformations and interactions, which mediate signaling. From a drug discovery perspective, the primary criteria for selecting a site of ligand intervention may be less a question of whether a site is technically orthosteric or allosteric but which site(s) provides the optimal pressure point. Understanding the interplay among these sites will be helpful here as well and is an area where computational investigations can play an important role [78]. The HUB site seems intuitively the most common, opening gambit—probably inherent in the GPCR architecture. While distal from the effector, the HUB seems to benefit from a leveraging mechanism (chop stick model). The inner most IC region, with its proximity to the incoming effector, may have advantages for inhibition of GPCR-effector interaction but may be difficult, albeit not inconceivable, to use as enhancers of GPCR-effector interactions by virtue of the potential for such a ligand to "get in its own way." As structural information with effectors continues to become more available, direct ligand interaction in this region may offer particular advantages for *biased inhibition* favoring specific effectors. The emerging structural information exemplified by the β_2AR /Gs and rhodopsin/β-arrestin structures may prove very effective here. Sites on the *external surface* of the transmembrane would seem at first to be more challenging as sites for small molecule binding because they are less enclosed, but could be advantageous in, e.g., *blocking dimerization*. At the other end of GPCR's TM region, proteins with significant interactions from above (EC direction) the HUB site, such as EC loops or N-terminus, may play a valuable role in the *kinetics* of ligand binding and have even been examined through X-ray structure determinations [79] to aid in rationally introducing desirable kinetic features into drug design. If we similarly consider the many non-HUB sites already determined (even for class A GPCRs) from the body of structural information available, they provide an increasing number of examples of *man-made operationally allosteric* sites. For the most part, ligands acting at these man-made sites were not a result of design. Rather, there were obtained through means such as in vitro screening and only subsequently characterized for their site of action. This is in contrast to the EC binding sites of non-class A GPCRs which offer examples of *natural operationally allosteric* sites as defined herein. Using the mounting body of structural information of the many potential sites/pressure points, an important future direction for structure-based drug discovery may be the targeting of these sites a priori, based on information from other targets or purely structural in nature, rather than after their discovery rooted in experimentally identified compounds/structures. *Moreover, as the structural information for GPCRs grows, we are moving toward a situation where we exploit this emerging structural knowledge to select the specific target sites of a given GPCR most likely to be effective for a given function/role.*

4.4 Multiple Sites Within and Between Proteins, Bivalent, Bitopic, Linked

An underlying theme in the search for a number of alternate modes of action of ligands is the incorporation of components of ligands acting at two sites into a single ligand. The various sites available for ligand modulation to be paired may be relegated to structure-based design in many cases. Pairing can occur within or between GPCRs. More traditional pairing of adjoining sites such as for opioid ligands having an "address" and "message" component now has available X-ray structures revealing the corresponding sites on the proteins [44–47]. As structural information becomes available it would be promising to expand the use of this structural information in the design of bitopic ligands linking together allosteric and orthosteric moieties such as was reported for M1 partial agonists [80]. Computational approaches to identifying possible sites where endogenous ligands may bind weakly and/or transiently as they migrate to the orthosteric sites provide a source for actually predicting new sites that can be paired with orthosteric sites in the targeting of bitopic ligands [81]. The dimerization of GPCRs offers the potential for ligands simultaneously binding to two GPCR monomers via linking of their binding moieties. Illustratively, structure-based models have accordingly been used to design bivalent ligands linking two oxytocin mimetics to bind to oxytocin receptor dimers with increased potency compared to the individual ligand [82]. However, to deploy such strategies most efficiently will require considerably more structural information regarding the dimers involved. Finally, it has even been suggested that bivalent ligands can be used to identify non-constitutive heterodimeric GPCR formation with novel physiological functions [83].

5 Concluding Remarks

The central physiological and pharmaceutical roles of GPCRs have generated an extensive history of investigations and understanding about these proteins and their actions. The start of detailed X-ray structure determinations of these proteins is now providing us with the ability to see and use GPCR structural information at an atomic level. Many hypotheses are now being validated and in some cases modified. The structural detail is effectively being deployed through computational methods to yield highly efficient drug discovery results for traditional strategies of protein targeting. More interestingly, this expanding information base affords opportunities to pursue novel design strategies and ways of thinking such as the use of ***man-made*** binding sites and ***operationally allosteric*** sites. The increasing supply of these structural studies promises to further advance our understanding and drug discovery capabilities.

References

1. Unwin PNT, Henderson R (1975) Molecular structure determination by electron microscopy of unstained crystalline specimens. J Mol Biol 94:425–440
2. Baldwin JM, Henderson R, Beckman E et al (1988) Images of purple membrane at 2.8A° resolution obtained by cryo-electron microscopy. J Mol Biol 202(3):585–591
3. Palczewski K, Kumasaka T, Hori T et al (2000) Crystal structure of rhodopsin: a G protein-coupled receptor. Science 289:739–745
4. Cherezov V, Rosenbaum DM, Hanson MA et al (2007) High-resolution crystal structure of an engineered human β2-adrenergic G protein-coupled receptor. Science 318:1258–1265
5. Rosenbaum DM, Cherezov V, Hanson MA et al (2007) GPCR engineering yields high-resolution structural insights into β2-adrenergic receptor function. Science 318:1266–1273
6. Wacker D, Wang S, McCorvy JD et al (2017) Crystal structure of an LSD-bound human serotonin receptor. Cell 168:377–389
7. Rasmussen SGF, DeVree BT, Zou Y et al (2011) Crystal structure of the β_2 adrenergic receptor–Gs protein complex. Nature 477:549–555

8. Rosenbaum DM, Zhang C, Lyons J et al (2011) Structure and function of an irreversible agonist-β_2 Adrenoceptor complex. Nature 469:236–240
9. Doré AS, Robertson N, Errey JC et al (2011) Structure of the adenosine A_{2A} receptor in complex with ZM241385 and the xanthines XAC and caffeine. Structure 19:1283–1293
10. Kang Y, Zhou XE, Gao X et al (2015) Crystal structure of rhodopsin bound to arrestin by femtosecond X-ray laser. Nature 523:561–567
11. Zheng Y, Qin L, Zacarías NVO et al (2016) Structure of CC chemokine receptor 2 with orthosteric and allosteric antagonists. Nature 540:458–461
12. Oswald C, Rappas M, Kean J et al (2016) Intracellular allosteric antagonism of the CCR9 receptor. Nature 540:462–465
13. Bernstein FC, Koetzle TF, Williams GJ et al (1977) The protein data bank: a computer-based archival file for macromolecular structures. J Mol Biol 112:535–542
14. Schrödinger Release 2014-2: Maestro, version 9.8, Schrödinger, LLC, New York, NY, 2014
15. Topiol S, Sabio M (2015) The role of experimental and computational structural approaches in 7TM drug discovery. Expert Opin Drug Discov 10:1071–1084
16. Hollenstein K, Kean J, Bortolato A et al (2013) Structure of class B GPCR corticotropin-releasing factor receptor 1. Nature 499:438–443
17. Siu FY, He M, de Graaf C et al (2013) Structure of the human glucagon class B G-protein-coupled receptor. Nature 499:444–449
18. Ballesteros JA, Weinstein H (1995) Integrated methods for the construction of three-dimensional models and computational probing of structure-function relations in G protein-coupled receptors. Methods Neurosci 25:366–428
19. Chien EYT, Liu W, Zhao Q et al (2010) Structure of the human dopamine D3 receptor in complex with a D2/D3 selective antagonist. Science 330:1091–1095
20. Wu H, Wang C, Gregory KJ et al (2014) Structure of a class C GPCR metabotropic glutamate receptor 1 bound to an allosteric modulator. Science 344:58–64
21. Doré AS, Okrasa K, Patel JC et al (2014) Structure of class C GPCR metabotropic glutamate receptor 5 transmembrane domain. Nature 511:557–562
22. Wang C, Wu H, Katritch V et al (2013) Structure of the human smoothened receptor bound to an antitumour agent. Nature 497:338–343
23. Wang C, Wu H, Evron T et al (2014) Structural basis for smoothened receptor modulation and chemoresistance to anticancer drugs. Nat Commun 5:4355.
24. Scheerer P, Park JH, Hildebrand PW et al (2008) Crystal structure of opsin in its G-protein-interacting conformation. Nature 455:497–502
25. Liu W, Chun E, Thompson AA et al (2012) Structural basis for allosteric regulation of GPCRs by sodium ions. Science 337:232–236
26. Topiol S, Sabio M (2016) 7TM X-ray structures for class C GPCRs as new drug-discovery tools. 1. mGluR5. Bioorg Med Chem Lett 26:484–494
27. Xu F, Wu H, Katritch V et al (2011) Structure of an agonist-bound human A_{2A} adenosine receptor. Science 332:322–327
28. Lebon G, Warne T, Edwards PC et al (2011) Agonist-bound adenosine A2A receptor structures reveal common features of GPCR activation. Nature 474:521–525
29. Howard MJ, Hughes RJ, Motulsky HJ et al (1987) Interactions of amiloride with alpha- and beta-adrenergic receptors: amiloride reveals an allosteric site on alpha 2-adrenergic receptors. Mol Pharmacol 32:53–58
30. Gutiérrez-de-Terán H, Massink A, Rodríguez D et al (2013) The role of a sodium ion binding site in the allosteric modulation of the A_{2A} adenosine G protein-coupled receptor. Structure 21:2175–2185
31. Lenselink EB, Beuming T, Sherman W et al (2014) Selecting an optimal number of binding site waters to improve virtual screening enrichments against the adenosine A_{2A} receptor. J Chem Inf Model 54:1737–1746
32. de Lera Ruiz M, Lim YH, Zheng J (2014) Adenosine A_{2A} receptor as a drug discovery target. J Med Chem 57:3623–3650
33. Sabbadin D, Ciancetta A, Moro S (2014) Perturbation of fluid dynamics properties of water molecules during G protein-coupled receptor–ligand recognition: the human A_{2A} adenosine receptor as a key study. J Chem Inf Model 54:2846–2855
34. Shimamura T, Shiroishi M, Weyand S et al (2011) Structure of the human histamine H_1 receptor complex with doxepin. Nature 475:65–70
35. Haga K, Kruse AC, Asada H et al (2010) Structure of the human M2 muscarinic acetylcholine receptor bound to an antagonist. 482:547–551
36. Jaakola V-P, Griffith MT, Hanson MA et al (2008) The 2.6 angstrom crystal structure of a human A_{2A} adenosine receptor bound to an antagonist. Science 322:1211–1217

37. Kruse AC, Hu J, Pan AC, Arlow DH et al (2012) Structure and dynamics of the M3 muscarinic acetylcholine receptor. Nature 482:552–556
38. Shao Z, Yin J, Chapman K et al (2016) High-resolution crystal structure of the human CB1 cannabinoid receptor. Nature 540:602–606
39. Zhang K, Zhang J, Gao Z-G (2014) Structure of the human P2Y12 receptor in complex with an antithrombotic drug. Nature 509:115–118
40. Warne T, Moukhametzianov R, Baker JG et al (2011) The structural basis for agonist and partial agonist action on a β_1-adrenergic receptor. Nature 469:241–244
41. Wang C, Jiang Y, Ma J et al (2013) Structural basis for molecular recognition at serotonin receptors. Science 340:610–614
42. Wacker D, Wang C, Katritch V et al (2013) Structural features for functional selectivity at serotonin receptors. Science 340:615–619
43. Kruse AC, Ring AM, Manglik A et al (2013) Activation and allosteric modulation of a muscarinic acetylcholine receptor. Nature 504:101–106
44. Manglik A, Kruse AC, Kobilka TS et al (2012) Crystal structure of the μ-opioid receptor bound to a morphinan antagonist. Nature 485:321–326
45. Granier S, Manglik A, Kruse AC et al (2012) Structure of the δ-opioid receptor bound to naltrindole. Nature 485:400–404
46. Wu H, Wacker D, Mileni M et al (2012) Structure of the human κ-opioid receptor in complex with JDTic. Nature 485:327–332
47. Thompson AA, Liu W, Chun E et al (2012) Structure of the nociceptin/orphanin FQ receptor in complex with a peptide mimetic. Nature 485:395–399
48. Hurst D, Grossfield A, Lynch D et al (2010) A lipid pathway for ligand binding is necessary for a cannabinoid G protein-coupled receptor. J Biol Chem 285:17954–17964
49. Schädel S, Heck MM, Maretzki D et al (2003) Ligand channeling within a G-protein-coupled receptor: the entry and exit of retinals in native opsin. J Biol Chem 278:24896–24903
50. Filipek S, Stenkamp R, Teller D et al (2003) G protein-coupled receptor rhodopsin: a prospectus. Annu Rev Physiol 65:851–879
51. Hanson MA, Roth CB, Jo E et al (2012) Crystal structure of a lipid G protein-coupled receptor. Science 335:851–855
52. Zhang C, Srinivasan Y, Arlow DH et al (2012) High-resolution crystal structure of human protease-activated receptor 1. Nature 492:387–392
53. Srivastava A, Yano JK, Hirozane Y et al (2014) High-resolution structure of the human GPR40 receptor bound to allosteric agonist TAK-875. Nature 513:124–127
54. Zhang D, Gao Z, Jacobson K et al (2015) Two disparate ligand-binding sites in the human $P2Y_1$ receptor. Nature 520:317–321
55. Rozenfeld R, Gomez I, Devi L (2011) Opioid receptor dimerization. In: Pasternak GW (ed) The opiate receptors, vol 23. Humana, New York, pp 407–437
56. Milligan G (2009) The role of dimerisation in the cellular trafficking of G-protein-coupled receptors. Curr Opin Pharmacol 10:1–7
57. Birdsall N (2010) Class A GPCR heterodimers: evidence from binding studies. Trends Pharmacol Sci 31:499–508
58. Gonza'lez-Maeso J, Ang RL, Yuen T et al (2008) Identification of a serotonin/glutamate receptor complex implicated in psychosis. Nature 452:93–98
59. Hanson MA, Cherezov V, Griffith MT et al (2008) A specific cholesterol binding site is established by the 2.8 a structure of the human β2-adrenergic receptor. Structure 16:897–905
60. Ruprecht JJ, Mielke T, Vogel R et al (2004) Electron crystallography reveals the structure of metarhodopsin I. EMBO J 23:3609–3620
61. Lee JY, Lyman E (2012) Predictions for cholesterol interaction sites on the A_{2A} adenosine receptor. J Am Chem Soc 134:16512–16515
62. Gupta K Donlan JAC Hopper JTS et al (2017) The role of interfacial lipids in stabilizing membrane protein oligomers. Nature 541:421–424
63. Kunishima N, Shimada Y, Tsuji Y et al (2000) Structural basis of glutamate recognition by a dimeric metabotropic glutamate receptor. Nature 407:971–977
64. Chappell MD, Li R, Smith SC et al (2016) Discovery of (1S,2R,3S,4S,5R,6R)-2-amino-3-[(3,4- difluorophenyl)sulfanylmethyl]-4-hydroxy-bicyclo[3.1.0]hexane-2,6- dicarboxylic acid hydrochloride (LY3020371·HCl): a potent, metabotropic glutamate 2/3 receptor antagonist with antidepressant-like activity. J Med Chem 59:10974–10993
65. Topiol S, Sabio M, Uberti M (2010) Exploration of structure-based drug design opportunities for mGluRs. Neuropharmacology 60:93–101
66. Byrne EFX, Sircar R, Paul Miller S et al (2016) Structural basis of smoothened regulation by its extracellular domains. Nature 535:517–522
67. Frimurer TM, Mende F, Graae A-S et al (2017) Model-based discovery of synthetic agonists for the Zn^{2+}–sensing G-protein-coupled

receptor 39 (GPR39) reveals novel biological functions. J Med Chem 60:886–898

68. Christopher JA, Aves SJ, Bennett KA et al (2015) Fragment and structure-based drug discovery for a class C GPCR: discovery of the mGlu5 negative allosteric modulator HTL14242 (3-Chloro-5-[6-(5-fluoropyridin-2-yl)pyrimidin-4-yl]benzonitrile). J Med Chem 58:6653–6664
69. Goudet C, Gaven F, Kniazeff J et al (2004) Heptahelical domain of metabotropic glutamate receptor 5 behaves like rhodopsin-like receptors. Proc Natl Acad Sci USA 101:378–383
70. Manglik A, Lin H, Aryal DK et al (2016) Structure-based discovery of opioid analgesics with reduced side effects. Nature 537:185–190
71. Kumar V, Bonifazi A, Ellenberger MP et al (2016) Highly selective dopamine D3 receptor (D3R) antagonists and partial agonists based on Eticlopride and the D3R crystal structure: new leads for opioid dependence treatment. J Med Chem 59:7634–7650
72. Saavedraa OM, Karilaa D, Brossarda D et al (2017) Design and synthesis of novel N-sulfonyl-2-indoles that behave as 5HT6 ligands with significant selectivity for D3 over D2 receptors. Bioorg Med Chem Lett 25:38–52
73. Coleman JA, Green EM, Gouaux E (2016) X-ray structures and mechanism of the human serotonin transporter. Nature 532:334–339
74. Topiol S, Bang-Andersen B, Sanchez C et al (2016) Exploration of insights, opportunities and caveats provided by X-ray structures of hSERT. Bioorg Med Chem Lett 26:5058–5064
75. Topiol S, Bang-Andersen B, Sanchez C et al (2016) X-ray structure based evaluation of analogs of citalopram: compounds with increased affinity and selectivity compared with R-citalopram for the allosteric site (S2) on hSERT. Bioorg Med Chem Lett 27:470–478
76. Bang-Andersen B, Ruhland T, Jørgensen M et al (2011) Discovery of 1-[2-(2,4-Dimethylphenylsulfanyl)phenyl]piperazine (Lu AA21004): a novel multimodal compound for the treatment of major depressive disorder. J Med Chem 54:3206–3221
77. Sun Y, Xu W, Fan N et al (2017) Design, synthesis and biological evaluation of (E)-3,4-dihydroxystyryl 4-acylaminophenethyl Sulfone, Sulfoxide derivatives as dual inhibitors of HIV-1 CCR5 and Integrase. Bioorg Med Chem 25:1076–1084
78. Bhattacharya S, Salomon-Ferrer R, Lee S et al (2016) Conserved mechanism of conformational stability and dynamics in G-protein-coupled receptors. J Chem Theory Comput 12:5575–5584
79. Segala E, Guo D, Cheng RKY et al (2016) Controlling the dissociation of ligands from the adenosine A_{2A} receptor through modulation of salt bridge strength. J Med Chem 59:6470–6479
80. Chen X, Klöckner J, Holze J et al (2015) Rational design of partial agonists for the muscarinic M1 acetylcholine receptor. J Med Chem 58:560–576
81. Fronik P, Gaiser BI, Pedersen DS (2017) Bitopic ligands and metastable binding sites: opportunities for G protein-coupled receptor (GPCR) medicinal chemistry. J Med Chem 60:4126–4134
82. Busnelli M, Kleinau G, Muttenthaler M et al (2016) Design and characterization of superpotent bivalent ligands targeting oxytocin receptor dimers via a channel-like structure. J Med Chem 59:7152–7166
83. Portoghese PS, Akgün E, Lunzer MM (2017) Heteromer induction: an approach to unique pharmacology? ACS Chem Neurosci 8:426–428

Chapter 2

Characterization of Ligand Binding to GPCRs Through Computational Methods

Silvana Vasile, Mauricio Esguerra, Willem Jespers, Ana Oliveira, Jessica Sallander, Johan Åqvist, and Hugo Gutiérrez-de-Terán

Abstract

The recent increase in available G protein-coupled receptor structures now contributes decisively to the structure-based ligand design. In this context, computational approaches in combination with medicinal chemistry and pharmacology are extremely helpful. Here, we provide an update on our structure-based computational protocols, used to answer key questions related to GPCR-ligand binding. All combined, these techniques can shed light on ligand binding modes, determine the molecular basis of conformational selection, for agonists and antagonists, as well as of subtype selectivity. To illustrate each of these questions, we will consider examples from existing projects on three families of class A (rhodopsin-like) GPCRs: one small-molecule (nucleotide-like) family, i.e., the adenosine receptors, and two peptide-binding receptors: neuropeptide-Y and angiotensin II receptors. The successful application of the same computational protocols to investigate this diverse group of receptor families gives an idea of the general applicability of our methodology in the characterization of GPCR-ligand binding.

Key words Homology modeling, Molecular dynamics, Free energy perturbation, Structure-based drug design

1 Introduction

The outstanding interest from the biopharmaceutical industry in G protein-coupled receptors (GPCRs) is usually illustrated with data that have not changed for decades: approximately one third of the current FDA-approved drugs target a GPCR to some extent, including many blockbusters in the pharmaceutical industry [1]. The oldest drugs in this group have been on the market even before the first receptor was cloned, as a result of a more traditional phenotypic screening. Since the late 1980s, the functional characterization of GPCRs through advances in molecular biology and

Electronic supplementary material: The online version of this article (https://doi.org/10.1007/978-1-4939-7465-8_2) contains supplementary material, which is available to authorized users.

Alexander Heifetz (ed.), *Computational Methods for GPCR Drug Discovery*, Methods in Molecular Biology, vol. 1705, https://doi.org/10.1007/978-1-4939-7465-8_2,

pharmacology has guided drug discovery in the field. With the new century, structural characterization of the targeted receptors started to be an accessible reality, allowing for the alluring perspective of GPCR structure-based drug design [2, 3]. The signature of GPCR ligand design and functional characterization has been a blend of biochemical studies, pharmacology, medicinal chemistry efforts, and computational modeling. In this framework, computer-aided ligand design for GPCRs has evolved from an era dominated by ligand-based techniques to the use of mature structure-based methods, such as virtual screening or free energy calculations [4, 5].

1.1 Structural Biology of GPCRs: Implications in Ligand Design

The superfamily of GPCRs is comprised of almost 900 members in humans and is clustered into five classes [6]. The most populated class A (rhodopsin-like) is further subdivided into four branches (α–δ), containing more than 50 families (defined by the nature of the endogenous ligand), approximately 300 olfactory receptors (sensitive to odorant molecules), and a number of orphans (for which no endogenous ligand has yet been identified). Nowadays, 20 of these class-A families have at least one structure solved, where a few privileged families stand out with several members and/or several conformations of the same receptor crystallized (i.e., opioids, muscarinic, adenosine). Lately, the structural biology map of GPCRs has been enriched with representatives of classes B, C, and F (*see* http://gpcrdb.org/structure/statistics). As opposed to the sequence diversity within the GPCR superfamily, the conserved topology of seven transmembrane helices (7TM) connected by three extracellular (EL) and three intracellular loops (IL) is an important aid in the 3D modeling of GPCRs. When none or very few structures were available, a popular strategy to build 3D models of GPCRs was based on ab initio modeling of the 7TM helices and subsequent packing onto the conserved 7TM topology. Though this method is still valid, the growing number of experimental GPCR structures now allows for the generation of high-quality homology-based 3D models [7–9]. As a result, both the experimentally determined structures and high-quality computer models are nowadays used in structure-based programs for ligand design [5].

Long before the crystal structure determination of GPCRs started to blossom, site-directed mutagenesis (SDM) was widely used to explore the binding sites of GPCRs, alone or in combination with computational modeling [10]. Here, one evaluates the experimental shift in binding affinity for one (or multiple) compounds upon a given mutation, which allows an indirect mapping of the residues that are important for binding of the molecule(s) of interest. Far from being an obsolete approach, SDM has proved itself very useful to map the precise protein-ligand interactions and is also applicable in cases where a crystal structure is available.

A paradigmatic example in real drug design projects is the systematic evaluation of an alanine scan of eight residue positions in the A_{2A} adenosine receptor on an array of ligands. This pipeline was designed by the company Heptares and denoted as biophysical mapping (BPM) [11]. Altogether, the accumulated data on structural and molecular biology on some receptors is allowing the structure-based discovery of novel compounds in the pharmaceutical industry [3].

1.2 Case Studies

We recently documented a combination of structure-based and ligand-based design techniques, as a strategy to assess the growth of a chemical library of adenosine receptors antagonists [12]. In this chapter, we provide an update on the related structure-based computational protocols, used in our lab to answer key questions related to GPCR-ligand binding: elucidation of ligand binding mode(s), determining subtype selectivity, molecular determinants of conformational selection for agonists and antagonists, and ligand optimization.

To illustrate each of these questions, we will consider examples from existing projects on three families of class-A (rhodopsin-like) GPCRs: one small-molecule (nucleotide-like) family, i.e., adenosine receptors (ARs), and two families of peptide binding receptors: neuropeptide-Y (NPY) and angiotensin II (AT) receptors (Fig. 1). The structural knowledge of these families is also diverse; with one representative subtype being crystallized for adenosine and angiotensin receptors[1] while no crystal structure is available for any NPY receptor. The application of the same computational protocols to this wide range of receptor families should give an idea of its general applicability in characterization of GPCR-ligand binding. We will here briefly describe each of the receptor families considered in this chapter.

1.3 Adenosine Receptors

Adenosine is the signaling molecule that activates four subtypes of class A GPCRs: A_1, A_{2A}, A_{2B} and A_3 adenosine receptors (ARs) [13]. These are highly demanded therapeutic targets, ubiquitously expressed in the human body and associated with several diseases such as several inflammatory processes (A_{2A} and A_3); respiratory pathological events such as allergic asthma (A_{2B} and A_3); vascular diseases (A_{2A}) as well as arrhythmias and stroke (A_1). Because of the widespread of the adenosine signaling system, and the high homology among the four ARs, the development of selective ligands is challenging. Traditionally, ligand design toward ARs has utilized ligand-based techniques, such as QSAR, as well as on modification of complex heterocycles with poor pharmacokinetic properties.

[1] During the processing of this manuscript one structure for the AT_2 receptor (5UHN) and 2 structures for the A_1 adenosine receptor (PDB codes 5N2S and 5UEN) were released.

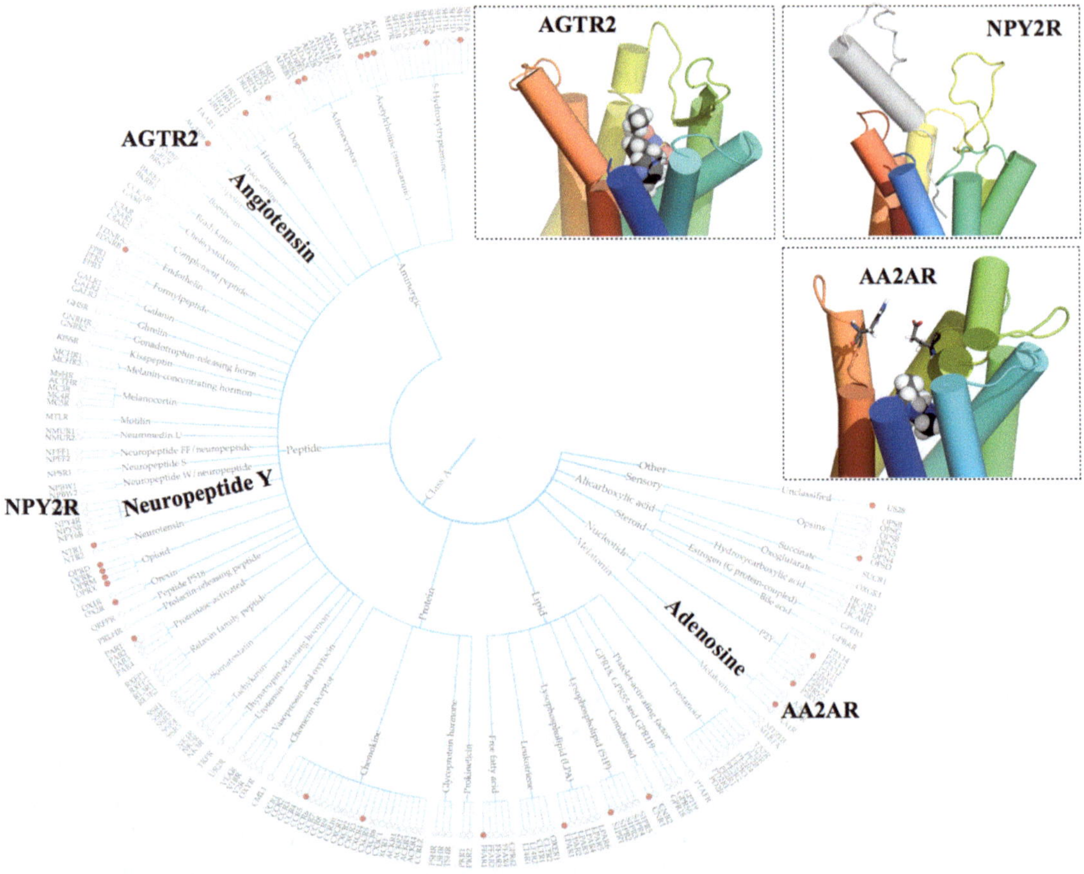

Fig. 1 Phylogenetic tree of GPCRs, indicating the location of the receptors considered in this chapter. Red dots denote families with at least one crystal structure. The *insets* show the 3D structure of the binding site of each receptor considered in this chapter, as obtained with the methodology here described

The field has dramatically changed with the crystallization of the A_{2A}AR, which stands as one of the best-characterized GPCRs at the structural level. The 12 crystal structures in the PDB include the inactive form co-crystallized with antagonists of three scaffolds: triazolopyrimidines (ZM241385), xantines (Caffeine, XAC), and triazines (T4G and T4E), as well as active-like conformations revealed in complex with adenosine and related agonists NECA and UK432097, and recently the fully active receptor in complex with a G-protein mimic [3].

The lack of X-ray structures of A_1, A_{2B}, and A_3 ARs and the high sequence similarity within the four subtypes prompt for the use of homology modeling techniques, as a basis to compute the determinants of high affinity and selectivity within the family members, as we described recently [12].

1.4 Neuropeptide Y Receptors

The next case study belongs to the neuropeptide-Y (NPY) system of signaling peptides and receptors mediating important physiological and behavioral processes, such as appetite regulation, anxiety, pain or learning, and memory control [14]. As a consequence, there is increasing interest from the pharmaceutical sector in the regulation of this system, in particular in the search of anti-obesity drugs [15]. Neuropeptide Y (NPY), peptide YY (PYY), and pancreatic polypeptide (PP) are each 36 residues peptides, arranged as a proline-rich N-terminus followed by a conserved α-helical central fold and an unwound amidated C-terminal pentapeptide. The three peptides are the natural ligands for the four Y receptors expressed in humans (Y_1, Y_2, Y_4 and Y_5; more subtypes are found in other species) [14]. While the modification of these peptides can lead to more selective agonists for any of the Y receptors, most antagonists are peptidomimetics of the C-terminal tail of the natural agonists.

During the past years, we have used and developed the computational protocols described here in the search for antagonists of Y_1 and agonists for the Y_2 receptors. With no crystal structure available for any of the Y receptors (Fig. 1), homology models of the two receptors were built, and used to define the antagonist and agonist binding modes respectively. The hY_1 receptor in complex with antagonist BIBP3226 was further used to develop our free energy perturbation (FEP) protocol, which we apply to characterize the effect of protein mutations and ligand SAR for different GPCRs [16]. As for the hY_2 receptor, the homology model of the active-like conformation was used to define the binding mode of the natural agonists NPY and PYY [17], in an example of receptor-peptide docking illustrated below.

1.5 Angiotensin Receptors

The renin-angiotensin system (RAS) produces hormonal peptides, which signal through the angiotensin receptors. The system is critical for cardiovascular control, impacting normal physiology and disease pathogenesis [18]. Although several biologically active peptides are generated by this system, its major actions are mediated by the peptide hormone angiotensin II (sequence DRVYIHPF) acting through its type 1 (AT_1) and type 2 (AT_2) receptors. Recent elucidation of crystal structures of human AT_1 receptor bound with the antagonists ZD7155 [19] and Olmesartan [20] facilitates discussion of the AT_1 receptor, though for the AT_2 receptor we still have to rely on homology models [21]. Furthermore, when studying interactions between agonists or antagonists these receptors need to be modeled in both their active-like and inactive conformations by means of homology-based modeling. In a collaborative study addressing the structural determinants of subtype and functional selectivity of AT receptors, we generated homology-based models of the two subtypes in the two relevant conformations [22]. These models were used to characterize the binding mode of the natural peptidic agonist ATII, and of four synthetic ligands that display different receptor subtype and conformational selectivities. The details of the multiple-template homology modeling and the characterization of selectivity will be discussed in further sections of this chapter.

2 Theory/Materials

2.1 Homology Modeling and MD Refinement with GPCR-ModSim

Despite the steady increase in the number of GPCR crystal structures, the reality is that the majority of receptors have yet unknown structure. In these cases, homology (or comparative) modeling remains one of the most important techniques for obtaining a reasonable 3D structural model to use in rational ligand design [7–9]. As a result, homology modeling is implemented in a number of specific web-servers, among which GPCR-ModSim emerges as a complete tool for the modeling and simulation of GPCRs [23, 24] (*see* **Note 1** for alternative solutions).

We will describe here the use of the last version of GPCR-ModSim [24], freely accessible at the web address http://gpcr-modsim.org (*see* **Note 2** about optional academic accounts). The only input required is a FASTA sequence or the UNIPROT code of the GPCR to be modeled. Thereafter, the modeling process consists of consecutive steps, which are illustrated below for the case of the AT_2 angiotensin receptor [22] (*see* **Note 3** for additional examples).

1. *Identification of the best template and generation of a template (s)/target pairwise sequence alignment.* This is done by performing a multiple sequence alignment (MSA) of the query sequence against a profile of available templates of known 3D structure. While in the case of a random protein this step involves a BLAST search against a large database of non-redundant protein sequences or available PDB structures, in GPCR-ModSim we restrict this search to a carefully curated structure-based profile-alignment of crystallized receptors. These are classified into three categories depending on receptor conformation: inactive (22 structures), active-like (8 structures), and fully active conformations (3 structures). The default criterion is to select the template with highest sequence identity in the transmembrane region (*see* **Note 4** for additional considerations). However, in cases of moderate sequence identity the overall structural similarity between the obtained model and the single template used can be artificially high, a problem that can be counterbalanced with the choice of additional templates [25, 26]. To use this option, a pairwise sequence identity (SI) is provided for each of the topological regions, aiding in the selection of the best template(s) for each region. Since the whole modeling process is based on the

accuracy of the assignment of topologically equivalent positions, the user can manually edit the MSA and produce a refined pairwise-sequence alignment. Here, one should account for the relatively infrequent distortions within the TM segments (which should be translated in terms of alignment gaps) and the high variability within loop regions.

2. *Model building*: the generation of the 3D structure of the query sequence is done by satisfying the set of spatial restraints, following the routines in Modeller [27], and dictated by the pairwise alignment with the *target* sequence. The models generated are ranked on the basis of a scoring function, and further evaluated for their stereochemical quality and other subjective criteria such as agreement with available mutagenesis data.
3. *Membrane insertion and Molecular Dynamics*: The resulting 3D–model can be further refined by molecular dynamics (MD) simulations, a process that can also be applied once a receptor-ligand complex has been generated (*see* Subheading 3.2.1). GPCR-ModSim uses a fully automated pipeline for membrane insertion and equilibration protocol called PyMemDyn [28]. The receptor (or the receptor-ligand complex) is automatically embedded into a pre-equilibrated lipid bilayer of POPC lipids in a way that its principal axis will be aligned with the vertical axis of the membrane. After the insertion in the membrane the whole system is placed in a hexagonal-prism shaped box, which is then solvated with water. If necessary the whole system is neutralized with counterions. Then a short steepest descent energy minimization is performed, before MD equilibration under periodic boundary conditions (PBC). The MD equilibration protocol consists of a recipe where position restraints on all heavy atoms are gradually released from a force constant of 1000 to 200 kJ/mol· nm^2 during 2.5 ns. Then another 2.5 ns simulation follows where NMR-style distance restraints are applied between pairs of residues, which have been identified to be interacting in a conserved way, forming an interhelical contact network as derived from the analysis of a non-redundant set of X-ray structures [28].

2.2 Ligand Docking

Once the 3D structure of the GPCR of interest is available, either from crystallography or from molecular modeling, the next step in the ligand design process is to define the binding mode of the ligand(s) of interest. The goal here is to describe the protein–ligand interactions at the physicochemical level and use this information to validate the structural model to guide further ligand optimization and establishment of the SAR. Depending on the chemical nature of the ligand (e.g., a small molecule or peptide), the number of ligands, and the quality of the 3D structural model of the receptor, the docking process can to a certain extent be automated. A number of docking algorithms are available today, which aim to identify the best binding mode for a given ligand. This is based on an exhaustive search of the rotational, translational, and internal degrees of freedom of the ligand around the binding site explored, in combination with the evaluation of the binding affinity of the candidate binding modes identified using a mathematical scoring function. The following software is used for GPCR-ligand docking in our lab:

- For small-molecule ligand projects, we mainly used the GOLD software [29] available (under license) from CCDC (*see* **Note 5** for alternative docking tools).
- The Protein Preparation Wizard workflow of the Schrödinger suite (Schrödinger LCC, New York, under license) is used to prepare the protein prior to the docking experiments [30].
- Peptide–protein docking can be considered a protein-protein problem. Thus, we selected the program HADDOCK for this purpose in one of our strategies. This program is available from the author's webpage (http://www.bonvinlab.org/software/haddock2.2/), where it can be freely downloaded and installed. However, the routines followed in our projects can also be reproduced using the latest web server edition at http://milou.science.uu.nl/services/HADDOCK2.2/. This docking algorithm is driven by constraints, originating from a list of "active" residues elaborated by the user for each of the two proteins. The program then generates a second list of residues defining the docking-surface contact of each protein ("passive" residues), and a list of "ambiguous distance restraints" is defined between the "active" residues of one protein and the "passive" residues from the other. During the docking search, an energy penalty is assigned for any of the ambiguous distance restraints not satisfied.
- Docking analysis, as well as manual docking adjustments that are part of our peptide-docking strategy, is performed with the molecular packages from the Schrödinger suite (through the Maestro graphical interface) and PyMOL (Version 1.4.1, free academic license) [30].

We will illustrate two different docking approaches in the next section with examples from docking of a series of small molecules in adenosine receptors and incremental peptide docking for the neuropeptide Y receptors. The files needed to reproduce some steps are available at the web address http://gpcr-modsim.org/tutorials.

2.3 Computation of Binding Free Energies

While docking provides a useful tool for the generation of possible binding poses, scoring functions generally fail to accurately predict the free energy of ligand binding. Therefore, rescoring with more elaborate methods for the estimation of free energy of binding is advised [31]. The most theoretically rigorous method is free energy perturbation (FEP), sometimes referred to as "computational alchemy" [32]. Here, one can estimate relative binding affinities of pairs of compounds or compare binding affinities between the wild-type (wt) and single-mutant versions of the same receptor. A historical limitation of FEP was that the compared complexes could not be too dissimilar in order to achieve accurate and converged results. An increase in computational power allowed overcoming these limitations, and we can affirm that nowadays FEP methodologies can be applied in the assessment of GPCR-ligand binding affinities in a systematic fashion. This is done in our lab from a double perspective: the evaluation of SAR for ligand series and, thanks to an appropriate redesign of the thermodynamic cycles involved, the estimation of SDM effects on ligand binding affinities. We will illustrate such applications from our extensive applications on the adenosine A_{2A} receptor [33, 34].

The FEP pipeline described here is implemented in our molecular dynamics package Q, which is available from the web address http://www.icm.uu.se/cbbi/aqvist-lab/q/. The current distribution of the program, Open source under the GPL Version 2 License, includes the necessary files to perform in silico SDM, based on the protocol recently described in detail in [16, 34]. The protocol applies a smooth and progressive annihilation of the atoms in the wt aminoacid sidechain to convert it into alanine (Ala-mutant). The transformation is by definition run in two parallel molecular dynamics (MD) simulations of the binding site, i.e., in presence (holo) and absence (apo) of the ligand. Then the binding free energy difference between the wt and Ala-mutant receptors can be calculated solving a standard thermodynamic cycle (left cycle in Fig. 2). The sidechain atoms are grouped following the definition of charge groups in the OPLS-AA forcefield (used in our MD simulations), and each group will undergo three consecutive transformations (a) annihilation of partial charges, (b) transformation of regular van der Waals (Lennard-Jones) potential to a soft-core potential to prevent singularities, and (c) annihilation of the soft-core potential. A given mutation is thus divided into a series of smaller subperturbations, each of which is in turn divided into 50 FEP windows (commonly known as λ-steps). Each λ-step is sampled with several short MD simulations. Using this scheme one can mutate basically any residue to alanine (proline is more difficult due to its cyclic backbone bonding). Moreover, by joining two thermodynamic cycles describing the reduction of a sidechain, for wt and mutant, to a common fragment (Ala-mutant), we can simulate the effect of any non-alanine mutation (*see* the extended cycle in Fig. 2). Since the number of amino acid residues is limited, the library provided in the Q distribution, containing all protocols for wt → Ala-mutant, is sufficient to sample the chemical space of any peptide or protein. The smooth annihilation ensures high

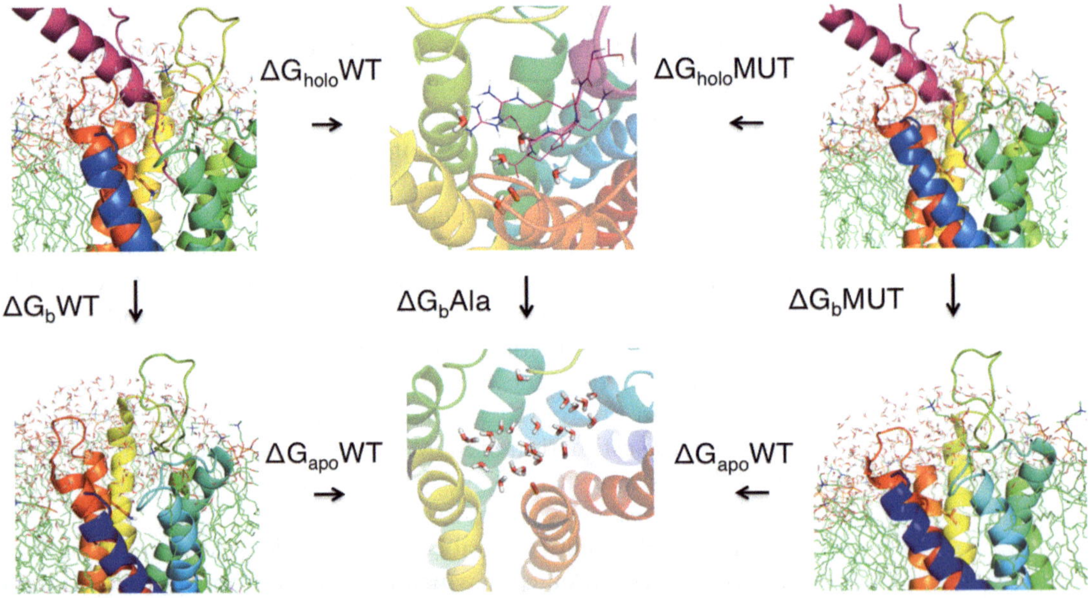

Fig. 2 Thermodynamic cycle used to perform in silico SDM. For a mutation of glutamine to leucine, one needs to perform four independent MD simulations (*horizontal arrows* in the figure) to transform each sidechain into alanine, both in the presence (*top panels*) and in the absence (*bottom panels*) of the ligand. The figure illustrates the quantification of the effect of Gln (wt) to Leu (mut) in the binding affinity of the PYY peptide into the Y_2 receptor (*vertical arrows*)

precision, fidelity, and a minimum hysteresis, which is the measure of convergent results.

The change in free energy between the final (*f*) and initial (*i*) states is calculated for each subperturbation by applying the FEP formula

$$\Delta G = G_f - G_i = -RT \sum_{m=1}^{n-1} \ln \langle \exp[-(U_{m+1} - U_m)/RT] \rangle_i$$

where

$$U_m = (1 - \lambda_m) U_i + \lambda_m U_f$$

Thus, concatenation of the energy change obtained upon each subperturbation gives the total free energy change for the given leg in the thermodynamic cycle (i.e., ΔG_{holo}WT, ΔG_{apo}WT, etc.). This analysis is performed for both the forward and the reverse (projected) pathways of the transformation, and the difference in the two values is used to assess the convergence of the simulations, by means of the hysteresis

$$\text{Hysteresis} = |\langle \Delta G_{fwd} \rangle - \langle \Delta G_{rev} \rangle|$$

where $\langle \Delta G \rangle$ are the average values of Gibbs free energies over several replicate MD trajectories (i.e., same conditions but different initial velocities) in the forward (fwd) and reverse (rev) directions. The total hysteresis is the sum of the hysteresis values for each subperturbation of the transformation taken into account.

We calculate the errors for the relative binding free energies as the standard error of the mean (s.e.m.).

The same stepwise annihilation of functional groups to a core scaffold is applied to study the SAR of ligands. However, as opposed to the residue mutations, the ligand chemical space is practically infinite, and the generation of a unified library or perturbation of functional groups is not possible. Here, one has to design the perturbation route to connect initial and ending states in each specific case.

3 Methods

3.1 Homology Modeling of the AT_2 Receptor

This example, extracted from our ligand design project for the AT_2 receptor, illustrates the capacity of both single template and multiple-template homology modeling. The full process has been adapted to be followed using the last version of GPCR-ModSim, and consists of the following steps:

- The native sequence for this receptor (P50052) can be downloaded from the Uniprot portal and manually edited in order to remove the long N-terminus (1–32) and C-terminus (336–360) fragments, because of the lack of templates for these regions among the crystallized GPCRs.
- The edited sequence is then uploaded to GPCR-ModSim. Select "Model a GPCR," and paste the edited sequence into the window. Leave the "inactive" templates as we will model the inactive conformation of the receptor in this example and press "Submit." The next window shows a range of options to select templates. We will illustrate both single and multiple template options in parallel modeling stages.
- *Option 1: Single-template modeling*. The server will offer as the best template the AT_1 receptor (PDB code 4YAY), which has a sequence overall/TM identity of 31/42% respectively. Click on "Model," select "10" number of models and click on "Submit." Once the process is finished, we will select the model with highest DopeHR score.
- *Option 2: multiple-template modeling*. In the "Model" window, one can select additional templates for different topological regions. The choice of templates is indicated in Table 1, and is based on dual criteria: the matrix of partial similarities provided by GPCR-ModSim, plus the consideration of phylogenetic relationships. The overall idea is to preserve the sequences with higher homologies for each topological region of the receptor.

Table 1
Templates selected, indicating the topological regions where each template is considered, for the multiple template homology modeling of the AT_2 inactive receptor

PDB code	Name	Regions
4YAY	hAGTR1	TM1-H8
4MBS	hCCR5	TM1-H8
4DJH	hOPRK	IL1
4DKL	mOPRM1	TM3
2Z73	sRHO	TM5
3VW7	hPAR1	TM6

Once the additional templates have been indicated, proceed to model the receptor as in the previous option.

- MD refinement: the homology model created with either option can then be submitted to our molecular dynamics (MD) equilibration protocol, by clicking on the corresponding frame of the selected model and select "Run MD." After 24 h, the output from MD simulations comes in a downloadable bundled compressed file (MD_output.tgz), including a PyMol script (load_gpcr.pml) to load the trajectory in PyMol, and two folders with detailed outputs (GROMACS format) and reports as plots in xmgrace formatted xvg files. Finally, a README file is included which explains the steps to isolate, minimize, and export a selected frame from the MD simulation.

The model built using "Option 2: multiple templates" is an adaptation to that recently reported by us, and will be used in the next sections to explain the selectivity of AT_2-specific ligands as compared to the crystallized AT_1 receptor [22] (*see* **Note 6** about active-like homology models).

3.2 Ligand Docking and Initial Screening

We illustrate here our docking protocols for small molecules as well as peptide docking to GPCRs. Small-molecule docking protocols are used to elucidate common binding mode of series of ligands in the adenosine receptors, and to identify the binding mode of AT_2 receptor agonists and antagonists. The peptide-docking protocols, on the other hand, allow us to identify the binding mode of the peptidic natural agonists of the AT receptors and NPY receptors.

3.2.1 Protein Preparation

Regardless of whether we are exploring the binding of a small molecule, a peptide, or a protein, the first step is to prepare the conformation of the receptor considered. As we have discussed, this might have been generated by homology modeling (see the AT_2

receptor modeled as illustrated above, or the neuropeptide Y receptors) if there is no crystal structure available. But even if this is the case, like for the A_{2A} adenosine receptor, we need to prepare the receptor structure for docking. In this example, we will use the inactive conformation of the A_{2A}AR with PDB code 4EIY (*see* **Note 7**).

The initial protein structure will be in the PDB format. Import this file into Maestro, and prepare it with the Protein Preparation Wizard workflow using default settings. At this stage, one has to pay attention to the protonation state of titratable residues and histidines, as well as the configuration of polar sidechains, and the crystallographic structural waters to retain. For the 4EIY structure, $His^{6.52}$250 and $His^{7.43}$278 should be modeled as neutral and protonated at Nδ1, based on our early works on this receptor [35]. The resulting structure is then saved in MOL2 format, and is ready to use in any of the following docking protocols (*see* file A2AAR_dock.mol2).

3.2.2 Small-Molecule Docking

- The ligands are drawn in the Maestro suite in 2D format, or alternatively imported into this program from a database (i.e., SDFile). Thereafter, a 3D structure is obtained with the LigPrep utility in the Schrödinger package considering the following options: the ionizable groups are protonated at physiological pH with the Epik extension, and all stereoisomers and tautomers are saved separately to be considered for parallel docking runs. A database of ligands is saved in SDF format (*see* file ligands_A2A.sdf).
- We check if there is evidence for considering water molecules as part of the binding site. If so, two parallel docking calculations are computed: one without any water molecule, and a second run where we include selected water molecules. The waters should have been extracted in the protein-preparation stage, saved as a separate PDB file, and considered for docking with the corresponding option in GOLD "toogle" option from GOLD Suite 5.2. With this flag, only the water molecules that improve the binding score are retained and considered for the binding (no water molecules are considered in this example).
- A typical docking run with GOLD in our lab considers the following parameters: each ligand is docked 20 times with default (high accuracy) genetic algorithm search parameters, using the scoring function Chemscore. The ligand is fully flexible, including the consideration of amide bond flipping and rotation of protein hydroxyl groups. We typically consider a sphere of 15 Å radius defining the binding pocket of the receptor, which in the AR projects is centered on the sidechain (CD1) of $Ile^{7.39}$. As a second example, in the AT projects a docking sphere of the same size is centered on an equidistant point between residues $K^{5.42}$, $R^{4.64}$, and $Y^{7.43}$.
- The selection of the docking pose is done by a combination of three criteria. (a) The binding mode proposed for a congeneric

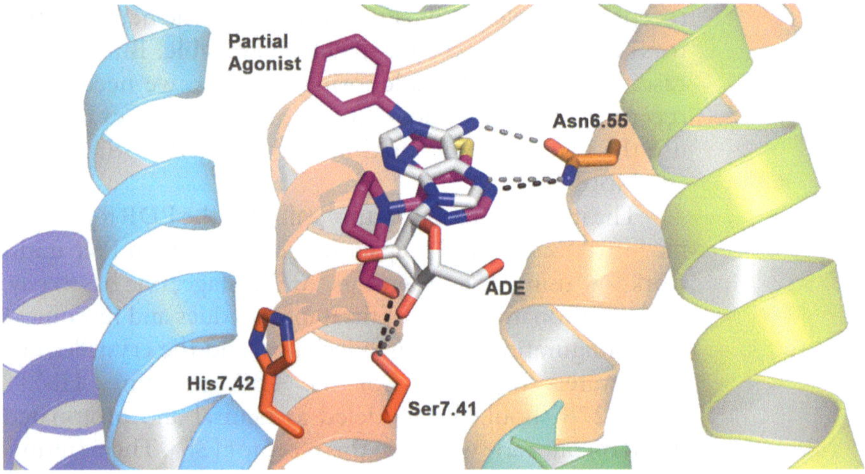

Fig. 3 Binding mode of the A_{2A}AR partial agonist used in this example [36] (*magenta*) in the crystal structure of the receptor (*rainbow color*, TM1-*blue* → TM7-*red*), superimposed on the crystallographic conformation of adenosine bound to the A_{2A}AR (*gray*)

series of ligands should be conserved, (b) good internal Chemscore ranking should be achieved (e.g., being within the top 25% poses) and (c) the population of the selected solution (according to a RMSD clustering criteria of 1 Å) should be as high as possible (*see* **Note 8** for additional considerations).

- MD equilibration of the complex. Optionally, the selected docked complex can be refined with the MD equilibration protocol in GPCR-ModSim. The receptor and ligand coordinates can be uploaded to the web server under the menu "Model a GPCR" → "Upload your GPCR pdb file" (*see* **Note 9** about the files needed).

The example that illustrates this protocol is the docking of 7-(Prolinol-*N*-yl)-2-phenylamino-thiazolo[5,4-*d*]pyrimidines to the A_{2A}AR, extracted from reference [36]. The results, which can be reproduced with the files provided, are summarized in Fig. 3. This conserved binding mode was found among the top three ranking poses in more than 78% of the compound series. In this pose, the ligand is anchored by a hydrogen bond interaction between the conserved $Asn^{6.55}$ and N4 of the ligand scaffold. Even if the docking is performed in the inactive structure of the receptor, it becomes evident by superposition with the agonist-like structure (PDB code 2YDO, in complex with adenosine) that the compounds bearing a 2-hydroxymethyl pyrolidine at R2 can mimic the interactions of the ribose in the active-like structure. Re-docking onto the active-like structure converged into the same binding pose, which reinforced the hypothesis of the hydroxyl mimicking. A hypothesis that was experimentally tested when the compounds were confirmed in vitro as partial agonists.

3.2.3 Peptide Docking

The protocols for peptide docking will be illustrated with the docking of the NPY peptide into the homology model built for the hY_2 receptor, as described in reference [17]. Two parallel strategies are considered:

Automated Docking with HADDOCK

This is a two-step strategy: first, the C-terminal dipeptide [CH_3C(O)-R35-Y36-NH_2] is docked into the TM crevice of the homology model, and the selected docking pose is used in a second stage to define distance restraints that guide the automated protein–protein docking.

- The 2D structure of the dipeptide is drawn with the help of the Maestro suite and the 3D version is obtained with LigPrep (Schrödinger LCC, New York, NY).
- Docking is performed with GOLD, using the same parameters as for small-molecule docking except for the following: sphere of 25 Å radius centered in the middle point between residues Thr2.61 and Gln6.55; 50 docking runs; clustering according to a 5 Å RMSD.
- Building of the initial structure of the NPY peptide: this is done with default homology modeling settings in Modeller, using as a template the structure of the aPP peptide (PDB code 2BF9, 53% seq ID), and selecting the best model according to the DOPE-HR scoring function.
- Docking with HADDOCK: based on mutagenesis data and the contacts estimated from the docking of the dipeptide with GOLD, residues Y36, R35, and R33 from the peptide and residues $Tyr^{2.64}$, $Tyr^{3.30}$, $Gln^{3.32}$, $Tyr^{5.43}$, $Asp^{6.59}$, and $Leu^{6.51}$ from the receptor are selected as active residues to define the "ambiguous distance restraints" guiding the docking. The best 200 structures are subjected to a rigid-body energy minimization (2000 runs) in explicit solvent, selecting DMSO to better represent the membrane environment of GPCRs.

Automated Docking of the C-Terminus and Manual Elongation

This is also a two-step strategy, where the automated docking is performed on a larger fragment (five residues) of the unwound C-terminal tail of the peptide, and the adjustment of the full α-helical structure of the NPY is done at a second stage by manual docking followed by geometry optimization and MD.

- The structure of the pentapeptide [$CH_3C(O)$-$^{32}TRQRY^{36}$-NH_2] is obtained as described previously (Subheading "Automated Docking with HADDOCK").
- The Induced Fit Docking in Schrödinger Suite 2011 (Schrödinger LCC, New York, NY) is used to dock the pentapeptide with default settings. The grid of 30x30x30 points is centered in the same point as described in the "Haddock" protocol.

- MD equilibration of the resulting top-scored pose, following the MD protocol in GPCR-ModSim as we also indicated in the "small molecule docking" section.
- The fragment of the NPY peptide with a defined secondary structure (residues 1–31) is then attached to the equilibrated pose of the ^{32}TRQRY^{36}-NH$_2$ pentapeptide using Maestro. The Y$_2$-NPY complex follows 1000 steps of energy minimization with Macromodel, with positional constraints to preserve the secondary structure of peptide and the protein (*see* **Note 10**).

3.3 Analyzing Subtype Selectivity

As stated in the introduction, the overall protein fold around the TM binding site is well conserved among GPCRs, though the meta-analysis of crystal structures reveals some differences in helical bending and orientation [37]. Earlier analyses revealed that the phylogenetic relationship within the GPCR superfamily could be reproduced by multiple sequence alignment of the pseudosequence comprising the residues in the binding crevice [38]. Such a pseudosequence alignment is the starting point for our exploration of selectivity issues. Here, one can map the ligand affinity data on receptor subtypes and receptor mutagenesis studies, with structural and sequence differences. A first inspection at this level might give some hints for the topologically equivalent positions that are responsible for selectivity, which can be further examined by the in silico mutagenesis protocol described in the next section.

In the angiotensin project, we modeled the two AT receptors in both the active-like and inactive conformations (*see* Subheading 3.1 and **Note 6**). Two pairs of agonists and antagonists, derived from the same chemical scaffold, were docked as explained in Subheading 3.2.1. A conserved binding mode was defined, which is shown in Fig. 4. The main anchoring points are salt-bridge interactions of the sulfonyl carbamate group with residues $K^{5.42}$ and $R^{4.64}$, conserved in all cloned angiotensin receptors. The imidazole ring, which is the only structural difference in selective compounds, is pointing toward the extracellular side of transmembrane regions TM1-TM2 and TM7. The pseudosequence alignment of the binding crevice (Fig. 4) revealed differences in the hydrophobic cluster composed by the residues $F/L^{2.53}$, $L^{2.57}$, $W^{2.60}$, $T/Y^{2.64}$, $Y^{2.65}$, $V/L^{3.32}$, $P^{7.36}$, and $I^{7.39}$ (note the notation AT_1/AT_2 for differing amino acid positions between the two receptors), as illustrated in Fig. 4.

3.4 Binding Free Energy Simulations: In Silico Mutagenesis and SAR

Our FEP protocol is here illustrated to reproduce the effect of a single-point mutation $H250^{6.52}N$ in the $A_{2A}AR$, which slightly favors the binding affinity of the agonist NECA, and is extracted from our extensive in silico characterization of this system [34].

- The receptor complex, which is obtained in this case from a crystal structure (PDB code 2YD0), is embedded in a lipid

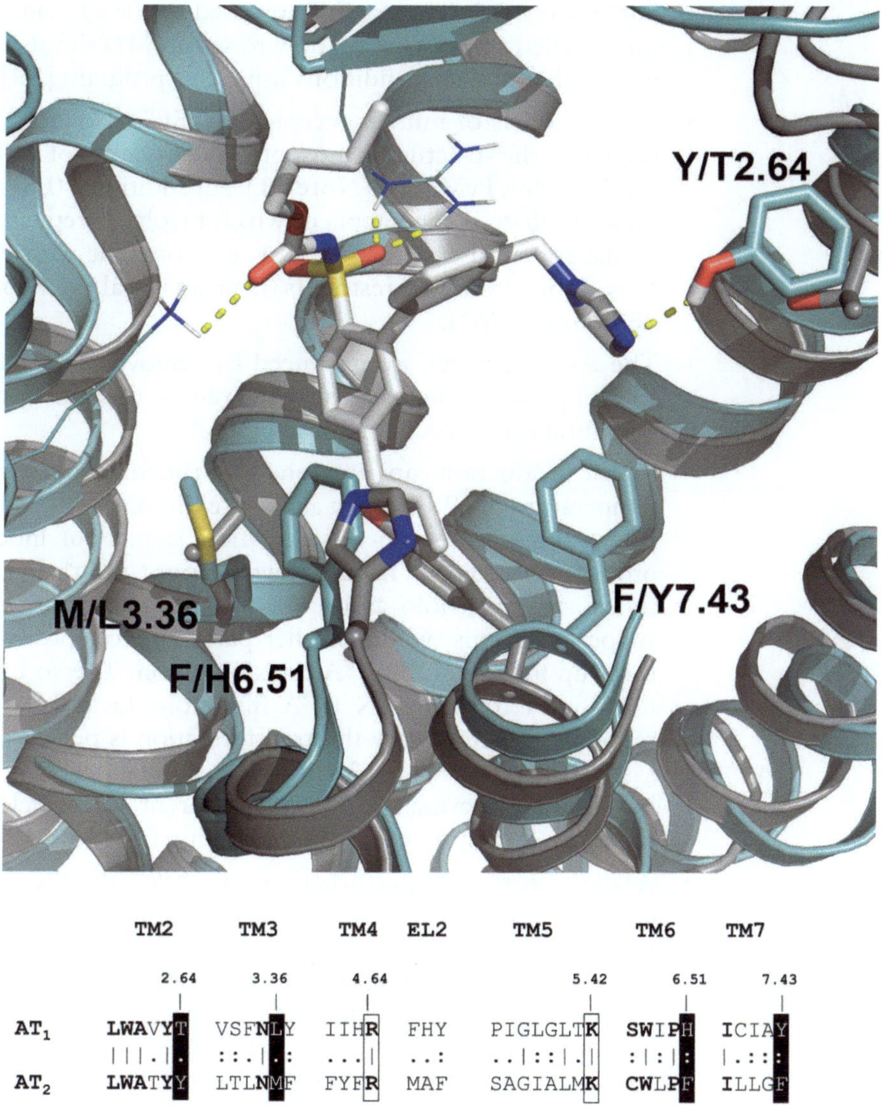

Fig. 4 The selective AT_2 antagonist (compound 2 in Ref. 22) docked to the AT_2 receptor model built in Subheading 3.1 is superimposed on the crystal structure of the AT_1 receptor (*dark gray*). The residues identified as selectivity hotspots are labeled and shown in sticks in the 3D picture, and identified in *black* on the pseudo-sequence alignment between AT_1 and AT_2 (*bottom*)

bilayer, solvated, and equilibrated using the PyMemDyn protocol implemented in GPCR-ModSim.

- A spherical system of 25 Å radius centered on the C2 adenine carbon in NECA is defined and used for MD simulations with the program Q. We used the same force field as above (OPLS-AA). Ionizable residues within 5 Å of the spherical boundaries are neutralized, while those within 20 Å from the center were assigned their most probable protonation state at pH 7. Protonation states of histidines should be kept as indicated in

Subheading 3.2. The reader is referred to the Q manual and to our detailed description on how to set up MD simulations with spherical boundary conditions using this program [39].

- Initial models of mutant receptor ($H250^{6.52}N$) are created by modeling the structurally most probable rotamer of the mutated residue using PyMol (*see* **Note 11**). An additional 0.61 ns equilibration phase is then applied, which involved stepwise heating of the spherical system to 298 K concomitant with release of heavy atom positional restraints (from an initial force constant of 25 kcal/mol/$Å^2$).
- The apo structures are produced by removing the ligand and solvating the created cavity with waters; thereafter, the same equilibration procedure should be applied.
- All the production runs are done with the following parameters: temperature of 298 K using a separate thermal bath coupling for solute and solvent and 1 fs MD time step. Four independent production runs are needed: two account for the $H250^{6.52}A$ transformation (holo and apo), which are divided into seven subperturbations, while another pair of holo/apo simulations is set up for the $N250^{6.52}A$ transformation. The so-called FEP files for H to A and N to A mutations (available in the Q repository) dictate how the transformation is performed. Each λ-step is sampled for 10–30 ns in MD simulations, which should be replicated several times with different initial velocities to obtain an appropriate statistical sampling (*see* **Note 12**).
- Analysis: for each perturbation, we calculate the associated change in free energy from the energy files (*.en) provided in Q, with a python script provided in the Q distribution. This script extracts and concatenates the energy change upon each subperturbation, to provide the free energy change for each leg in the thermodynamic cycle (*see* Fig. 2). Calculated relative binding free energies $\left(\Delta\Delta G_{\text{bind}}^{\text{calc}}\right)$ are obtained from a series of small, convergent FEP calculations as $\Delta\Delta G_{\text{bind}}^{\text{calc}} = \left(\Delta G_{\text{holo}}^{\text{wt}} - \Delta G_{\text{apo}}^{\text{wt}}\right) - \left(\Delta G_{\text{holo}}^{\text{mut}} - \Delta G_{\text{apo}}^{\text{mut}}\right)$. The s.e.m. and hysteresis values are calculated from independent replicates as indicated in Subheading 2.

4 Notes

1. For a complete list of tools and servers, we refer to the GPCR database GPCRdb (http://gpcrdb.org/), which contains a number of GPCR-dedicated modeling web servers, which together with GPCR-ModSim are partners of the European Network for GPCRs GLISTEN (COST action CM1207): GPCRM [40], GPCR-SSFE [25], and GOMoDo [41].

2. GPCR-ModSim is completely free and anonymous through the option "Model a GPCR." However, academic users can create a free account, where they can store their projects and navigate through the folders that store the alignments, models, and MD trajectories, which can be accessed and modified (i.e., creating subbranches of the project. etc.) during a limited time.
3. Other examples concerning the use of GPCR-ModSim can be found in the tutorial of the webserver, which includes modeling of the orexin-2 receptor (http://gpcr-modsim.org/static/docs/tutorial.pdf) and the MD simulation of the A_{2A}AR in the complex with the antagonist caffeine (https://www.youtube.com/channel/UCQP49VirtLEPZQ264qkLodg/videos). In addition, our participation in the last edition of the GPCR-Dock competition provides an example on the successful use of GPCR-ModSim to model a GPCR-antagonist complex [9].
4. Other criteria to consider when selecting the best template include: (a) proximity of the sequences in the phylogenetic tree; this sometimes reveals common structural areas, as is the case for peptide binding receptors, which share common secondary structural elements within the EL regions, or aminergic receptors, which have a common binding site motif. (b) The sequence identity of topological regions (i.e., TM or loops) or similarities in the binding pocket regions. In general, manual adjustment of alignments is possible (and often advisable), and can be done by downloading, editing, and uploading the modified alignment, or using the JalView visualizer implemented in GPCR-ModSim.
5. Alternatively we have also used the GLIDE docking software, from the Schrödinger suite [30], using default settings and a grid box typically of 30x30x30 Å, using and selecting the best pose as ranked by GLIDE-SP.
6. Using a similar procedure one can build active-like conformations of the two angiotensin receptors, as reported in reference [22]. This way, not only subtype but also functional selectivity can be analyzed for similar ligands that differ in their agonist/antagonist profile.
7. While the homology model is a minimized structure that does not need further refinement, a crystal structure of a GPCR (as is the case of 4EIY) requires some treatment. Fragments added for crystallization purposes are removed (i.e., the BrIL protein fused to the N-terminus) and the missing loops are remodeled. This is particularly important if one wants to run MD of the complex generated during docking, while this step can be avoided otherwise.
8. In the AT projects, since the ligands are designed as mimics of the HPF C-terminal fragment of tripeptide the natural agonist,

ATII, the criteria have been substituted by a reliable superposition to this fragment of ATII as docked into the receptor.

9. The corresponding parameters for the ligand must be obtained by the user. One strategy is to generate OPLS-AA parameters with the Schrödinger suite (academic license FFLD), and use the python script that we provide in the web server to translate them into the GROMACS format needed.

10. A docking pose of the PYY peptide can be easily created from the Y_2-NPY complex, by mutation of the residues that differ between the two peptides with Maestro, followed by the same energy minimization protocol described for NPY.

11. If more than one side chain rotamer could be modeled, all possibilities were subjected to MD simulation and the most stable one was selected as the initial rotamer.

12. The system is usually more sensible to the first subperturbation, in particular in polar sidechains, since this step induces annihilation of charges from the equilibrated structure, therefore sampling is increased to 30 ps per λ-step; analogously, the last subperturbation includes annihilation of the Cβ and replacement to H (in the case of a pure Ala mutation), which again needs a higher sampling (in this case we set to 20 ps / λ-step). In the remaining subperturbations, we found that 10 ps/ λ-step is enough, but this default can be changed by the user. Depending on the sidechain, the transformation to Ala considers 4–9 subperturbations, thus the total sampling per replicate simulation is 3.5 to 6 ns. Considering ten replicate simulations, each leg of the thermodynamic cycle is sampled for around 50 ns.

References

1. Santos R, Ursu O, Gaulton A et al (2017) A comprehensive map of molecular drug targets. Nat Rev Drug Discov 16:19–34. https://doi.org/10.1038/nrd.2016.230
2. Stevens RC, Cherezov V, Katritch V et al (2013) The GPCR network: a large-scale collaboration to determine human GPCR structure and function. Nat Rev Drug Discov 12:25–34. https://doi.org/10.1038/nrd3859
3. Jazayeri A, Andrews SP, Marshall FH (2016) Structurally enabled discovery of adenosine A2A receptor antagonists. Chem Rev 117:21–37. https://doi.org/10.1021/acs.chemrev.6b00119
4. Kooistra AJ, Roumen L, Leurs R et al (2013) From Heptahelical bundle to hits from the haystack: structure-based virtual screening for GPCR ligands. Methods Enzymol 522:279–336. https://doi.org/10.1016/B978-0-12-407865-9.00015-7
5. Rodriguez D, Gutierrez-de-Teran H (2013) Computational approaches for ligand discovery and design in class-a G protein-coupled receptors. Curr Pharm Des 19:2216–2236
6. Fredriksson R, Lagerström MC, Lundin L-G, Schiöth HB (2003) The G-protein-coupled receptors in the human genome form five main families. Phylogenetic analysis, paralogon groups, and fingerprints. Mol Pharmacol 63:1256–1272. https://doi.org/10.1124/mol.63.6.1256
7. Michino M, Abola E, GPCR Dock 2008 Participants et al (2009) Community-wide assessment of GPCR structure modelling and ligand docking: GPCR dock 2008. Nat Rev Drug Discov 8:455–463. https://doi.org/10.1038/nrd2877

8. Kufareva I, Rueda M, Katritch V et al (2011) Status of GPCR modeling and docking as reflected by community-wide GPCR dock 2010 assessment. Structure 19:1108–1126. https://doi.org/10.1016/j.str.2011.05.012
9. Kufareva I, Katritch V, Stevens RC et al (2014) Advances in GPCR modeling evaluated by the GPCR dock 2013 assessment: meeting new challenges. Structure 22:1120–1139. https://doi.org/10.1016/j.str.2014.06.012
10. Kristiansen K (2004) Molecular mechanisms of ligand binding, signaling, and regulation within the superfamily of G-protein-coupled receptors: molecular modeling and mutagenesis approaches to receptor structure and function. Pharmacol Ther 103:21–80. https://doi.org/10.1016/j.pharmthera.2004.05.002
11. Zhukov A, Andrews SP, Errey JC et al (2011) Biophysical mapping of the adenosine A2A receptor. J Med Chem 54:4312–4323. https://doi.org/10.1021/jm2003798
12. Gutierrez-de-Teran H, Keränen H, Azuaje J et al (2015) Computer-aided design of GPCR ligands. Methods Mol Biol 1272:271–291. https://doi.org/10.1007/978-1-4939-2336-6_19
13. Chen J-F, Eltzschig HK, Fredholm BB (2013) Adenosine receptors as drug targets – what are the challenges? Nat Rev Drug Discov 12:265–286. https://doi.org/10.1038/nrd3955
14. Michel MC, Beck-Sickinger A, Cox H et al (1998) XVI. International Union of Pharmacology recommendations for the nomenclature of neuropeptide Y, peptide YY, and pancreatic polypeptide receptors. Pharmacol Rev 50:143–150
15. Zhang L, Bijker MS, Herzog H (2011) The neuropeptide Y system: pathophysiological and therapeutic implications in obesity and cancer. Pharmacol Ther 131:91–113. https://doi.org/10.1016/j.pharmthera.2011.03.011
16. Boukharta L, Gutierrez-de-Teran H, Aqvist J (2014) Computational prediction of alanine scanning and ligand binding energetics in G-protein coupled receptors. PLoS Comput Biol 10:e1003585. https://doi.org/10.1371/journal.pcbi.1003585
17. Xu B, Fällmar H, Boukharta L et al (2013) Mutagenesis and computational Modeling of human G protein-coupled receptor Y2 for neuropeptide Y and peptide YY. Biochemistry 52:7987–7998
18. Stegbauer J, Coffman TM (2011) New insights into angiotensin receptor actions: from blood pressure to aging. Curr Opin Nephrol Hypertens 20:84–88. https://doi.org/10.1097/MNH.0b013e3283414d40
19. Zhang H, Unal H, Gati C et al (2015) Structure of the angiotensin receptor revealed by serial femtosecond crystallography. Cell 161:833–844. https://doi.org/10.1016/j.cell.2015.04.011
20. Zhang H, Unal H, Desnoyer R et al (2015) Structural basis for ligand recognition and functional selectivity at angiotensin receptor. J Biol Chem 290:29127–29139. https://doi.org/10.1074/jbc.M115.689000
21. Zhang H, Han GW, Batyuk A, Ishchenko A, White KL, Patel N, Sadybekov A, Zamlynny B, Rudd MT, Hollenstein K, Tolstikova A, White TA, Hunter MS, Weierstall U, Liu W, Babaoglu K, Moore EL, Katz RD, Shipman JM, Garcia-Calvo M, Sharma S, Sheth P, Soisson SM, Stevens RC, Katritch V, Cherezov V (2017) Structural basis for selectivity and diversity in angiotensin II receptors. Nature 544:327–332
22. Sallander J, Wallinder C, Hallberg A et al (2016) Structural determinants of subtype selectivity and functional activity of angiotensin II receptors. Bioorg Med Chem Lett. https://doi.org/10.1016/j.bmcl.2015.10.084
23. Rodriguez D, Bello X, Gutierrez-de-Teran H (2012) Molecular modelling of G protein-coupled receptors through the web. Mol Inform 31:334–341. https://doi.org/10.1002/minf.201100162
24. Esguerra M, Siretskiy A, Bello X et al (2016) GPCR-ModSim: a comprehensive web based solution for modeling G-protein coupled receptors. Nucleic Acids Res 44:W455–W462. https://doi.org/10.1093/nar/gkw403
25. Worth CL, Kreuchwig A, Kleinau G, Krause G (2011) GPCR-SSFE: a comprehensive database of G-protein-coupled receptor template predictions and homology models. BMC Bioinform 12:185. https://doi.org/10.1186/1471-2105-12-185
26. Mobarec JC, Sanchez R, Filizola M (2009) Modern homology modeling of G-protein coupled receptors: which structural template to use? J Med Chem 52:5207–5216. https://doi.org/10.1021/jm9005252
27. Webb B, Sali A (2014) Comparative protein structure modeling using modeller. Curr Protoc Bioinformatics 47:5–6. https://doi.org/10.1002/0471250953.bi0506s47
28. Gutierrez-de-Teran H, Bello X, Rodriguez D (2013) Characterization of the dynamic events of GPCRs by automated computational simulations. Biochem Soc Trans 41:205–212. https://doi.org/10.1042/BST20120287

29. Verdonk ML, Cole JC, Hartshorn MJ et al (2003) Improved protein-ligand docking using GOLD. Proteins 52:609–623. https://doi.org/10.1002/prot.10465
30. Schrödinger L (2012) Schrödinger Suite 2012. doi: https://doi.org/10.3389/fphar.2015.00011/full
31. Nervall M, Hanspers P, Carlsson J et al (2008) Predicting binding modes from free energy calculations. J Med Chem 51:2657–2667. https://doi.org/10.1021/jm701218j
32. Brandsdal BO, Osterberg F, Almlöf M et al (2003) Free energy calculations and ligand binding. Adv Protein Chem 66:123–158
33. Keränen H, Gutierrez-de-Teran H, Aqvist J (2014) Structural and energetic effects of A2A adenosine receptor mutations on agonist and antagonist binding. PLoS One 9:e108492. https://doi.org/10.1371/journal.pone.0108492
34. Keränen H, Aqvist J, Gutierrez-de-Teran H (2015) Free energy calculations of a (2A) adenosine receptor mutation effects on agonist binding. Chem Commun (Camb) 51:3522–3525. https://doi.org/10.1039/c4cc09517k
35. Rodriguez D, Pineiro A, Gutierrez-de-Teran H (2011) Molecular dynamics simulations reveal insights into key structural elements of adenosine receptors. Biochemistry 50:4194–4208. https://doi.org/10.1021/bi200100t
36. Bharate SB, Singh B, Kachler S et al (2016) Discovery of 7-(Prolinol-N-yl)-2-phenylamino-thiazolo[5,4-d]pyrimidines as novel non-nucleoside partial agonists for the A2A adenosine receptor: prediction from molecular Modeling. J Med Chem 59:5922–5928. https://doi.org/10.1021/acs.jmedchem.6b00552
37. Venkatakrishnan AJ, Deupi X, Lebon G et al (2013) Molecular signatures of G-protein-coupled receptors. Nature 494:185–194. https://doi.org/10.1038/nature11896
38. Surgand J-S, Rodrigo J, Kellenberger E, Rognan D (2006) A chemogenomic analysis of the transmembrane binding cavity of human G-protein-coupled receptors. Proteins 62:509–538. https://doi.org/10.1002/prot.20768
39. Gutierrez-de-Teran H, Aqvist J (2012) Linear interaction energy: method and applications in drug design. In: Barron R (ed) Computational drug discovery and design. Springer, New York, pp 305–323
40. Latek D, Pasznik P, Carlomagno T, Filipek S (2013) Towards improved quality of GPCR models by usage of multiple templates and profile-profile comparison. PLoS One 8: e56742. https://doi.org/10.1371/journal.pone.0056742
41. Sandal M, Duy TP, Cona M et al (2013) GOMoDo: a GPCRs online modeling and docking webserver. PLoS One 8:e74092. https://doi.org/10.1371/journal.pone.0074092

Chapter 3

Breakthrough in GPCR Crystallography and Its Impact on Computer-Aided Drug Design

Antonella Ciancetta and Kenneth A. Jacobson

Abstract

Recent crystallographic structures of G protein-coupled receptors (GPCRs) have greatly advanced our understanding of the recognition of their diverse agonist and antagonist ligands. We illustrate here how this applies to A_{2A} adenosine receptors (ARs) and to $P2Y_1$ and $P2Y_{12}$ receptors (P2YRs) for ADP. These X-ray structures have impacted the medicinal chemistry aimed at discovering new ligands for these two receptor families, including receptors that have not yet been crystallized but are closely related to the known structures. In this Chapter, we discuss recent structure-based drug design projects that led to the discovery of: (a) novel A_3AR agonists based on a highly rigidified (N)-methanocarba scaffold for the treatment of chronic neuropathic pain and other conditions, (b) fluorescent probes of the ARs and $P2Y_{14}$R, as chemical tools for structural probing of these GPCRs and for improving assay capabilities, and (c) new more drug-like antagonists of the inflammation-related $P2Y_{14}$R. We also describe the computationally enabled molecular recognition of positive (for A_3AR) and negative ($P2Y_1$R) allosteric modulators that in some cases are shown to be consistent with structure-activity relationship (SAR) data. Thus, computational modeling has become an essential tool for the design of purine receptor ligands.

Key words Adenosine receptor, P2Y receptor, Structure-based drug design, X-ray crystallography, Nucleosides, Nucleotides

1 Introduction

GPCR ligands represent 33% of the small-molecule drugs that target major protein families [1]. The study of GPCR structure and function has been revolutionized as a result of new X-ray crystallographic findings and correlation with the structure activity relationships (SAR) and pharmacology of their ligands [2–5]. We present purine receptors as examples of GPCR families that have benefited enormously from these new high-resolution receptor structures.

Purinergic signaling is an element in the control of many human physiological functions. The signaling interactions associated with extracellular purines and pyrimidines, loosely referred to as the "purinergic signalome," consist of twelve G protein-

Alexander Heifetz (ed.), *Computational Methods for GPCR Drug Discovery*, Methods in Molecular Biology, vol. 1705,
https://doi.org/10.1007/978-1-4939-7465-8_3, © Springer Science+Business Media LLC 2018

coupled receptors (GPCRs), i.e., activated by adenosine (four adenosine receptors, ARs) and nucleotides (eight P2YRs), and seven distinct subunits of multimeric ATP-gated P2X ion channels [6, 7]. Among the GPCRs, the ARs respond principally to adenine nucleosides and the P2YRs respond to adenine and uracil nucleotides. These receptors can be activated or blocked with small-molecule modulators, some of which have entered clinical trials or been approved as diagnostic probes and agents for treating chronic diseases. In addition to orthosteric ligands, i.e., binding in the same site as native ligand, allosteric modulators, which bind at a separate site, are being explored for the purine receptors.

Furthermore, inhibition, by small molecules, of the enzymes that regulate the levels of the native AR agonists (Fig. 1a) and native P2YR agonists (Fig. 1b) and inhibitors of nucleoside transporters adds another dimension to the exogenous control of this system [8].

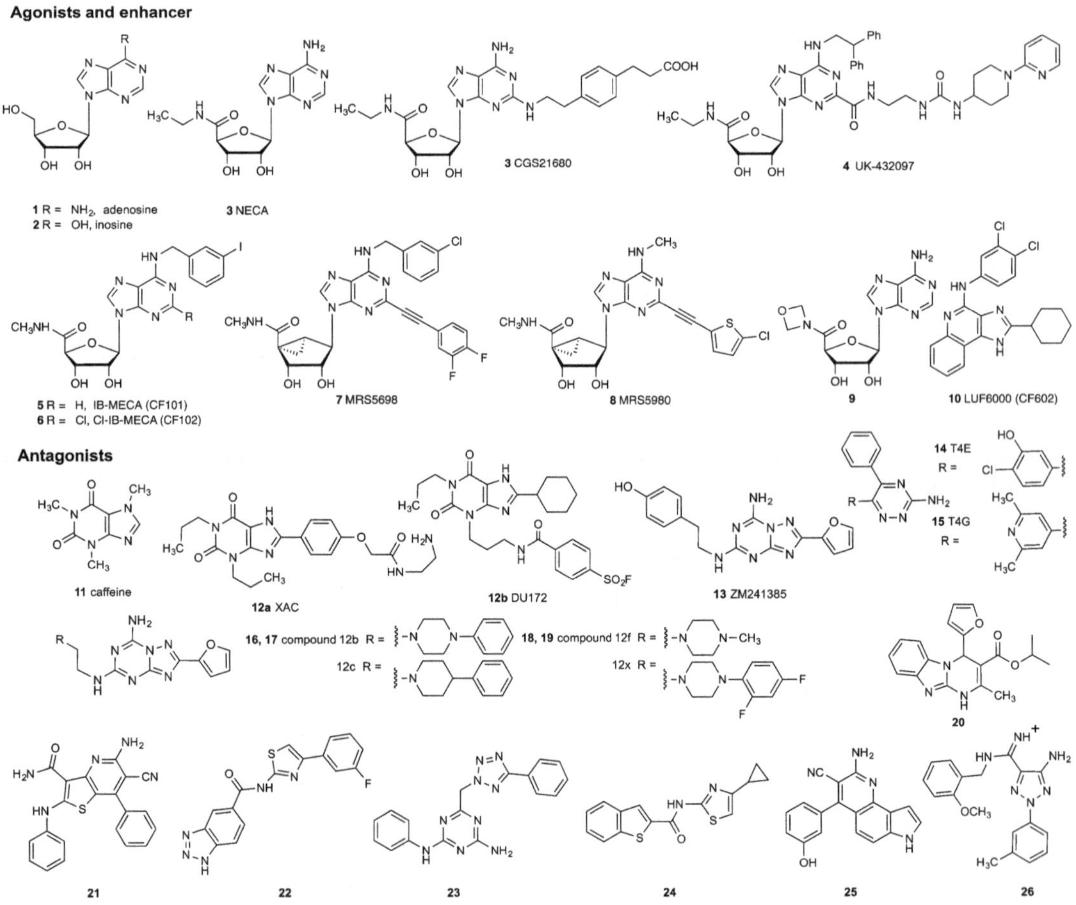

Fig. 1 Structures of ligands described in the text for the: (**a**) AR and (**b**) P2YR families. The native agonists are adenosine (**1**) and inosine (**2**), to a lesser extent, for ARs and ADP (**26**), ATP (**27**), UDP-glucose (**30**), and UDP and UTP (not shown) for the P2YRs

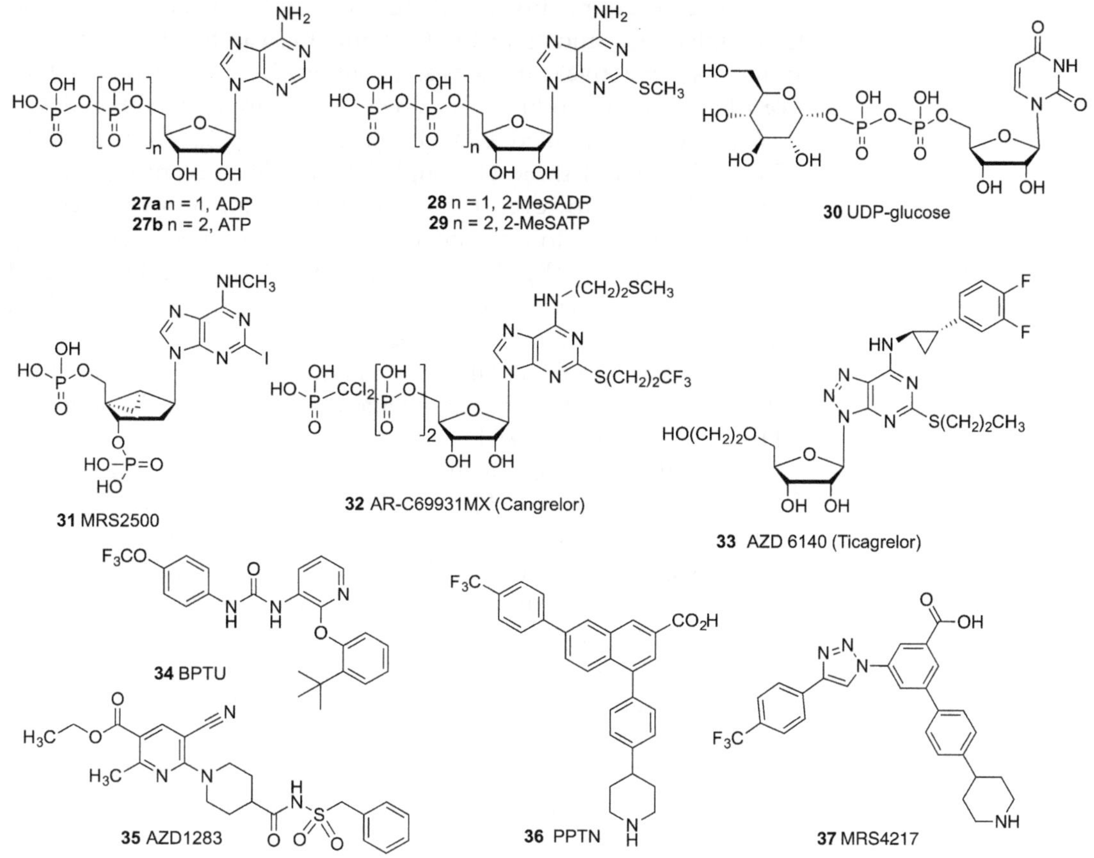

Fig. 1 (continued)

Some of these extracellular enzymes are: CD39 (ectonucleoside triphosphate diphosphohydrolase 1, ENTPD1), which hydrolyzes ATP **27b** to ADP **27a** and to AMP; CD73 (ecto-5′-nucleotidase, 5′-NT), which hydrolyzes AMP to adenosine **1**, and adenosine deaminase, which converts adenosine to inosine **2**. Adenosine kinase that forms AMP is an intracellular enzyme, but it tends to reduce the pool of available adenosine both inside and outside the cell, because of nucleoside transporters such as the equilibrative ENT1 that allow adenosine to cross the cell membrane [9]. The overall importance of the purinergic signalome is consistent with its being well conserved throughout evolution as indicated by the phylogenetic relationships of the receptor family members [10, 11]. We use convergent approaches, i.e., medicinal chemical, pharmacological, and structural, to discover new agonists, antagonists, and inhibitors to modulate the purinergic signalome (Fig. 1) [12–14].

There are four subtypes of ARs: G_i-coupled A_1 and A_3ARs and G_s-coupled A_{2A} and A_{2B}ARs. Caffeine **11**, the most widely consumed psychoactive drug, acts as a competitive, nonselective inhibitor of adenosine binding at the ARs. There are two subfamilies of P2YRs: five G_q-coupled P2Y_1-like receptors (P2Y_1, P2Y_2, P2Y_4, P2Y_6, P2Y_{11}) and three G_i-coupled P2Y_{12}-like receptors (P2Y_{12}, P2Y_{13}, P2Y_{14}). However, the diversity of endogenous nucleotide agonists does not follow the division of subtypes based on G protein coupling. P2Y_2, P2Y_4, P2Y_6, and P2Y_{14}Rs can be activated by uracil nucleotides, while P2Y_1, P2Y_2, P2Y_{11}, P2Y_{12}, and P2Y_{13} are activated by adenine nucleotides. The only subtype that is associated with well-validated therapeutics is P2Y_{12}R, for which antagonists act as antithrombotic agents by blocking the action of ADP **27a** on platelets [15].

Since 2007, numerous high-resolution X-ray structures of the human (h) A_{2A}AR and P2Y_1 and P2Y_{12}Rs have been reported (Table 1) [16, 17, 20, 23, 29–32]. Recently, the structure of the hA_1AR was also solved [19]. These structural advances have opened up new opportunities to rationally design ligands, both by structural enhancement of known agonists and antagonists and through the in silico screening and modeling of novel chemotypes. Research based on these GPCR X-ray structures has also led to advances regarding the other two AR subtypes that have not yet been crystallized but are closely related to the known A_{2A}AR and A_1AR structures. Similarly, closely related homology models are predictive of ligand recognition at P2YR subtypes other than P2Y_1 and P2Y_{12}Rs. Chemical tools for the crystallized receptors and related subtypes, such as high affinity fluorescent probes, were designed with the aid of molecular modeling and applied to drug discovery, including homology modeling, docking, and molecular dynamics (MD) simulations. The importance of molecular modeling in the discovery of new purinergic ligands has therefore increased in recent years.

2 Materials

2.1 X-Ray Structures of Complexes of Purine Receptors

The pace of reports on X-ray crystallographic GPCR structures, which are membrane-bound, has recently accelerated due to methodological advances [33–36]. Although it is not yet a routine process, and not amenable to the high-throughput methods applied to X-ray crystallography of soluble proteins, the several hundred structures, corresponding to dozens of GPCRs in different complexes, have already turned the tide in drug discovery approaches for this important superfamily of drug targets. We analyze the current state of knowledge of purine receptors as an illustration.

Table 1
Reported X-ray crystal structures of purine receptors (all human) and methods used in the determination

Receptor, ligand	Method,[a] Resolution (Å)	PBD ID:	References
Adenosine $A_{2A}AR$ agonists			
Adenosine (**1**)	StaR (5), 3.0	2YDO	Lebon et al. [16]
NECA (**3**)	StaR (5), 2.6	2YDV	Lebon et al. [16]
UK432097 (**4**)	T4L(IL3)-ΔC, 2.71	3QAK	Xu et al. [17]
CGS21680 (**3**)	StaR (5), 2.6	4UG2	Lebon et al. [18]
	StaR (5), 2.6	4UHR	Lebon et al. [18]
Adenosine A_1AR antagonist[c]			
DU172 (**12b**)	bRIL(IL3)-ΔC (9)[b], 3.2	5UEN	Glukhova et al. [19]
Adenosine $A_{2A}AR$ antagonists			
ZM241385 (**13**)	T4L(IL3), 2.6	3EML	Jaakola et al. [20]
	StaR (8), 3.3	3PWH	Doré et al. [21]
	Fab2838 (1), 2.7	3VG9	Hino et al. [22]
	Fab2838 (1), 3.1	3VGA	Hino et al. [22]
	bRIL(IL3)-ΔC (3), 1.8	4EIY	Liu et al. [23]
	StaR2-bRIL(IL3) (11), 1.72	5IU4	Segala et al. [24]
Caffeine (**11**)	StaR (8), 3.6	3RFM	Doré et al. [21]
XAC (**12a**)	StaR (8), 3.31	3REY	Doré et al. [21]
T4E (**14**)	StaR (8), 3.34	3UZC	Congreve et al. [25]
T4G (**15**)	StaR (8), 3.27	3UZA	Congreve et al. [25]
12c (**17**)	StaR2-bRIL(IL3) (10), 1.9	5IU7	Segala et al. [24]
12f (**18**)	StaR2-bRIL(IL3) (12), 2.0	5IU8	Segala et al. [24]
12b (**16**)	StaR2-bRIL(IL3) (12), 2.2	5IUA	Segala et al. [24]
12x (**19**)	StaR2-bRIL(IL3) (10), 2.1	5IUB	Segala et al. [24]
Cmpd-1 (**26**)	bRIL(IL3), 3.5	5UIG	Sun et al. [26]
Adenosine $A_{2A}AR$ – other			
Engineered G protein	StaR (18), 3.4	5G53	Carpenter et al. [27]
na	SAD/XFEL (3), 2.5	5K2A	Batyuk et al. [28]
na	MR/XFEL (3), 2.5	5K2B	Batyuk et al. [28]
na	SAD/XFEL (3), 1.9	5K2C	Batyuk et al. [28]
na	MR/XFEL (3), 1.9	5K2D	Batyuk et al. [28]

(continued)

Table 1
(continued)

Receptor, ligand	Method,[a] Resolution (Å)	PBD ID:	References
$P2Y_1R$ antagonists			
MRS2500 (**31**)	rub(IL3) (1), 2.7	4XNW	D. Zhang et al. [29]
BPTU (**34**)	bRIL(N-term) (1), 2.2	4XNV	D. Zhang et al. [29]
$P2Y_{12}R$ antagonist			
AZD1283 (**35**)	bRIL(IL3) (4), 2.62	4NTJ	K. Zhang et al. [30]
$P2Y_{12}R$ agonists			
2-MeSADP (**28**)	bRIL(IL3) (4), 2.5	4PXZ	K. Zhang et al. [31]
2-MeSATP (partial agonist, **29**)	bRIL(IL3) (4), 3.1	4PY0	K. Zhang et al. [31]

[a] Construct or stabilization method. Number of mutations, if present, is given in *parentheses*, and if StaR or fusion construct (protein and location). T4L, cysteine-free phage T4 lysozyme; bRIL, thermostabilized apocytochrome b_{562}RIL (e.g., A23-L128); ΔC, truncated C-terminus. Fab, antibody Fab fragment; N-term, N-terminus; rub, M1–E54 of rubredoxin; XFEL, X-ray free-electron laser; MR, molecular replacement method; sulfur phasing with SAD, single-wavelength anomalous dispersion; na, not applicable

[b] Construct is substituted with N159A (glycosylation site) and residues of the A_{2A}AR at 220–228 ("to optimize bRIL insertion sites") [19]

[c] A_1 and A_{2A} receptor structures (StaRs stabilized with b562RIL) in complex with three xanthine antagonists were recently reported by Chen et al. [100]

2.1.1 X-ray Structures of the A_1AR and A_{2A}AR

The first structure of any purine receptor was an A_{2A}AR complex with potent antagonist 4-[2-[7-amino-2-(2-furyl)-1,2,4-triazolo [1,5-*a*][1, 3, 5]triazin-5-yl-amino]ethyl]phenol (ZM241385, **13**) reported in 2007 [20]. The overall ligand arrangement was roughly perpendicular to the membrane plane, in contrast to other GPCR structures. This general orientation of the various ligands as stretching from the pharmacophore binding site toward the outer surface applies to various ligands, such as the A_{2A}AR structure in complex with the high affinity xanthine amine congener (XAC, **12a**) antagonist [21]. This ligand arrangement—roughly parallel to the TMs (transmembrane domains)—was anticipated by earlier modeling and by the many chain-functionalized analogues of agonists and antagonists, which suggested that the distal tethered portions of the molecules were exposed to the medium. There is more freedom of substitution on the distal portions, because they are not limited by the steric constraints of the pharmacophore binding site [14, 37, 38].

Coordination of triazolo-triazine ZM241385 **13** and other antagonists in the A_{2A}AR binding site (Fig. 2a) occurs through both H-bonding and interactions with hydrophobic sidechains, such as Leu6.51 (using Ballesteros-Weinstein convention for amino acid numbering in the TMs [39]). The heterocyclic ring forms π–π stacking with Phe168 (EL2). A conserved Asn6.55 (Asn253 in the A_{2A}AR) forms bidentate H-bonds with the

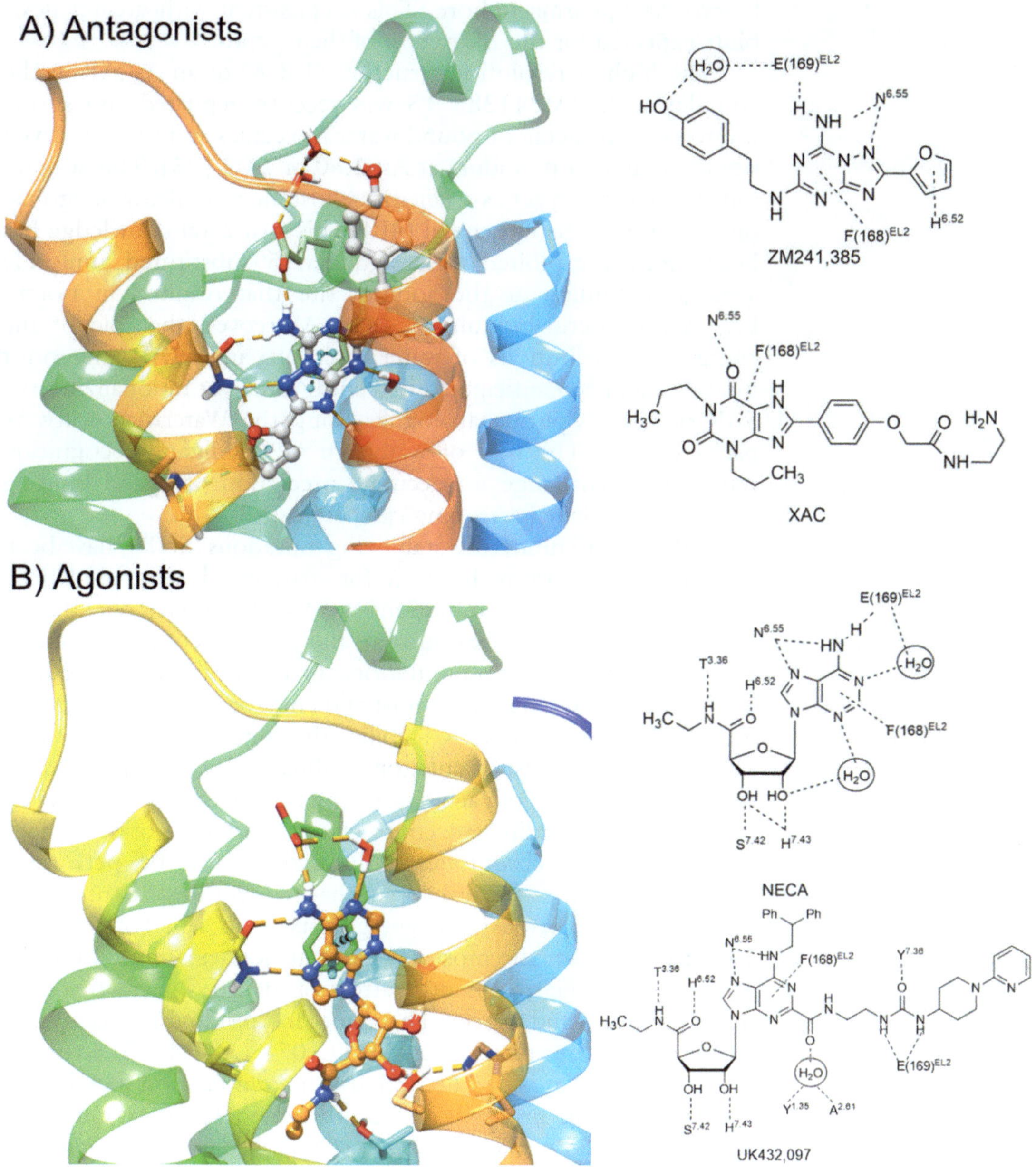

Fig. 2 *Left panel*: High-resolution X-ray crystallographic structures of the $hA_{2A}AR$ [23] with (**a**) ZM241385 (**13**) and (**b**) NECA [16] (**3**). *Right panel*: Contacts between four cocrystallized ligands and the $hA_{2A}AR$ are depicted

exocyclic amine and a triazole ring nitrogen atom, and the exocyclic amine also H-bonds with Glu169 in EL2 (Fig. 2b). There are differences in the position of the 2-(4-hydroxyphenyl)ethyl side-chain of ZM241385 between different reported structures [16, 30], which suggests that this moiety, which approaches the receptor's exofacial side, has more conformational freedom than the heterocyclic pharmacophore. This is consistent with greater flexibility expected for the EL region of the receptor.

The highest resolution structure (1.8 Å) of an $A_{2A}AR$ in the complex with ZM241385 **13** was recently reported and shows unprecedented detail of bound water molecules and the conserved site for sodium ion binding at Asp2.50 in TM2 [23]. This sodium ion is proposed to act as a negative allosteric modulator of agonist binding at GPCRs in general [40]. This structural knowledge has been recently exploited to design new 5′-substituted amiloride analogues binding at the sodium site that resulted in potent $hA_{2A}AR$ allosteric modulators [41]. Moreover, the role of the energetic contributions of water molecules in $A_{2A}AR$ antagonist binding had been anticipated in a study aimed at rationalizing the SAR of triazolylpurine analogues using the WaterMap software package [42]. The role of waters in GPCR ligand recognition represents a challenge for accurate prediction of ligand binding mode during virtual screening campaigns [43].

GPCRs with numerous stabilizing mutations (StaRs) have been developed as a core technology for structural biology and drug discovery [44]. These mutations are usually placed outside the binding site to avoid disruption of the drug-receptor interaction. Using multiple stabilizing mutations, many GPCRs can be crystallized with a wide range of agonist and antagonist ligands. Separate mutation sets were found to stabilize the conformation needed for binding either agonist or antagonist; thus, the StaRs can be customized for discovery of each ligand type. Furthermore, biophysical mapping of the binding site contour for a given class of congeneric ligands, such as chromones and triazines, can be performed [32, 45, 46]. Biophysical mapping has led to the structure-based optimization of new $A_{2A}AR$ antagonists, such as a 1,2,4-triazine T4E **14**, of which the most advanced analogues are on a translational path for the treatment of attention-deficit/hyperactivity disorder (ADHD) and cancer. This effort has resulted in a candidate molecule for cancer co-therapy HTL1071 (AZD4635, structure not disclosed) that is in a Phase 1 trial in combination with durvalumab, targeting programmed death ligand-1, in patients with advanced solid malignancies (ClinicalTrials.gov identifier: NCT02740985). The StaRs are sufficiently stable to be studied in biophysical experiments at room temperatures, including incorporation into HDL particles for biosensor studies [47] or covalently immobilization on gold surfaces or in a lipid bilayer for measuring binding using surface plasmon resonance [48]. StaRs may also be used for NMR studies, specifically target-immobilized NMR screening (TINS) for fragment-based drug discovery (FBDD) of very small molecules that bind to the $A_{2A}AR$ with relatively low affinity [49]. These biophysical techniques have been useful in the discovery of new $A_{2A}AR$ ligands, despite the presence of mutations in the receptor. Other $A_{2A}AR$ antagonists are being developed for treating Parkinson's disease (PD) [20].

An advanced crystallographic technique that was recently applied to the $A_{2A}AR$ in the absence of bound ligand is native phasing of X-ray data determined with a pulsed free-electron laser (XFEL) [28]. The data is collected using serial femtosecond crystallography (SFX) using microcrystals that are delivered in a continuous hydrated stream and subsequently destroyed by the beam.

The conformational differences between the multiple agonist-bound and antagonist-bound structures of the $A_{2A}AR$ present a consistent picture of the reorganization of the orthosteric site to accommodate nucleosides [16, 17]. Furthermore, the predicted binding of nucleosides that vary greatly in efficacy at the A_3AR, particularly after ribose modification [50, 51], provides insights into residues implicated in the activation process.

We have collaborated with Ray Stevens and colleagues in the structural characterization of an agonist-bound $A_{2A}AR$ structure, antagonist-bound $P2Y_1R$ structures as well as agonist- and antagonist-bound $P2Y_{12}R$ structures containing stabilizing fusion proteins in the third intracellular loop (IL3) or at the N-terminus [17, 29–31]. The receptor constructs contain few if any mutations in the TM regions, so they tend to preserve the native structures.

This first X-ray structure of an agonist-bound $A_{2A}AR$ used a bulky adenosine agonist with extended C2 and N^6 groups on the adenine moiety (UK432097, 6-(2,2-diphenylethylamino)-9-((2R,3R,4S,5S)-5-(ethylcarbamoyl)-3,4-dihydroxytetrahydrofuran-2-yl)-*N*-(2-(3-(1-(pyridin-2-yl)piperidin-4-yl)ureido)ethyl)-9*H*–purine-2-carboxamide, **4**) [17]. This substituent combination on the adenine and ribose moieties sufficiently stabilized the receptor complex as indicated by a rise in T_m (melting temperature). The **4**-$A_{2A}AR$ structure and a subsequent report on structurally simpler agonists, **1** and **3**, bound to $A_{2A}AR$ StaRs [16, 18] feature a deep hydrophilic pocket. This forms the ribose moiety's binding site, while the upper regions of the binding site where the nucleobase resides are largely surrounded by hydrophobic residues (Fig. 2b). The hydrophobic regions accommodate the adenine moiety and its typically bulky hydrophobic substituents at the C2 and N^6 positions (Figs. 2b and 3 left). Among the interactions that stabilize the bound nucleobase is a π–π interaction between Phe168 (EL2) and the heterocyclic ring, which is conserved for diverse AR ligands (Fig. 2b). However, the adenine moiety also has polar interactions with the $A_{2A}AR$; similar to the coordination of ZM241385, the sidechain of conserved Asn6.55 is H-bonded to both the adenine exocyclic secondary amine and N7. The various adenine substitutions, particularly at C2 and N^6 positions, are often responsible for the subtype selectivity of the nucleosides, and the ribose with its multiple H-bonding groups is required for AR activation. His7.43

acts as an H-bond donor to the ribose 2′-hydroxyl group (Fig. 2b). The latter interaction and the proximity of the $A_{2A}AR$ residue Thr3.36 to the ribose moiety of bound agonists were determined earlier using a neoceptor approach, which is based on complementary structural changes of the ligand and its targeted receptor [53, 54], having now been confirmed in the X-ray structures. Both of these residues and the residues coordinating the adenine ring have identical or homologous functionality among the ARs. Thus, there is much conservation of the recognition pattern for agonists across the AR family.

However, there are some differences in interactions of nucleoside ligands between the $A_{2A}AR$ and A_3AR that account for pharmacological differences of such ligands, with respect to their affinity, selectivity, and efficacy. For example, His6.52 in $A_{2A}AR$, which H-bonds to the 5′-carbonyl group of NECA **3** (Fig. 2b), is replaced with Ser6.52 and is not in direct contact with docked nucleosides in the A_3AR. The role of Thr3.36 in $A_{2A}AR$, which H-bonds to the 5′-amide NH group of NECA, is maintained in the A_3AR. The chemical "removal" of this NH, by alkylation or truncation of the amide, leads to A_3AR nucleoside antagonists, possibly suggesting a role for TM3 in A_3AR activation [17, 27]. Some water molecules deep in the binding site of the antagonist-bound $A_{2A}AR$ are displaced by the ribose moiety of agonists when bound, which contributes to the favorable energetics of agonist binding. Unlike various other GPCRs, the native agonist binds to the several subtypes (except $A_{2B}AR$) with near nanomolar affinity. AR activation by agonists is thought to involve a side-chain rotation of Trp6.48, which, however, has not been captured in the X-ray structures [16, 27, 55]. The agonist-bound and agonist-bound $A_{2A}AR$ structures, including the surrounding membrane, were subjected to unbiased molecular dynamics and metadynamics simulations to predict conformational transitions upon activation, including rotamers of Trp6.48 [56].

In general, the ribose or ribose-like moiety of adenosine is involved in AR activation and increase affinity at the four ARs, and the substituted adenine moiety mainly determines the subtype selectivity. These two domains—hydrophilic for ribose and hydrophobic for adenine were recognized even in early AR modeling based on rhodopsin [57]. This suggests that the binding site of adenosine derivatives in the ARs can be divided into separate functional domains: ribose as the message (facilitating the receptor activation) moiety and adenine (directing the ligand to a receptor or receptor subtype) as the address moiety [37]. The same distinction has been applied to peptide GPCR ligands, i.e., they are conceptually divided into separate address and message sequences.

An engineered G protein segment (mini-G_s) was used to stabilize the $A_{2A}AR$ (NECA **3** complex) in an active-like state [27]. The observed conformational changes associated with binding of the mini-G_s as a GDP complex were similar to but not identical to those observed for the β_2 adrenergic receptor active state, which was stabilized by a nanobody mimic of G_s protein. The most dramatic change in the overall structure in the NECA-A_{2A}AR-mini-G_s complex was a shift of the cytoplasmic end of TM6 away from the receptor core by 14 Å with respect to the inactive state, with only slight changes for TMs 5 and 7. Furthermore, TMs 3, 5, and 7 underwent rotations. Thus, in the earlier active-intermediate state with only bound agonist UK432097 **4**, the conformational changes in the overall receptor structure, with respect to the cytosolic side, were underestimated [17]; in opsin, the corresponding outward movement of TM6 was 6–7 Å.

Most of the above approaches help define the recognition of ligands in the orthosteric binding site. In addition, allosteric modulators are well explored for A_1AR and A_3AR, but the case for A_{2A}AR remains unresolved [26]. Bitopic ligands of the A_1AR that bridge the allosteric and orthosteric binding site have been reported. At the A_{2A}AR, a bitopic antagonist **26**, which additionally antagonizes the N-methyl D-aspartate (NMDA) receptor subtype 2B, was cocrystallized with the hA_{2A}AR and shown to access the orthosteric site as well as a distal binding region on A_{2A}AR (potentially an allosteric site) [58]. This defined a previously uncharacterized binding pocket of the A_{2A}AR that could be exploited for allosteric modulation. Antagonizing both A_{2A}AR and NMDA$_{2B}$R with a single bifunctional compound might be a fruitful approach to treating PD.

Until recently, the A_{2A}AR was the only AR with a structure determined. However, a recent report featured the A_1AR structure in complex with an irreversibly bound antagonist DU172 (**12b**) [19]. The ligand's reactive fluorosulfonyl group was linked to the hydroxyl group of Tyr7.36. The extracellular cavity was more exposed than in the A_1AR due to a distinct conformation of EL2, and a secondary binding pocket suggests that it could accommodate both orthosteric and allosteric ligands.

2.1.2 X-ray Structures of the P2Y$_1$R and P2Y$_{12}$R

Unlike AR molecular modeling, early P2YR molecular modeling based on the structure of bovine rhodopsin and other templates was less successful in predicting the position and key interactions of agonists and antagonists [59]. Ligands that have been cocrystallized in P2YR X-ray structures include both agonists and antagonists (Fig. 1b). The subsequent P2Y$_1$ and P2Y$_{12}$R X-ray structures [29–31], representing each of the two P2YR subfamilies, brought many surprises, i.e., features that were unlike any GPCR structures previously determined (Figs. 3 and 4). Comparison of agonist-bound and antagonist-bound P2Y$_{12}$R indicates unprecedented structural plasticity in the outer TM portions and the extracellular loops (Fig. 4). There is a major difference in conformation needed to

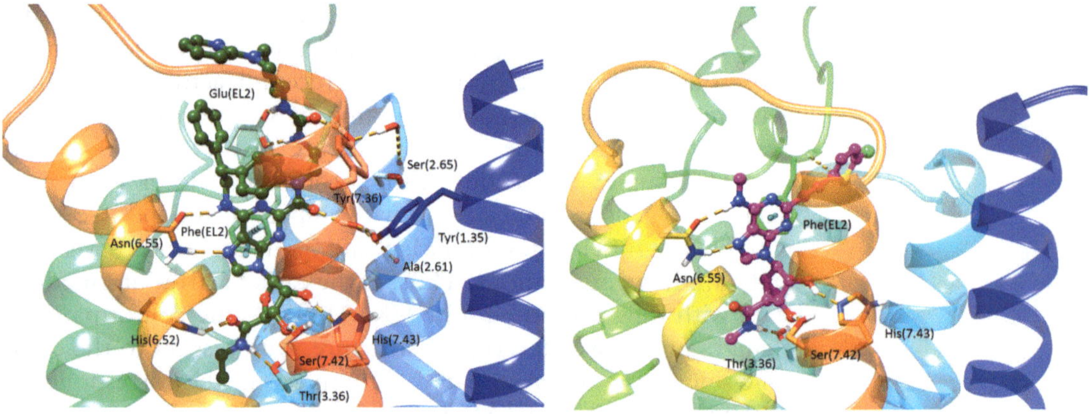

Fig. 3 High-resolution X-ray crystallographic structure of the $hA_{2A}AR$ with high affinity agonist UK432097 (**4**) bound (*left*, *green carbon atoms*) [17], and comparison with a hybrid homology model of the hA_3AR with potent C2-arylethynyl agonist MRS5980 (**8**) bound (*right*, *magenta carbon atoms*) [52]

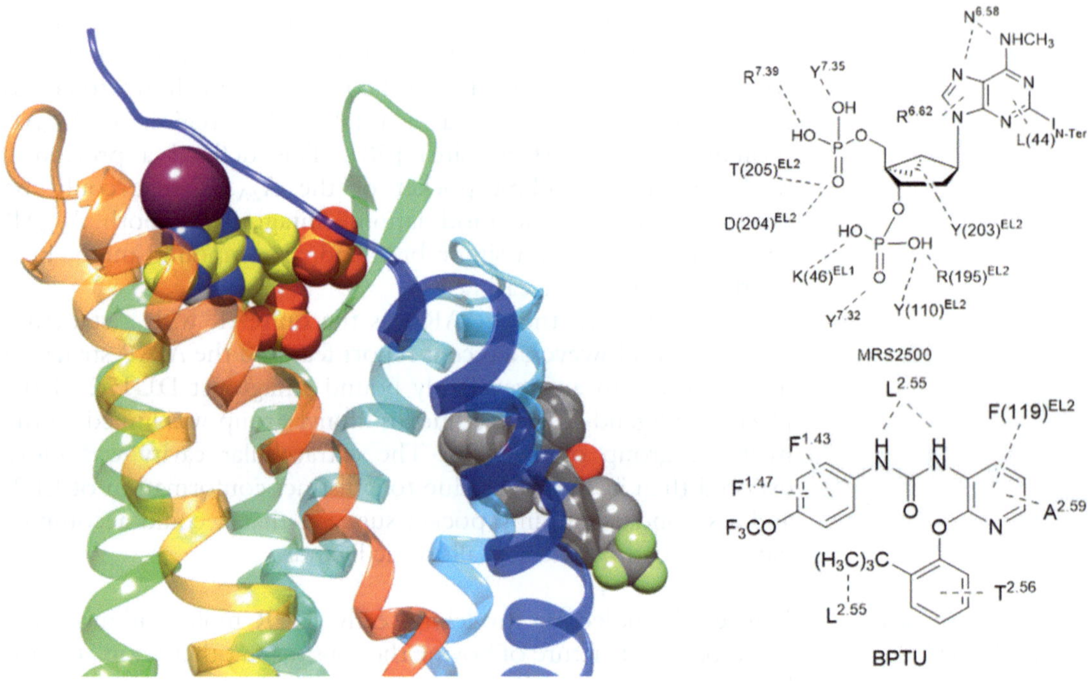

Fig. 4 *Left panel*: High-resolution X-ray crystallographic structures [29] of the $hP2Y_1R$ showing the binding sites for orthosteric antagonist MRS2500 (**31**) (*yellow carbon atoms*) and allosteric antagonist (NAM) BPTU (**34**) (*gray carbon atoms*). *Right panel*: Contacts between two cocrystallized ligands and the $hP2Y_1R$

bind nonnucleotide antagonists (AZD1283, ethyl 6-(4-((benzylsulfonyl)carbamoyl)piperidin-1-yl)-5-cyano-2-methylnicotinate, **35**) [60] and nucleotide partial agonists (such as 2-methylthio-ATP, 2-MeSATP **29**) [30, 31]. One of the most unusual features of the

AZD1283-P2Y$_{12}$R complex is the apparent lack of a disulfide bridge between EL2 and TM3 (otherwise conserved), but the presence of a disulfide between TM7 and the N-terminus (conserved for P2YRs). The outer TM portions and the ELs are also rich in positively charged Lys and Arg residues. Consequently, these residues close around the nucleotide ligands in both complexes with 2-methylthio-ADP (full agonist, **28**) and 2-methylthio-ATP, and the conserved disulfide bridge (TM3 to EL2) is present. The variation in the presence of this disulfide bond between different complexes suggests that this bond is dynamic in the P2Y$_{12}$R.

Although P2Y$_1$R, like P2Y$_{12}$R, is activated by ADP, it belongs to a structurally distinct subset of the rhodopsin-like GPCR δ-branch. Structural comparison of the P2Y$_1$R with bound nucleotide (orthosteric) and nonnucleotide (allosteric) antagonists indicates that completely different residue sets are involved, i.e., no amino acids were shared by the two sites (Fig. 5) [29]. The orthosteric nucleotide antagonist MRS2500 **31** binds at a location involving the ELs that is more external than most small molecule orthosteric ligands of other GPCRs. Positively charged EL residues coordinate the phosphate groups at the 5′ and 3′-positions of the bisphosphate antagonist (Fig. 5). The overall effect of bound MRS2500 **31** is to constrain two receptor domains, i.e., TM1-TM4 and TM5-TM7, which would be expected to have a relative movement during activation. The negative allosteric modulator BPTU (1-(2-(2-(*tert*-butyl)phenoxy)pyridin-3-yl)-3-

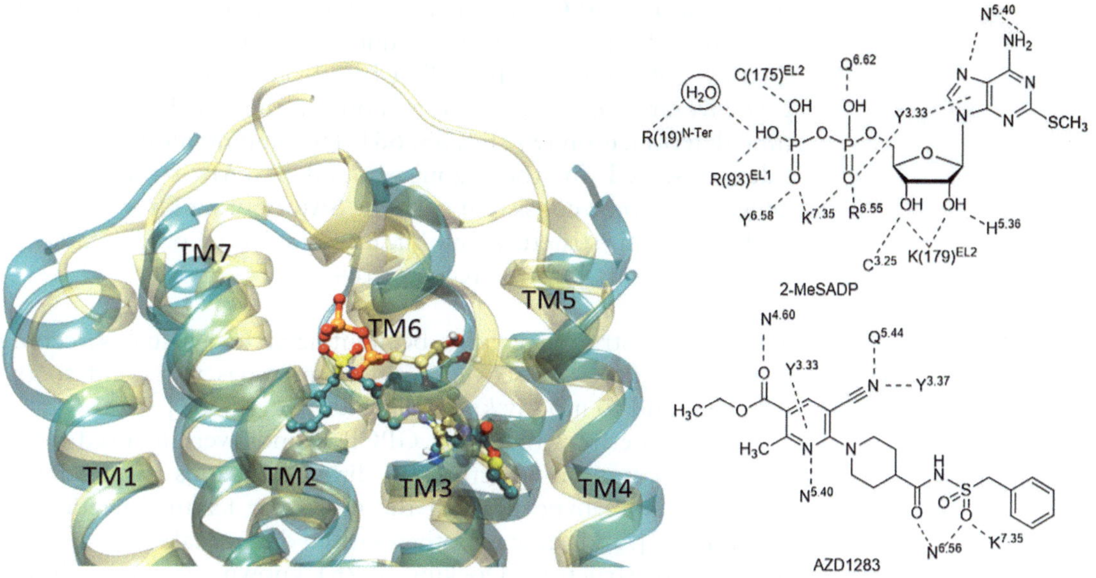

Fig. 5 *Left panel*: Superimposition between X-ray crystallographic structures of the hP2Y$_{12}$R in complex with 2-MeSADP (**28**) (*yellow ribbon* representation for the receptor and *yellow carbon atoms* for the ligand) [31] and AZD1283 (**35**) (*dark cyan* ribbon representation for the receptor and *dark cyan* carbon atoms for the ligand) [30]. *Right panel*: Contacts between two cocrystallized ligands and the hP2Y$_{12}$R

(4-(trifluoromethoxy)phenyl)urea, **34**) [61] surprisingly bound to the outer $P2Y_1R$ surface in contact with the hydrophobic phospholipid membrane components and with residues in TMs 1–3. The binding site was shallow with the urea group of BPTU forming H-bonds with a TM2 backbone carbonyl group (Fig. 5). The placement of this allosteric antagonist would constrain relative TM3 movement with respect to TM2, which is also characteristic of GPCR activation [17]. The mechanism for the propagation of the conformational effects of **34** binding to other $P2Y_1R$ regions might involve the characteristic interactions of conserved hydrophobic and aromatic residues in TMs 3, 5, and 6 of various GPCRs [62]. The molecular recognition of other negative $P2Y_1R$ allosteric modulators can be modeled and is shown to be consistent with SAR data [29]. Curiously, the $P2Y_1R$ conformations present in complexes with the allosteric and orthosteric antagonists are nearly identical, and no agonist-bound structure is yet reported.

3 Methods

3.1 Structure-Based Medicinal Chemistry of the ARs

Early molecular modeling and site-directed mutagenesis of the ARs based on the bovine rhodopsin structure was relatively successful in predicting the position of agonists and antagonists in the orthosteric binding site and their interacting residues, as was subsequently validated in the X-ray structures of antagonist-bound and later agonist-bound A_{2A}ARs. The ARs are members of Family A rhodopsin-like GPCRs, which is very close in overall structure to rhodopsin itself, although the sequence homology is low. Consequently, AR molecular modeling based on a rhodopsin template gave favorable results prior to and during the initial A_{2A}AR structural determination [37, 54, 55, 63]. The binding mode of nucleosides observed in the agonist-bound A_{2A}AR structures was generalized to be consistent with the SAR of other known ligands, leading to their structural modification guided by predicted favorable interactions with the receptor [64].

3.1.1 Medicinal Chemistry of the A_{2A}AR

The A_{2A}AR structures have been utilized for virtual screening (VS) campaigns to discover novel chemotypes that bind to the ARs [65] and to modify known ligands [61, 65–67].

Previous efforts to model GPCR ligands were focused on an overlay of common structures in different ligands to generate a pharmacophore hypothesis of the molecular features needed for binding [68], and these hypotheses often did not take into account the receptor structure. Docking of 751 known antagonists in the A_{2A}AR structure to create a refined pharmacophore model improved the predictive ability of subsequent quantitative SAR statistical models [69]. The pharmacophore hypothesis was validated with new test set of 29 A_{2A}AR antagonists related to ZM241385 **13**.

Kinetic parameters of binding, as well as affinity, may be studied using modeling approaches. The dissociation kinetics of antagonists derived from ZM241385, i.e., **16–19**, has been studied by X-ray crystallography of A_{2A}AR StaRs [24]. A salt bridge involving EL residues of the A_{2A}AR controlled the antagonist dissociation kinetics of analogues that had chemical modifications at the terminal phenol, as is present in ZM241385. Metadynamics investigations revealed that the salt bridge was readily broken in the simulations for ligands exhibiting short residence times, whereas it was maintained for ligands with longer residence times. The X-ray structures of the ligand-receptor complexes highlighted that long residence time ligands established stabilizing interactions with His264 (EL3) that were not detected instead for ligands with shorter residence times.

The EL structure of a given GPCR is not only important for ligand affinity and binding kinetics, but can be utilized as the site of covalent modification by reactive, bitopic ligands. A_{2A}AR agonists that modify the receptor irreversibly by acylating a specific residue, Lys153, in EL2 were designed through docking to the agonist-bound receptor structure [70]. The terminal position of the C2 chain contained a chemically reactive group, i.e., an active ester, and its proximity to Lys153 (EL2) facilitated this covalent modification of the receptor.

A_{2A}AR structures have guided the design of A_3AR agonists and antagonists as well.

3.1.2 Medicinal Chemistry of the A_3AR

Originally, it was thought that A_3AR antagonists might have broader application in the clinic than agonists, because of proinflammatory effects associated with acute agonist administration [71]. A_3AR antagonists have been proposed for the treatment of glaucoma or kidney fibrosis [71–73]. However, it is now recognized that A_3AR agonists have shown efficacy in various animal models of inflammatory disease, cancer and chronic neuropathic pain [71, 74–76]. Thus, the improvement of the affinity and selectivity of A_3AR agonists has clear clinical relevance. Furthermore, prototypical agonists IB-MECA **5** and Cl-IB-MECA **6** have demonstrated safety and efficacy in Phase 2 and 1/2 clinical trials, respectively, and are now progressing to more advances trials in rheumatoid arthritis, psoriasis, hepatocellular carcinoma, and non-alcoholic steatohepatitis (NASH) [71, 75].

The three-dimensional GPCR structures and related homology models have been used for improving the affinity and selectivity of known ligands as needed. For example, the design of A_3AR agonists has benefited from iterative cycles of ligand docking to A_3AR homology models followed by synthesis of new ligand analogues and model refinement. Differences between the structures of the AR subtypes are identified in the modeling process and can be used advantageously to increase selectivity.

The ribose moiety is the core of AR agonists that establishes the spatial relationship of important ligand recognition elements. We have substituted ribose in A_3AR agonists with a variety of sterically constrained, bicyclic rings, to mimic the conformation of native tetrahydrofuryl ring of ribose when bound to the receptor. In particular, novel A_3AR agonists are being refined, based on a highly rigidified (*N*)-methanocarba (a [3.1.0]bicyclohexane) scaffold, with computational approaches based on A_{2A}AR X-ray structures as essential components, leading to nM binding affinities, exceptionally high selectivities and improved in vivo efficacy [13]. For example, MRS5698 7 is one such selective agonist that maintains consistent affinity across species ($K_i = 3$ nM at hA_3AR and mouse A_3AR). A hybrid A_3AR model, based on the active or active-like structures of different GPCRs, was found most suitable for these new agonists [13]. High specificity (~10,000-fold selectivity for the A_3AR) and clean ADME-tox and off-target properties were achieved.

In addition to the ongoing clinical trials of A_3AR agonists for cancer and inflammation, they also have potential in pain treatment [76]. Novel A_3AR agonists for pain control were designed and screened using an in vivo phenotypic model, which reflected both pharmacokinetic and pharmacodynamic parameters. This close coupling of structure-based ligand design and in vivo testing permitted multiple parameters to be enhanced during the choice of synthetic target molecules. Activation of the A_3AR in peripheral neurons, spinal cord, and brain by highly specific C2-arylalkynyl agonists, e.g., MRS5980 **8**, as well as less selective A_3AR agonists was found to reduce chronic neuropathic pain in vivo. This protection was dependent on $GABA_A$ receptor modulation, oxidative pathways, astrocytic activation, and cytokine levels in the spinal cord.

Although the AR binding site has two domains for ribose and adenine, there are examples of compensation for the loss of an otherwise important recognition element in one domain by features of a different domain of the ligand. For 4′-truncated A_1AR agonists, a particular N^6-dicyclopropylmethyl bound precisely in two subpockets of the N^6 region to compensate for the absence of receptor binding stabilization at the 5′ position and activated the receptor [77]. In the MRS5980 **8** chemical series of A_3AR agonists, the N^6 methylamino group could be group truncated to H or substituted with CH_3 with retention of high affinity [52]. Thus, the exocyclic amine of adenosine derivatives was not essential for high affinity and selectivity at the A_3AR, although the Asn6.55 sidechain in A_3AR models is similar to its position with more conventional agonists docked (Fig. 3) [52]. The other stabilizing groups on these truncated analogues of MRS5980, e.g., at 5′ and C2, could partly compensate for the lack of H-bond stabilization between the exocyclic amine and this residue.

Biased agonists for a given GPCR display a preference for one or more signaling pathways [78, 79], and such an action has been explored for AR agonists. Biased agonism is based on the concept that a GPCR active state is associated with multiple conformations, and different activated conformations a given GPCR depend on which agonist is bound. More importantly, each conformation in theory has its own profile of preferred signaling, such as G-protein dependent vs. arrestin-dependent signaling. This is based on a particular receptor conformation resulting from agonist binding that is more effective in the preferred pathway compared to others. Biased A_3AR agonism was detected in (*N*)-methanocarba analogues [80]. Conformational plasticity of the A_3AR, especially with respect to the orientation of TM2, was proposed to accommodate bulky groups and correlate with signaling bias within this agonist class. Within a set of rigid C2-arylalkynyl and C2-polyarylalkynyl derivatives, the degree of bias for A_3AR-dependent cell survival was directly proportional to the length of this substituent on the agonist. Hypothetically, this bias would correspond to the degree of outward displacement of the upper TM2 portion in the A_3AR agonist-bound state. Thus, conformational A_3AR plasticity was proposed to accommodate bulky groups and correlate with signaling bias within this class of agonists. However, it is to be noted that nucleoside A_3AR antagonists with extended C2 substituents would also require TM2 displacement in the inactive receptor state [72]. Consistent with the prediction of an outward movement of TM2 to accommodate known A_3AR ligands, the recently determined structure of an antagonist-bound A_1AR features TM2 displaced in the same direction by ~5 Å [19], relative to the $A_{2A}AR$ structure. Like the A_3AR, the A_1AR has only one disulfide bridge in the EL region; thus, TM2 is not constrained as in the $A_{2A}AR$.

3.1.3 Virtual Screening to Discover AR Ligands in General

In silico screening of diverse chemical libraries has identified numerous novel chemotypes for GPCR modulation. In silico screening has identified chemically diverse ligands at each of the AR subtypes. This structure-based drug design approach has been applied using the $A_{2A}AR$ X-ray structures resulting in typical hit rates of up to 40% with K_i values $\leq 10\ \mu M$ [65, 74, 81]. In many cases, the hits displayed not only $A_{2A}AR$ affinity but also affinity at the closely related A_1AR and/or A_3AR. Thus, VS is a means of discovering novel chemotypes for binding to other AR family members because of the close structural homology.

Furthermore, nearly all of the hit molecules in VS of ARs were found to be AR antagonists, using as a template either an antagonist-bound $A_{2A}AR$ structure (as expected) or an agonist-bound structure (unexpected) [74, 81, 82]. This illustrated the close structural tolerance needed for AR activation and that non-ribosides were highly unlikely to be suitable. Instead, the challenge of discovering novel agonists for the AR family was approached by screening within a set of ~7000 diverse nucleobases available commercially [65]. The screening was accomplished by first virtually attaching the ribose moiety before docking in an agonist-bound A_{2A}AR structure—and finally by chemically synthesizing the ribosides from the nucleobase hits. In this manner, the pool of candidate molecules in the database was greatly expanded. Because the number of pre-formed ribosides in the databases was quite limited, the resulting hit rate for riboside products that activated one or more AR subtypes was greatly enhanced.

Modeling approaches have also facilitated the design of tool compounds for drug discovery, such as high affinity fluorescent ligands. Selective fluorescent agonist and antagonist probes of the A_3AR have been reported [83]. For example, agonist MRS5218 (structure not shown) binds to the A_3AR with high affinity (Ki 17 nM). The fluorophore (AlexaFluor488), attached through a functionalized alkynyl chain at C2, is thought to participate in the recognition when bound to the receptor's outer loop region. Thus, the same functionalized AR ligand can vary enormously in affinity, with both increases and decreases with respect to the parent ligand. The affinity can depend strongly on the nature of the tethered fluorophore, because this moiety can contribute to the affinity by interacting with the receptor's EL region.

Agonist discovery at other AR subtypes has also benefited from the high-resolution A_{2A}AR structures, in some cases unexpectedly through lateral hits. The structure-guided chemical modification of AR agonists at the adenosine 5′-amide position resulted in an oxetane derivative **9** that achieved high affinity at the A_1AR instead of the A_{2A}AR, and this gain of function was interpreted in terms of a different A_1AR geometry in the ribose region that allowed H-bonding of conserved Asn5.42 to the oxetane ether [66]. The bulky presence of the Trp5.46 sidechain in A_1AR, compared to Cys5.42 in A_{2A}AR, creates a cavity for the oxetane ring to lodge between TM5 and TMs 3 and 4. The binding region of the A_1AR around the N^6 group was predicted to accommodate two hydrophobic groups, e.g., dicyclopropyl, consistent with the previously explored recognition of α-branched substituents demonstrating a diastereomeric preference [77].

Ligand docking at an A_{2B}AR homology model based on an antagonist-bound A_{2A}AR structure has aided the design of 3,4-dihydropyrimidin-2(1*H*)-ones, such as **20** (K_i $hA_{2B}AR = 3.49$ nM), as novel antagonist chemotypes that display high selectivity [84]. Compounds **21** and **24** arose from A_{2A}AR screens and were found to bind selectively to the A_1AR [81, 85]. Compounds **22** and **23** were identified in a screen using A_1AR homology models and bound selectively to the A_3AR [74]. Compound **25** was identified using the agonist-bound

$A_{2A}AR$, but was found to be a mixed antagonist at the A_1AR and $A_{2A}AR$ [86]. Thus, there is considerable activity of screening hits at related receptors. In silico screening and structure-based design strategies have also led to the identification of nonnucleoside atypical partial agonists of the ARs based on a 7-prolinol-thiazolo [5,4-*d*]-pyrimidine scaffold [87]. 2-Amino-3-cyanopyridines appeared as hits in VS based on agonist-bound $A_{2A}ARs$, but they were AR antagonists [86], unlike related 2-amino-3,5-dicyanopyridines that act as atypical AR agonists [88]. Docking studies helped in rationalizing the binding of these compounds at the $A_{2A}AR$ and in identifying surrogates for the ribose ring that afforded receptor activation. Indeed, a 2-furylmethanol moiety in the 2-amino-3-cyanopyridine series was predicted to establish H-bond interaction with Ser 7.42, whereas the 2-hydroxylmethyl pyrrolidine moiety in the 7-prolinol-thiazolo[5,4-*d*]-pyrimidine series was predicted to establish H-bond interaction with His 7.43.

3.1.4 A_3AR Allosteric Modulators

Nearly all of the AR X-ray structures reported contained orthosteric ligands, but modeling has been applied to the binding of allosteric ligands as well. Several classes of nitrogen heterocyclic molecules have been shown to be positive allosteric modulators (PAMs) for the A_3AR, but no X-ray structures to firmly establish their binding site on the receptor. One such allosteric enhancer is *N*-(3,4-dichloro-phenyl)-2-cyclohexyl-1*H*-imidazo[4,5-*c*]quinolin-4-amine (LUF6000, **10**). Nevertheless, site-directed mutagenesis and modeling approaches, including Supervised Molecular Dynamics (SuMD), have attempted to determine the residues involved in their allosteric binding vs. the residues needed for orthosteric binding [89]. In the SuMD study, LUF6000 was found to engage in interactions with a putative meta-binding site located at the interface between EL2 and the upper region of TM5 and TM6, prior to reaching the orthosteric site. A π–π stacking interaction established with Phe168 (EL2) triggered the allosteric modulator to reach the orthosteric binding site occupied by the agonist adenosine (**1**). In the final ternary complex, LUF6000 established hydrophobic contacts with residues in the upper region of the orthosteric binding site, thus acting as "pocket cap."

3.2 Structure-Based Medicinal Chemistry of the P2YRs

$P2Y_{12}R$ antagonists display potent anti-thrombotic activity by preventing the action of ADP on the platelet surface, which is the basis of blockbuster drugs such as clopidopgrel. $P2Y_1R$ antagonists, such as the orthosteric antagonist MRS2500 **31** of 0.8 nM affinity, also have potent anti-thrombotic activity by antagonizing ADP at platelet receptors, but this therapeutic concept has not yet resulted in clinical trials [90]. Thus, there is continuing interest in designing ligands of the ADP-activated $P2Y_1R$ and $P2Y_{12}R$ as antithrombotic agents and also to explore clinical potential of other P2YR ligands. Although there is not yet an X-ray structure for any of the other

P2YRs, much insight can be gained from homology modeling, as we illustrate below for the $P2Y_{14}R$. We have used P2YR structures to understand the recognition of numerous antagonists at these receptors [91, 92].

3.2.1 Medicinal Chemistry of the $P2Y_1R$

Because there is no agonist-bound $P2Y_1R$ structure, the extension of the P2YR structures to binding of diverse agonists is best justified for the $P2Y_{12}R$. Nevertheless, a $P2Y_1R$ nucleotide agonist 2-MeSADP **28** was found to dock in a similar position, at least with respect to adenine and ribose moieties, as the nucleotide orthosteric antagonist **31**. For allosteric antagonists, such as BPTU **34**, the docking of other urea and related derivatives [15] that were reported to have the same antagonistic effect on $P2Y_1R$ explains some observed SAR in this series. These allosteric antagonists are predominantly hydrophobic molecules, consistent with the required passage through the phospholipid bilayer in order to reach the structurally defined binding site. Attempts to introduce polar groups while retaining $P2Y_1R$ affinity were only partially successful [93].

The environments surrounding the hydrophilic nucleotide MRS2500 **31** binding site of the $P2Y_1R$ structure and the hydrophobic allosteric binding site have very different properties and would require separate treatment in VS. An in silico screen was performed using the MRS2500-$P2Y_1R$ structure to identify compounds from medicinal plants related to Chinese traditional medicines that might have potential as antithrombotic drugs [94].

3.2.2 Medicinal Chemistry of the $P2Y_{12}R$

The first step in extending the reported P2YR structures to novel ligands and subsequently to other P2YR subtypes was an effort to explain the known ligand SAR. Representative $P2Y_{12}R$ ligands from different chemical classes were docked in the structures [58]. The $P2Y_{12}R$ X-ray structures with nucleotides bound (2-MeSADP and the corresponding triphosphate) differ greatly in conformation from the complex with nonnucleotide antagonist AZD1283 **35** bound (Fig. 4). Therefore, the question arose as to which structure could best serve as a suitable template for modeling the binding of various known $P2Y_{12}R$ ligands. The nucleotide complex(es) were found to be a suitable template for various agonists and diverse antagonists, many of which contain negatively charged groups. These diverse anionic groups were predicted to interact with Lys7.35—similar to its interaction with the partial negative charge of a sulfonyl oxygen of **35** in its $P2Y_{12}R$ complex (Fig. 4). The binding of nucleotide Cangrelor **32**, which is now an approved antithrombotic drug, was well accommodated in the same orientation as 2-MeSADP **28** in its $P2Y_{12}R$ complex (Fig. 6 left). The adenine moiety forms $\pi-\pi$ stacking with Tyr3.33, suggesting that nonaromatic nucleobase substitution is not possible at $P2Y_{12}R$ (Fig. 4). Although bulkier than the corresponding

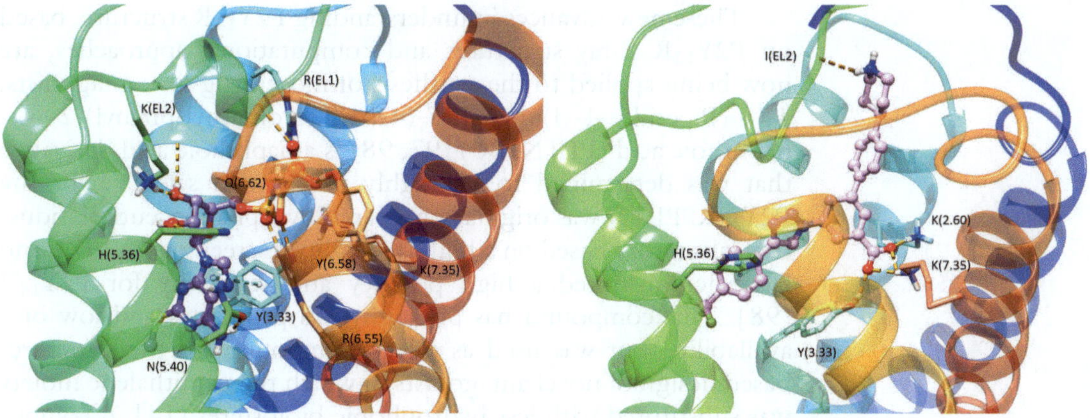

Fig. 6 High-resolution X-ray crystallographic structures of the $hP2Y_{12}R$ with agonist 2-MeSADP (**28**) bound (*purple carbon atoms*, left), and comparison with a homology model of the $hP2Y_{14}R$ with potent nonnucleotide antagonist MRS4217 (**37**) bound (*pink carbon atoms*, right) [14]

substitutions in **28**, the C2 and N^6 substituents fit in small pockets delimited, respectively, by TMs 3, 4, and 5 and at the base of the binding site toward TM6. The binding of Ticagrelor **33**, which was the first competitive $P2Y_{12}R$ antagonist [95] to be approved as an antithrombotic drug, required a $P2Y_{12}R$ hybrid model to accommodate the extended C2 and N^6 groups. However, SuMD simulation did not predict the proposed position of **33** in the orthosteric binding site of $P2Y_{12}R$ [58].

3.2.3 Medicinal Chemistry of the $P2Y_{14}R$

The $P2Y_{14}R$ is activated by UDP-sugars and is the focus of structure-based ligand design studies. Homology modeling based on the closely related nucleotide-$P2Y_{12}R$ structure (with 45% sequence identity), docking and MD predicted the position of the nucleotide $P2Y_{14}R$ agonists [96]. The uridine moiety of UDP-glucose **30** is bound in the same region as the adenosine moiety of ADP **27a** in the $P2Y_{12}R$ (Fig. 6 left), and a similar analogy applies to the 5′-diphosphate moieties. The uracil ring is located in a smaller hydrophobic region of the binding site than the corresponding region in $P2Y_{12}R$, which is consistent with this receptor's strong preference for uracil nucleotides. The glucose moiety is predicted to bind in the second subpocket of the bifurcated orthosteric binding site of $P2Y_{14}R$, which is present yet vacant in the $P2Y_{12}R$. Because of its predicted position facing the ELs, this glucose moiety is amenable to structural modification without impairing receptor binding and is used as an attachment site to produce a high affinity fluorescent agonist of $P2Y_{14}R$, MRS4183 (structure not shown) [96]. In that fluorescent analogue, a boron-dipyrromethene TR (BODIPY TR) fluorophore is coupled by amide linkage to the distal carboxylate of UDP-glucuronic acid.

These new advances in understanding $P2Y_{14}R$ structure, based on $P2Y_{12}R$ X-ray structures and computational approaches, are now being applied to the synthesis of more drug-like antagonists. 4-(4-(Piperidin-4-yl)phenyl)-7-(4-(trifluoromethyl)phenyl)-2-naphthoic acid (PPTN, **36**) [97, 98] is a naphthoic acid derivative that was determined to be highly potent and selective for the $P2Y_{14}R$. PPTN was originally prepared by a pharmaceutical industrial laboratory based on a high throughput screening hit [97], and later demonstrated a high potency and selectivity for $P2Y_{14}R$ [98]. This compound has poor physical properties and low bioavailability, but was used as a lead compound for the structure-based design of novel antagonists in which the naphthalene moiety was substituted with less hydrophobic bioiosteres [14]. A computational pipeline was used to compare $P2Y_{14}R$ recognition of proposed analogues and led to the identification of 4′-(piperidin-4-yl)-5-(4-(4-(trifluoromethyl)phenyl)-1*H*-1,2,3-triazol-1-yl)-[1,1-′-biphenyl]-3-carboxylic acid (MRS4217, **37**), which was only six-fold less potent than PPTN in a fluorescence-based binding assay in whole cells. Docking and MD simulation suggested that the compound was able to establish an H-bond network with Lys2.60, Lys7.35 and Tyr3.33 along with a π–π tacking interaction with His5.36 (Fig. 6 right). These interactions were stably maintained during 30 ns of MD simulations.

Chemical tools for structural probing of these GPCRs and improving assay capabilities, such as fluorescent probes of the inflammation-related $P2Y_{14}R$ [96, 99], were designed with the aid of docking and MD. This was especially important for drug discovery at the $P2Y_{14}R$, because there are no high affinity radioligands. The high affinity fluorescent antagonist MRS4174 (structure not shown, K_i 0.08 nM) was designed as a PPTN analogue, in which a fluorophore (AlexaFluor488) is strategically tethered through the alkylation of the piperidine ring's amino group. The location of the functionalized chain was planned at this position based on a prediction from docking of the parent antagonist, which featured the piperidine moiety exposed to the extracellular medium. The affinity gain with respect to PPTN is predicted to be from polar interactions of the charged fluorophore moiety with specific amino acids of the $P2Y_{14}R$ ELs.

4 Notes

Recent breakthroughs in computational modeling approaches have impacted the medicinal chemistry aimed at discovering new ligands for adenosine and P2Y receptors. Purine receptor structures and an interdisciplinary approach have enabled the elucidation of their biological role, the conceptualization of future therapeutics, and novel ligand discovery. Computational approaches based on $A_{2A}AR$ X-ray structures guided the identification and refinement of novel A_3AR agonists for treating chronic neuropathic pain, on the basis of a highly rigidified (*N*)-methanocarba scaffold. Molecular docking and MD aided the design of fluorescent AR and $P2Y_{14}R$ probes as chemical tools for structural exploration of these GPCRs and for improving assay capabilities. Computational approaches based on $P2Y_{12}R$ structures guided the design of more drug-like antagonists of the inflammation-related $P2Y_{14}R$. The molecular recognition of positive (for A_3AR) and negative ($P2Y_1R$) allosteric modulators has also been modeled to shed light on ARs and PYRs allosteric modulation. Thus, computational modeling based on physically determined structures is now an essential tool for GPCR ligand design.

Acknowledgments

We acknowledge support from the NIDDK, NIH Intramural Research Program.

References

1. Santos R, Ursu O, Gaulton A et al (2016) A comprehensive map of molecular drug targets. Nat Rev Drug Discov 16:19–34. https://doi.org/10.1038/nrd.2016.230
2. Mason JS, Bortolato A, Weiss DR et al (2013) High end GPCR design: crafted ligand design and druggability analysis using protein structure, lipophilic hotspots and explicit water networks. In Silico Pharmacol 1:23. https://doi.org/10.1186/2193-9616-1-23
3. Tautermann CS (2014) GPCR structures in drug design, emerging opportunities with new structures. Bioorg Med Chem Lett 24:4073–4079. https://doi.org/10.1016/j.bmcl.2014.07.009
4. Rodríguez D, Ranganathan A, Carlsson J (2015) Discovery of GPCR ligands by molecular docking screening: novel opportunities provided by crystal structures. Curr Top Med Chem 15:2484–2503
5. Kooistra AJ, Vischer HF, McNaught-Flores D et al (2016) Function-specific virtual screening for GPCR ligands using a combined scoring method. Sci Rep 6:28288. https://doi.org/10.1038/srep28288
6. Burnstock G (2016) Short- and long-term (trophic) purinergic signalling. Philos Trans R Soc B Biol Sci 371:20150422. https://doi.org/10.1098/rstb.2015.0422
7. Cronstein BN, Sitkovsky M (2016) Adenosine and adenosine receptors in the pathogenesis and treatment of rheumatic diseases. Nat Rev Rheumatol 13:41–51. https://doi.org/10.1038/nrrheum.2016.178
8. Zimmermann H, Zebisch M, Sträter N (2012) Cellular function and molecular structure of ecto-nucleotidases. Purinergic Signal 8:437–502. https://doi.org/10.1007/s11302-012-9309-4
9. Boison D (2013) Adenosine kinase: exploitation for therapeutic gain. Pharmacol Rev 65:906–943. https://doi.org/10.1124/pr.112.006361
10. Schöneberg T, Hermsdorf T, Engemaier E et al (2007) Structural and functional evolution of the $P2Y_{12}$-like receptor group. Purinergic Signal 3:255–268. https://doi.org/10.1007/s11302-007-9064-0
11. Verkhratsky A, Burnstock G (2014) Biology of purinergic signalling: its ancient evolutionary roots, its omnipresence and its multiple functional significance. BioEssays 36:697–705. https://doi.org/10.1002/bies.201400024
12. Toti KS, Osborne D, Ciancetta A et al (2016) South (S)- and north (N)-methanocarba-7-deazaadenosine analogues as inhibitors of human adenosine kinase. J Med Chem 59:6860–6877. https://doi.org/10.1021/acs.jmedchem.6b00689
13. Tosh DK, Deflorian F, Phan K et al (2012) Structure-guided design of A_3 adenosine

receptor-selective nucleosides: combination of 2-arylethynyl and bicyclo[3.1.0]hexane substitutions. J Med Chem 55:4847–4860. https://doi.org/10.1021/jm300396n

14. Junker A, Balasubramanian R, Ciancetta A et al (2016) Structure-based design of 3-(4-aryl-1 *H* –1,2,3-triazol-1-yl)-biphenyl derivatives as $P2Y_{14}$ receptor antagonists. J Med Chem 59:6149–6168. https://doi.org/10.1021/acs.jmedchem.6b00044
15. Conroy S, Kindon N, Kellam B, Stocks MJ (2016) Drug-like antagonists of P2Y receptors—from lead identification to drug development. J Med Chem 59:9981–10005. https://doi.org/10.1021/acs.jmedchem.5b01972
16. Lebon G, Warne T, Edwards PC et al (2011) Agonist-bound adenosine A_{2A} receptor structures reveal common features of GPCR activation. Nature 474:521–525. https://doi.org/10.1038/nature10136
17. Xu F, Wu H, Katritch V et al (2011) Structure of an agonist-bound human A_{2A} adenosine receptor. Science 332:322–327. https://doi.org/10.1126/science.1202793
18. Lebon G, Edwards PC, Leslie AGW, Tate CG (2015) Molecular determinants of CGS21680 binding to the human adenosine A_{2A} receptor. Mol Pharmacol 87:907–915. https://doi.org/10.1124/mol.114.097360
19. Glukhova A, Thal DM, Nguyen AT et al (2017) Structure of the adenosine A_1 receptor reveals the basis for subtype selectivity. Cell 168:867–877.e13. https://doi.org/10.1016/j.cell.2017.01.042
20. Jaakola V-P, Griffith MT, Hanson MA et al (2008) The 2.6 angstrom crystal structure of a human A_{2A} adenosine receptor bound to an antagonist. Science 322:1211–1217. https://doi.org/10.1126/science.1164772
21. Doré AS, Robertson N, Errey JC et al (2011) Structure of the adenosine A_{2A} receptor in complex with ZM241385 and the xanthines XAC and caffeine. Structure 19:1283–1293. https://doi.org/10.1016/j.str.2011.06.014
22. Hino T, Arakawa T, Iwanari H et al (2012) G-protein-coupled receptor inactivation by an allosteric inverse-agonist antibody. Nature 482:237–240. https://doi.org/10.1038/nature10750
23. Liu W, Chun E, Thompson AA et al (2012) Structural basis for allosteric regulation of GPCRs by sodium ions. Science 337:232–236. https://doi.org/10.1126/science.1219218
24. Segala E, Guo D, Cheng RKY et al (2016) Controlling the dissociation of ligands from the adenosine A_{2A} receptor through modulation of salt bridge strength. J Med Chem 59:6470–6479. https://doi.org/10.1021/acs.jmedchem.6b00653
25. Congreve M, Andrews SP, Doré AS et al (2012) Discovery of 1,2,4-triazine derivatives AS adenosine A_{2A} antagonists using structure based drug design. J Med Chem 55:1898–1903. https://doi.org/10.1021/jm201376w
26. Jacobson KA, Gao ZG (2017) Chapter 11: Allosteric modulators of adenosine, P2Y and P2X receptors. In: Doller D (ed) Allosterism in drug discovery (RSC drug discovery series no. 56), pp 247–270. https://doi.org/10.1039/9781782629276
27. Carpenter B, Nehmé R, Warne T et al (2016) Structure of the adenosine A_{2A} receptor bound to an engineered G protein. Nature 536:104–107. https://doi.org/10.1038/nature18966
28. Batyuk A, Galli L, Ishchenko A et al (2016) Native phasing of x-ray free-electron laser data for a G protein-coupled receptor. Sci Adv 2: e1600292–e1600292. https://doi.org/10.1126/sciadv.1600292
29. Zhang D, Gao Z-G, Zhang K et al (2015) Two disparate ligand-binding sites in the human $P2Y_1$ receptor. Nature 520:317–321. https://doi.org/10.1038/nature14287
30. Zhang K, Zhang J, Gao Z-G et al (2014) Structure of the human $P2Y_{12}$ receptor in complex with an antithrombotic drug. Nature 509:115–118. https://doi.org/10.1038/nature13083
31. Zhang J, Zhang K, Gao Z-G et al (2014) Agonist-bound structure of the human $P2Y_{12}$ receptor. Nature 509:119–122. https://doi.org/10.1038/nature13288
32. Jazayeri A, Andrews SP, Marshall FH (2017) Structurally enabled discovery of adenosine A_{2A} receptor antagonists. Chem Rev 117:21–37. https://doi.org/10.1021/acs.chemrev.6b00119
33. Zou Y, Weis WI, Kobilka BK (2012) N-terminal T4 lysozyme fusion facilitates crystallization of a G protein coupled receptor. PLoS One 7:e46039. https://doi.org/10.1371/journal.pone.0046039
34. Katritch V, Cherezov V, Stevens RC (2013) Structure-function of the G protein–coupled receptor superfamily. Annu Rev Pharmacol Toxicol 53:531–556. https://doi.org/10.1146/annurev-pharmtox-032112-135923
35. Steyaert J, Kobilka BK (2011) Nanobody stabilization of G protein-coupled receptor conformational states. Curr Opin Struct Biol

21:567–572. https://doi.org/10.1016/j.sbi.2011.06.011

36. Ghosh E, Kumari P, Jaiman D, Shukla AK (2015) Methodological advances: the unsung heroes of the GPCR structural revolution. Nat Rev Mol Cell Biol 16:69–81. https://doi.org/10.1038/nrm3933
37. Ivanov AA, Barak D, Jacobson KA (2009) Evaluation of homology modeling of G-protein-coupled receptors in light of the A_{2A} adenosine receptor crystallographic structure. J Med Chem 52:3284–3292. https://doi.org/10.1021/jm801533x
38. Jacobson KA (2009) Functionalized congener approach to the design of ligands for G protein-coupled receptors (GPCRs). Bioconjug Chem 20:1816–1835. https://doi.org/10.1021/bc9000596
39. Ballesteros JA, Weinstein H (1995) [19] integrated methods for the construction of three-dimensional models and computational probing of structure-function relations in G protein-coupled receptors. Methods Neurosci Elsevier:366–428
40. Katritch V, Fenalti G, Abola EE et al (2014) Allosteric sodium in class a GPCR signaling. Trends Biochem Sci 39:233–244. https://doi.org/10.1016/j.tibs.2014.03.002
41. Massink A, Louvel J, Adlere I et al (2016) 5′-substituted Amiloride derivatives as allosteric modulators binding in the sodium ion pocket of the adenosine A_{2A} receptor. J Med Chem 59:4769–4777. https://doi.org/10.1021/acs.jmedchem.6b00142
42. Higgs C, Beuming T, Sherman W (2010) Hydration site thermodynamics explain SARs for triazolylpurines analogues binding to the A_{2A} receptor. ACS Med Chem Lett 1:160–164. https://doi.org/10.1021/ml100008s
43. Lenselink EB, Beuming T, Sherman W et al (2014) Selecting an optimal number of binding site waters to improve virtual screening enrichments against the adenosine A_{2A} receptor. J Chem Inf Model 54:1737–1746. https://doi.org/10.1021/ci5000455
44. Magnani F, Serrano-Vega MJ, Shibata Y et al (2016) A mutagenesis and screening strategy to generate optimally thermostabilized membrane proteins for structural studies. Nat Protoc 11:1554–1571. https://doi.org/10.1038/nprot.2016.088
45. Langmead CJ, Andrews SP, Congreve M et al (2012) Identification of novel adenosine A_{2A} receptor antagonists by virtual screening. J Med Chem 55:1904–1909. https://doi.org/10.1021/jm201455y
46. Gutiérrez-de-Terán H, Sallander J, Sotelo E (2017) Structure-based rational design of adenosine receptor ligands. Curr Top Med Chem 17:40–58
47. Segala E, Errey JC, Fiez-Vandal C et al (2015) Biosensor-based affinities and binding kinetics of small molecule antagonists to the adenosine A_{2A} receptor reconstituted in HDL like particles. FEBS Lett 589:1399–1405. https://doi.org/10.1016/j.febslet.2015.04.030
48. Bocquet N, Kohler J, Hug MN et al (2015) Real-time monitoring of binding events on a thermostabilized human A_{2A} receptor embedded in a lipid bilayer by surface plasmon resonance. Biochim Biophys Acta Biomembr 1848:1224–1233. https://doi.org/10.1016/j.bbamem.2015.02.014
49. Chen D, Errey JC, Heitman LH et al (2012) Fragment screening of GPCRs using biophysical methods: identification of ligands of the adenosine A_{2A} receptor with novel biological activity. ACS Chem Biol 7:2064–2073. https://doi.org/10.1021/cb300436c
50. Gao ZG, Kim SK, Biadatti T et al (2002) Structural determinants of A_3 adenosine receptor activation: nucleoside ligands at the agonist/antagonist boundary. J Med Chem 45:4471–4484
51. Toti KS, Moss SM, Paoletta S et al (2014) Synthesis and evaluation of N^6-substituted apioadenosines as potential adenosine A_3 receptor modulators. Bioorg Med Chem 22:4257–4268. https://doi.org/10.1016/j.bmc.2014.05.036
52. Tosh DK, Ciancetta A, Warnick E et al (2016) Purine (*N*)-methanocarba nucleoside derivatives lacking an exocyclic amine as selective A_3 adenosine receptor agonists. J Med Chem 59:3249–3263. https://doi.org/10.1021/acs.jmedchem.5b01998
53. Gao Z-G, Duong HT, Sonina T et al (2006) Orthogonal activation of the reengineered A_3 adenosine receptor (neoceptor) using tailored nucleoside agonists. J Med Chem 49:2689–2702. https://doi.org/10.1021/jm050968b
54. Jacobson KA, Ohno M, Duong HT et al (2005) A neoceptor approach to unraveling microscopic interactions between the human A_{2A} adenosine receptor and its agonists. Chem Biol 12:237–247. https://doi.org/10.1016/j.chembiol.2004.12.010
55. Kim S-K, Gao Z-G, Jeong LS, Jacobson KA (2006) Docking studies of agonists and antagonists suggest an activation pathway of the A_3 adenosine receptor. J Mol Graph

Model 25:562–577. https://doi.org/10.1016/j.Jmgm.2006.05.004

56. Li J, Jonsson AL, Beuming T, Shelley JC, Voth GA (2103) Ligand-dependent activation and deactivation of the human adenosine A_{2A} receptor. J Am Chem Soc 135:8749--8759. https://doi.org/10.1021/ja404391q

57. Kim JH, Wess J, van Rhee AM et al (1995) Site-directed mutagenesis identifies residues involved in ligand recognition in the human A_{2a} adenosine receptor. J Biol Chem 270:13987–13997

58. Sun B, Bachhawat P, Chu MLH et al (2017) Crystal structure of the adenosine A_{2A} receptor bound to an antagonist reveals a potential allosteric pocket. Proc Natl Acad Sci U S A 114:2066–2071. https://doi.org/10.1073/pnas.1621423114

59. Paoletta S, Sabbadin D, von Kügelgen I et al (2015) Modeling ligand recognition at the $P2Y_{12}$ receptor in light of X-ray structural information. J Comput Aided Mol Des 29:737–756. https://doi.org/10.1007/s10822-015-9858-z

60. Bach P, Boström J, Brickmann K et al (2013) Synthesis, structure–property relationships and pharmacokinetic evaluation of ethyl 6-aminonicotinate sulfonylureas as antagonists of the $P2Y_{12}$ receptor. Eur J Med Chem 65:360–375. https://doi.org/10.1016/j.ejmech.2013.04.007

61. Chao H, Turdi H, Herpin TF et al (2013) Discovery of 2-(Phenoxypyridine)-3-phenylureas as small molecule $P2Y_1$ antagonists. J Med Chem 56:1704–1714. https://doi.org/10.1021/jm301708u

62. Rosenbaum DM, Zhang C, Lyons JA et al (2011) Structure and function of an irreversible agonist-β_2 adrenoceptor complex. Nature 469:236–240. https://doi.org/10.1038/nature09665

63. Kim S-K, Jacobson KA (2007) Three-dimensional quantitative structure—activity relationship of nucleosides acting AT the A_3 adenosine receptor: analysis of binding and relative efficacy. J Chem Inf Model 47:1225–1233. https://doi.org/10.1021/ci600501z

64. Deflorian F, Kumar TS, Phan K et al (2012) Evaluation of molecular modeling of agonist binding in light of the crystallographic structure of an agonist-bound A_{2A} adenosine receptor. J Med Chem 55:538–552. https://doi.org/10.1021/jm201461q

65. Rodríguez D, Chakraborty S, Warnick E et al (2016) Structure-based screening of uncharted chemical space for atypical adenosine receptor agonists. ACS Chem Biol 11:2763–2772. https://doi.org/10.1021/acschembio.6b00357

66. Tosh DK, Phan K, Gao Z-G et al (2012) Optimization of adenosine 5′-carboxamide derivatives as adenosine receptor agonists using structure-based ligand design and fragment screening. J Med Chem 55:4297–4308. https://doi.org/10.1021/jm300095s

67. Paoletta S, Tosh DK, Finley A et al (2013) Rational design of sulfonated A_3 adenosine receptor-selective nucleosides as pharmacological tools to study chronic neuropathic pain. J Med Chem 56:5949–5963. https://doi.org/10.1021/jm4007966

68. Jacobson KA, Costanzi S, Paoletta S (2014) Computational studies to predict or explain G protein-coupled receptor polypharmacology. Trends Pharmacol Sci 35:658–663. https://doi.org/10.1016/j.tips.2014.10.009

69. Bacilieri M, Ciancetta A, Paoletta S et al (2013) Revisiting a receptor-based pharmacophore hypothesis for human A_{2A} adenosine receptor antagonists. J Chem Inf Model 53:1620–1637. https://doi.org/10.1021/ci300615u

70. Moss SM, Jayasekara PS, Paoletta S et al (2014) Structure-based design of reactive nucleosides for site-specific modification of the A_{2A} adenosine receptor. ACS Med Chem Lett 5:1043–1048. https://doi.org/10.1021/ml5002486

71. Borea PA, Varani K, Vincenzi F et al (2014) The A_3 adenosine receptor: history and perspectives. Pharmacol Rev 67:74–102. https://doi.org/10.1124/pr.113.008540

72. Tosh DK, Paoletta S, Phan K et al (2012) Truncated nucleosides as A_3 adenosine receptor ligands: combined 2-arylethynyl and bicyclohexane substitutions. ACS Med Chem Lett 3:596–601. https://doi.org/10.1021/ml300107e

73. Nayak A, Chandra G, Hwang I et al (2014) Synthesis and anti-renal fibrosis activity of conformationally locked truncated 2-hexynyl-N^6-substituted-(N)-methanocarbanucleosides as A_3 adenosine receptor antagonists. J Med Chem 57:1344–1354. https://doi.org/10.1021/jm4015313

74. Kolb P, Phan K, Gao Z-G et al (2012) Limits of ligand selectivity from docking to models: in silico screening for A_1 adenosine receptor antagonists. PLoS One 7:e49910. https://doi.org/10.1371/journal.pone.0049910

75. Fishman P, Bar-Yehuda S, Liang BT, Jacobson KA (2012) Pharmacological and therapeutic

effects of A_3 adenosine receptor agonists. Drug Discov Today 17:359–366. https://doi.org/10.1016/j.drudis.2011.10.007

76. Janes K, Symons-Liguori A, Jacobson KA, Salvemini D (2016) Identification of A_3 adenosine receptor agonists as novel non-narcotic analgesics. Br J Pharmacol 173:1253–1267. https://doi.org/10.1111/bph.13446
77. Tosh DK, Paoletta S, Deflorian F et al (2012) Structural sweet spot for A_1 adenosine receptor activation by truncated (N)-methanocarba nucleosides: receptor docking and potent anticonvulsant activity. J Med Chem 55:8075–8090. https://doi.org/10.1021/jm300965a
78. Strachan RT, Sun JP, Rominger DH et al (2014) Divergent transducer-specific molecular efficacies generate biased agonism at a G protein-coupled receptor (GPCR). J Biol Chem 289:14211–14224. https://doi.org/10.1074/jbc.M114.548131
79. Galandrin S, Onfroy L, Poirot MC et al (2016) Delineating biased ligand efficacy at 7TM receptors from an experimental perspective. Int J Biochem Cell Biol 77:251–263. https://doi.org/10.1016/j.biocel.2016.04.009
80. Baltos J-A, Paoletta S, Nguyen ATN et al (2016) Structure-activity analysis of biased agonism at the human adenosine A_3 receptor. Mol Pharmacol 90:12–22. https://doi.org/10.1124/mol.116.103283
81. Katritch V, Jaakola V-P, Lane JR et al (2010) Structure-based discovery of novel chemotypes for adenosine A_{2A} receptor antagonists. J Med Chem 53:1799–1809. https://doi.org/10.1021/jm901647p
82. Carlsson J, Yoo L, Gao Z-G et al (2010) Structure-based discovery of A_{2A} adenosine receptor ligands. J Med Chem 53:3748–3755. https://doi.org/10.1021/jm100240h
83. Kozma E, Gizewski ET, Tosh DK et al (2013) Characterization by flow cytometry of fluorescent, selective agonist probes of the A_3 adenosine receptor. Biochem Pharmacol 85:1171–1181. https://doi.org/10.1016/j.bcp.2013.01.021
84. el Maatougui A, Azuaje J, González-Gómez M et al (2016) Discovery of potent and highly selective A_{2B} adenosine receptor antagonist chemotypes. J Med Chem 59:1967–1983. https://doi.org/10.1021/acs.jmedchem.5b01586
85. Ranganathan A, Stoddart LA, Hill SJ, Carlsson J (2015) Fragment-based discovery of subtype-selective adenosine receptor ligands from homology models. J Med Chem 58:9578–9590. https://doi.org/10.1021/acs.jmedchem.5b01120
86. Rodríguez D, Gao Z-G, Moss SM et al (2015) Molecular docking screening using agonist-bound GPCR structures: probing the A_{2A} adenosine receptor. J Chem Inf Model 55:550–563. https://doi.org/10.1021/ci500639g
87. Bharate SB, Singh B, Kachler S et al (2016) Discovery of 7-(prolinol-N-yl)-2-phenylaminothiazolo[5,4-d]pyrimidines as novel non-nucleoside partial agonists for the A_{2A} adenosine receptor: prediction from molecular modeling. J Med Chem 59:5922–5928. https://doi.org/10.1021/acs.jmedchem.6b00552
88. Louvel J, Guo D, Soethoudt M et al (2015) Structure-kinetics relationships of Capadenoson derivatives as adenosine A_1 receptor agonists. Eur J Med Chem 101:681–691. https://doi.org/10.1016/j.ejmech.2015.07.023
89. Deganutti G, Cuzzolin A, Ciancetta A, Moro S (2015) Understanding allosteric interactions in G protein-coupled receptors using supervised molecular dynamics: a prototype study analysing the human A_3 adenosine receptor positive allosteric modulator LUF6000. Bioorg Med Chem 23:4065–4071. https://doi.org/10.1016/j.bmc.2015.03.039
90. Wong PC, Watson C, Crain EJ (2016) The $P2Y_1$ receptor antagonist MRS2500 prevents carotid artery thrombosis in cynomolgus monkeys. J Thromb Thrombolysis 41:514–521. https://doi.org/10.1007/s11239-015-1302-7
91. Conroy S, Kindon N, Kellam B, Stocks MJ (2016) Nucleotides acting at P2Y receptors: connecting structure and function. J Med Chem 59:9981–10005. https://doi.org/10.1021/acs.jmedchem.5b01972
92. Jacobson KA, Paoletta S, Katritch V et al (2015) Nucleotides acting at P2Y receptors: connecting structure and function. Mol Pharmacol 88:220–230. https://doi.org/10.1124/mol.114.095711
93. Hu CH, Qiao JX, Han Y et al (2014) 2-amino-1,3,4-thiadiazoles in the 7-hydroxy-N-neopentyl spiropiperidine indolinyl series as potent $P2Y_1$ receptor antagonists. Bioorg Med Chem Lett 24:2481–2485. https://doi.org/10.1016/j.bmcl.2014.04.011
94. Yi F, Sun L, L-j X, Peng Y et al (2017) In silico approach for anti thrombosis drug discovery: $P2Y_1R$ structure-based TCMs screening.

Front Pharmacol 7:531. https://doi.org/10.3389/fphar.2016.00531

95. Hoffmann K, Lutz DA, Straßburger J et al (2014) Competitive mode and site of interaction of ticagrelor at the human platelet $P2Y_{12}$ -receptor. J Thromb Haemost 12:1898–1905. https://doi.org/10.1111/jth.12719
96. Kiselev E, Balasubramanian R, Uliassi E et al (2015) Design, synthesis, pharmacological characterization of a fluorescent agonist of the $P2Y_{14}$ receptor. Bioorg Med Chem Lett 25:4733–4739. https://doi.org/10.1016/j.bmcl.2015.08.021
97. Gauthier JY, Belley M, Deschênes D et al (2011) The identification of 4,7-disubstituted naphthoic acid derivatives as UDP-competitive antagonists of $P2Y_{14}$. Bioorg Med Chem Lett 21:2836–2839. https://doi.org/10.1016/j.bmcl.2011.03.081
98. Barrett MO, Sesma JI, Ball CB et al (2013) A selective high-affinity antagonist of the $P2Y_{14}$ receptor inhibits UDP-glucose-stimulated chemotaxis of human neutrophils. Mol Pharmacol 84:41–49. https://doi.org/10.1124/mol.113.085654
99. Kiselev E, Barrett MO, Katritch V et al (2014) Exploring a 2-naphthoic acid template for the structure-based design of $P2Y_{14}$ receptor antagonist molecular probes. ACS Chem Biol 9:2833–2842. https://doi.org/10.1021/cb500614p
100. Cheng RKY, Segala E, Robertson N et al (2017) Structures of human A1 and A2A adenosine receptors with xanthines reveal determinants of selectivity. Structure, 25:1275–1285. https://doi.org/10.1016/j.str.2017.06.012

Chapter 4

A Structural Framework for GPCR Chemogenomics: What's In a Residue Number?

Márton Vass, Albert J. Kooistra, Stefan Verhoeven, David Gloriam, Iwan J.P. de Esch, and Chris de Graaf

Abstract

The recent surge of crystal structures of G protein-coupled receptors (GPCRs), as well as comprehensive collections of sequence, structural, ligand bioactivity, and mutation data, has enabled the development of integrated chemogenomics workflows for this important target family. This chapter will focus on cross-family and cross-class studies of GPCRs that have pinpointed the need for, and the implementation of, a generic numbering scheme for referring to specific structural elements of GPCRs. Sequence- and structure-based numbering schemes for different receptor classes will be introduced and the remaining caveats will be discussed. The use of these numbering schemes has facilitated many chemogenomics studies such as consensus binding site definition, binding site comparison, ligand repurposing (e.g. for orphan receptors), sequence-based pharmacophore generation for homology modeling or virtual screening, and class-wide chemogenomics studies of GPCRs.

Key words G protein-coupled receptors, GPCRs, Crystal structures, Chemogenomics, Drug discovery, Ligand repurposing, Numbering schemes, Mutations

1 Crystal Structures Reveal Common and Specific Features of GPCRs

G protein-coupled receptors (GPCRs) are the largest family of cell membrane embedded proteins. GPCRs recognize and bind endogenous ions, small molecules such as neurotransmitters, lipids, carbohydrates, nucleotides, amino acids, and taste and odorant molecules as well as larger molecules such as peptide hormones and proteins. Upon binding these molecules from the extracellular space, a signal is relayed to the intracellular space by activating various signaling pathways [1, 2]. Because of their abundance, specific expression patterns, biological function, and druggability at least one-third of all marketed drugs act on GPCRs [3–6], and these receptors are implicated in many medical conditions including immunological, cardiovascular, and neurological disorders, cancer and obesity. There are 825 human GPCRs including

Alexander Heifetz (ed.), *Computational Methods for GPCR Drug Discovery*, Methods in Molecular Biology, vol. 1705, https://doi.org/10.1007/978-1-4939-7465-8_4, © Springer Science+Business Media LLC 2018

olfactory receptors [7], however, surprisingly only 108 of them are already exploited targets by approved drugs and 66 additional targets are being evaluated in clinical trials. Furthermore, 100–160 GPCRs (depending on the definition) are currently undercharacterized (so-called orphan receptors) with no or only a few known ligands and unknown endogenous ligands, which are potentially exploitable drug targets with novel mechanisms of action [8–10]. GPCRs are classified into the classes A-F for all human and nonhuman GPCRs [11, 12] (A: Rhodopsin-like receptors divided into 19 subfamilies, B: Secretin receptor family divided into B1: Secretin receptors, B2: Adhesion receptors, B3: Insect Methuselah-like proteins, C: Metabotropic glutamate/pheromone receptors divided into five subfamilies, D: Fungal mating pheromone receptors, E: Slime mold cAMP receptors, F: Frizzled/Smoothened receptors, and Taste2, which are now considered as their own class), or according to the alternative GRAFS (Glutamate, Rhodopsin, Adhesion, Frizzled, Secretin and additionally Taste2) system [13, 14] for the human receptors.

GPCRs feature a conserved seven transmembrane spanning helical fold connected by intra- and extracellular loops. In the past two decades, advances in membrane protein engineering and structure elucidation have facilitated an outset followed by an exponential increase in the number of solved GPCR crystal structures. Altogether over 213 crystal structures are now available (25 September 2017) for 43 unique receptors (47 including orthologs) belonging to all of the human classes (A-C, F) [15–20] and families, with the exception of the Taste type 2 and Adhesion families (Fig. 1). These crystal structures comprise GPCRs in complex with small molecule, peptide, and protein ligands, activity modulating (sodium) ions, and stabilizing membrane lipids (cholesterol), in active and inactive states, and also in complex with signaling proteins or surrogates, as shown in Fig. 1. The structures have greatly enhanced our understanding of GPCR protein structure, ligand binding, and signaling [38, 39]. Moreover, they have revealed shared as well as receptor-specific information on ligand binding sites [40–42] and conformational changes during receptor activation and signaling protein interactions [43]. The transmembrane domain is responsible for endogenous ligand binding and signal transduction in class A receptors as well as comprising part of the binding site of peptide ligands in class B1 receptors. Class B2, C, and F receptors furthermore feature large extracellular domains involved in protein-protein interactions or ligand binding, whereas the transmembrane domain may be targeted by allosteric modulators in these receptors [44, 45].

Given the large body of comparative research carried out on GPCRs, the need arose for a framework for referring to specific structural elements over all GPCRs, GPCR classes, subfamilies, or unique receptors. Structural alignment is feasible for the

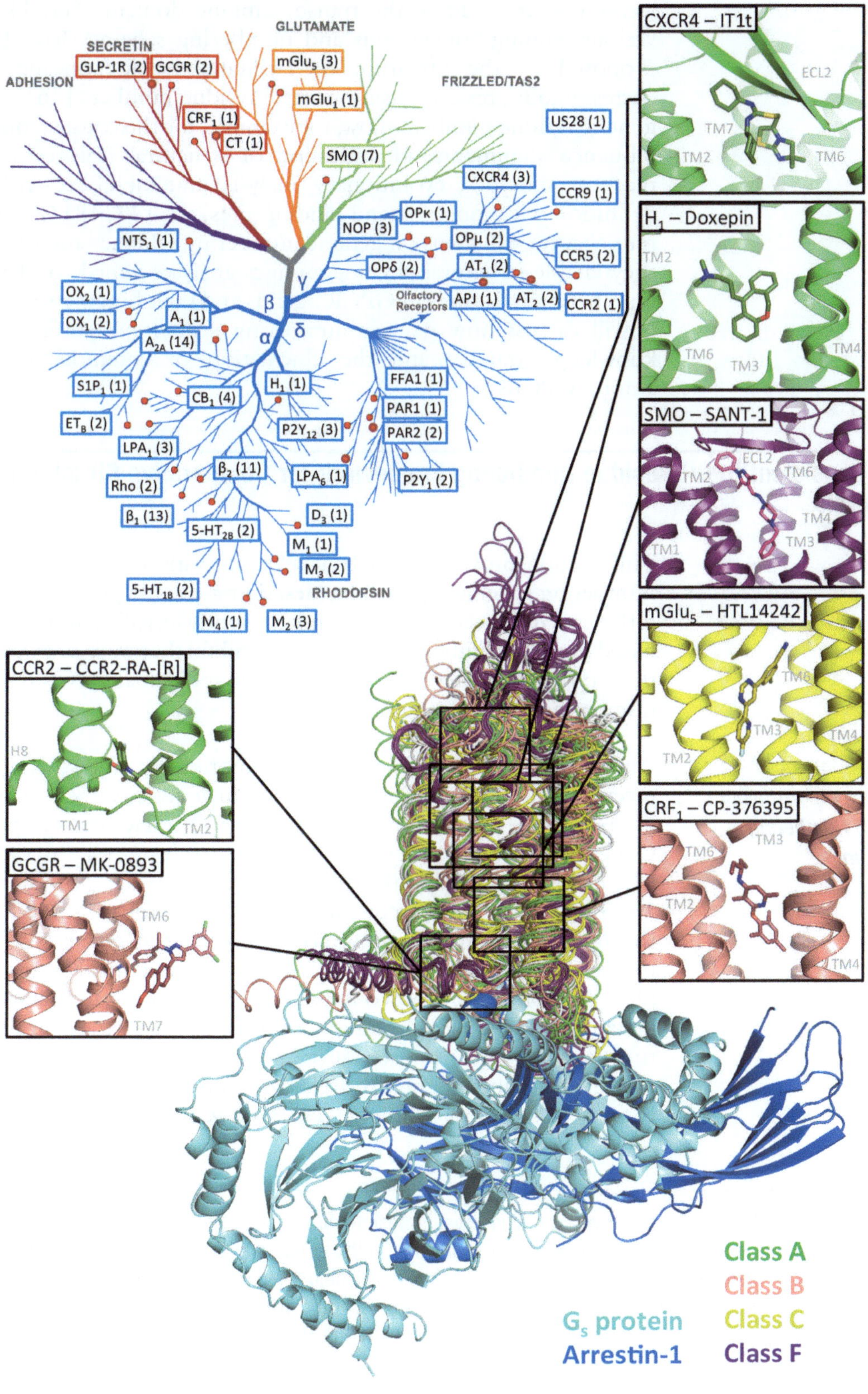

Fig. 1 GPCR phylogenetic tree with crystallized receptors and number of publicly available crystal structures with unique ligands indicated. The alignment of representative class A and all class B, C and F GPCR X-ray

well-conserved fold of the transmembrane domain (Fig. 1) and various naming conventions and numbering schemes have been proposed in the literature to facilitate the identification of corresponding residues using generic residue numbers [46]. Such generic residue numbers allow, for example, for the construction of sequence alignments, identification of structural features across receptor subtypes, evolutionary analysis, mutant effect analysis, definition of canonical ligand binding sites, etc. [47, 48]. The current chapter will focus on generic numbering schemes and the way they enable structural chemogenomics analyses aimed to identify links between features of GPCR ligands and their binding sites, and it will discuss how generic numbering schemes have facilitated knowledge transfer and the elucidation of common features among GPCRs.

2 Consistent GPCR Residue Numbering Framework for Comparative Structural Analysis

This section provides an overview of the various sequence-based numbering schemes applied for describing residue positions in the different GPCR classes and the more recently introduced structure-based generic numbering schemes. It will furthermore draw attention to the advantages and caveats of the individual GPCR residue numbering schemes.

2.1 Sequence Conservation-Based GPCR Numbering Schemes

Class A or Rhodopsin-like GPCRs comprise the largest family of receptors including opsins, aminergic, nucleotide, carboxylic acid, lipid, opioid, chemokine, hormone, and olfactory receptors. Generic numbering schemes were first introduced for this family. The Ballesteros-Weinstein (BW) numbering [49] was introduced in 1995 and is the most frequently used scheme for class A GPCRs throughout the literature. The BW numbering scheme is based on the sequence alignment of all class A GPCR sequences and the identification of the most conserved residue in each transmembrane helix (TMs). A generic residue number consists of the number of the helix (1–7), a dot separator, and a two-digit number. The two-digit number denotes the relative position (upstream or downstream) to the most conserved TM residue, which is denoted with

Fig. 1 (continued) structures reveals the conserved transmembrane heptahelical fold as well as large differences in the extra- and intracellular loop regions and in the structures co-crystallized in complex with G_s-protein or Arrestin-1. Moreover, the diversity of ligand binding sites is also highlighted. The aligned structures are Rhodopsin (PDB: 1F88 [21], 5DGY [22] co-crystallized with Arrestin-1), β_1 (PDB: 2YOO [23]), β_2 (PDB: 3SN6 [24] co-crystallized with trimeric G_s-protein), H_1 (PDB: 3RZE [25]), CXCR4 (PDB: 3ODU [26]), CCR2 (PDB: 5T1A [27]), GCGR (PDB: 4L6R [28], 5EE7 [29]), CRF_1 (PDB: 4K5Y [30], 4Z9G [31]), $mGlu_1$ (PDB: 4OR2 [32]), $mGlu_5$ (PDB: 4OO9 [33], 5CGC [34], 5CGD [34]), SMO (PDB: 4JKV [35], 4N4W [19], 4O9R [36], 4QIM [19], 4QIN [19], 5L7D [37] truncated before S190, 5L7I [37])

50. $R^{3.50}$, for example, is the most-conserved residue located in TM3 (it is conserved in 94% of the class A GPCRs) and is part of the so-called D[E]RY motif involved in an ionic lock stabilizing the inactive receptor state. $C^{3.25}$ is also located in TM3, but 25 positions before $R^{3.50}$ and is also highly conserved (in 87% of class A GPCRs and 83% in all GPCRs), and forms a characteristic disulfide bridge with a cysteine located in extracellular loop 2. For a complete overview of conserved positions in the different GPCR classes, *see* Table 1.

Alternative sequence-based numbering schemes have been proposed by Oliveira [50], and Baldwin [51, 52] and Schwartz [53, 54] (BS) a few years earlier but they did not gain such widespread usage in the literature. These numbering schemes attempted to assign similar numbers to the amino acids located at the same height of the membrane based on the structure of bacteriorhodopsin (PDB: 1BRD [55]), bovine [56], and frog rhodopsin [57] determined by electron cryo-microscopy. Oliveira et al. used an iterative profile alignment method that was based on knowledge about the function of key residues to align 225 rhodopsin-like GPCRs and in each helix the number of the most conserved residue is a multiple of 10 while the start of the helix is at the same time as close as possible to 100 times the number of the helix. Baldwin and Schwartz used an alignment of 105 sequences and defined each helix to include 26 amino acids based on hydrophobicity profile analysis. The central position numbers (13 and 14) of each helix were assigned to the middle of the region that had the most sites that could be in contact with lipid. In these numbering schemes the orders of TM2, TM4, and TM6 sequences are reversed and furthermore the variations in length and inclination of the helices, which were uncovered later by the novel crystal structures, render them inefficient to use. The original format of these schemes also differs, the Oliveira numbering does not use a separator, while the BS numbering uses roman numerals for the TM helix and a colon as a separator, for example, $R^{3.50}$ according to the BW scheme is denoted as 340 according to Oliveira and III:26 according to the BS scheme.

2.2 Comparison of Conserved Residues and Motifs within GPCR Classes

Numbering schemes using a similar logic to the BW numbering have later been proposed by Wootten for class B [58], Pin for class C [59], and Wang for class F GPCRs [19]. However, the reference positions for these numbering schemes are chosen to be the most conserved residues within each of these classes and therefore differ from the ones used by the class A-specific BW scheme. Although the reference residues in the Wootten numbering were derived from the sequence alignment of the B1/secretin subclass, they are also the most conserved residues for five TMs of the B2/adhesion receptor TM helices and highly conserved in the remaining two TMs [60]. Class F is a small class of receptors and in cases where a

Table 1

Comparison of Ballesteros-Weinstein reference positions (X.50) for representative class A/Rhodopsin-like (bovine rhodopsin, bRho), class B1/Secretin-like (glucagon receptor, GCGR), class B2/adhesion (GPR56 or ADGRG1), class C/Glutamate (metabotropic glutamate receptor 1, $mGlu_1$), and class F/Frizzled (smoothened, SMO) receptors. Receptor-specific UniProt numbers are colored gray, and their Ballesteros-Weinstein residue number appended by the corresponding GPCR class letter are shown in superscript. In addition to Ballesteros-Weinstein reference positions, cysteines in the conserved disulfide bridge ($C^{3.25a/f}/C^{3.29b/c}$ and $C^{45.50}$) are indicated. Most conserved residues (at the X.50 position) are marked cyan, whereas conserved residues (>45%) between GPCR families that are not considered conserved reference residues for a specific family are marked green. Receptors where the most conserved residues (at the X.50 position) are not conserved are listed in the last column

Segment	Class A Rhodopsin (bRho)	Class B1 Secretin (GCGR)	Class B2 Adhesion (GPR56)	Class C Glutamate ($mGlu_1$)	Class F Frizzled (SMO)	Not conserved in		
TM1	$G51^{1.46a}$	$S152^{1.50b}$	$S414^{1.50b}$	$G603^{1.50c}$	$T241^{1.43f}$	ADGRA1 ADGRA2 ADGRA3 CELSR2	TAS1R1 TAS1R3 GPR158 GPR179	
	$N55^{1.50a}$	$L156^{1.54b}$	$C418^{1.54b}$	$T607^{1.54c}$	$T245^{1.47f}$	ACKR1 GPR141 GPR142 GPR148 GPR171		
	$T58^{1.53a}$	$A159^{1.57b}$	$T421^{1.57b}$	$V610^{1.57c}$	$T248^{1.50f}$	-		
TM2	$L76^{2.43a}$	$H177^{2.50b}$	$H446^{2.50b}$	$C631^{2.39c}$	$L267^{2.44f}$	CELSR2 ADGRB1 ADGRB2 ADGRB3 ADGRF1 ADGRG2 ADGRG4 ADGRG6 ADGRG7 ADGRV1		
	$A82^{2.49a}$	$S183^{2.56b}$	$A452^{2.56b}$	$G637^{2.45c}$	$C273^{2.50f}$	-		
	$D83^{2.50a}$	$F184^{2.57b}$	$V453^{2.57b}$	$I638^{2.46c}$	$F274^{2.51f}$	$GnRH_1$ NTS_2 NK_1 OPN1SW ACKR1 ACKR2 CCRL2 GPBAR GPR26 GPR33 GPR78 GPR82 GPR141 GPR146 GPR149 GPR153 GPR161 GPR162 GPR176 LGR4 LGR5 LGR6		
	$V87^{2.54a}$	$A188^{2.61b}$	$D457^{2.61b}$	$Y642^{2.50c}$	$S278^{2.55f}$	CaSR $GPRC_6$ $GABA_{B1}$ TAS1R1 TAS1R3 GPRC5A GPRC5B GPRC5C GPRC5D		
TM3	$C110^{3.25a}$	$C224^{3.29b}$	$C475^{3.29b}$	$C657^{3.29c}$	$C314^{3.25f}$	$MC_{1\text{-}5}$ $LPA_{1\text{-}3}$ $S1P_{1\text{-}5}$ $CB_{1,2}$ GPR3 GPR6 GPR12 GPR20 GPR82 GPR88 GPR139 GPR142 GPR146 GPR149 GPR156 MAS1 MAS1L MRGPRD MRGPRE MRGPRF MRGPRG MRGPRX1 MRGPRX2 MRGPRX3 MRGPRX4 GPRC5A GPRC5D		
	$L131^{3.46a}$	$E245^{3.50b}$	$E496^{3.50b}$	$K678^{3.50c}$	$L335^{3.46f}$	CELSR3 ADGRA1 ADGRA2 ADGRA3 ADGRF1 ADGRF2 ADGRF3 ADGRF4 ADGRF5 ADGRG7 ADGRV1	TAS1R1 TAS1R2 TAS1R3 GPRC5A GPRC5B GPRC5C GPRC5D	
	$R135^{3.50a}$	$L249^{3.54b}$	$L500^{3.54b}$	$I682^{3.54c}$	$W339^{3.50f}$	$C5a_2$ RXFP1 RXFP2 ACKR1 ACKR2 DP_1 GPR1 GPR78 GPR146 GPR148 GPR149 GPR151 GPR160 LGR6 MRGPRE	-	
TM4	$W161^{4.50a}$	$W272^{4.50b}$	$W524^{4.50b}$	$I714^{4.40c}$	$W365^{4.50f}$	DP_1 EP_1 EP_2 EP_4 FP IP GPR132 GPR139 GPR142 LGR4 LGR5 LGR6	ADGRA1-3 ADGRE1-5 ADGRF1-5 ADGRL1-4	-
	$P171^{4.60a}$	$W282^{4.60b}$	$V534^{4.60b}$	$L724^{4.50c}$ (64% W)	$I375^{4.60f}$	$mGlu_1$ $mGlu_5$ TAS1R2 TAS1R3 GPR158 GPR179 GPRC5A GPRC5D		
ECL2	$C187^{45.50}$	$C294^{45.50}$	$C562^{45.50}$	$C746^{45.50}$	$C390^{45.50}$	$MC_{1\text{-}5}$ GPR20 GPR139 GPR142 GPR148 MAS1 MAS1L MRGPRD MRGPRE MRGPRF MRGPRG MRGPRX1 MRGPRX2 MRGPRX3 MRGPRX4 GPRC5A GPRC5D (possibly others)		
TM5	$V210^{5.45a}$	$P310^{5.42b}$	$L577^{5.42b}$	$Y759^{5.46c}$	$P407^{5.50f}$	-		
	$P215^{5.50a}$	$A314^{5.46b}$	$V581^{5.46b}$	$L763^{5.50c}$	$V411^{5.54f}$	$MC_{1\text{-}5}$ RXFP1 RXFP2 XCR1 ACKR1 FSH LH TSH $LPA_{1\text{-}3}$ $S1P_{1\text{-}5}$ $CB_{1,2}$ PAF DP_1 $EP_{1\text{-}4}$ FP IP TP $P2Y_{12\text{-}14}$ GPR3 GPR6 GPR12 GPR22 GPR26 GPR27 GPR34 GPR39 GPR75 GPR82 GPR87 GPR141 GPR153 GPR160 GPR162 GPR171 LGR4 LGR5 LGR6 MAS1L MRGPRE MRGPRG MRGPRX1 MRGPRX2 MRGPRX3 MRGPRX4	GPR158	
	$I219^{5.54a}$	$N318^{5.50b}$	$N585^{5.50b}$	$C767^{5.54c}$	$G415^{5.58f}$	ADGRA1 ADGRA2 ADGRG3 ADGRV1 CELSR2		
TM6	$L262^{6.45a}$	$G359^{6.50b}$	$G620^{6.50b}$	$C795^{6.47c}$	$V463^{6.44f}$	ADGRA1 ADGRA2 ADGRA3 ADGRB1 ADGRB2 ADGRB3 ADGRV1 CELSR1 CELSR2 CELSR3		
	$W265^{6.48a}$	$E362^{6.53b}$	$W623^{6.53b}$	$W798^{6.50c}$	$T466^{6.47f}$	$GABA_{B1}$ $GABA_{B2}$ TAS1R2 GPR156 GPR158 GPR179		
	$P267^{6.50a}$	$V364^{6.55b}$	$L625^{6.55b}$	$A800^{6.52c}$	$S468^{6.49f}$	GPR148 MRGPRF		
	$Y268^{6.51a}$	$F365^{6.56b}$	$I626^{6.56b}$	$F801^{6.53c}$	$C469^{6.50f}$	-		
TM7	$A299^{7.46a}$	$G393^{7.50b}$	$G651^{7.50b}$	$V823^{7.40c}$	$F526^{7.46f}$	ADGRA3 ADGRF2		
	$P303^{7.50a}$	$A397^{7.54b}$	$F655^{7.54b}$	$L827^{7.44c}$	$I530^{7.50f}$	$PKR_{1,2}$ LPA_2 $MT_{1,2}$ OXGR GPR18 GPR21 GPR35 GPR50 GPR52 GPR55 GPR139 GPR141 GPR142 GPR148 GPR160	-	
	$M308^{7.55a}$	$L403^{7.60b}$	$M661^{7.60b}$	$P833^{7.50c}$	$W535^{7.55f}$	CaSR		

helix had more than one fully conserved position, the one closest to the class A BW numbering was used by Wang et al. as the reference position. Since these numbering schemes share the same formatting, it was proposed to use the class designation as a lower case letter after the residue number if there is possible ambiguity [16].

Different GPCR classes share reference positions in TM1 ($S^{1.50b}/G^{1.50c}$), TM3 ($R^{3.50a}/W^{3.50f}$, $E^{3.50b}/K^{3.50c}$), TM4 ($W^{4.50a50b50f}$), TM5 ($P^{5.50a}/L^{5.50c}$), and TM7 ($P^{7.50a}/I^{7.50f}$), as shown in Table 1. The most conserved residue in TM3 in class B GPCRs, for example, is $E^{3.50b}$ that structurally aligns with the conserved $K^{3.50c}$ in class C GPCRs and corresponds to positions 3.46a and 3.46f in class A and class F GPCRs, respectively. For class F GPCRs the most conserved residue in TM3 is $W^{3.50f}$, which structurally overlaps with the class A $R^{3.50a}$. Consequently, the conserved cysteine in the TM3-ECL2 cysteine bridge gets the number $C^{3.25a}/C^{3.25f}$ in class A and F GPCRs, and $C^{3.29b}/C^{3.29c}$ in class B and C GPCRs. The most conserved residues in the seven transmembrane helices are also often part of larger conserved motifs in the different receptor classes and are implicated in signal transduction. Classes A ($CW^{6.48a}xP^{6.50a}$), B1 ($W^{6.53b}$), and C ($W^{6.50c}$) for example share a conserved tryptophan residue that is part of the so-called toggle switch that is implicated in the activation of class A GPCRs [61]. The rearrangement of conserved hydrophobic contacts between hydrophobic residues in TM3 (including $I/L/M^{3.46a}$) and TM6 (including $L/V/I^{6.37a}$) connects the toggle switch with the ionic lock between TM3 ($D[E]R^{3.50a}Y$) and TM6 ($D/E^{6.30a}$) and the intracellular end of TM7 ($NP^{7.50a}xxY^{7.53a}$) along the activation pathways of class A GPCRs [62]. Motifs with a different composition but possibly a similar function have been identified in class B, C, and F receptors as well and are shown in Fig. 2. The ionic locks in Class B ($H^{2.50b}$ and $E^{3.50b}$) and class C ($K^{3.50c}$ and $E^{6.35c}$) GPCRs for example form H-bond networks with conserved residues in TM7 ($FQG^{7.50b}xxVxxxY^{7.75b}$, $FxP^{7.50c}KxY^{7.53c}$) that have been proposed to facilitate conformational changes of TM7 associated with receptor activation in these GPCR classes [16].

2.3 From Gapless to Structure-Based GPCR Sequence Numbering

Although the sequence conservation between the different GPCR classes is low, the structural fold is well conserved and the elucidation of multiple crystal structures from all the major classes has made it possible to construct structure-based cross-class sequence alignments. Such alignments are currently easily available in an interactive manner in GPCRdb [66, 67]. Although TM helices are known to deviate laterally, and by their tilt angles and kinks in the helices, local helix alignments are still possible and do not complicate generic numbering attempts. However, the former assumption that sequences should always be aligned without gaps within the TM helices is no longer valid. Bulges and constrictions have been identified that are localized to one π-helical (additional

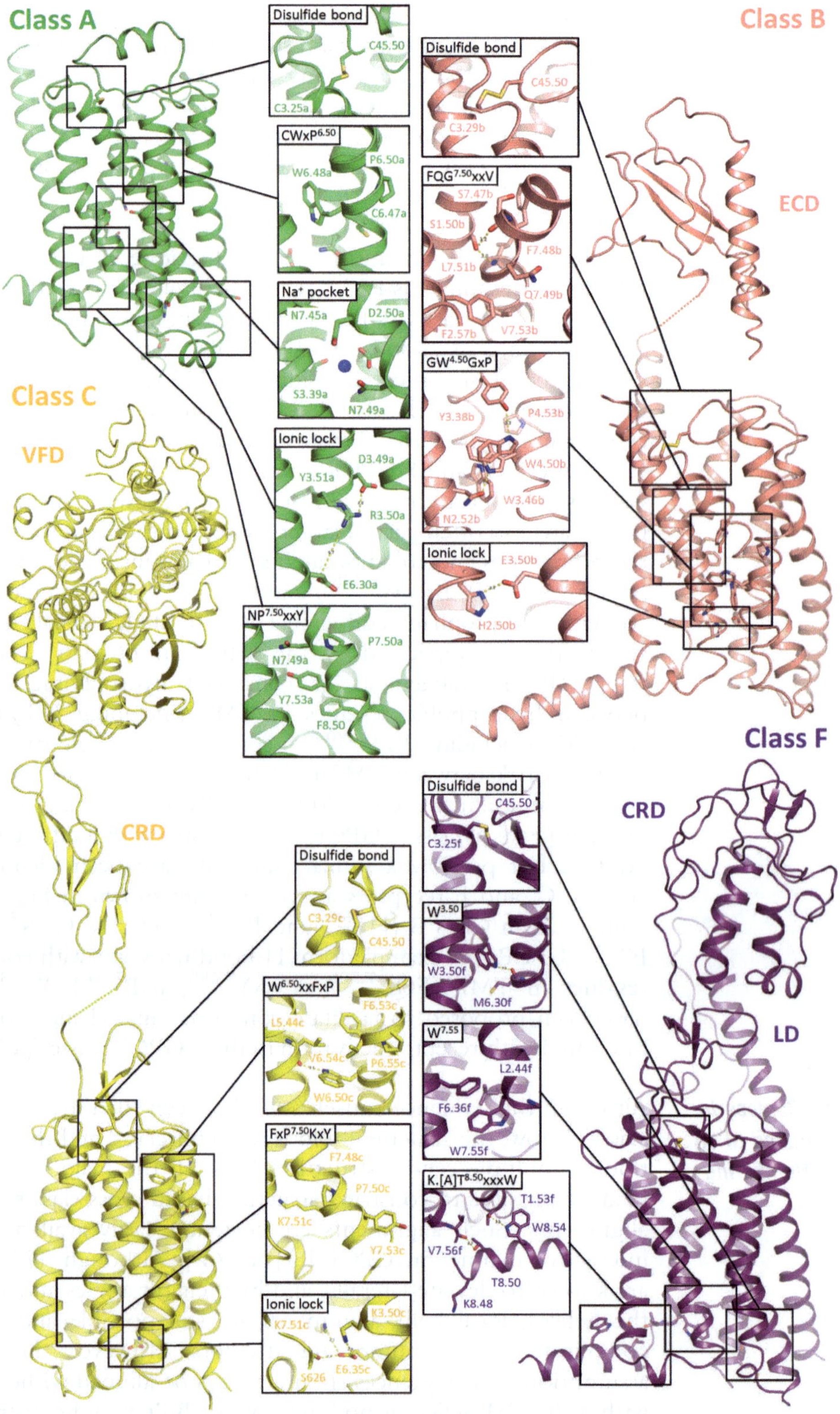

Fig. 2 Conserved structural motifs in class A, B, C, and F GPCRs: β_1 (PDB: 2Y00 [23]), GCGR (PDB: 4L6R [28] for TMD, 4ERS [63] for ECD, 5EE7 [29] for alternative $W^{4.50b}$ rotamer), mGlu$_1$ (PDB: 4OR2 [32] for TMD, 1EWK

residue) or 3_{10}-helical turn (absent residue), respectively. This results in shifted generic numbering for receptors containing the bulges and constrictions relative to the receptors that have standard α-helices, thereby offsetting the comparison of residue positions for such receptors. Isberg et al. identified nine bulges and six constrictions in all TMs except in TM3 [46]. These were found across all classes A, B, C, and F; and two constrictions in TM4 and TM7 were even found in both classes A and B. Some helical distortions affect only a specific receptor subtype, whereas others are shared by the majority of receptors in a GPCR class. In the same paper, a new structure-based generic numbering scheme was proposed that corrects for the bulges and constrictions by comparing the affected structures to standard α-helices, and assigns the same number as the preceding residue followed by the digit 1 for the most protruding residue in the bulge and skips a number for the constrictions. This method also future-proofs the residue numbers in case more distortions are discovered later. To distinguish it from the former numbering schemes the character "x" is used as a separator and may be used together with the sequence-based numbering. The usage of both is recommended as the structure-based numbering can only be transferred to non-crystallized receptors through sequence motif matching. In the current overview, we have assigned the structure-based number in case these are different from the BW number. The examples of the use of this structure-based GPCR residue numbering scheme are described below and presented in Fig. 3.

Bulges and constrictions also affect which amino acids are projected toward the core of the TM domain and toward the membrane. In some cases, this also affects the composition of the binding pocket and it is very important to take this fact into consideration for homology modeling or chemogenomic studies on GPCRs. Figure 3 shows several examples of distortions around experimentally determined binding modes of GPCR ligands. For example, a bulge in TM2 denoted by the structure-based generic number 2x551 is present in all the aminergic GPCR crystal structures and in a few peptide receptors as well, but it is not present in, for example, chemokine receptors and thus orienting $D^{2.63x63a}$ to the minor binding pocket of CXCR4. TM4 shows constrictions in several class A receptors and, for example, in the histamine H_1 receptor (H_1R) this directly affects the orientation of $W^{4.56x57a}$ into the major aminergic binding pocket forming aromatic stacking interaction with the butterfly-shaped tricyclic system of doxepin and other similar H_1R ligands. A bulge located in TM5 in the

Fig. 2 (continued) [64] for VFD, 2E4U [65] for CRD from $mGlu_3$), SMO (PDB: 5L7D [37]). *TMD* transmembrane domain, *ECD* extracellular domain, *CRD* cysteine-rich domain, *VFD* Venus flytrap domain, *LD* linker domain

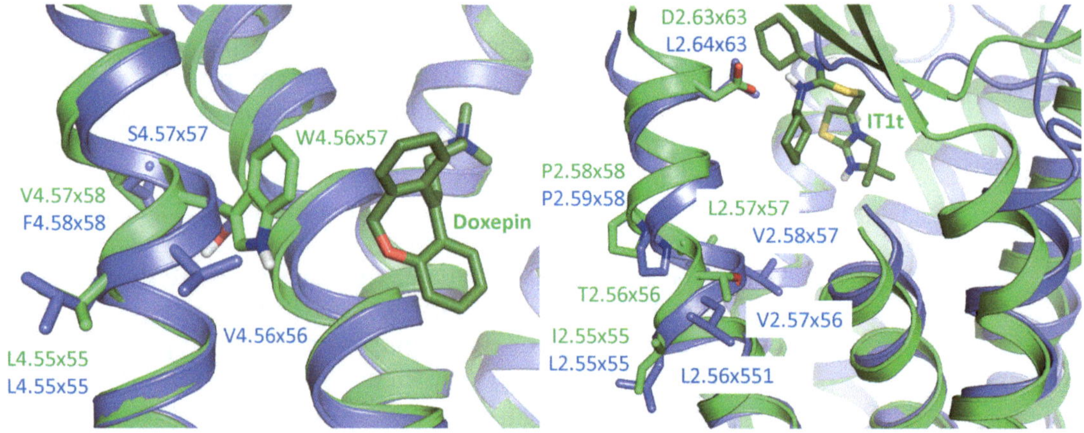

BW (A)	52	53	54	55	-	56	57	58	59
	52	53	54	55	56	57	58	59	60
GPCRDB (A)	52	53	54	55	56	57	58	59	60
H_1	L	S	F	L	-	W	V	I	P
β_1	I	S	A	L	V	S	F	L	P

BW (A)	52	53	54	55	-	56	57	58	59
	52	53	54	55	56	57	58	59	60
GPCRDB (A)	52	53	54	55	551	56	57	58	59
CXCR4	L	F	V	I	-	T	L	P	F
β_1	V	M	G	L	L	V	V	P	F

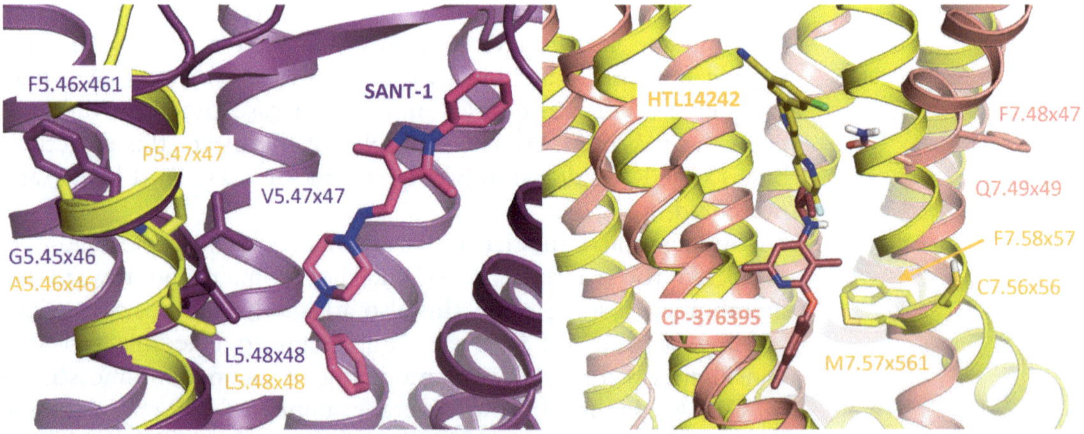

Wang (F)	43	44	45	46	47	48	49	50	51
	44	45	46	-	47	48	49	50	51
GPCRDB (F)	44	45	46	461	47	48	49	50	51
Pin (C)	39	40	41	42	43	44	45	46	47
	40	41	42	-	43	44	45	46	47
GPCRDB (C)	40	41	42	421	43	44	45	46	47
SMO	R	A	G	F	V	L	A	P	I
$mGlu_1$	V	V	A	-	P	L	G	Y	N

Wootten (B)	47	48	-	49	50	...	55	56	-	57	58
	46	47	48	49	50	...	55	56	57	58	59
GPCRDB (B)	46	47	48	49	50	...	55	56	561	57	58
Pin (C)	37	38	-	39	40	...	45	46	-	47	48
	36	37	38	39	40	...	45	46	47	48	49
GPCRDB (C)	37	38	39	40	41	...	46	47	471	48	49
CRF_1	S	F	-	Q	G	...	V	F	-	Y	C
$mGlu_5$	V	S	L	S	A	...	G	C	M	F	V

Fig. 3 Bulges and constrictions in GPCR crystal structures determining the structure of ligand binding sites: H_1 (PDB: 3RZE [25]), CXCR4 (PDB: 3ODU [26]), β_1 (PDB: 2Y00 [23]), SMO (PDB: 4N4W [19]), $mGlu_1$ (PDB: 4OR2 [32]), CRF_1 (PDB: 4Z9G [31]), $mGlu_5$ (PDB: 5CGD [34]). Class specific continuous and gapped structure-based generic numbers are indicated by colored/*gray* and *black numbers*, respectively

class F smoothened receptor orients $F^{5.46x461f}$ toward the membrane and projects the smaller $V^{5.47x47f}$ toward the binding pocket of one of the cocrystallized ligands, SANT-1. Finally, though the list is longer, both a constriction and a bulge in TM7 cause this helix to adopt very different shapes in class C receptors affecting the binding sites of allosteric modulators located deep in the transmembrane bundle. The constriction between 7x47b and 7x49b is conserved in class A and B receptors but not in class C receptors. Furthermore, the bulge M/L^{7x471c} in mGlu receptors cause the extracellular half of TM7 to be perpendicular to the membrane and the intracellular half tilting outward, while in class A and B receptors it is the opposite. The binding site of the cocrystallized ligands in the CRF_1 structure and the $mGlu_{1/5}$ structures is located next to the kink in TM7 explaining the strict SAR of mGlu allosteric modulators.

2.4 Residue Numbering in Loops and Helix 8

On top of the numbering of the transmembrane helices now helix 8 and ICL1/2 and ECL1/2—which contain conserved secondary structure elements—are also numbered in GPCRdb denoting the most conserved residue with 50 and upstream and downstream residues by continuously decreasing and increasing numbers. For example, the conserved cysteine bridge in ECL2 gets the number 45x50 (45 for the loop between TM4 and TM5) and the well-conserved phenylalanine in H8 gets the number 8x50. ICL1 often features a single-turn helix, ICL2 a double-turn helix, and ECL1 a bend (aromatic-proline-hydrophobic). However, amino acids outside these conserved motifs can still only be referred to by their UniProt sequence numbers. The generic GPCR numbering scheme has also inspired the definition of numbering schemes for other transmembrane protein families [68, 69].

2.5 Assigning and Comparing Different Residue Numbering Schemes

All of the aforementioned numbering schemes are available in GPCRdb [66, 67] through the web interface or the GPCRdb REST API. GPCRdb also provides a service to assign generic numbers to GPCR structures uploaded by the user. Furthermore, the various numbering schemes returned by the GPCRdb REST API are easily retrieved and utilized in chemogenomics workflows in KNIME [70] using the 3D-e-Chem GPCRdb KNIME nodes [71, 72] which are discussed in more detail in Subheading 3.6.

3 Applications of GPCR Residue Numbering to Structural Chemogenomics Studies

3.1 Consensus Binding Site Definitions and Ligand Repurposing

After the elucidation of the first rhodopsin crystal structure in 2000 there was a surge of GPCR chemogenomics studies primarily aimed at comparative modeling and binding site characterization of other GPCRs and defining common motifs or motifs conveying selectivity in the receptors. The identification of the common binding site in different GPCR classes was one of the most common endeavors. For example, Bondensgaard et al. docked previously identified privileged GPCR scaffolds (i.e., molecular substructures that occur frequently in active ligands of different receptor subtypes, Fig. 5) to rhodopsin-based homology models of their respective targets to identify common binding site features of class A receptors [73]. They furthermore extracted conservation entropy information of a sequence alignment of 111 class A GPCRs and identified a highly conserved hydrophobic core in GPCRs. The authors identified a binding pocket of 35 residues using the BW numbering and three conserved aromatic residues ($F^{5.47a}$, $F^{6.44a}$, and $W^{6.48a}$) involved in the binding of the investigated privileged scaffolds. From the new crystal structures, we know that $W^{6.48a}$ indeed contacts most of the ligands bound in the TM binding pocket but the two other residues are generally located below the major binding pocket. Residues $F^{5.47a}$ and $F^{6.44a}$ do not interact with any of the class A GPCR crystal structures that have been reported so far, but have been shown to play an important role in the activation mechanism of class A GPCRs [38, 43, 62].

In another study a consensus binding pocket was identified with 22 amino acids given by their BS numbers and possible ligand interaction features (hydrophobic, aromatic, charged, polar) were encoded in a fixed length fingerprint to facilitate binding site similarity assessment of class A GPCRs by the physicogenetics method [74, 75]. The binding site similarity analysis revealed the $AT_{1/2}$ receptors as evolutionary distinct receptors but possessing a similar binding pocket to the DP_2 receptor (also known as CRTH2), which was the target of investigation. Therefore, a set of angiotensin ligands was screened against DP_2 and also a commercial library was screened against a pharmacophore model derived from the rhodopsin-based DP_2 homology model and $AT_{1/2}$ mutagenesis and SAR data mapped on the homology model. Using the repurposing methodology, two AT_1 ligands were found to possess micromolar activity at DP_2, whereas screening of the commercial library yielded several nanomolar ligands (Fig. 5). The anchor point the authors identified for the negatively charged tetrazole ligands was K V:08 (BS numbering) or $K^{5.42x43a}$ (BW and generic numbering) which is close in the AT_1 crystal structures (Fig. 6, PDB: 4YAY [77] and 4ZUD [76]) to the tetrazole ring but $R^{4.64x65a}$ is the residue directly interacting with it, which is also conserved in the DP_2 receptor. Several amino acids of the binding pocket were correctly predicted by the study ($S^{3.29a}$, $F^{3.32a}$, $F^{3.33a}$, $W^{6.48a}$, $Y^{6.51a}$, $F^{7.43x42a}$ using BW and generic numbers) but some others were located deeper in the TM bundle, which are known to contact ligands in other receptors but not in the AT_1 crystal structures. As Frimurer and Högberg later point out the BW and BS numbering schemes were gaining similar popularity and it was a matter of preference of the authors which scheme was used in publications.

They also gave a numbering conversion table for the 22 amino acids constituting the consensus binding site [75]. Using the same physicogenetics method the MCHR1 receptor was identified to possess similar features in its binding site as the dopamine D_2 and D_3 receptors and therefore dopamine antagonist scaffolds were utilized in identifying novel MCHR1 ligands with $D^{3.32a}$ hypothesized to interact with the basic ligand moieties similarly to aminergic receptors [75, 78]. Finally, CCR2 was found to have similar features in its binding site to serotonin receptors with $E^{7.39x38a}$ anchoring the positively charged ligands, which was verified by the recently published CCR2 crystal structure.

The consensus binding pocket approach was extended to class C receptors in a similar analysis as the previous ones but using a robust alignment algorithm based on functional amino acid conservation indices, able to cope with the low sequence similarities between GPCR classes [79, 80]. Using this method an alignment between bovine rhodopsin and rat $mGlu_1$ and $mGlu_5$ and human CaSR TM helices was possible and a consensus binding pocket of 35 residues given by their class A BW numbers. Converting the residue maps of the consensus binding pocket to pharmacophore models allowed the prediction of ligand binding modes and the identification of key interacting residues for mutational studies. The authors correctly predicted that ligands like EM-TBPC (such as the cocrystallized FITM) interact with $T^{7.32x33c}$ in a more extracellular pocket in $mGlu_1$ than the acetylene ligands MPEP (such as the cocrystallized Mavoglurant, PDB: 4OO9 [33], Fig. 4) deep in the TM bundle in $mGlu_5$ contacted by e.g., $P^{3.40c}$, $Y^{3.44c}$, $L^{5.44c}$, $W^{6.50c}$, and $F^{6.53c}$. Although the ligands that have been later cocrystallized with the human $mGlu_1$ and $mGlu_5$ are not exactly the ones studied there, they are structurally similar and the identified interacting residues are also in contact with the cocrystallized ligands (except for $N^{5.37c}$).

Surgand et al. provided a systematic and detailed overview of the TM sequence alignments of all non-olfactory receptors and compared binding sites based clustering into 22 clusters to the full sequence-based phylogenetic tree for all the GPCR classes [136]. They analyzed the composition of binding sites for all the GPCR clusters and related it to the physico-chemical parameters of their known ligands and receptor mutagenesis studies. Furthermore, one of the first attempts was made to relate orphan GPCRs to similar, well-studied GPCRs and to propose ligand repurposing for orphans (this approach is further discussed in the following section). GPR88, for example, was identified to be close to dopamine D_1 and D_5 receptors when considering the TM domain, but clustered with class C GPCRs based on the 30 residues designated by the authors to form the binding site. The binding site definition was used by the authors to construct a fixed-length protein-ligand fingerprint (PLFP) consisting of 240 bits

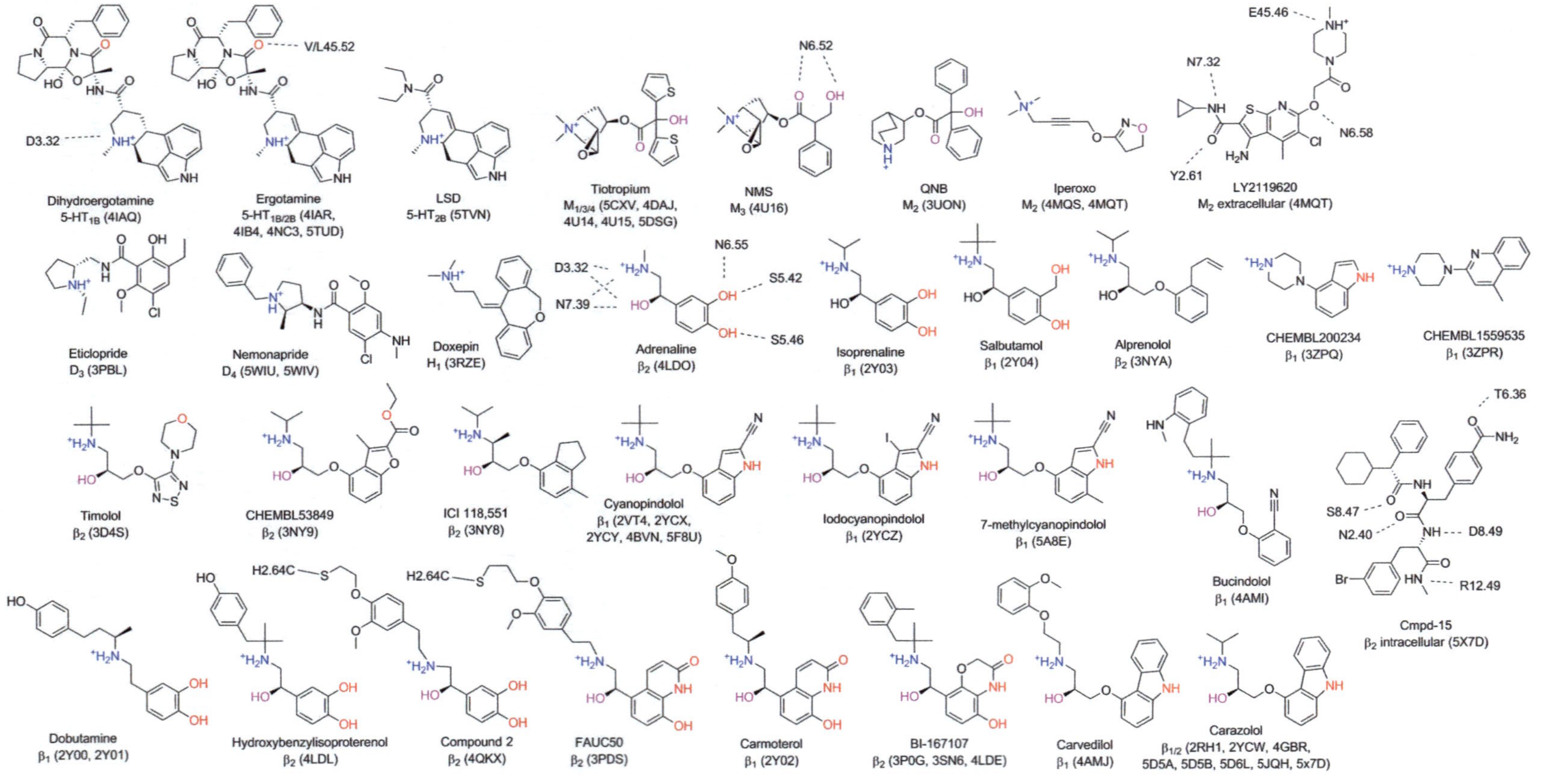

Fig. 4 (**a–d**) Co-crystallized ligands in GPCR X-ray structures [19, 21, 23–27, 29, 30, 32–37, 76, 77, 81–135]. Polar and covalent interactions with binding pocket residues are indicated. H-bonds conserved within receptor subfamilies are indicated with colored atoms (*red, blue, magenta*)

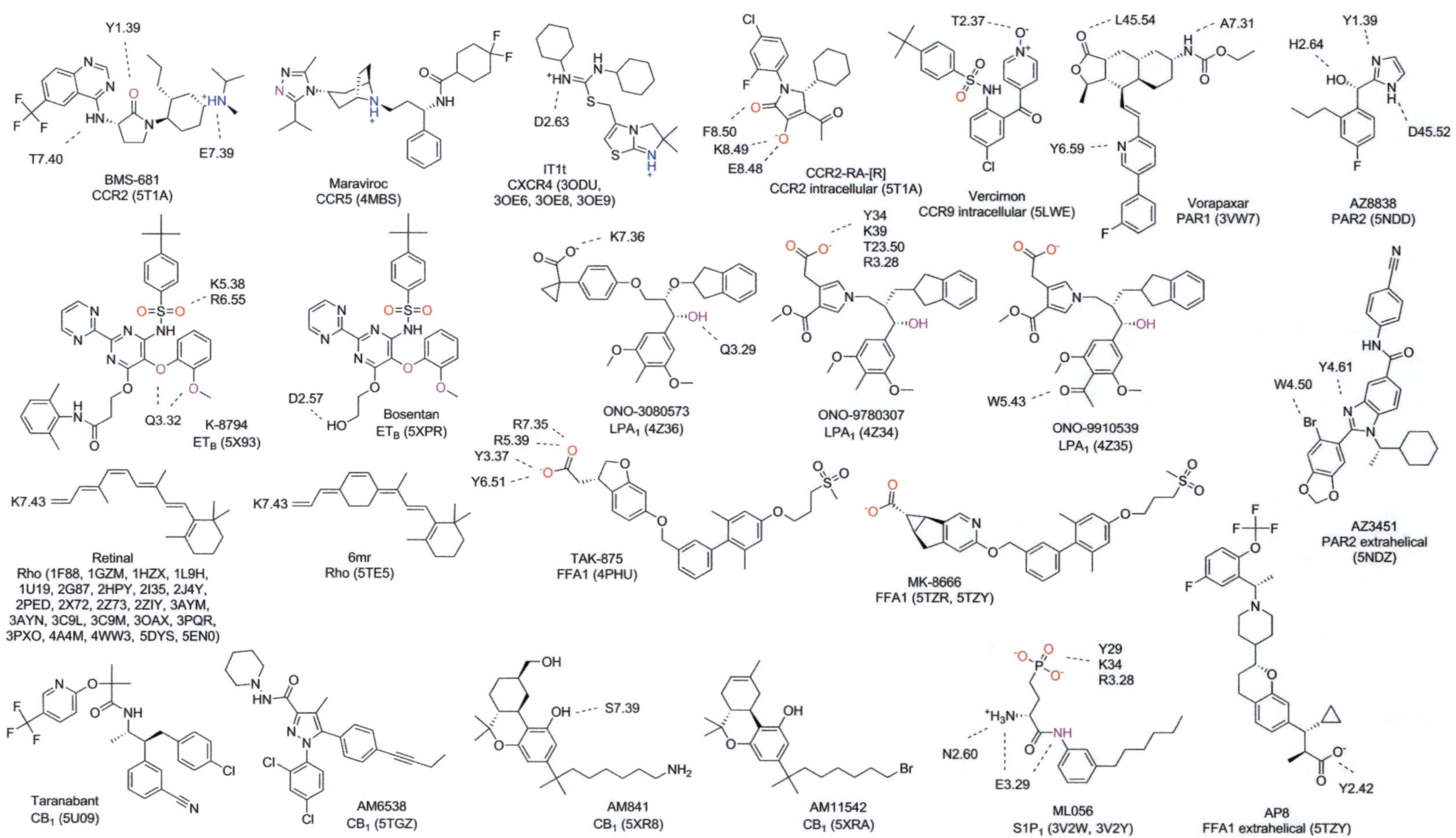

Fig. 4 (continued)

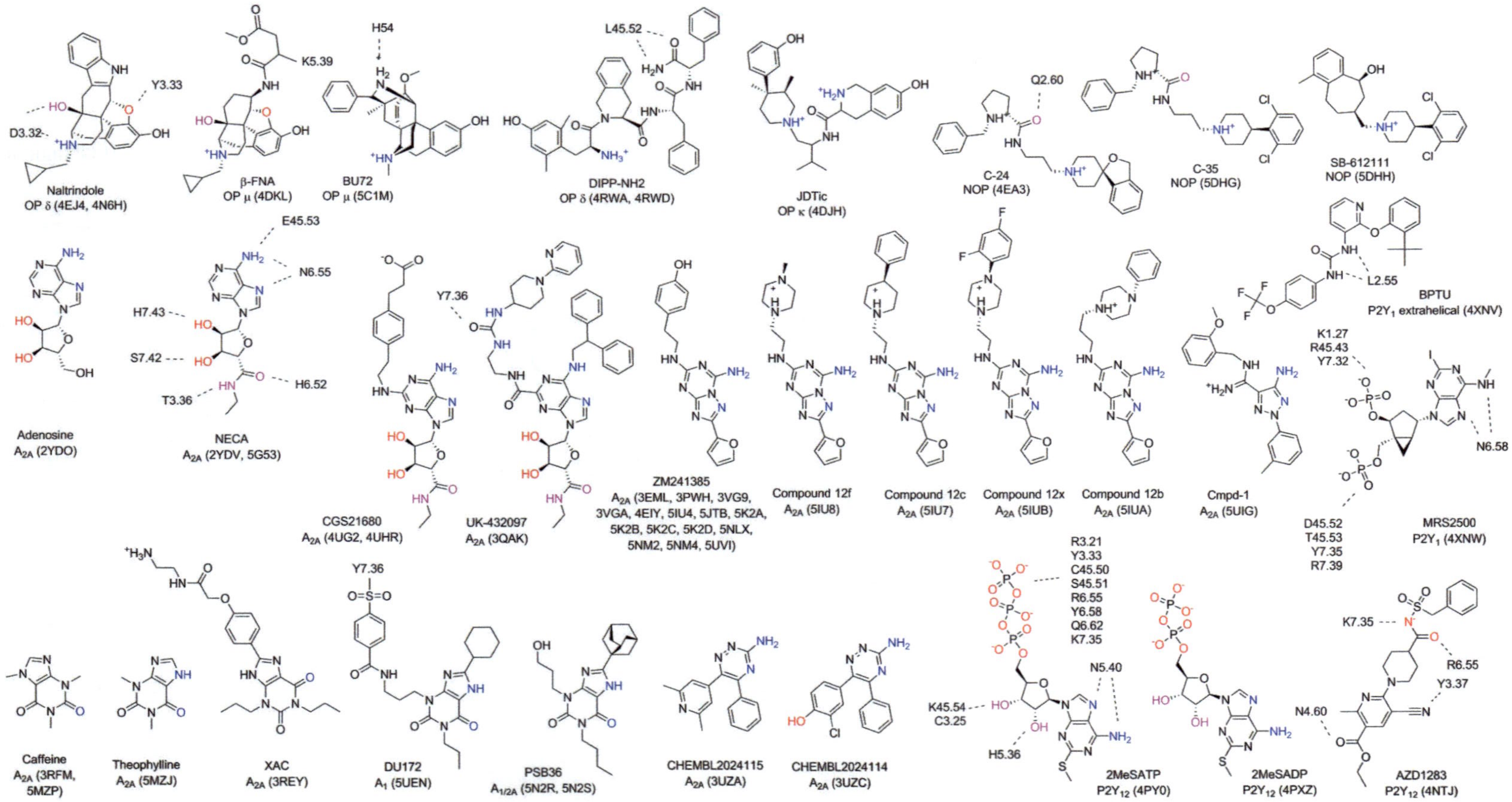

Fig. 4 (continued)

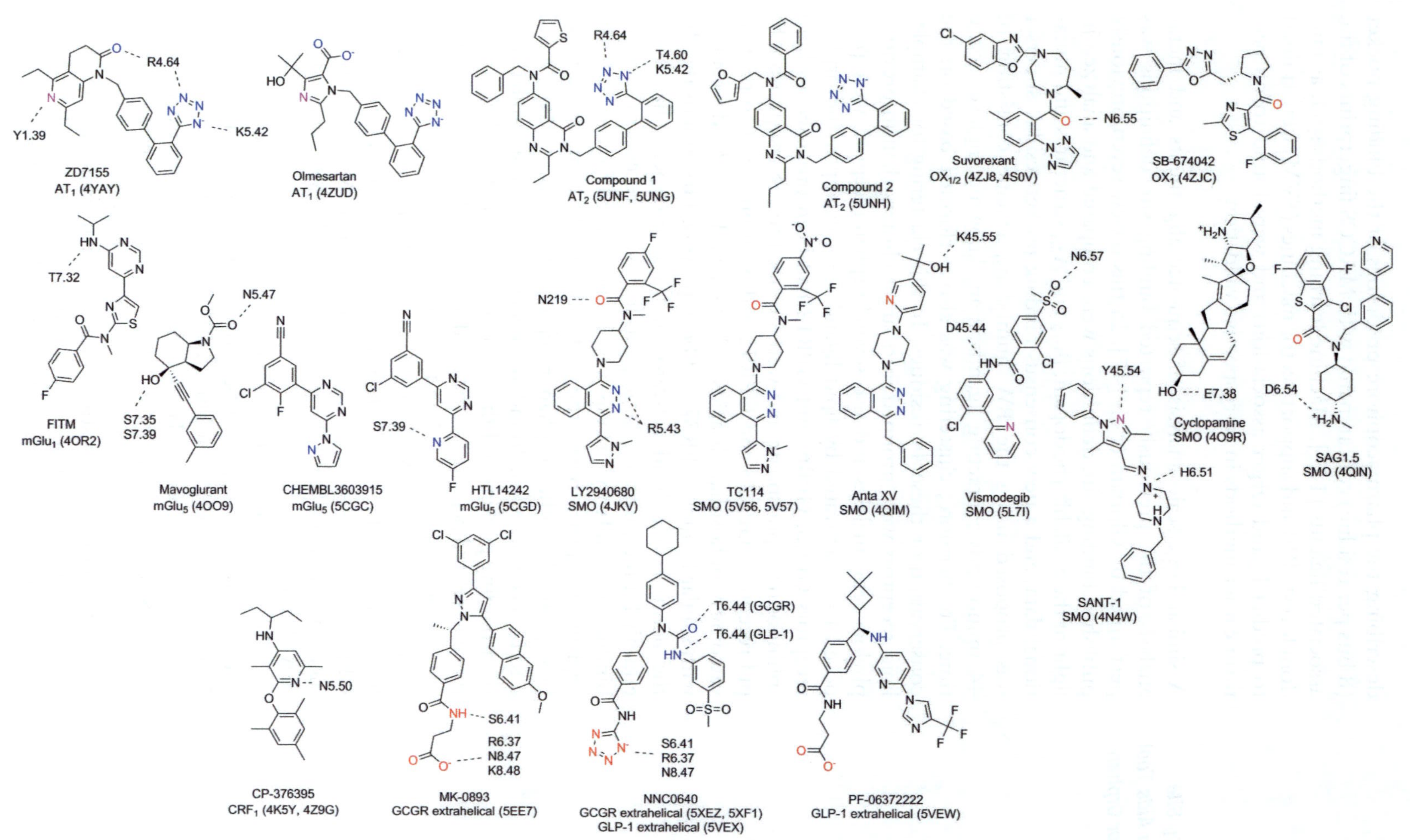

Fig. 4 (continued)

describing the pharmacophore properties of the binding pocket (8 bits per residue) concatenated with MACCS fingerprints of the associated ligands [137]. Machine learning methods such as random forest (RF) and support vector machines (SVM) were trained to predict ligand-target associations and were found to outperform earlier methods in retrospective validation.

3.2 Binding Site Comparison Aids Tool Discovery for Orphan Receptors

A similar large-scale analysis of sequence alignments and meta-analysis of the previously reported binding site definitions was performed by Gloriam et al. [47]. In this review seven previously published binding site definitions were compared and analyzed in light of the available rhodopsin, $\beta_{1/2}$ and A_{2A} structures and mutational data, and a new consensus binding pocket residue superset was proposed using the BW numbering scheme consisting of 44 amino acids contacting ligands in any known complexes at the time. Furthermore, clustering was also performed based on the conservation or these 44 residues. In this clustering for example, lipid receptors were grouped into larger clusters than in previous phylogenetic analyses and also several orphan receptors were proposed to be activated by lipid ligands. For example, the orphan receptors GPR3, GPR6, and GPR12 had been proposed to bind sphingosine 1-phosphate and were grouped to the lysophospholipid receptors. GPR23 and GPR92 had been proposed to be activated by lysophosphatidic acids and grouped close to lipid receptors with similar ligands. GPR37 had been proposed to be activated by the neuropeptide head activator and was grouped adjacent to endothelin receptors in a large peptide binding receptor cluster. A binding site similarity analysis was also performed for the GPR139 orphan receptor over the 44 identified contact residues [138]. Besides other orphan receptors, the human $MC_{2/4}$ and TRH_1 receptors were found to possess the highest similarity in their binding pockets and therefore peptide ligands of these receptors and related peptides were tested against GPR139 (Fig. 5). The peptide hormones β-MSH and ACTH and several truncated products of the latter (α-MSH, α-$MSH_{1–9}$, α-$MSH_{1–10}$, and the core tetrapeptide α-$MSH_{6–9}$ = HFRW) were indeed found to activate GPR139 with 0.3–6 μM EC_{50}. Furthermore, α-$MSH_{1–9}$ was predicted to originate from the pre-pro-protein POMC via a putative new cleavage site.

The class C GPRC6A receptor was recently discovered and deorphanized and found to be activated by basic L-α-amino acids L-Arg, L-Lys, and L-ornithine [139]. Another ligand repurposing study from the closely related CaSR receptor identified Calindol and NPS2143 as nonselective negative allosteric modulators of GPRC6A binding in the TM domain anchored by $E^{7.32x33c}$ [140]. Chemogenomics studies revealed that while the similarity of the GPRC6A receptor to class A receptors based on the full TM region is low, it is substantially similar in its binding site. This

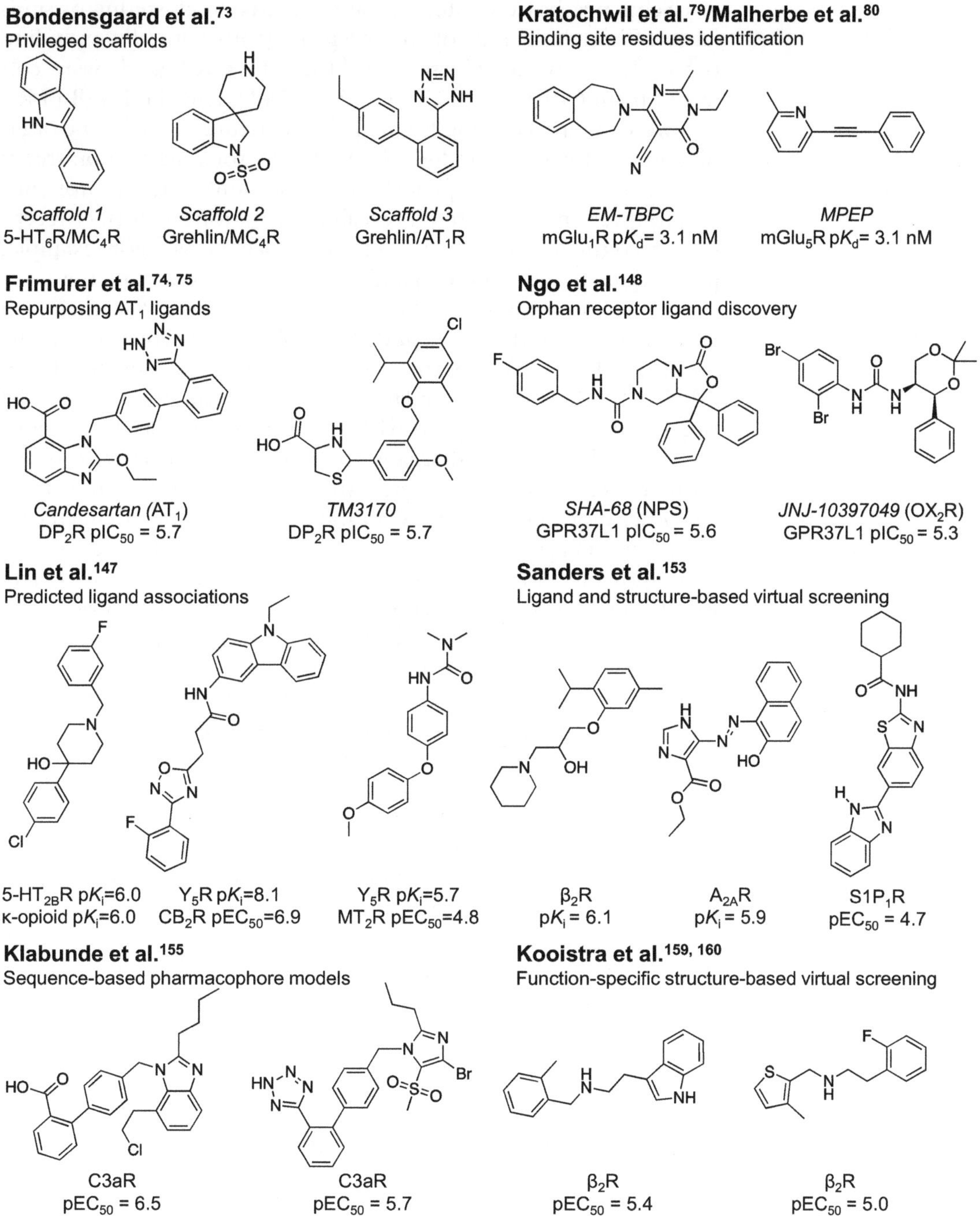

Fig. 5 Selected examples of ligands used in, or identified by structural chemogenomics studies as discussed in Subheading 3. Where available the affinities or potencies of the ligands for the targeted GPCRs are provided

finding allowed repurposing of the privileged 3-substituted 2-phenyl-indole scaffold from class A receptors to the class C GPRC6A receptor and the discovery of micromolar but selective GPRC6A negative allosteric modulators [141].

Sequence-based clustering was compared to clustering by similarity of known ligands of the receptors (thus naturally excluding orphans) by van der Horst et al. [142]. GPCR ligands were collected from ChEMBL [143], GLIDA [144], and the KiDB [145] and were compared using frequent substructure mining. The organization of the resulting ligand-based receptor classification tree is overall similar to the sequence-based one, however, receptor subfamilies were more scattered revealing evolutionary distinct receptors recognizing similar ligands and vice versa. For example, purinergic and adenosine ligands cluster together, while the muscarinic M_2 receptor and histamine receptors cluster together with chemokine, opioid, and peptide receptors. In another study, the Similarity Ensemble Approach (SEA) [146] was used to cluster GPCRs by their ligand similarities to obtain the pharmacological organization of the receptors with a similar outcome as the previous study [147]. The general organization of the map was similar to the sequence-based one but the muscarinic receptors shifted away from the other biogenic amine GPCRs and toward the chemokine receptors, β-adrenergic receptors separated from the α-adrenergic receptors moving closer to the cannabinoid lipid receptors and melatonin receptors, cysteinyl leukotriene, and leukotriene B4 receptors clustered together with lipid GPCRs, and so on. Several new ligand associations from this map were experimentally tested and verified such as between 5-HT_{2B} and the κ-opioid receptor, and between the NPY_5 and the CB_2 and MT_2 receptors (Fig. 5).

With the elucidation of crystal structures from a number of different class A receptors another approach incorporating also a scaling by the interaction strength of each contact residue with cocrystallized ligands was used to update the pharmacological organization map of GPCRs [148]. In this method termed CoIN-Pocket crystal structures of 27 unique class A GPCRs were analyzed and a superset of 61 residue positions given by their GPCRdb numbers were used to define the class A TM binding pocket. It was successful in recapitulating several previously known GPCR associations such as the similarity between, among others, sst_5 and various biogenic amine receptors, the MT_2 and NPY_5 receptors, the muscarinic receptors and MCH_1, and between CCR5 and the δ opioid receptors. The method was used to identify pharmacological neighbors of the orphan GPR37L1 receptor (also called Endothelin B receptor-like protein 2) and it was found that endothelin receptors were in fact not good candidates for ligand repurposing studies. Important ligand contacting residues were predicted to be $E^{3.32a}$, $V^{3.33a}$, $E^{6.51a}$, $N^{6.55a}$, $G^{7.35x34a}$, $Q^{7.39x38a}$, and $F^{7.43x42a}$ for GPR37L1, thus comprising a more acidic binding site instead of the more basic one found in closely related receptors. Instead, ligands of the pharmacologically related $OX_{1/2}$, $BB_{1\text{–}3}$ and the NPS receptors were purchased and tested against GPR37L1 with three of the ten ligands displaying inverse agonistic effects in a concentration-dependent manner (Fig. 5). Thus, it can be seen that chemogenomics methods can aid pharmacological tool discovery for studying orphan GPCRs.

The currently available 213 GPCR crystal structures reveal various binding pockets for small-molecule ligands (*see* Fig. 1). Most of them are located between the transmembrane helices, which allows for the definition of a consensus transmembrane binding pocket based on experimentally observed contacts between residues and small-molecule ligands. A comparison of such a binding site definition using the current GPCRdb protein-ligand interaction annotations and previous binding site definitions is shown in Table 2, while ligand-contacting residues in noncanonical binding sites and class B, C, and F receptors are shown in Table 3. Figure 4 lists all the small-molecule ligands cocrystallized with GPCRs and important polar interactions between the ligands and pocket residues. However, given the large variations already observed in the small-molecule binding pockets in GPCRs this list is likely to further expand with the elucidation of novel GPCR structures.

3.3 Sequence-Based Pharmacophore Methods

A large-scale alignment of 13,324 sequences from UniProt [149] and Ensembl [150] covering most of the species homologues of the human set of GPCRs was used to predict residues involved in ligand binding for any G protein-coupled receptor using the ss-TEA method [151]. The subfamily-specific two entropy analysis (ss-TEA) method identifies ligand binding residues by comparing the conservation level of a residue position within a subfamily to the level of conservation for this residue outside the subfamily. Residues highly conserved within but not outside of a subfamily are likely to be involved in ligand binding or other specific functions of the receptors within the subfamily. The method was tested on ten cases with structural ligand binding information or abundant mutational data and was found to correlate well with the known information on ligand binding residues. The FFAR1 and CCR5 GPCR crystal structures have since been elucidated and the key interacting residues predicted by ss-TEA were indeed found to interact with cocrystallized ligands: $R^{5.39x40a}$, $N^{6.55a}$ and $R^{7.35x34a}$ with TAK-875 in FFAR1 (Fig. 4, PDB: 4PHU [109]), and $Y^{1.39}$, $W^{2.60a}$, $Y^{3.32a}$ and $E^{7.39x38a}$ with Maraviroc in CCR5 (Fig. 4, PDB: 4MBS [110]) using the BW and generic numbering.

The large-scale alignment and entropy analysis were later used in the Snooker method for automatic homology model creation, binding site selection, and pharmacophore generation for the prediction of ligand binding modes in GPCRs with yet no experimentally determined structure [152]. The usefulness of the method was demonstrated in binding mode reproduction of known beta-2-adrenergic receptor ligands, the prospective prediction of the D_3 receptor– Eticlopride complex in the community-wide GPCR DOCK 2010 assessment, and in retrospective virtual screening

Table 2
Consensus transmembrane binding pocket definitions for class A receptors. The number of crystal structures in which amino acids at the specific positions (based on the GPCRdb structure-based numbers) are contacting the cocrystallized small molecule are indicated out of a total of 134 class A GPCR crystal structures (not including extrahelical and intracellular binding pockets or structures with peptide/protein ligands; contacts defined by GPCRdb). Positions defined as comprising the consensus transmembrane binding pocket in previous studies are marked with an X (data taken from Refs. 24 and 77)

	5-HT1B receptor	5-HT2B receptor	M1 receptor	M2 receptor	M3 receptor	M4 receptor	β1-adrenoceptor	β2-adrenoceptor	D3 receptor	H1 receptor	AT1 receptor	δ receptor	κ receptor	μ receptor	NOP receptor	OX1 receptor	OX2 receptor	PAR1	CCR2	CCR5	CXCR4	FFA1 receptor	LPA1 receptor	S1P1 receptor	CB1 receptor	A1 receptor	A2A receptor	P2Y1 receptor	P2Y12 receptor	Rhodopsin	Sum	ref. 78	GSK – ref. 31	CNRS	7TM Pharma	Novo Nordisk	Roche	Novartis	Bio-Focus	BCM-UCSF
1x27																			1												1									
1x30																			1												1									
1x31																			1		1							1			3									
1x32																					1										1									
1x34																			1												1									
1x35											1								1				3		2	1	2				10	X	X	X				X		
1x39											2								1	1	4				2	1					11	X	X	X			X	X		
2x53					1			4			2		1												2					9	19	X	X					X	X	X
2x54																									2						2							X		X
2x56													1			2	1								2						6	X	X	X				X	X	
2x57											2												3	2	2					5	14	X	X	X			X			X
2x59																2	1														3	X							X	
2x60							2	2	1		2	2	1	1	3	2	1		1	1	4		3	2	1	1	11				41	X	X	X			X	X	X	
2x61																							3		2		1				6	X	X							
2x63	2						4	5			2		1		3	2	1		1	1	4						8	1			35	X	X					X	X	
2x64							3	2			2				1				1		4					1	5				19	X	X	X				X	X	
2x65																									1						1									
23x50								2					1	1	3	2	1				4		2	2							18									
3x21																													2		2	X								
3x25																								2					2		4								X	X
3x26																						1									1	X							X	
3x27																						1									1									
3x28	2	3					15	17	1				1	1	3	2	1				4		2	2						18	72	X	X	X	X		X	X	X	
3x29	2	3					14	15	1		2	2	1	1	3	2	1		1	1	4	1	3	2	2		1		2	14	78	X	X	X	X		X	X	X	
3x30																						1							3		4	X								
3x32	2	3	1	3	4	1	17	17	1	1	2	4	1	2	3	2	1		1	1	4		3	2	2	1	18			23	120	X	X	X	X	X	X	X	X	
3x33	2	3	1	3	4	1	17	17	1	1	2	4	1	2	3	2	1	1	1	1		1	3	2	2	1	24		3	24	128	X	X	X	X	X	X	X	X	X
3x34																													3		3	X								
3x35													1												1						2					X	X	X	X	X
3x36	2	3	1	3	4	1	17	17	1	1	2	4	1	2	3	2	1			1			3	2		1	7			21	100	X	X	X	X	X	X	X		X
3x37	2	3		3	4	1	12	16	1	1	2				2			1				1					5		3	24	81	X	X		X	X	X	X		
3x40	2		1	3	3	1				1																	5			18	34	X	X	X		X	X	X		
4x53																													1		1								X	
4x54																						1									1									
4x56	2	2																											3	4	11	X	X	X	X		X	X		
4x57			1	3	4	1				1								1				1							3		15	X	X		X			X	X	
4x58																						1									1	X								
4x60																		1				1							3		5	X	X	X	X				X	
4x61											2					2	1	1				1									7	X	X							
4x62																						1									1									
4x64																													2		2									
4x65											2					2	1		1												6									
45x50							1	5					1		3				1		4								2		17									
45x51	2	2					1	6				1			1			1	1		4	1					5	1	1	19	46									
45x52	2	3					17	17	1		2	2			2	2	1	1			3	1				1	25	1		24	105									
5x36		1										1		1						1		1							3		8	X							X	
5x37		1																								1	17				19	X							X	
5x39		3					17	16										1		1		1	1						3		43	X	X	X				X		
5x40		3	1	1	3	1	14	13	1			4	1	2		2	1			1		1	3		1	1	25		3	3	85	X	X	X	X	X		X	X	X
5x43	2	3	1	1	4	1	17	17	1	1	1	4	1	2	3	2	1			1			3	2	2		22		3	24	119	X	X	X	X	X	X	X	X	
5x44		3	1	1	4	1	17	17	1	1		1												2					1	23	73	X	X	X	X		X	X	X	
5x461	2	3	1	3	4	1	17	17	1	1					3												4			8	65	X	X	X	X	X	X	X	X	
5x47			1	3	4	1				1														2			2			24	38	X	X				X	X	X	
6x44										1																				16	17	X	X	X		X	X	X	X	
6x48	2	3	1	3	4	1	15	17	1	1	1	4	1	2	3	2	1			1			3	2	2	1	23			24	118	X	X	X	X	X	X	X	X	X
6x51	2	3	1	3	4	1	17	17	1	1		4	1	2	3	2	1			1	3	1	3	2	2	1	25		1	24	126	X	X	X	X		X	X	X	
6x52	2	3	1	3	4	1	17	17	1	1		4	1	2	3	2							3	2			24			24	115	X	X	X	X	X	X	X	X	
6x54																		1					3			1	1		1		7	X							X	
6x55	2	3		1	4	1	17	17	1	1		4	1	2	3	2	1	1		1		1	3	2	1	1	25		3	7	105	X	X	X	X	X	X	X	X	
6x56									1			1																			2									
6x58	2	2										4		1				1								1	1	1	3		16	X	X		X					
6x59		2										2						1		1											6	X	X		X				X	
6x61																												1			1	X								
6x62																		1										1	2		4	X								
7x27																												1			1	X								
7x30																		1									3	1	1		6	X								
7x31	2	3						3										1									17	1	2		29	X								
7x33																		1											1		2	X							X	
7x34	2	3					5	17	1			4	1	2				1		1		1	3		2	1	25	1	3		73	X	X	X	X	X		X	X	
7x35	2	3					5	6			2				3				1		1		3	2	2	1	17	1	1		50	X						X	X	
7x36																			1												1									
7x37																							3						1		4	X					X	X		
7x38	2	3	1	3	4	1	17	17	1	1	2	4	1	2	3	2	1		1	1	4		3	2	2	1	25	1		23	128	X	X	X	X	X	X	X	X	
7x39							4	2											1	1					2					4	14	X				X	X	X		
7x41			1	3	4	1						2	1	1		2	1						3		2		8			12	41	X	X		X	X	X	X	X	
7x42	2	3	1	3	4	1	17	17	1	1	2	4	1	2	3	2	1		1	1				2	2	1	9			24	105	X	X	X	X	X	X	X	X	
7x46																									2						2		X			X		X	X	

Table 3
Ligand contacting residues of noncanonical binding pockets in class A GPCRs and all class B, C, and F GPCR crystal structures. The number of crystal structures in which amino acids at the specific positions (based on the GPCRdb structure-based numbers) are contacting the cocrystallized small molecule are indicated (contacts defined by GPCRdb)

	2x60a	2x63a	2x64a	3x33a	45x51	6x58a	7x31a	7x34a	7x35a	7x38a
M2 receptor allosteric	1	1	1	1	1	1	1	1	1	1

	1x43a	1x47a	2x55a	2x56a	2x58a	2x59a	23x52	3x24a	3x27a	3x28a	3x31a
P2Y1 receptor extrahelical	1	1	1	1	1	1	1	1	1	1	1

	1x53a	1x56a	1x57a	1x60a	12x49	12x51	2x37a	2x39a	2x40a	2x43a	2x44a	3x46a	3x50a	5x65a	6x31a	6x32a	6x33a	6x34a	6x35a	6x36a	6x37a	6x38a	6x39a	6x40a	7x52a	7x53a	7x56a	8x47	8x48	8x49	8x50	8x51	8x53
CCR2 intracellular	1	1	1					1		1		1	1			1	1			1	1					1		1	1	1	1	1	1
CCR9 intracellular	1	1	1	1	1	1	1	1	1	1	1	1	1								1			1		1		1	1	1	1		1

	1x57b	1x60b	1x61b	1x64b	12x49	12x51	2x44b	2x46b	2x47b	2x50b	2x51b	3x50b	3x54b	5x61b	6x36b	6x37b	6x38b	6x39b	6x40b	6x41b	6x42b	6x43b	6x44b	6x45b	7x56b	7x57b	7x60b	8x47	8x48	8x49	8x50	8x51	8x53
GCGR intracellular														1	1	1		1	1	1		1	1		1		1	1	1	1			

	2x57b	3x43b	3x44b	3x47b	3x48b	3x50b	3x51b	3x54b	5x46b	5x47b	5x50b	5x51b	5x53b	5x54b	5x57b	6x42b	6x45b	6x46b	6x49b	6x50b	6x53b	7x49b	7x56b
CRF1	2	2	2	2	2	2	2	1	2	2	2	2	2	2	2	2	2	2	2	2	2	1	2

	2x45c	2x46c	2x49c	2x52c	2x56c	3x32c	3x33c	3x36c	3x37c	3x39c	3x40c	3x43c	3x44c	4x43c	45x52	5x40c	5x43c	5x44c	5x47c	5x51c	6x46c	6x49c	6x50c	6x53c	6x57c	7x29c	7x30c	7x33c	7x36c	7x37c	7x40c	7x41c	7x44c
mGlu1					1	1	1	1	1		1		1	1	1	1	1	1	1	1	1	1	1	1	1	1	1	1	1	1	1		
mGlu5	2	3	3	2				3	1	3	3	3	3			3	1	3	1	1		3	3	3				3	3	3	3	3	3

	1x32f	1x36f	2x51f	2x58f	3x32f	3x36f	3x37f	3x40f	45x51	5x44f	5x47f	5x51f	6x44f	6x47f	6x51f	6x54f	6x58f	6x61f	6x62f	6x65f	7x35f	7x38f	7x41f	7x42f	7x45f	7x48f
Smoothened	4	3	1	4	1	2	1	1	4	5	1	1	1	1	5	5	5	4	5	5	2	6	4	4	3	1

against 15 diverse GPCRs. Snooker pharmacophore models were successfully used in prospective virtual screening against the β_2, A_{2A}, and $S1P_1$ receptors affording 6, 18, and 3 novel active molecules (Fig. 5) against these targets, respectively [153]. The proposed ligand interacting amino acids $R^{3.28a}$, $E^{3.29a}$, $Y^{5.39a}$, and $L^{6.55a}$ (BW and generic numbering) for $S1P_1$ were confirmed by the experimentally determined complex structure with the ligand ML056. Finally, the Snooker method was also extended to the class of taste 2 receptors [154]. A structure-based pharmacophore of the hTAS2R39 was built using Snooker, and revealed putative interaction of flavonoid agonists and blockers with the residues $N^{3.36a}$, $T^{5.45a}$, $F^{6.55a}$, $Q^{7.35}$, and $N^{7.39a}$ (BW and generic numbering). As there are no bitter taste receptor structures solved yet, the validation of these predictions remains to be elucidated.

Another sequence-to-pharmacophore method called Pharma was developed by Klabunde et al. [155]. In this method, pharmacophore features originating for clusters of amino acids in 13 reference GPCR crystal structures or homology models called chemoprints were identified and a sequence alignment of

270 GPCRs by the BW numbering allowed for the creation of sequence-derived 3–7 point pharmacophores for all of the studied receptors. This method was prospectively applied to the complement component 3a receptor 1 (C3AR1), for which the pharmacophore model was based on an AT_1 homology model resulting in four hits with 0.3–3 μM EC_{50}. The important pharmacophore features from $V^{3.32a}$, $S^{3.29a}$, $K^{5.42x43a}$ and $W^{6.48}$ for AT_1 were correctly defined in retrospect based on the AT_1 crystal structures (Fig. 6, PDB: 4YAY [77], 4ZUD [76]), however, $F^{5.47a}$ was predicted to contact the ligand but later it was found to be located deeper in the TM bundle facing the membrane, and other important interactions with $Y^{1.39a}$ and $R^{4.64x65a}$ were missed by the method. The lower resolution of pharmacophore-based models, however, still affords useful models for virtual screening as shown by successful applications of the Snooker and Pharma methods.

Gloriam et al. developed a pharmacophore generation method based on the available protein-ligand interaction information in crystallized GPCRs [156, 157]. First, the cocrystallized ligands from all class A GPCR structures were fragmented and specific fragments were assigned to the amino acid(s) they are interacting with. Then for any non-crystallized GPCR the closest homolog with a crystal structure is selected and the fragments from other X-ray structures are superimposed on this template (after correcting residue positions for helix bulges and constrictions). Finally, the pharmacophore points associated with the superimposed fragments are clustered and representatives are selected to create the final pharmacophore. The fragment sets are available in GPCRdb [66, 67]. The method was validated in retrospective virtual screening against the H_1 and H_3 receptors, for which good ROC AUC values (0.88 and 0.82 respectively) were achieved.

3.4 Structural Determinants of Ligand Functional Effects

Not only the prediction of ligand binding but also of the functional effects of specific ligands is of great interest for drug discovery. Chemogenomics methods were also used to uncover specific GPCR features responsible for the agonistic and antagonistic effects. Wichard et al. used mutual information analysis of coupled GPCR sequence and ligand descriptor data to extract features in agonist and antagonist ligands responsible for their specific functional effects, significant molecular features in GPCRs for recognizing these ligands, and furthermore for their interactions with specific G protein coupling partners [158]. Positions extracted for selective agonistic effects in the helices were mainly located between TM1, TM2, and TM3 (e.g., 1.31a, 1.36a, 2.40a, 2.44a, 3.33a), while selective antagonistic sensitive positions were located mainly between TM5 and TM6 (e.g., 5.36x37a, 5.40x41a, 5.57a). Ligand descriptors were not so sensitive for separating antagonists and agonists but presence of alcohols and phenols, and larger polar surface area were related to agonists while, for example, a larger

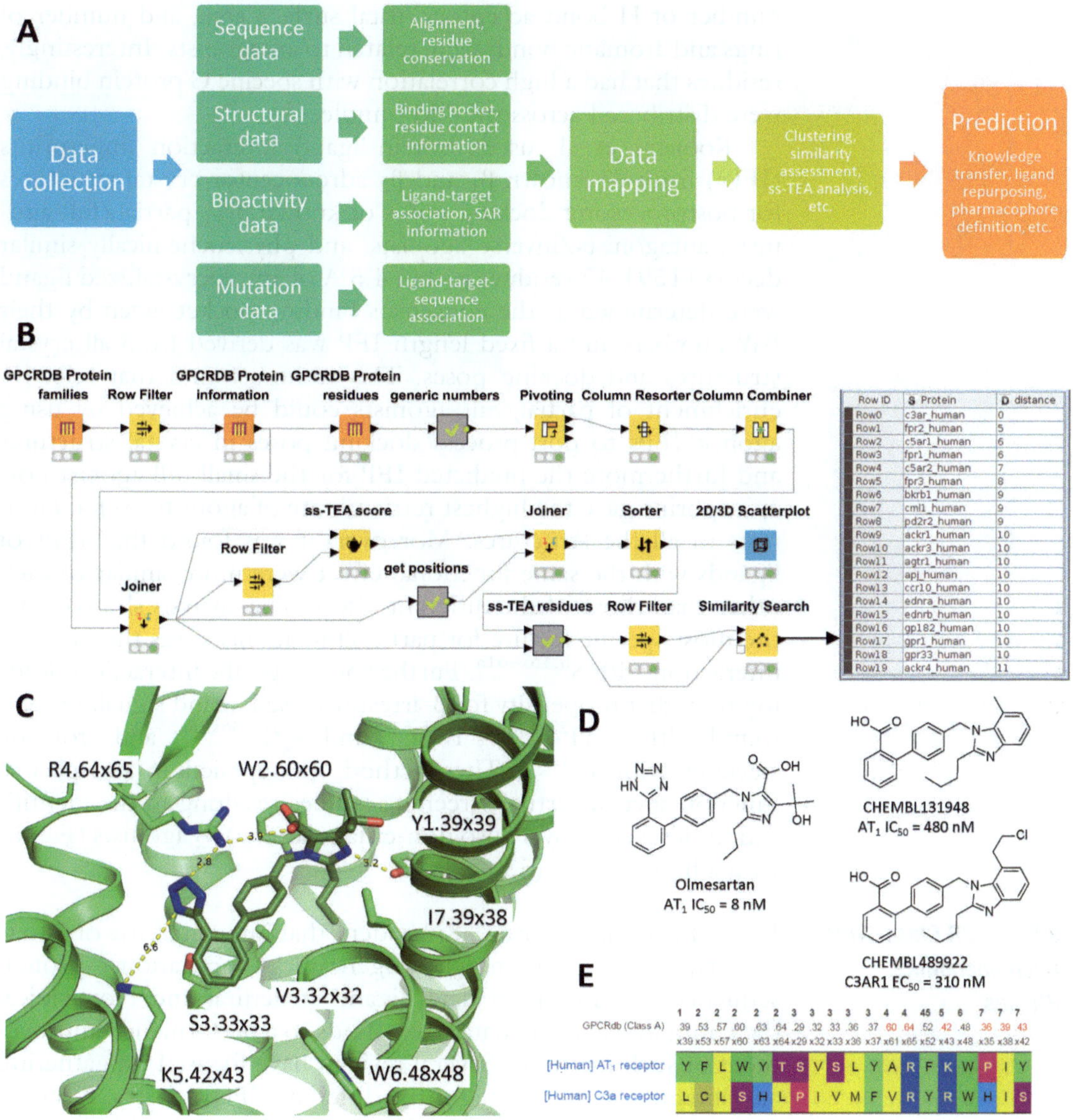

Fig. 6 (**a**) General chemogenomics workflow scheme and (**b**) a specific workflow designed using the KNIME analytics platform [70] that exploits and integrates heterogeneous data sources for the prediction of GPCR-ligand interactions. The KNIME analytical workflow makes use of the 3D-e-Chem GPCRdb KNIME nodes [71, 72] to collect phylogenetic and sequence information from GPCRdb on all class A GPCRs, then uses the structure-based generic numbering scheme to construct a sequence alignment (Pivoting node), and performs an ss-TEA analysis using the 3D-e-Chem ss-TEA score KNIME node for the prediction of hot spots in the C3AR1 receptor. Finally, the top 15 predicted hot spot residues are selected for sequence comparison (Similarity Search node) and a list of related receptors is returned of which the top 20 are shown, including the crystallized AT_1 receptor also identified to be similar in terms of binding pocket composition to C3AR1 in Ref. 53. (**c**) The AT_1 receptor binding pocket with co-crystallized Olmesartan (PDB: 4ZUD [76]). (**d**) Bioactivities of the co-crystallized Olmesartan, the most potent C3AR1 ligand identified in Ref. 53 and the most similar AT_1 ligand from ChEMBL. (**e**) Alignment of binding pocket residues of AT_1 and C3AR1 from GPCRdb based on the structure-based numbering scheme

number of H-bond acceptors, total surface area, and number of rings and aromatic bonds were related to antagonists. Interestingly, residues that had a high correlation with specific G protein binding were distributed across the TM bundle.

Kooistra et al. used protein-ligand interaction fingerprints (IFPs) from 31 known β_1 and β_2 adrenoceptor crystal structures for postprocessing docking poses of known $\beta_{1/2}$ partial/full agonists, antagonists/inverse agonists, and physicochemically similar decoys [159]. 47 residues within 4.5 Å of any cocrystallized ligand were determined as the consensus binding pocket given by their BW numbers and a fixed length IFP was derived from all crystal structures and docking poses. The authors found that selective enrichment of partial/full agonists could be achieved by using agonist IFPs to post-process docking poses in crystal structures and furthermore the predicted IFP for the small full agonist norepinephrine gave the highest retrieval rate of agonists over antagonists in all the structures. Moreover, it was found that IFPs of ligands with the same functional effect were more similar to each other than those of the different effect. The analysis showed that the most specific feature for partial/full agonists was the H-bond interaction with $S^{5.46x461a}$. Furthermore, specific interactions leading to higher propensity for β-arrestin-biased ligand signaling were found with $L/H^{2.64x63a}$, $D^{45.51}$, and $V/I^{7.36x35a}$, and aromatic stacking with $W^{3.28a}$. The method was applied in prospective function-specific virtual screening studies resulting in the identification of several novel (also non-ethanolamine) β_2 agonists (Fig. 5) with a high hit-rate (53%) [160].

3.5 GPCR Class-Wide Chemogenomics Studies

From the recent literature it is evident that integrated use of different types of data in chemogenomics applications—protein sequence and structural data, ligand structural and biochemical data, mutational effect data, etc.—leads to a performance improvement of such methods. Several reviews have focused on gathering and integrating these data for the different GPCR classes. A review on aminergic GPCRs combined ligand affinity data, receptor mutagenesis studies, amino acid sequence, and high-resolution structural analyses of GPCR-ligand interactions to highlight correlations and differences between ligand similarity and ligand binding site similarity of different aminergic receptors [48]. The study analyzes the composition of the major bioamine and the minor allosteric binding site, the latter of which is exploited by appendages of dualsteric ligands and allosteric ligands experimentally determined for the muscarinic M_2 receptor but postulated for many other aminergic receptors as well. The authors review homology modeling and virtual screening efforts against aminergic receptors and collected a large body of mutational effect data (1420 single-point mutations for 128 individual amino acid positions given by their BW numbering) for the histamine receptor subfamily and the crystallized aminergic receptors and related these data to protein-ligand interaction features. Finally, by comparing binding site and ligand similarity and receptor-ligand selectivity profiles, the authors report affinity and selectivity cliffs in aminergic ligands.

Integrated chemogenomics analyses for class B1 secretin-like receptors were performed by Hollenstein and De Graaf et al. [16]. While extracellular domains (ECDs) of these receptors in complex with peptide ligands had been available since 2007, there was a delay in obtaining a structure of the TMD of any class B receptor owing to the inherent instability of these receptors and the limited availability of small-molecule ligands capable of stabilizing them. The authors analyze the crystal structures of GCGR and CRF_1 which have a pronounced opening toward the extracellular side. In the CRF_1 structure the ligand CP-376395 was surprisingly located very deep in the TM domain anchored by the conserved $N^{5.50b}$ and in the GCGR structure solved later in complex with the antagonist MK-0893 an even more unexpected extra-helical binding mode was revealed, located outside the 7TM-bundle on the intracellular side of TM6 anchored by $R^{6.37b}$, $S^{6.41b}$ and $N^{8.47b}$ from the transition between TM7 and H8 [29]. In the review the authors also give a detailed overview of the ECDs and their peptide ligands as well. Furthermore, the elucidation of both ECD and TMD crystal structures of the receptors as well as incorporation of mutagenesis and photo-crosslinking data allowed the construction of a full GCGR-glucagon model in which an extended flexible conformation of the first seven residues of glucagon is proposed to reach deep into the TMD binding pocket [28]. Finally, the authors discuss potential druggability of the elucidated class B1 GPCR binding pockets. Combined structure-based comparative modeling, site-specific mutational studies (66 novel mutants complementing 76 mutants from literature), and molecular dynamics simulations facilitated the construction of a full-length GLP-1R bound to multiple truncated or mutated variants of the peptide ligands GLP-1 and exendin-4 [161]. The model was validated by concerted mutations of the receptor and ligands, and molecular dynamics simulations of the wild-type and mutant systems revealed conserved and receptor-specific ligand interaction hot spots in the binding modes of GLP-1R—GLP-1 and GCGR—glucagon complexes. Furthermore, the relative flexibility of the TMD and the ECD (swinging and rotational motion) was shown to accommodate binding of the different peptide ligand variants demonstrating the complexity of ligand recognition by class B GPCRs.

Despite the breakthroughs in the elucidation of GPCR crystal structures, a structure of a member of the class B2 adhesion GPCR family has not yet been solved. Nijmeijer et al. recently presented a comparative sequence and structure-based analysis of the TM domain of adhesion GPCRs [60]. Conserved sequence motifs present both in class B1 and B2 receptors suggest that their TM domains share a similar structural fold and that the conserved residues in both families may be involved in similar intermolecular interaction networks. Adhesion GPCRs consist of a large ECD, and a TM domain followed by an intracellular domain (ICD). The ECD contains a GPCR autoproteolysis inducing (GAIN) domain that cleaves it to an N-terminal fragment (NTF) and a membrane-spanning C-terminal fragment (CTF) that are non-covalently associated. The β-13 strand (or Stachel) sequence, which is the remaining part of the GAIN domain after cleavage, has been recently shown to be a tethered agonist sequence capable of activating adhesion GPCRs. Several conserved motifs between adhesion and other GPCRs were analyzed. Since they are most similar to secretin-like receptors, the class B Wootten numbering scheme was proposed to be used also for adhesion GPCRs. For example, the transmission switch including $W^{6.53b}$ is conserved in adhesion GPCRs corresponding to the highly conserved $W^{6.48a}$ in the rhodopsin family and $W^{6.50c}$ in glutamate family GPCRs. Bulges in TM4 induced by the GW/$Y^{4.50b}$GxP motif, in TM5 by $P^{5.42b}$ and in TM7 by $G^{7.50b}$ suggest a similar helix arrangement to secretin-like receptors. At the intracellular side of the TM domain, $H^{2.50b}$ and $E^{3.50b}$ are conserved in secretin family GPCRs and are present in most adhesion GPCRs, suggesting the presence of a putative ionic lock in this family as well. However, alternative polar/ionic networks are also observed in some adhesion receptor subtypes. Finally, the authors analyze the conservation of the TM binding site residues and the extra-helical one found in GCGR and conclude that the druggability of these sites is probably lower than in class A or B1 receptors.

Gloriam et al. have reported integrated chemogenomics analyses of the TM domain of class C GPCRs [18, 162]. Allosteric modulators binding the TM domain have been discovered for all non-orphan class C GPCR families. The authors collected 1670 single-point mutation data points covering 99 TM and several extracellular loop positions and analyzed sequence conservation, the available $mGlu_{1/5}$ crystal structures, and ligand information. Although all crystal structures were solved in complex with negative allosteric modulators (NAMs), mutagenesis data suggests that NAMs and PAMs bind a common allosteric site in all class C GPCRs. Most mutations have similar influence on PAM and NAM effects, and modeling suggests that functional effects may be determined by subtle variations in the structure such as disturbing the water network around $Y^{3.44c}$, $T^{6.46c}$, and $S^{7.39x40c}$. Not only the mutation of TM residues but that of several residues in extracellular loop 2 also had a significant effect on allosteric modulator binding. Finally, as only the mutation of $F^{1.46c}$ out of 22 positions had a significant effect on the ligand CPPHA while not affecting other PAMs or NAMs, this suggests the presence of a distinct, possibly extra-helical binding site for this ligand in $mGlu_{1/5}$ receptors.

3.6 Structural Cheminformatics Workflows to Integrate and Analyze GPCR Chemogenomics Data

The combination of cheminformatics and bioinformatics tools can facilitate the systematic analysis of GPCR chemogenomics data, as exemplified by the KNIME nodes and data analytics workflows that enable efficient data mining from established structural (PDB [163]) and bioactivity (ChEMBL [143]) databases as well as customized G Protein-Coupled Receptor (GPCRdb [66, 67]) focused data resources. It should be noted that for the efficient use of such data resources these have to comply with the FAIR data principles (the data should be Findable, Accessible, Interoperable, and Reusable) [164]. The GPCRdb KNIME nodes [71, 72] facilitate the extraction and analysis of structure-based annotation of GPCR sequence alignments of 14805 sequences of 414 receptor subtypes and of 3547 species, analysis of 203 GPCR crystal structures and GPCR-ligand interactions, and 28126 mutational data points, stored in the GPCRdb repository (http://gpcrdb.org, accessed 10 March 2017) [66]. Current GPCRDB KNIME node functionalities include: (a) The extraction of protein family information, including the protein names and classifications of all GPCRs based on class, ligand type, subfamily, subtype (GPCRDB Protein families node). (b) The retrieval of source, species, and sequence data from UniProt identifiers or protein family identifier (GPCRDB Protein information node). (c) The retrieval of residues and numbering schemes (GPCRDB Protein residues node). (d) The retrieval of experimental GPCR structures with literature references, PDB codes, and ligands (GPCRDB Structures and Structures of a protein nodes). (e) The retrieval of single-point mutations in GPCRs, including the sequence position, mutation, ligand, assay type, mutation effect, protein expression information, and publication reference (GPCRDB Mutations of a protein node). (f) The sequence numbers of amino acid residues interacting with ligands in the specified PDB entry (GPCRDB Structure-ligand interactions node). (g) The sequence identity and similarity of a query receptor versus a set of receptors, based on the full sequence or a specified set of residues (GPCRDB Protein similarity node). A general integrated chemogenomics workflow scheme and a specific KNIME workflow utilizing GPCRdb and 3D-e-Chem resources are shown in Fig. 6. The latter workflow can be used to obtain all GPCR sequences stored in GPCRdb, create an alignment based on the structure-based numbering scheme, perform a double entropy analysis (ss-TEA) [151] of C3AR1 receptor orthologs against all other GPCRs, and assess the similarity of C3AR1 with other GPCRs based on the hot spot residues identified by ss-TEA identifying overlapping neighbors with the study based on binding pocket composition in Ref. 53.

4 Conclusions

The increasing amount of structural information on GPCRs combined with comprehensive sequence, ligand bioactivity, and mutation data facilitates the construction of integrated chemogenomics workflows to gain new insight into this important target family. Crystal structures from class A, B1, C, and F GPCRs revealed a well-conserved transmembrane fold but large variations in extracellular domain structures, loop regions, and ligand binding pockets. Cross-family and cross-class studies have pinpointed the need for a common reference framework for referring to specific structural elements and conserved or specific residues across all GPCRs, GPCR classes, subfamilies, or unique receptors. Various class-specific numbering schemes have been proposed over the years, of which the Ballesteros-Weinstein scheme became the primarily used scheme for class A GPCRs and formed the basis for the similar Wootten, Pin, and Wang schemes for class B, C, and F GPCRs, respectively. However, these schemes used consecutive numbering, whereas the bulges and constrictions discovered in transmembrane helices of specific receptors urged the construction of a structure-based numbering system by GPCRdb that is more suitable for the comparison of structurally equivalent residues. The recent efforts to standardize the residue numbering schemes based on the increasing insights in GPCR structure have facilitated many chemogenomics studies such as consensus binding site definitions, binding site comparisons, ligand repurposing for orphan receptors, sequence-based pharmacophore generation for homology modeling or virtual screening, and GPCR class-wide analysis studies. Moreover, standardized resources such as GPCRdb and automated data retrieval and analysis using the 3D-e-Chem cheminformatics tools in the popular workflow management software KNIME enable the efficient integration of different data types in chemogenomics workflows. Elucidation of adhesion, taste, and olfactory receptor crystal structures, full-length receptor crystal structures, and novel ligand binding modes will probably further impact the use of numbering schemes in chemogenomics studies in the future.

Acknowledgments

Netherlands eScience Center/NWO (3D-e-Chem, grant 027.014.201, to C.d.G). M. V., A. J. K., D. G., I. J. P. d. E. and C. d. G. participate in the COST Action CM1207 (GLISTEN). M. V., D. G., and C. d. G. participate in the GPCR Consortium (gpcrconsortium.org). Vignir Isberg and Christian Munk (GPCRdb (http://gpcrdb.org), University of Copenhagen) are acknowledged for useful discussions on the developments of the GPCRdb KNIME nodes.

References

1. Pierce KL, Premont RT, Lefkowitz RJ (2002) Seven-transmembrane receptors. Nat Rev Mol Cell Biol 3(9):639–650. https://doi.org/10.1038/nrm908
2. Stevens RC, Cherezov V, Katritch V, Abagyan R, Kuhn P, Rosen H, Wuthrich K (2013) The GPCR network: a large-scale collaboration to determine human GPCR structure and function. Nat Rev Drug Discov 12 (1):25–34. https://doi.org/10.1038/nrd3859
3. Overington JP, Al-Lazikani B, Hopkins AL (2006) How many drug targets are there? Nat Rev Drug Discov 5(12):993–996. https://doi.org/10.1038/nrd2199
4. Rask-Andersen M, Masuram S, Schioth HB (2014) The druggable genome: evaluation of drug targets in clinical trials suggests major shifts in molecular class and indication. Annu Rev Pharmacol Toxicol 54:9–26. https://doi.org/10.1146/annurev-pharmtox-011613-135943
5. Garland SL (2013) Are GPCRs still a source of new targets? J Biomol Screen 18 (9):947–966. https://doi.org/10.1177/1087057113498418
6. Hauser AS, Attwood MM, Rask-Andersen M, Schiöth HB, Gloriam DE (2017) Trends in GPCR drug discovery: new agents, targets and indications. Nat Rev Drug Discov. In press
7. Lv X, Liu J, Shi Q, Tan Q, Wu D, Skinner JJ, Walker AL, Zhao L, Gu X, Chen N, Xue L, Si P, Zhang L, Wang Z, Katritch V, Liu ZJ, Stevens RC (2016) In vitro expression and analysis of the 826 human G protein-coupled receptors. Protein Cell 7(5):325–337. https://doi.org/10.1007/s13238-016-0263-8
8. Chung S, Funakoshi T, Civelli O (2008) Orphan GPCR research. Br J Pharmacol 153 (Suppl 1):S339–S346. https://doi.org/10.1038/sj.bjp.0707606
9. Roth BL, Kroeze WK (2015) Integrated approaches for genome-wide interrogation of the Druggable non-olfactory G protein-coupled receptor superfamily. J Biol Chem 290(32):19471–19477. https://doi.org/10.1074/jbc.R115.654764
10. Davenport AP, Alexander SP, Sharman JL, Pawson AJ, Benson HE, Monaghan AE, Liew WC, Mpamhanga CP, Bonner TI, Neubig RR, Pin JP, Spedding M, Harmar AJ (2013) International Union of Basic and Clinical Pharmacology. LXXXVIII. G protein-coupled receptor list: recommendations for new pairings with cognate ligands. Pharmacol Rev 65(3):967–986. https://doi.org/10.1124/pr.112.007179
11. Attwood TK, Findlay JB (1994) Fingerprinting G-protein-coupled receptors. Protein Eng 7(2):195–203
12. Kolakowski LF Jr (1994) GCRDb: a G-protein-coupled receptor database. Receptors Channels 2(1):1–7
13. Fredriksson R, Lagerstrom MC, Lundin LG, Schioth HB (2003) The G-protein-coupled receptors in the human genome form five main families. Phylogenetic analysis, paralogon groups, and fingerprints. Mol Pharmacol 63(6):1256–1272. https://doi.org/10.1124/mol.63.6.1256
14. Nordstrom KJ, Sallman Almen M, Edstam MM, Fredriksson R, Schioth HB (2011) Independent HHsearch, Needleman--Wunsch-based, and motif analyses reveal the overall hierarchy for most of the G protein-coupled receptor families. Mol Biol Evol 28 (9):2471–2480. doi:https://doi.org/10.1093/molbev/msr061
15. Tautermann CS (2014) GPCR structures in drug design, emerging opportunities with new structures. Bioorg Med Chem Lett 24 (17):4073–4079. https://doi.org/10.1016/j.bmcl.2014.07.009
16. Hollenstein K, de Graaf C, Bortolato A, Wang MW, Marshall FH, Stevens RC (2014) Insights into the structure of class B GPCRs. Trends Pharmacol Sci 35(1):12–22. https://doi.org/10.1016/j.tips.2013.11.001
17. Leach K, Gregory KJ (2016) Molecular insights into allosteric modulation of class C G protein-coupled receptors. Pharmacol Res. https://doi.org/10.1016/j.phrs.2016.12.006
18. Harpsoe K, Boesgaard MW, Munk C, Brauner-Osborne H, Gloriam DE (2016) Structural insight to mutation effects uncover a common allosteric site in class C GPCRs. Bioinformatics. https://doi.org/10.1093/bioinformatics/btw784
19. Wang C, Wu H, Evron T, Vardy E, Han GW, Huang XP, Hufeisen SJ, Mangano TJ, Urban DJ, Katritch V, Cherezov V, Caron MG, Roth BL, Stevens RC (2014) Structural basis for smoothened receptor modulation and chemoresistance to anticancer drugs. Nat Commun 5:4355. https://doi.org/10.1038/ncomms5355
20. McCabe JM, Leahy DJ (2015) Smoothened goes molecular: new pieces in the hedgehog

signaling puzzle. J Biol Chem 290 (6):3500–3507. https://doi.org/10.1074/jbc.R114.617936

21. Palczewski K, Kumasaka T, Hori T, Behnke CA, Motoshima H, Fox BA, Le Trong I, Teller DC, Okada T, Stenkamp RE, Yamamoto M, Miyano M (2000) Crystal structure of rhodopsin: a G protein-coupled receptor. Science 289(5480):739–745
22. Zhou XE, Gao X, Barty A, Kang Y, He Y, Liu W, Ishchenko A, White TA, Yefanov O, Han GW, Xu Q, de Waal PW, Suino-Powell KM, Boutet S, Williams GJ, Wang M, Li D, Caffrey M, Chapman HN, Spence JC, Fromme P, Weierstall U, Stevens RC, Cherezov V, Melcher K, HE X (2016) X-ray laser diffraction for structure determination of the rhodopsin-arrestin complex. Sci Data 3:160021. https://doi.org/10.1038/sdata.2016.21
23. Warne T, Moukhametzianov R, Baker JG, Nehme R, Edwards PC, Leslie AG, Schertler GF, Tate CG (2011) The structural basis for agonist and partial agonist action on a beta (1)-adrenergic receptor. Nature 469 (7329):241–244. https://doi.org/10.1038/nature09746
24. Rasmussen SG, DeVree BT, Zou Y, Kruse AC, Chung KY, Kobilka TS, Thian FS, Chae PS, Pardon E, Calinski D, Mathiesen JM, Shah ST, Lyons JA, Caffrey M, Gellman SH, Steyaert J, Skiniotis G, Weis WI, Sunahara RK, Kobilka BK (2011) Crystal structure of the beta2 adrenergic receptor-Gs protein complex. Nature 477(7366):549–555. https://doi.org/10.1038/nature10361
25. Shimamura T, Shiroishi M, Weyand S, Tsujimoto H, Winter G, Katritch V, Abagyan R, Cherezov V, Liu W, Han GW, Kobayashi T, Stevens RC, Iwata S (2011) Structure of the human histamine H1 receptor complex with doxepin. Nature 475 (7354):65–70. https://doi.org/10.1038/nature10236
26. Wu B, Chien EY, Mol CD, Fenalti G, Liu W, Katritch V, Abagyan R, Brooun A, Wells P, Bi FC, Hamel DJ, Kuhn P, Handel TM, Cherezov V, Stevens RC (2010) Structures of the CXCR4 chemokine GPCR with small-molecule and cyclic peptide antagonists. Science 330(6007):1066–1071. https://doi.org/10.1126/science.1194396
27. Zheng Y, Qin L, Zacarias NV, de Vries H, Han GW, Gustavsson M, Dabros M, Zhao C, Cherney RJ, Carter P, Stamos D, Abagyan R, Cherezov V, Stevens RC, Ijzerman AP, Heitman LH, Tebben A, Kufareva I, Handel TM (2016) Structure of CC chemokine receptor 2 with orthosteric and allosteric antagonists. Nature 540 (7633):458–461. https://doi.org/10.1038/nature20605
28. Siu FY, He M, de Graaf C, Han GW, Yang D, Zhang Z, Zhou C, Xu Q, Wacker D, Joseph JS, Liu W, Lau J, Cherezov V, Katritch V, Wang MW, Stevens RC (2013) Structure of the human glucagon class B G-protein-coupled receptor. Nature 499(7459):444–449. https://doi.org/10.1038/nature12393
29. Jazayeri A, Dore AS, Lamb D, Krishnamurthy H, Southall SM, Baig AH, Bortolato A, Koglin M, Robertson NJ, Errey JC, Andrews SP, Teobald I, Brown AJ, Cooke RM, Weir M, Marshall FH (2016) Extra-helical binding site of a glucagon receptor antagonist. Nature 533(7602):274–277. https://doi.org/10.1038/nature17414
30. Hollenstein K, Kean J, Bortolato A, Cheng RK, Dore AS, Jazayeri A, Cooke RM, Weir M, Marshall FH (2013) Structure of class B GPCR corticotropin-releasing factor receptor 1. Nature 499(7459):438–443. https://doi.org/10.1038/nature12357
31. Dore AS, Bortolato A, Hollenstein K, Cheng RK, Read RJ, Marshall FH (2017) Decoding Corticotropin-releasing factor receptor type 1 crystal structures. Curr Mol Pharmacol. https://doi.org/10.2174/1874467210666170110114727
32. Wu H, Wang C, Gregory KJ, Han GW, Cho HP, Xia Y, Niswender CM, Katritch V, Meiler J, Cherezov V, Conn PJ, Stevens RC (2014) Structure of a class C GPCR metabotropic glutamate receptor 1 bound to an allosteric modulator. Science 344(6179):58–64. https://doi.org/10.1126/science.1249489
33. Dore AS, Okrasa K, Patel JC, Serrano-Vega M, Bennett K, Cooke RM, Errey JC, Jazayeri A, Khan S, Tehan B, Weir M, Wiggin GR, Marshall FH (2014) Structure of class C GPCR metabotropic glutamate receptor 5 transmembrane domain. Nature 511 (7511):557–562. https://doi.org/10.1038/nature13396
34. Christopher JA, Aves SJ, Bennett KA, Dore AS, Errey JC, Jazayeri A, Marshall FH, Okrasa K, Serrano-Vega MJ, Tehan BG, Wiggin GR, Congreve M (2015) Fragment and structure-based drug discovery for a class C GPCR: discovery of the mGlu5 negative allosteric modulator HTL14242 (3-Chloro-5-[6-(5-fluoropyridin-2-yl)pyrimidin-4-yl]benzonitrile). J Med Chem 58(16):6653–6664. https://doi.org/10.1021/acs.jmedchem.5b00892

35. Wang C, Wu H, Katritch V, Han GW, Huang XP, Liu W, Siu FY, Roth BL, Cherezov V, Stevens RC (2013) Structure of the human smoothened receptor bound to an antitumour agent. Nature 497(7449):338–343. https://doi.org/10.1038/nature12167
36. Weierstall U, James D, Wang C, White TA, Wang D, Liu W, Spence JC, Bruce Doak R, Nelson G, Fromme P, Fromme R, Grotjohann I, Kupitz C, Zatsepin NA, Liu H, Basu S, Wacker D, Han GW, Katritch V, Boutet S, Messerschmidt M, Williams GJ, Koglin JE, Marvin Seibert M, Klinker M, Gati C, Shoeman RL, Barty A, Chapman HN, Kirian RA, Beyerlein KR, Stevens RC, Li D, Shah ST, Howe N, Caffrey M, Cherezov V (2014) Lipidic cubic phase injector facilitates membrane protein serial femtosecond crystallography. Nat Commun 5:3309. https://doi.org/10.1038/ncomms4309
37. Byrne EF, Sircar R, Miller PS, Hedger G, Luchetti G, Nachtergaele S, Tully MD, Mydock-McGrane L, Covey DF, Rambo RP, Sansom MS, Newstead S, Rohatgi R, Siebold C (2016) Structural basis of smoothened regulation by its extracellular domains. Nature 535(7613):517–522. https://doi.org/10.1038/nature18934
38. Venkatakrishnan AJ, Deupi X, Lebon G, Tate CG, Schertler GF, Babu MM (2013) Molecular signatures of G-protein-coupled receptors. Nature 494(7436):185–194. https://doi.org/10.1038/nature11896
39. Piscitelli CL, Kean J, de Graaf C, Deupi X (2015) A molecular Pharmacologist's guide to G protein-coupled receptor crystallography. Mol Pharmacol 88(3):536–551. https://doi.org/10.1124/mol.115.099663
40. Munk C, Harpsoe K, Hauser AS, Isberg V, Gloriam DE (2016) Integrating structural and mutagenesis data to elucidate GPCR ligand binding. Curr Opin Pharmacol 30:51–58. https://doi.org/10.1016/j.coph.2016.07.003
41. Katritch V, Fenalti G, Abola EE, Roth BL, Cherezov V, Stevens RC (2014) Allosteric sodium in class a GPCR signaling. Trends Biochem Sci 39(5):233–244. https://doi.org/10.1016/j.tibs.2014.03.002
42. Gimpl G (2016) Interaction of G protein coupled receptors and cholesterol. Chem Phys Lipids 199:61–73. https://doi.org/10.1016/j.chemphyslip.2016.04.006
43. Tehan BG, Bortolato A, Blaney FE, Weir MP, Mason JS (2014) Unifying family a GPCR theories of activation. Pharmacol Ther 143 (1):51–60. https://doi.org/10.1016/j.pharmthera.2014.02.004
44. Lagerstrom MC, Schioth HB (2008) Structural diversity of G protein-coupled receptors and significance for drug discovery. Nat Rev Drug Discov 7(4):339–357. https://doi.org/10.1038/nrd2518
45. Conn PJ, Christopoulos A, Lindsley CW (2009) Allosteric modulators of GPCRs: a novel approach for the treatment of CNS disorders. Nat Rev Drug Discov 8(1):41–54. https://doi.org/10.1038/nrd2760
46. Isberg V, de Graaf C, Bortolato A, Cherezov V, Katritch V, Marshall FH, Mordalski S, Pin JP, Stevens RC, Vriend G, Gloriam DE (2015) Generic GPCR residue numbers – aligning topology maps while minding the gaps. Trends Pharmacol Sci 36 (1):22–31. https://doi.org/10.1016/j.tips.2014.11.001
47. Gloriam DE, Foord SM, Blaney FE, Garland SL (2009) Definition of the G protein-coupled receptor transmembrane bundle binding pocket and calculation of receptor similarities for drug design. J Med Chem 52 (14):4429–4442. https://doi.org/10.1021/jm900319e
48. Kooistra AJ, Kuhne S, de Esch IJ, Leurs R, de Graaf C (2013) A structural chemogenomics analysis of aminergic GPCRs: lessons for histamine receptor ligand design. Br J Pharmacol 170(1):101–126. https://doi.org/10.1111/bph.12248
49. Ballesteros JA, Weinstein H (1995) Integrated methods for the construction of three-dimensional models and computational probing of structure-function relations in G protein-coupled receptors. Methods Neurosci 25:366–428
50. Oliveira L, Paiva ACM, Vriend GJ (1993) A common motif in G-protein-coupled seven transmembrane helix receptors. J Comput Aided Mol Des 7(6):649–658
51. Baldwin JM (1993) The probable arrangement of the helices in G protein-coupled receptors. EMBO J 12(4):1693–1703
52. Baldwin JM, Schertler GF, Unger VM (1997) An alpha-carbon template for the transmembrane helices in the rhodopsin family of G-protein-coupled receptors. J Mol Biol 272 (1):144–164. https://doi.org/10.1006/jmbi.1997.1240
53. Schwartz TW (1994) Locating ligand-binding sites in 7TM receptors by protein engineering. Curr Opin Biotechnol 5 (4):434–444

54. Schwartz TW, Gether U, Schambye HT, Hjorth SA (1995) Molecular mechanism of action of non-peptide ligands for peptide receptors. Curr Pharm Des 1(3):325–342
55. Henderson R, Baldwin JM, Ceska TA, Zemlin F, Beckmann E, Downing KH (1990) Model for the structure of bacteriorhodopsin based on high-resolution electron cryo-microscopy. J Mol Biol 213 (4):899–929. https://doi.org/10.1016/S0022-2836(05)80271-2
56. Schertler GF, Villa C, Henderson R (1993) Projection structure of rhodopsin. Nature 362(6422):770–772. https://doi.org/10.1038/362770a0
57. Unger VM, Hargrave PA, Baldwin JM, Schertler GF (1997) Arrangement of rhodopsin transmembrane alpha-helices. Nature 389 (6647):203–206. https://doi.org/10.1038/38316
58. Wootten D, Simms J, Miller LJ, Christopoulos A, Sexton PM (2013) Polar transmembrane interactions drive formation of ligand-specific and signal pathway-biased family B G protein-coupled receptor conformations. Proc Natl Acad Sci U S A 110 (13):5211–5216. https://doi.org/10.1073/pnas.1221585110
59. Pin JP, Galvez T, Prezeau L (2003) Evolution, structure, and activation mechanism of family 3/C G-protein-coupled receptors. Pharmacol Ther 98(3):325–354
60. de Graaf C, Nijmeijer S, Wolf S, Ernst OP (2016) 7TM Domain structure of adhesion GPCRs. Handb Exp Pharmacol 234:43–66. https://doi.org/10.1007/978-3-319-41523-9_3
61. Trzaskowski B, Latek D, Yuan S, Ghoshdastider U, Debinski A, Filipek S (2012) Action of molecular switches in GPCRs – theoretical and experimental studies. Curr Med Chem 19(8):1090–1109
62. Venkatakrishnan AJ, Deupi X, Lebon G, Heydenreich FM, Flock T, Miljus T, Balaji S, Bouvier M, Veprintsev DB, Tate CG, Schertler GF, Babu MM (2016) Diverse activation pathways in class a GPCRs converge near the G-protein-coupling region. Nature 536 (7617):484–487. https://doi.org/10.1038/nature19107
63. Koth CM, Murray JM, Mukund S, Madjidi A, Minn A, Clarke HJ, Wong T, Chiang V, Luis E, Estevez A, Rondon J, Zhang Y, Hotzel I, Allan BB (2012) Molecular basis for negative regulation of the glucagon receptor. Proc Natl Acad Sci U S A 109 (36):14393–14398. https://doi.org/10.1073/pnas.1206734109
64. Kunishima N, Shimada Y, Tsuji Y, Sato T, Yamamoto M, Kumasaka T, Nakanishi S, Jingami H, Morikawa K (2000) Structural basis of glutamate recognition by a dimeric metabotropic glutamate receptor. Nature 407(6807):971–977. https://doi.org/10.1038/35039564
65. Muto T, Tsuchiya D, Morikawa K, Jingami H (2007) Structures of the extracellular regions of the group II/III metabotropic glutamate receptors. Proc Natl Acad Sci U S A 104 (10):3759–3764. https://doi.org/10.1073/pnas.0611577104
66. Isberg V, Mordalski S, Munk C, Rataj K, Harpsoe K, Hauser AS, Vroling B, Bojarski AJ, Vriend G, Gloriam DE (2016) GPCRdb: an information system for G protein-coupled receptors. Nucleic Acids Res 44(D1): D356–D364. https://doi.org/10.1093/nar/gkv1178
67. Munk C, Isberg V, Mordalski S, Harpsoe K, Rataj K, Hauser AS, Kolb P, Bojarski AJ, Vriend G, Gloriam DE (2016) GPCRdb: the G protein-coupled receptor database – an introduction. Br J Pharmacol 173 (14):2195–2207. https://doi.org/10.1111/bph.13509
68. Lee J, Sands ZA, Biggin PC (2016) A numbering system for MFS transporter proteins. Front Mol Biosci 3:21. https://doi.org/10.3389/fmolb.2016.00021
69. Randolph AL, Mokrab Y, Bennett AL, Sansom MS, Ramsey IS (2016) Proton currents constrain structural models of voltage sensor activation. elife 5. https://doi.org/10.7554/eLife.18017
70. Berthod M, Cebron N, Dill F, Gabriel T, Kötter T, Meinl T, Ohl P, Sieb C, Thiel K, Wiswedel B (2007) KNIME: the Konstanz information miner. Studies in classification, data analysis, and knowledge organization. Springer, Berlin, Heidelberg
71. Verhoeven S, Kooistra AJ, Vass M, McGuire R, Ritschel T, de Graaf C (2017) 3D-e-Chem/knime-gpcrdb: v1.1.0. Zenodo. http://doi.org/10.5281/zenodo.240491
72. McGuire R, Verhoeven S, Vass M, Vriend G, De Esch IJ, Lusher SJ, Leurs R, Ridder L, Kooistra AJ, Ritschel T, de Graaf C (2017) 3D-E-Chem-VM: structural cheminformatics research infrastructure in a freely available virtual machine. J Chem Inf Model. https://doi.org/10.1021/acs.jcim.6b00686
73. Bondensgaard K, Ankersen M, Thogersen H, Hansen BS, Wulff BS, Bywater RP (2004) Recognition of privileged structures by G-protein coupled receptors. J Med Chem

47(4):888–899. https://doi.org/10.1021/jm0309452

74. Frimurer TM, Ulven T, Elling CE, Gerlach LO, Kostenis E, Hogberg T (2005) A physicogenetic method to assign ligand-binding relationships between 7TM receptors. Bioorg Med Chem Lett 15(16):3707–3712. https://doi.org/10.1016/j.bmcl.2005.05.102
75. Frimurer TM, Hogberg T (2011) Drug design of GPCR ligands using physicogenetics and chemogenomics – principles and case studies. Curr Top Med Chem 11 (15):1882–1901
76. Zhang H, Unal H, Desnoyer R, Han GW, Patel N, Katritch V, Karnik SS, Cherezov V, Stevens RC (2015) Structural basis for ligand recognition and functional selectivity at angiotensin receptor. J Biol Chem 290 (49):29127–29139. https://doi.org/10.1074/jbc.M115.689000
77. Zhang H, Unal H, Gati C, Han GW, Liu W, Zatsepin NA, James D, Wang D, Nelson G, Weierstall U, Sawaya MR, Xu Q, Messerschmidt M, Williams GJ, Boutet S, Yefanov OM, White TA, Wang C, Ishchenko A, Tirupula KC, Desnoyer R, Coe J, Conrad CE, Fromme P, Stevens RC, Katritch V, Karnik SS, Cherezov V (2015) Structure of the angiotensin receptor revealed by serial femtosecond crystallography. Cell 161(4):833–844. https://doi.org/10.1016/j.cell.2015.04.011
78. Receveur JM, Bjurling E, Ulven T, Little PB, Norregaard PK, Hogberg T (2004) 4-Acylamino-and 4-ureidobenzamides as melanin-concentrating hormone (MCH) receptor 1 antagonists. Bioorg Med Chem Lett 14 (20):5075–5080. https://doi.org/10.1016/j.bmcl.2004.07.077
79. Kratochwil NA, Malherbe P, Lindemann L, Ebeling M, Hoener MC, Muhlemann A, Porter RH, Stahl M, Gerber PR (2005) An automated system for the analysis of G protein-coupled receptor transmembrane binding pockets: alignment, receptor-based pharmacophores, and their application. J Chem Inf Model 45(5):1324–1336. https://doi.org/10.1021/ci050221u
80. Malherbe P, Kratochwil N, Muhlemann A, Zenner MT, Fischer C, Stahl M, Gerber PR, Jaeschke G, Porter RH (2006) Comparison of the binding pockets of two chemically unrelated allosteric antagonists of the mGlu5 receptor and identification of crucial residues involved in the inverse agonism of MPEP. J Neurochem 98(2):601–615. https://doi.org/10.1111/j.1471-4159.2006.03886.x
81. Cherezov V, Rosenbaum DM, Hanson MA, Rasmussen SG, Thian FS, Kobilka TS, Choi HJ, Kuhn P, Weis WI, Kobilka BK, Stevens RC (2007) High-resolution crystal structure of an engineered human beta2-adrenergic G protein-coupled receptor. Science 318 (5854):1258–1265. https://doi.org/10.1126/science.1150577
82. Chien EY, Liu W, Zhao Q, Katritch V, Han GW, Hanson MA, Shi L, Newman AH, Javitch JA, Cherezov V, Stevens RC (2010) Structure of the human dopamine D3 receptor in complex with a D2/D3 selective antagonist. Science 330(6007):1091–1095. https://doi.org/10.1126/science.1197410
83. Chrencik JE, Roth CB, Terakado M, Kurata H, Omi R, Kihara Y, Warshaviak D, Nakade S, Asmar-Rovira G, Mileni M, Mizuno H, Griffith MT, Rodgers C, Han GW, Velasquez J, Chun J, Stevens RC, Hanson MA (2015) Crystal structure of antagonist bound human Lysophosphatidic acid receptor 1. Cell 161(7):1633–1643. https://doi.org/10.1016/j.cell.2015.06.002
84. Christopher JA, Brown J, Dore AS, Errey JC, Koglin M, Marshall FH, Myszka DG, Rich RL, Tate CG, Tehan B, Warne T, Congreve M (2013) Biophysical fragment screening of the beta1-adrenergic receptor: identification of high affinity arylpiperazine leads using structure-based drug design. J Med Chem 56(9):3446–3455. https://doi.org/10.1021/jm400140q
85. Congreve M, Andrews SP, Dore AS, Hollenstein K, Hurrell E, Langmead CJ, Mason JS, Ng IW, Tehan B, Zhukov A, Weir M, Marshall FH (2012) Discovery of 1,2,4-triazine derivatives as adenosine a (2A) antagonists using structure based drug design. J Med Chem 55(5):1898–1903. https://doi.org/10.1021/jm201376w
86. Dore AS, Robertson N, Errey JC, Ng I, Hollenstein K, Tehan B, Hurrell E, Bennett K, Congreve M, Magnani F, Tate CG, Weir M, Marshall FH (2011) Structure of the adenosine a(2A) receptor in complex with ZM241385 and the xanthines XAC and caffeine. Structure 19(9):1283–1293. https://doi.org/10.1016/j.str.2011.06.014
87. Fenalti G, Zatsepin NA, Betti C, Giguere P, Han GW, Ishchenko A, Liu W, Guillemyn K, Zhang H, James D, Wang D, Weierstall U, Spence JC, Boutet S, Messerschmidt M, Williams GJ, Gati C, Yefanov OM, White TA, Oberthuer D, Metz M, Yoon CH, Barty A, Chapman HN, Basu S, Coe J, Conrad CE, Fromme R, Fromme P, Tourwe D, Schiller PW, Roth BL, Ballet S, Katritch V, Stevens

RC, Cherezov V (2015) Structural basis for bifunctional peptide recognition at human delta-opioid receptor. Nat Struct Mol Biol 22(3):265–268. https://doi.org/10.1038/nsmb.2965

88. Glukhova A, Thal DM, Nguyen AT, Vecchio EA, Jorg M, Scammells PJ, May LT, Sexton PM, Christopoulos A (2017) Structure of the adenosine A1 receptor reveals the basis for subtype selectivity. Cell 168(5):867–877.e813. https://doi.org/10.1016/j.cell.2017.01.042
89. Granier S, Manglik A, Kruse AC, Kobilka TS, Thian FS, Weis WI, Kobilka BK (2012) Structure of the delta-opioid receptor bound to naltrindole. Nature 485(7398):400–404. https://doi.org/10.1038/nature11111
90. Haga K, Kruse AC, Asada H, Yurugi-Kobayashi T, Shiroishi M, Zhang C, Weis WI, Okada T, Kobilka BK, Haga T, Kobayashi T (2012) Structure of the human M2 muscarinic acetylcholine receptor bound to an antagonist. Nature 482(7386):547–551. https://doi.org/10.1038/nature10753
91. Hanson MA, Cherezov V, Griffith MT, Roth CB, Jaakola VP, Chien EY, Velasquez J, Kuhn P, Stevens RC (2008) A specific cholesterol binding site is established by the 2.8 A structure of the human beta2-adrenergic receptor. Structure 16(6):897–905. https://doi.org/10.1016/j.str.2008.05.001
92. Hanson MA, Roth CB, Jo E, Griffith MT, Scott FL, Reinhart G, Desale H, Clemons B, Cahalan SM, Schuerer SC, Sanna MG, Han GW, Kuhn P, Rosen H, Stevens RC (2012) Crystal structure of a lipid G protein-coupled receptor. Science 335(6070):851–855. https://doi.org/10.1126/science.1215904
93. Hua T, Vemuri K, Pu M, Qu L, Han GW, Wu Y, Zhao S, Shui W, Li S, Korde A, Laprairie RB, Stahl EL, Ho JH, Zvonok N, Zhou H, Kufareva I, Wu B, Zhao Q, Hanson MA, Bohn LM, Makriyannis A, Stevens RC, Liu ZJ (2016) Crystal structure of the human cannabinoid receptor CB1. Cell 167(3):750–762.e714. https://doi.org/10.1016/j.cell.2016.10.004
94. Huang W, Manglik A, Venkatakrishnan AJ, Laeremans T, Feinberg EN, Sanborn AL, Kato HE, Livingston KE, Thorsen TS, Kling RC, Granier S, Gmeiner P, Husbands SM, Traynor JR, Weis WI, Steyaert J, Dror RO, Kobilka BK (2015) Structural insights into micro-opioid receptor activation. Nature 524(7565):315–321. https://doi.org/10.1038/nature14886
95. Jaakola VP, Griffith MT, Hanson MA, Cherezov V, Chien EY, Lane JR, Ijzerman AP, Stevens RC (2008) The 2.6 Angstrom crystal structure of a human A2A adenosine receptor bound to an antagonist. Science 322(5905):1211–1217. https://doi.org/10.1126/science.1164772
96. Kruse AC, Hu J, Pan AC, Arlow DH, Rosenbaum DM, Rosemond E, Green HF, Liu T, Chae PS, Dror RO, Shaw DE, Weis WI, Wess J, Kobilka BK (2012) Structure and dynamics of the M3 muscarinic acetylcholine receptor. Nature 482(7386):552–556. https://doi.org/10.1038/nature10867
97. Kruse AC, Ring AM, Manglik A, Hu J, Hu K, Eitel K, Hubner H, Pardon E, Valant C, Sexton PM, Christopoulos A, Felder CC, Gmeiner P, Steyaert J, Weis WI, Garcia KC, Wess J, Kobilka BK (2013) Activation and allosteric modulation of a muscarinic acetylcholine receptor. Nature 504(7478):101–106. https://doi.org/10.1038/nature12735
98. Lebon G, Edwards PC, Leslie AG, Tate CG (2015) Molecular determinants of CGS21680 binding to the human adenosine A2A receptor. Mol Pharmacol 87(6):907–915. https://doi.org/10.1124/mol.114.097360
99. Lebon G, Warne T, Edwards PC, Bennett K, Langmead CJ, Leslie AG, Tate CG (2011) Agonist-bound adenosine A2A receptor structures reveal common features of GPCR activation. Nature 474(7352):521–525. https://doi.org/10.1038/nature10136
100. Manglik A, Kruse AC, Kobilka TS, Thian FS, Mathiesen JM, Sunahara RK, Pardo L, Weis WI, Kobilka BK, Granier S (2012) Crystal structure of the micro-opioid receptor bound to a morphinan antagonist. Nature 485(7398):321–326. https://doi.org/10.1038/nature10954
101. Miller RL, Thompson AA, Trapella C, Guerrini R, Malfacini D, Patel N, Han GW, Cherezov V, Calo G, Katritch V, Stevens RC (2015) The importance of ligand-receptor conformational pairs in stabilization: spotlight on the N/OFQ G protein-coupled receptor. Structure 23(12):2291–2299. https://doi.org/10.1016/j.str.2015.07.024
102. Moukhametzianov R, Warne T, Edwards PC, Serrano-Vega MJ, Leslie AG, Tate CG, Schertler GF (2011) Two distinct conformations of helix 6 observed in antagonist-bound structures of a beta1-adrenergic receptor. Proc Natl Acad Sci U S A 108(20):8228–8232. https://doi.org/10.1073/pnas.1100185108
103. Oswald C, Rappas M, Kean J, Dore AS, Errey JC, Bennett K, Deflorian F, Christopher JA,

Jazayeri A, Mason JS, Congreve M, Cooke RM, Marshall FH (2016) Intracellular allosteric antagonism of the CCR9 receptor. Nature 540(7633):462–465. https://doi.org/10.1038/nature20606

104. Ring AM, Manglik A, Kruse AC, Enos MD, Weis WI, Garcia KC, Kobilka BK (2013) Adrenaline-activated structure of beta2-adrenoceptor stabilized by an engineered nanobody. Nature 502(7472):575–579. https://doi.org/10.1038/nature12572
105. Rosenbaum DM, Zhang C, Lyons JA, Holl R, Aragao D, Arlow DH, Rasmussen SG, Choi HJ, Devree BT, Sunahara RK, Chae PS, Gellman SH, Dror RO, Shaw DE, Weis WI, Caffrey M, Gmeiner P, Kobilka BK (2011) Structure and function of an irreversible agonist-beta(2) adrenoceptor complex. Nature 469(7329):236–240. https://doi.org/10.1038/nature09665
106. Sato T, Baker J, Warne T, Brown GA, Leslie AG, Congreve M, Tate CG (2015) Pharmacological analysis and structure determination of 7-Methylcyanopindolol-bound beta1-adrenergic receptor. Mol Pharmacol 88(6):1024–1034. https://doi.org/10.1124/mol.115.101030
107. Segala E, Guo D, Cheng RK, Bortolato A, Deflorian F, Dore AS, Errey JC, Heitman LH, Ijzerman AP, Marshall FH, Cooke RM (2016) Controlling the dissociation of ligands from the adenosine A2A receptor through modulation of salt bridge strength. J Med Chem 59(13):6470–6479. https://doi.org/10.1021/acs.jmedchem.6b00653
108. Shao Z, Yin J, Chapman K, Grzemska M, Clark L, Wang J, Rosenbaum DM (2016) High-resolution crystal structure of the human CB1 cannabinoid receptor. Nature. https://doi.org/10.1038/nature20613
109. Srivastava A, Yano J, Hirozane Y, Kefala G, Gruswitz F, Snell G, Lane W, Ivetac A, Aertgeerts K, Nguyen J, Jennings A, Okada K (2014) High-resolution structure of the human GPR40 receptor bound to allosteric agonist TAK-875. Nature 513(7516):124–127. https://doi.org/10.1038/nature13494
110. Tan Q, Zhu Y, Li J, Chen Z, Han GW, Kufareva I, Li T, Ma L, Fenalti G, Li J, Zhang W, Xie X, Yang H, Jiang H, Cherezov V, Liu H, Stevens RC, Zhao Q, Wu B (2013) Structure of the CCR5 chemokine receptor-HIV entry inhibitor maraviroc complex. Science 341(6152):1387–1390. https://doi.org/10.1126/science.1241475
111. Thal DM, Sun B, Feng D, Nawaratne V, Leach K, Felder CC, Bures MG, Evans DA, Weis WI, Bachhawat P, Kobilka TS, Sexton PM, Kobilka BK, Christopoulos A (2016) Crystal structures of the M1 and M4 muscarinic acetylcholine receptors. Nature 531(7594):335–340. https://doi.org/10.1038/nature17188
112. Thompson AA, Liu W, Chun E, Katritch V, Wu H, Vardy E, Huang XP, Trapella C, Guerrini R, Calo G, Roth BL, Cherezov V, Stevens RC (2012) Structure of the nociceptin/orphanin FQ receptor in complex with a peptide mimetic. Nature 485(7398):395–399. https://doi.org/10.1038/nature11085
113. Thorsen TS, Matt R, Weis WI, Kobilka BK (2014) Modified T4 lysozyme fusion proteins facilitate G protein-coupled receptor crystallogenesis. Structure 22(11):1657–1664. https://doi.org/10.1016/j.str.2014.08.022
114. Wacker D, Fenalti G, Brown MA, Katritch V, Abagyan R, Cherezov V, Stevens RC (2010) Conserved binding mode of human beta2 adrenergic receptor inverse agonists and antagonist revealed by X-ray crystallography. J Am Chem Soc 132(33):11443–11445. https://doi.org/10.1021/ja105108q
115. Wacker D, Wang S, McCorvy JD, Betz RM, Venkatakrishnan AJ, Levit A, Lansu K, Schools ZL, Che T, Nichols DE, Shoichet BK, Dror RO, Roth BL (2017) Crystal structure of an LSD-bound human serotonin receptor. Cell 168(3):377–389. e312. https://doi.org/10.1016/j.cell.2016.12.033
116. Wang C, Jiang Y, Ma J, Wu H, Wacker D, Katritch V, Han GW, Liu W, Huang XP, Vardy E, McCorvy JD, Gao X, Zhou XE, Melcher K, Zhang C, Bai F, Yang H, Yang L, Jiang H, Roth BL, Cherezov V, Stevens RC, HE X (2013) Structural basis for molecular recognition at serotonin receptors. Science 340(6132):610–614. https://doi.org/10.1126/science.1232807
117. Warne T, Edwards PC, Leslie AG, Tate CG (2012) Crystal structures of a stabilized beta1-adrenoceptor bound to the biased agonists bucindolol and carvedilol. Structure 20(5):841–849. https://doi.org/10.1016/j.str.2012.03.014
118. Warne T, Serrano-Vega MJ, Baker JG, Moukhametzianov R, Edwards PC, Henderson R, Leslie AG, Tate CG, Schertler GF (2008) Structure of a beta1-adrenergic G-protein-coupled receptor. Nature 454(7203):486–491. https://doi.org/10.1038/nature07101
119. Weichert D, Kruse AC, Manglik A, Hiller C, Zhang C, Hubner H, Kobilka BK, Gmeiner P

(2014) Covalent agonists for studying G protein-coupled receptor activation. Proc Natl Acad Sci U S A 111(29):10744–10748. https://doi.org/10.1073/pnas.1410415111

120. Wu H, Wacker D, Mileni M, Katritch V, Han GW, Vardy E, Liu W, Thompson AA, Huang XP, Carroll FI, Mascarella SW, Westkaemper RB, Mosier PD, Roth BL, Cherezov V, Stevens RC (2012) Structure of the human kappa-opioid receptor in complex with JDTic. Nature 485(7398):327–332. https://doi.org/10.1038/nature10939
121. Xu F, Wu H, Katritch V, Han GW, Jacobson KA, Gao ZG, Cherezov V, Stevens RC (2011) Structure of an agonist-bound human A2A adenosine receptor. Science 332 (6027):322–327. https://doi.org/10.1126/science.1202793
122. Yin J, Babaoglu K, Brautigam CA, Clark L, Shao Z, Scheuermann TH, Harrell CM, Gotter AL, Roecker AJ, Winrow CJ, Renger JJ, Coleman PJ, Rosenbaum DM (2016) Structure and ligand-binding mechanism of the human OX1 and OX2 orexin receptors. Nat Struct Mol Biol 23(4):293–299. https://doi.org/10.1038/nsmb.3183
123. Yin J, Mobarec JC, Kolb P, Rosenbaum DM (2015) Crystal structure of the human OX2 orexin receptor bound to the insomnia drug suvorexant. Nature 519(7542):247–250. https://doi.org/10.1038/nature14035
124. Zhang C, Srinivasan Y, Arlow DH, Fung JJ, Palmer D, Zheng Y, Green HF, Pandey A, Dror RO, Shaw DE, Weis WI, Coughlin SR, Kobilka BK (2012) High-resolution crystal structure of human protease-activated receptor 1. Nature 492(7429):387–392. https://doi.org/10.1038/nature11701
125. Zhang D, Gao ZG, Zhang K, Kiselev E, Crane S, Wang J, Paoletta S, Yi C, Ma L, Zhang W, Han GW, Liu H, Cherezov V, Katritch V, Jiang H, Stevens RC, Jacobson KA, Zhao Q, Wu B (2015) Two disparate ligand-binding sites in the human P2Y1 receptor. Nature 520(7547):317–321. https://doi.org/10.1038/nature14287
126. Zhang J, Zhang K, Gao ZG, Paoletta S, Zhang D, Han GW, Li T, Ma L, Zhang W, Muller CE, Yang H, Jiang H, Cherezov V, Katritch V, Jacobson KA, Stevens RC, Wu B, Zhao Q (2014) Agonist-bound structure of the human P2Y12 receptor. Nature 509 (7498):119–122. https://doi.org/10.1038/nature13288
127. Zhang K, Zhang J, Gao ZG, Zhang D, Zhu L, Han GW, Moss SM, Paoletta S, Kiselev E, Lu W, Fenalti G, Zhang W, Muller CE, Yang H, Jiang H, Cherezov V, Katritch V, Jacobson KA, Stevens RC, Wu B, Zhao Q (2014) Structure of the human P2Y12 receptor in complex with an antithrombotic drug. Nature 509(7498):115–118. https://doi.org/10.1038/nature13083
128. Liu X, Ahn S, Kahsai AW, Meng KC, Latorraca NR, Pani B, Venkatakrishnan AJ, Masoudi A, Weis W, Dror RO, Chen X, Lefkowitz RJ, Kobilka BK (2017) Mechanism of intracellular allosteric β 2AR antagonist revealed by X-ray crystal structure. Nature 548(7668):480–484. https://doi.org/10.1038/nature23652
129. Cheng RKY, Fiez-Vandal C, Schlenker O, Edman K, Aggeler B, Brown DG, Brown GA, Cooke RM, Dumelin CE, Doré AS, Geschwindner S, Grebner C, Hermansson NO, Jazayeri A, Johansson P, Leong L, Prihandoko R, Rappas M, Soutter H, Snijder A, Sundström L, Tehan B, Thornton P, Troast D, Wiggin G, Zhukov A, Marshall FH, Dekker N (2017) Structural insight into allosteric modulation of proteaseactivated receptor 2. Nature 545(7652):112–115. https://doi.org/10.1038/nature22309
130. Lu J, Byrne N, Wang J, Bricogne G, Brown FK, Chobanian HR, Colletti SL, Di Salvo J, Thomas-Fowlkes B, Guo Y, Hall DL, Hadix J, Hastings NB, Hermes JD, Ho T, Howard AD, Josien H, Kornienko M, Lumb KJ, Miller MW, Patel SB, Pio B, Plummer CW, Sherborne BS, Sheth P, Souza S, Tummala S, Vonrhein C, Webb M, Allen SJ, Johnston JM, Weinglass AB, Sharma S, Soisson SM (2017) Structural basis for the cooperative allosteric activation of the free fatty acid receptor GPR40. Nat Struct Mol Biol 24 (7):570–577. https://doi.org/10.1038/nsmb.3417
131. Hua T, Vemuri K, Nikas SP, Laprairie RB, Wu Y, Qu L, Pu M, Korde A, Jiang S, Ho JH, Han GW, Ding K, Li X, Liu H, Hanson MA, Zhao S, Bohn LM, Makriyannis A, Stevens RC, Liu ZJ (2017) Crystal structures of agonist-bound human cannabinoid receptor CB1. Nature 547(7664):468–471. https://doi.org/10.1038/nature23272
132. Shihoya W, Nishizawa T, Yamashita K, Inoue A, Hirata K, Kadji FMN, Okuta A, Tani K, Aoki J, Fujiyoshi Y, Doi T, Nureki O (2017) X-ray structures of endothelin ETB receptor bound to clinical antagonist bosentan and its analog. Nat Struct Mol Biol 24(9):758–764. https://doi.org/10.1038/nsmb.3450
133. Zhang H, Han GW, Batyuk A, Ishchenko A, White KL, Patel N, Sadybekov A, Zamlynny B, Rudd MT, Hollenstein K, Tolstikova A,

White TA, Hunter MS, Weierstall U, Liu W, Babaoglu K, Moore EL, Katz RD, Shipman JM, Garcia-Calvo M, Sharma S, Sheth P, Soisson SM, Stevens RC, Katritch V, Cherezov V (2017) Structural basis for selectivity and diversity in angiotensin II receptors. Nature 544(7650):327–332. https://doi.org/10.1038/nature22035

134. Zhang H, Qiao A, Yang D, Yang L, Dai A, de Graaf C, Reedtz-Runge S, Dharmarajan V, Zhang H, Han GW, Grant TD, Sierra RG, Weierstall U, Nelson G, Liu W, Wu Y, Ma L, Cai X, Lin G, Wu X, Geng Z, Dong Y, Song G, Griffin PR, Lau J, Cherezov V, Yang H, Hanson MA, Stevens RC, Zhao Q, Jiang H, Wang MW, Wu B (2017) Structure of the full-length glucagon class B G-protein-coupled receptor. Nature 546(7657):259–264. https://doi.org/10.1038/nature22363

135. Song G, Yang D, Wang Y, de Graaf C, Zhou Q, Jiang S, Liu K, Cai X, Dai A, Lin G, Liu D, Wu F, Wu Y, Zhao S, Ye L, Han GW, Lau J, Wu B, Hanson MA, Liu ZJ, Wang MW, Stevens RC (2017) Human GLP- 1 receptor transmembrane domain structure in complex with allosteric modulators. Nature 546 (7657):312–315. https://doi.org/10.1038/nature22378

136. Surgand JS, Rodrigo J, Kellenberger E, Rognan D (2006) A chemogenomic analysis of the transmembrane binding cavity of human G-protein-coupled receptors. Proteins 62 (2):509–538. https://doi.org/10.1002/prot.20768

137. Weill N, Rognan D (2009) Development and validation of a novel protein-ligand fingerprint to mine chemogenomic space: application to G protein-coupled receptors and their ligands. J Chem Inf Model 49 (4):1049–1062. https://doi.org/10.1021/ci800447g

138. Nohr AC, Shehata MA, Hauser AS, Isberg V, Mokrosinski J, Andersen KB, Farooqi IS, Pedersen DS, Gloriam DE, Brauner-Osborne H (2017) The orphan G protein-coupled receptor GPR139 is activated by the peptides: adrenocorticotropic hormone (ACTH), alpha-, and beta-melanocyte stimulating hormone (alpha-MSH, and beta-MSH), and the conserved core motif HFRW. Neurochem Int 102:105–113. https://doi.org/10.1016/j.neuint.2016.11.012

139. Wellendorph P, Hansen KB, Balsgaard A, Greenwood JR, Egebjerg J, Brauner-Osborne H (2005) Deorphanization of GPRC6A: a promiscuous L-alpha-amino acid receptor with preference for basic amino acids. Mol Pharmacol 67(3):589–597. https://doi.org/10.1124/mol.104.007559

140. Faure H, Gorojankina T, Rice N, Dauban P, Dodd RH, Brauner-Osborne H, Rognan D, Ruat M (2009) Molecular determinants of non-competitive antagonist binding to the mouse GPRC6A receptor. Cell Calcium 46 (5–6):323–332. https://doi.org/10.1016/j.ceca.2009.09.004

141. Gloriam DE, Wellendorph P, Johansen LD, Thomsen AR, Phonekeo K, Pedersen DS, Brauner-Osborne H (2011) Chemogenomic discovery of allosteric antagonists at the GPRC6A receptor. Chem Biol 18 (11):1489–1498. https://doi.org/10.1016/j.chembiol.2011.09.012

142. van der Horst E, Peironcely JE, Ijzerman AP, Beukers MW, Lane JR, van Vlijmen HW, Emmerich MT, Okuno Y, Bender A (2010) A novel chemogenomics analysis of G protein-coupled receptors (GPCRs) and their ligands: a potential strategy for receptor de-orphanization. BMC Bioinformatics 11:316. https://doi.org/10.1186/1471-2105-11-316

143. Bento AP, Gaulton A, Hersey A, Bellis LJ, Chambers J, Davies M, Kruger FA, Light Y, Mak L, McGlinchey S, Nowotka M, Papadatos G, Santos R, Overington JP (2014) The ChEMBL bioactivity database: an update. Nucleic Acids Res 42(Database issue): D1083–D1090. https://doi.org/10.1093/nar/gkt1031

144. Okuno Y, Tamon A, Yabuuchi H, Niijima S, Minowa Y, Tonomura K, Kunimoto R, Feng C (2008) GLIDA: GPCR – ligand database for chemical genomics drug discovery – database and tools update. Nucleic Acids Res 36 (Database issue):D907–D912. https://doi.org/10.1093/nar/gkm948

145. Roth BL, Lopez E, Beischel S, Westkaemper RB, Evans JM (2004) Screening the receptorome to discover the molecular targets for plant-derived psychoactive compounds: a novel approach for CNS drug discovery. Pharmacol Ther 102(2):99–110. https://doi.org/10.1016/j.pharmthera.2004.03.004

146. Keiser MJ, Roth BL, Armbruster BN, Ernsberger P, Irwin JJ, Shoichet BK (2007) Relating protein pharmacology by ligand chemistry. Nat Biotechnol 25(2):197–206. https://doi.org/10.1038/nbt1284

147. Lin H, Sassano MF, Roth BL, Shoichet BK (2013) A pharmacological organization of G protein-coupled receptors. Nat Methods 10 (2):140–146. https://doi.org/10.1038/nmeth.2324

148. Ngo T, Ilatovskiy AV, Stewart AG, Coleman JL, McRobb FM, Riek RP, Graham RM, Abagyan R, Kufareva I, Smith NJ (2017) Orphan receptor ligand discovery by pick-pocketing pharmacological neighbors. Nat Chem Biol 13(2):235–242. https://doi.org/10.1038/nchembio.2266
149. The UniProt C (2017) UniProt: the universal protein knowledgebase. Nucleic Acids Res 45 (D1):D158–D169. https://doi.org/10.1093/nar/gkw1099
150. Yates A, Akanni W, Amode MR, Barrell D, Billis K, Carvalho-Silva D, Cummins C, Clapham P, Fitzgerald S, Gil L, Giron CG, Gordon L, Hourlier T, Hunt SE, Janacek SH, Johnson N, Juettemann T, Keenan S, Lavidas I, Martin FJ, Maurel T, McLaren W, Murphy DN, Nag R, Nuhn M, Parker A, Patricio M, Pignatelli M, Rahtz M, Riat HS, Sheppard D, Taylor K, Thormann A, Vullo A, Wilder SP, Zadissa A, Birney E, Harrow J, Muffato M, Perry E, Ruffier M, Spudich G, Trevanion SJ, Cunningham F, Aken BL, Zerbino DR, Flicek P (2016) Ensembl 2016. Nucleic Acids Res 44(D1):D710–D716. https://doi.org/10.1093/nar/gkv1157
151. Sanders MP, Fleuren WW, Verhoeven S, van den Beld S, Alkema W, de Vlieg J, Klomp JP (2011) ss-TEA: entropy based identification of receptor specific ligand binding residues from a multiple sequence alignment of class a GPCRs. BMC Bioinformatics 12:332. https://doi.org/10.1186/1471-2105-12-332
152. Sanders MP, Verhoeven S, de Graaf C, Roumen L, Vroling B, Nabuurs SB, de Vlieg J, Klomp JP (2011) Snooker: a structure-based pharmacophore generation tool applied to class a GPCRs. J Chem Inf Model 51(9):2277–2292. https://doi.org/10.1021/ci200088d
153. Sanders MP, Roumen L, van der Horst E, Lane JR, Vischer HF, van Offenbeek J, de Vries H, Verhoeven S, Chow KY, Verkaar F, Beukers MW, McGuire R, Leurs R, Ijzerman AP, de Vlieg J, de Esch IJ, Zaman GJ, Klomp JP, Bender A, de Graaf C (2012) A prospective cross-screening study on G-protein-coupled receptors: lessons learned in virtual compound library design. J Med Chem 55 (11):5311–5325. https://doi.org/10.1021/jm300280e
154. Roland WS, Sanders MP, van Buren L, Gouka RJ, Gruppen H, Vincken JP, Ritschel T (2015) Snooker structure-based pharmacophore model explains differences in agonist and blocker binding to bitter receptor hTAS2R39. PLoS One 10(3):e0118200. https://doi.org/10.1371/journal.pone.0118200
155. Klabunde T, Giegerich C, Evers A (2009) Sequence-derived three-dimensional pharmacophore models for G-protein-coupled receptors and their application in virtual screening. J Med Chem 52(9):2923–2932. https://doi.org/10.1021/jm9001346
156. Fidom K, Isberg V, Hauser AS, Mordalski S, Lehto T, Bojarski AJ, Gloriam DE (2015) A new crystal structure fragment-based pharmacophore method for G protein-coupled receptors. Methods 71:104–112. https://doi.org/10.1016/j.ymeth.2014.09.009
157. Frandsen IO, Boesgaard MW, Fidom K, Hauser AS, Isberg V, Bräuner-Osborne H, Wellendorph P, Gloriam DE (2017) Identification of Histamine H3 Receptor Ligands Using a New Crystal Structure Fragmentbased Method. Sci Rep. 7(1):4829. https://doi.org/10.1038/s41598-017-05058-w.
158. Wichard JD, Ter Laak A, Krause G, Heinrich N, Kuhne R, Kleinau G (2011) Chemogenomic analysis of G-protein coupled receptors and their ligands deciphers locks and keys governing diverse aspects of signalling. PLoS One 6(2):e16811. https://doi.org/10.1371/journal.pone.0016811
159. Kooistra AJ, Leurs R, de Esch IJ, de Graaf C (2015) Structure-based prediction of G-protein-coupled receptor ligand function: a beta-adrenoceptor case study. J Chem Inf Model 55(5):1045–1061. https://doi.org/10.1021/acs.jcim.5b00066
160. Kooistra AJ, Vischer HF, McNaught-Flores-D, Leurs R, de Esch IJ, de Graaf C (2016) Function-specific virtual screening for GPCR ligands using a combined scoring method. Sci Rep 6:28288. https://doi.org/10.1038/srep28288
161. Yang D, de Graaf C, Yang L, Song G, Dai A, Cai X, Feng Y, Reedtz-Runge S, Hanson MA, Yang H, Jiang H, Stevens RC, Wang MW (2016) Structural determinants of binding the seven-transmembrane domain of the glucagon-like Peptide-1 receptor (GLP-1R). J Biol Chem 291(25):12991–13004. https://doi.org/10.1074/jbc.M116.721977
162. Harpsoe K, Isberg V, Tehan BG, Weiss D, Arsova A, Marshall FH, Brauner-Osborne H, Gloriam DE (2015) Selective negative allosteric modulation of metabotropic glutamate receptors - a structural perspective of ligands and mutants. Sci Rep 5:13869. https://doi.org/10.1038/srep13869
163. Berman HM, JW ZF, Gilliland G, Bhat TN, Weissig H, Shindyalov IN, Bourne PE (2000)

The protein data bank. Nucleic Acids Res 28:235–242

164. Wilkinson MD, Dumontier M, Aalbersberg IJ, Appleton G, Axton M, Baak A, Blomberg N, Boiten JW, da Silva Santos LB, Bourne PE, Bouwman J, Brookes AJ, Clark T, Crosas M, Dillo I, Dumon O, Edmunds S, Evelo CT, Finkers R, Gonzalez-Beltran A, Gray AJ, Groth P, Goble C, Grethe JS, Heringa J, t Hoen PA, Hooft R, Kuhn T, Kok R, Kok J, Lusher SJ, Martone ME, Mons A, Packer AL, Persson B, Rocca-Serra P, Roos M, van Schaik R, Sansone SA, Schultes E, Sengstag T, Slater T, Strawn G, Swertz MA, Thompson M, van der Lei J, van Mulligen E, Velterop J, Waagmeester A, Wittenburg P, Wolstencroft K, Zhao J, Mons B (2016) The FAIR guiding principles for scientific data management and stewardship. Sci Data 3:160018. https://doi.org/10.1038/sdata.2016.18

Chapter 5

GPCR Homology Model Generation for Lead Optimization

Christofer S. Tautermann

Abstract

The vast increase of recently solved GPCR X-ray structures forms the basis for GPCR homology modeling to atomistic accuracy. Nowadays, homology models can be employed for GPCR-ligand optimization and have been reported as invaluable tools for drug design in the last few years. Elucidation of the complex GPCR pharmacology and the associated GPCR conformations made clear that different homology models have to be constructed for different activation states of the GPCRs. Therefore, templates have to be chosen accordingly to their sequence homology as well as to their activation state. The subsequent ligand placement is nontrivial, as some recent X-ray structures show very unusual ligand binding sites and solvent involvement, expanding the space of the putative ligand binding site from the generic retinal binding pocket to the whole receptor. In the present study, a workflow is presented starting from the selection of the target sequence, guiding through the GPCR modeling process, and finishing with ligand placement and pose validation.

Key words GPCR, Homology models, Lead optimization, Docking

1 Introduction

Homology modeling of GPCRs used to be a quite approximate task until a decade ago, because dark state rhodopsin was the only template available [1]. Advanced methods have been developed to overcome the shortcomings of the very narrow, buried binding site of retinal to make best use of the rhodopsin template [2]. The situation changed a lot, when the structure of β2 was reported in 2007 [3, 4] because several important features of GPCR structures already became obvious. While the orientation and arrangement of the 7 transmembrane helix (7TM) region was quite similar in the structures, the geometry of the extracellular loops (EL) as well as the ligand binding sites were quite different. For the next upcoming structure ($A_{2A}R$) in 2008 a prospective competition for the scientific community was set up to predict the ligand/receptor complex structure [5]. The results showed that the prediction of the geometry of the 7TM region was quite satisfying, but the prediction of ligand binding was very hard and the EL regions

Alexander Heifetz (ed.), *Computational Methods for GPCR Drug Discovery*, Methods in Molecular Biology, vol. 1705,
https://doi.org/10.1007/978-1-4939-7465-8_5,

were not satisfying at all. Two more contests were held in 2010 and 2013 to predict the structures of CXCR4, D3 and SMO, 5-HT_{2B}, and 5-HT_{1B} receptors respectively [6, 7]. In a nutshell, for the class A receptors the overall topology and structure of the 7TM region was predicted quite well. When it comes to the ligand binding pose and the prediction of the structure of EL2 the performance was much worse, unless a very close template (like β2 for D3) was available. Recent quantitative investigations confirmed the strong model quality dependence on the distance to the template [8]. Altogether these three GPCR structure prediction exercises showed that even the world-leading groups in GPCR modeling had a hard time to predict a quantitatively correct ligand/receptor structure for a distant receptor.

Lead optimization (LO) is a late and decisive phase in preclinical research of a drug. During LO the overall profile of a compound class is optimized for multiple parameters, such as selectivity, pharmacokinetics, potency, toxicity, and pharmacodynamics, to name a few. Therefore the optimization of affinity, which is discussed as the most essential parameter in most studies, is only *one* goal among others within LO phases. The other parameters obviously depend on the ligand structure as well and can directly be addressed by ligand modifications. Therefore, the knowledge of the ligand/receptor complex structure is of high interest, especially when modifications of the ligand have to be performed while preserving a high affinity to the receptor. One quite common goal is to decrease the clogP of a ligand, which is in many cases directly related to an increased metabolic stability and solubility [9]. If the structure of the ligand/receptor complex is not known, it is not clear which ligand positions allow higher polarity without being detrimental to the affinity. Such optimization efforts often become very demanding, because at the end drugs with a balanced profile are required to be moved into clinical development.

Having said this, it is not surprising that LO support demands high-quality ligand binding poses in a quite accurate binding site of a receptor [10, 11]. However, it is not hopeless to get a decent homology model for a GPCR target supported by the structure revolution taking place over the last few years. Since 2012 every year on average more than six new receptors have been structurally solved and have been reported to the RSBC Protein Data Bank [12] and the coverage of the GPCR phylogenetic tree increases steadily. This means that for many GPCRs of interest at least one appropriate template is available and the model generation may be quite accurate in many cases. The very rich pharmacology of GPCRs, however, adds another layer of complexity to the modeling problem. It has been realized that often quite subtle differences in the binding site of inactive and active state GPCRs are most important for ligand binding [13]. In several cases agonists and inverse agonists (or likewise PAMs and NAMs or ligands with different biases) are structurally highly related, but the pharmacology displayed at the receptor is contrary [14]. Thus, very small changes in the binding site can lead to unwanted effects and slightly incorrect modeled pockets and binding modes can generally result in wrong hypotheses. Therefore, experimental validation of ligand/ receptor complexes is crucial—be it by mutagenesis, extensive SAR, or a combination of these approaches. There is a strong difference in the required accuracy and quality of homology models used for LO vs. models which are used for virtual screening (VS). The objective of a VS campaign is to enrich active compounds in a small fraction of a very large library. This means that also filtering out inactive compounds will lead to enrichment over random selection. This in turn makes also approximate homology models well-suited tools for VS, because also the approximate shape of the binding pocket can reject most molecules which do not fit [15]. For LO support the requirement is a correct atomistic description of ligand/receptor interactions and the ability to suggest ligand modifications in order to optimize various parameters (as mentioned above). Therefore, the second obstacle concerns the correct placement of the ligand into the receptor. In a very insightful study by Beuming and Sherman plain docking was shown to be useful for X-ray structures of GPCRs, although the involvement of water in ligand binding, such as in $A_{2A}R$, leads to significantly worse docking performance [16]. When docking into homology models, the performance of docking was rather poor, yielding less than 20% of the poses to be useful for further modeling efforts. In further studies the importance of water has been thoroughly investigated, showing that the GPCR ligands often replace clusters of water which are in an unfavorable state of free energy in the binding site [17].

There are several reports in the literature, where GPCR homology models have been successfully used for ligand optimization. In recent studies Heifetz et al. were able to optimize and delineate the very subtle selectivity of ligands on serotonin receptors by employing accurate models generated through a hierarchical modeling approach [18, 19]. In another study by the same research group on the orexin receptor family, receptor models have been generated and validated based on a set of site-directed mutagenesis experiments [20]. Moreover, a new method for assessing GPCR model quality has been introduced—going far beyond the usual assessment methods of general homology models [21]. In GPCRs a network of conserved residues is known to connect the seven helices [22]. Based on the conservation of these contacts, the quality and integrity of models or MD-snapshots have been performed.

In addition to the mentioned examples of GPCR modeling in LO, there are several reviews that describe methods and examples of recent successful studies [23, 24]. In the following, a general method for generating accurate homology models and ligand binding modes which suffice the need of lead optimization is described in detail.

2 Methods

2.1 General Requirements and Resources

2.1.1 Required Computational Resources and Programming Skills

The procedure described herein requires the presence of a high performance computer environment, ideally based on a Linux operating system. Shell scripting skills as well as basic programming skills are advantageous, but not a prerequisite. The sound knowledge of protein sequence and structure datatypes is highly recommended.

2.1.2 GPCR Resources in the Internet—Data and Web Services

GPCR resources in the Internet have become crucial for the daily life of researchers in the field—for sequences, structures, and pharmacology several well-curated databases exist which are well kept up to date. In addition, for many steps in the procedures described herein, Internet resources (web services) are available, which can perform similar or identical tasks. The resources listed neither provide an exhaustive overview, nor are only limited to GPCRs, but most common tools are described. (For a more thorough overview excellent reviews are available [25].) In many cases all the databases cross-link to more detailed databases, but these are often too detailed for usual GPCR modeling efforts.

1. UniProt [26] (http://www.uniprot.org/): "The mission of UniProt is to provide the scientific community with a comprehensive, high-quality, and freely accessible resource of protein sequence and functional information" is stated on the homepage. This is the gold-standard of databases to retrieve the correct protein sequence for a protein of interest.
2. GPCRdb [27] (http://gpcrdb.org/): The GPCRdb is a resource which combines databases such as repositories of GPCR sequences, structures, and mutagenesis data with tools which enable the user to generate sequence alignments, phylogenetic trees, and snake plots, allows structure retrieval, template selection, and enable the prediction of mutagenesis sites.
3. IUPHAR/BPS Guide to PHARMACOLOGY [28] (http://www.guidetopharmacology.org/): This is a tertiary database of expert curated data of pharmacological targets and the substances that act on them. This database is extremely useful to get a very condensed overview about a target and its most important ligands.

4. ChEMBL [29] (https://www.ebi.ac.uk/chembl/): This is a literature curated chemical database of bioactive molecules with drug-like properties. The retrieval of several active compounds from ChEMBL can be very useful to validate a receptor model or derive a pharmacophore model.
5. RSBC Protein Data Bank [12] (http://www.rcsb.org/): The generic repository containing all published protein X-ray structures. Some basic web-services such as similarity searches are also available.
6. GPCR-ModSim [30] (http://open.gpcr-modsim.org/): This is a web-service for computational modeling and simulation of GPCRs. It is a multi-step procedure which takes a sequence as input, chooses the best templates, models the GPCR, assesses the models by the internal conserved contacts, and eventually performs a molecular dynamics (MD) simulation on the best model. In principle, this pipeline can substitute the whole manual model generation process and it is recommended to use this web-service in addition to manual modeling efforts as a complementary approach.

2.2 Software

The modeling process is a multi-step process and for every step a specialized software tool is available. There are also companies (such as Schrödinger inc.—https://www.schrodinger.com/) offering integrated software packages, which are able to perform everything starting from the sequence alignment to the model generation and ligand docking steps and the subsequent MD simulation within one environment. The huge advantage of such a one-stop-shop solution is that the user does not need to worry about file formats and input files and in addition these programs are optimized for usability. However, there are reasons to keep the multi-stage modeling process modular by using different tools for different modeling steps. One important reason is the high degree of specialization of various tools, where the expert user can choose myriads of different settings, which may increase the result quality. In the following various tools are presented for the different modeling steps.

1. Tool for sequence alignment and modification: To get a first good guess for an alignment the GPCRdb web-service [31] can be used. For further refinement a tool should be used which allows applying alignment position restraints and manually modifying and moving residues. The molecular operating environment (MOE) by Chemical Computing Group (https://www.chemcomp.com/) can be used.
2. Generating homology models: A plethora of homology modeling programs is available. One state-of-the-art standard program is Modeller, [32] where homology models are built by

obeying spatial restraints derived from the templates. As input files the templates in pdb format as well as the sequence alignment in the pir format are required.

3. Structural clustering of models: Homology modeling programs do generate a set of models which may be ranked by an internal energy score. Clustering based on structural features such as the χ-angles of the putative binding site or the RMSD of selected amino acids helps to identify representative models which enter further investigation. RMSD-based clustering can be directly done within MOE, clustering of χ-angles or other structural parameters is preferentially done in MATLAB (http://www.mathworks.com/).
4. Water placement and energy assessment: The putative binding sites should be assessed by the propensity of hosting ordered networks of water. The WaterMap [33] method was one of the first tools, which was able to perform water placement and classification into "happy" and "unhappy" waters. The method is based on short MD simulations and the analysis of water mobility in the binding site. A complementary approach is 3D–RISM [34] which is based on the rigid protein structure and water placements and energy assessment are done through statistically modeling of the solvent.
5. Detection of consensus ligand features: The Cresset suite of programs (http://www.cresset-group.com/) allows the ligand-based derivation of the bioactive conformation by generating pharmacophores and QSAR models.
6. Ligand docking: Several docking programs are available, the performance of the state-of-the-art tools, such as Glide, [35] Gold, [36] and Autodock [37], is similar [38]. In the present study Gold is used as docking tool.
7. A software suite covering broadly most required steps is provided by the Schrödingers suite of programs (https://www.schrodinger.com/), thus allowing an easier kick-start to non-expert modelers. Most importantly, sequence alignment, homology modeling, water placement, pharmacophore generation, ligand docking, and MD simulation can be triggered from one common frontend.

3 Methods

The overall procedure is displayed in Fig. 1 and can be divided into three main steps: The generation of the model, the preparation of ligand data, and the combination of both, i.e., the placement of the ligands into the receptor ensuring the compatibility of ligand-derived SAR.

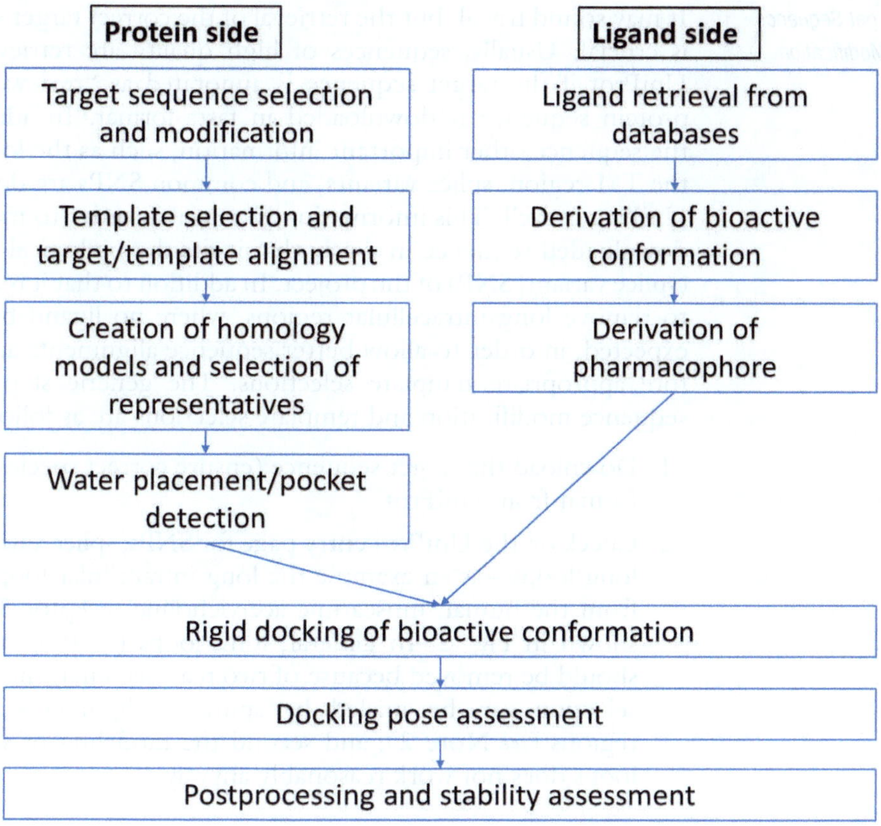

Fig. 1 Workflow to generate GPCR homology models useful for lead optimization

3.1 Generation and Preparation of Representative Homology Models

The most common method to generate GPCR models is by homology modeling. In this procedure the target sequence is aligned to a homologous protein, of which the structure is known. The main assumption of this approach is that the structure of a protein family is more conserved than its sequence, which in turn means that sequentially homologous proteins are of similar structure. The so-called twilight zone, where this assumption starts to break down is usually reported to be around 25–30% sequence identity of target and template [39]. For GPCRs this threshold is even lower because all so far reported crystal structures share a common fold; however, the sequence conservation drops as low as 10% (*see* **Note 1**). What makes GPCRs so special is the presence of conserved sequence motifs in the TM domains [22] and conserved packing patterns between the helices. Based on these observations, the sequence alignment and model assessment can be reliably done for quite distantly related receptors still yielding good results.

3.1.1 Target Sequence Selection/Modification

It may sound trivial, but the retrieval of the correct target sequence is crucial. Usually, sequences of high quality are retrieved from UniProt if the target sequence is annotated as "reviewed." The protein sequence is downloaded in fasta-format. In addition to the sequence other important information, such as the location of the TM-region, splice variants, and common SNPs are denoted in UniProt as well. This information is crucial in order to modify the downloaded sequence in a way that it fits the biological rationale (splice variant, SNP) of the project. In addition to that it is advisable to remove long intracellular regions, where no ligand binding is expected, in order to allow better sequence alignments and therefore appropriate template selections. The generic steps of this sequence modification and template selections are as follows:

1. Download the target sequence (ensure correct species) in fasta format from UniProt.
2. Check on the UniProt entry page for SNPs, splice variants, and long loops—as an example the long intracellular loop 3 (IL3) from the human muscarinic acetylcholine receptor 3 (M3) is shown in Fig. 2. In general, long loops (>30 amino acids) should be removed because of two reasons: First, the template selection can be misled by spurious alignments in these regions (*see* **Note 2**), and second the modeling of such long loops does not work reasonably anyway.

3.1.2 GPCR Template Selection and Template/Target Alignment

1. Search for templates: The easiest method to select best templates is a web-service provided by GPCRdb [31]. Under "Receptors" > "Template selection" the user pastes the UniProt identifier of the target protein and the most closely related

Topology

Feature key	Position(s)	Description	Actions	Graphical view	Length
Topological domain	1 – 67	Extracellular By similarity	Add BLAST		67
Transmembrane	68 – 91	Helical; Name=1 By similarity	Add BLAST		24
Topological domain	92 – 104	Cytoplasmic By similarity	Add BLAST		13
Transmembrane	105 – 130	Helical; Name=2 By similarity	Add BLAST		26
Topological domain	131 – 142	Extracellular By similarity	Add BLAST		12
Transmembrane	143 – 164	Helical; Name=3 By similarity	Add BLAST		22
Topological domain	165 – 184	Cytoplasmic By similarity	Add BLAST		20
Transmembrane	185 – 206	Helical; Name=4 By similarity	Add BLAST		22
Topological domain	207 – 229	Extracellular By similarity	Add BLAST		23
Transmembrane	230 – 252	Helical; Name=5 By similarity	Add BLAST		23
Topological domain	253 – 491	Cytoplasmic By similarity	Add BLAST		239
Transmembrane	492 – 514	Helical; Name=6 By similarity	Add BLAST		23
Topological domain	515 – 526	Extracellular By similarity	Add BLAST		12
Transmembrane	527 – 546	Helical; Name=7 By similarity	Add BLAST		20
Topological domain	547 – 590	Cytoplasmic By similarity	Add BLAST		44

Fig. 2 Predicted topology for M3 in UniProt. The very long IL3 at positions 253–491 is easily to be spotted

templates are presented for download. This procedure only works for unchanged human UniProt sequences as a query. If the user prefers to look for templates for a modified GPCR sequence, a BLAST sequence search in the RSBC Protein Data Bank is most straightforward. In addition, one has to pay attention to choose the correct activation states of the templates (*see* **Note 4**). Generally, templates with a sequence identity of >20% and the correct activation state are useful. If the similarity drops even lower, then the employment of more templates has proven to be beneficial [40] (*see* **Notes 1** and **2**).

2. Template preparation: GPCR X-ray structures very often carry fusion protein such as T4L or BRIL, which have shown to be one of the key success factors for crystallization [41]. For homology modeling these constructs are not desirable and therefore the removal of all extra-domains is recommended. Easiest is the comparison of the UniProt sequence of the template receptor with the X-ray sequence. This immediately identifies the regions of discrepancy (as shown in Fig. 3) and the additional domains can be removed manually in a modeling program.

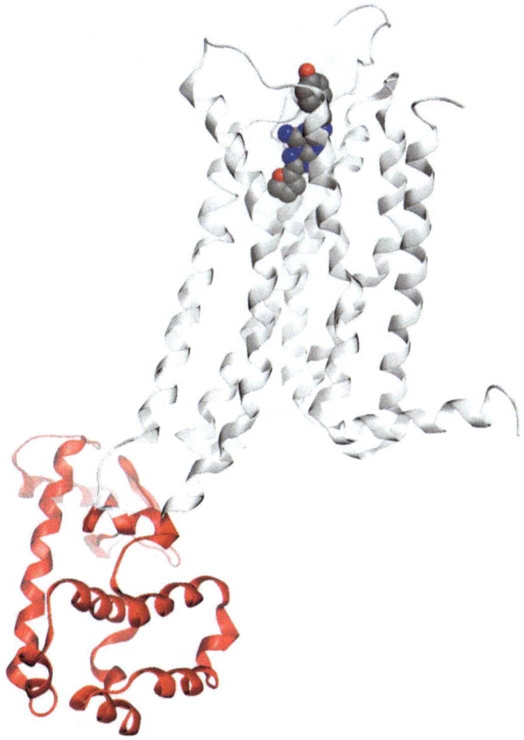

Fig. 3 T4L fused structure of the $A_{2A}R$ (pdb code: 3EML)—by sequence alignment the T4L region can easily be identified (colored in *red*)

3. Template/Target alignment: Once the templates have been prepared, the alignment should be done by a tool which allows manual intervention. In a first round, a combined structural (important for the X-ray structures) and similarity-based alignment is done. Such functionality is implemented in MOE and beyond that manual alignment constraints can be applied. It is essential to ensure that the known conserved features in GPCRs [22] are aligned properly.

3.1.3 Creation of Homology Models and Selection of Representatives

Once the target/template(s) alignment is done, the homology modeling step can be performed by many different tools. In this chapter the Modeller program, which is one of the most widely used tools for such purposes, is used. The generation of models per se is not a difficult task for the user, but the consecutive selection of the "best" models may become very demanding. Usually, a set of protein-geometry scores is used in order to rank the resulting homology models, where structural properties such as dihedrals, rotamers, and clashes are taken into account. These scores are not designed to predict the usefulness of a binding site for ligand docking. Therefore, they are only applied to remove the worst models from potential next steps because of severe deficits in their geometry. The best way to identify models for docking is to try to cover a large area of the conformational space, and this means that clustering based on geometrical features is a viable way forward. With this approach the number of models should be reduced to a tractable number (<50). In the following the individual steps are described.

1. Homology model generation: The template/target (s) alignment is saved in the pir format. The pdb files of the targets have to exactly correspond to the sequences in the .pir file. Modeller requires a python-based script-like input file— and a good tutorial (https://salilab.org/modeller/tutorial/basic.html) gives an idea of the input file for model building. Most importantly, the automodel() function is very powerful, yet easy to employ, and producing good results. This keeps the complexity of the input files to a minimum. It is recommended to generate a large number of models (>200), especially if various templates are used (*see* **Note 1**). Modeller writes the final structures in pdb format.
2. Homology model assessment: During the model building process various internal quality scores from the modeling tool should be calculated, such as, e.g., the common DOPE score. The bottom 10% of the ranked models should be directly discarded.
3. Homology model clustering: The task of identifying structurally diverse models for ligand docking implicitly assumes that the putative ligand binding site is more or less known. Recently solved GPCR structures show that the generic class A ligand

binding site close to the extracellular surface is not the only druggable site in GPCRs. The sites on the intracellular GPCR/G-protein interface have been reported and inhibitors are also reported to bind outside of the GPCR helical bundle. For the sake of simplicity, we assume that the coarse location of the ligand binding pocket is known. If no satisfying results are obtained, the whole procedure should be repeated with other potential binding sites (which are identified by tools like the SiteFinder application in MOE). If more than one template is employed, clustering based on the backbone RMSD (and also the χ-angles) of the amino acids lining the putative binding sites is recommended. If only one template is used, clustering of the χ-angles in the binding site is sufficient. Clustering is done in MATLAB, employing the k-means clustering algorithm. From each cluster 1–2 representatives with a good overall DOPE score are selected for further processing.

4. Check for conserved motifs and interactions: GPCRs are known to possess a number of conserved interactions between the transmembrane helices [22]. High-quality homology models have to show a large fraction of these interactions, otherwise something has gone wrong in the modelling procedure. Ideally, these checks are done in an automated way by smart scripts as described in the literature, or they can also be done by hand. Only the models with the highest numbers of conserved interactions should be kept for final optimization.
5. Protonation and optimization: The final models have to be protonated and the tautomeric forms of the histidines have to be checked, a procedure that is implemented in most modeling packages. Within MOE the "prepare protein" procedure corrects chain breaks, unusual geometric features, and ionizes/protonates employing a generalized Born methodology. Finally, the models should be minimized by putting restraints on the heavy atoms to relax all remaining major clashes.

3.1.4 Water Placement and Ligand Binding Site Detection

Ligand binding to a protein causes the replacement of solvent molecules by the ligand. This means that the ligand has to have a lower free energy of binding compared to the replaced water molecules. It has been shown that water molecules in GPCR binding sites have distinctly different free energies of binding, and ligands usually displace patches of water which are in an unfavorable energetic state, the sometimes called "unhappy" waters. Therefore, the detection of patches of unhappy water will help to identify potential locations of the ligand binding site. Different methods for water placement and energy assessment have been reported, [17] herein we refer to 3D–RISM, which is applied to a CCR3 model [42] and the result is shown in Fig. 4. Alternatively to the energetic assessment of water molecules in the binding site, plain methods for

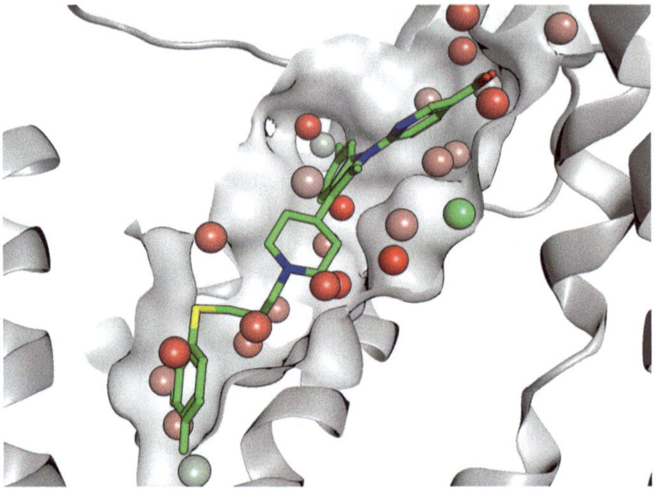

Fig. 4 Homology model of CCR3 with docked lead compound [42]. Water placement by RISM as implemented in MOE. *Red* spheres indicate energetically unfavorable water positions

pocket detection can be used, such as PocketPicker [43] or the site finder tool as implemented in MOE (*see* **Note 3**). The advantage of water energy assessment methods is that they are also useful when a pocket is much bigger than the ligand and the ligand only partly displaces water molecules.

3.2 Ligand Preparation and Overlay

Generating high-quality homology models is a prerequisite for LO support, but it is not sufficient. The placement of the ligands is also very critical to enable ligand design. Plain ligand docking into homology models is not reliable; therefore, a procedure based on an ensemble of ligands is employed to yield better ligand poses.

3.2.1 Ligand Retrieval and Generation of Bioactive Conformations

The goal of this paragraph is to derive a bioactive conformation of the ligands. The best starting points are highly rigid and potent ligands, where only few conformations can be adopted. To generate the bioactive conformations for all chemical classes potent representatives are overlaid with the first guess of the bioactive conformation. Tools to do such overlays are included in all major modeling packages. Importantly, the overlay must not induce strain in any of the structures (*see* **Note 5**).

3.2.2 Generation of Pharmacophores

Once the bioactive conformation is generated, more ligands spanning a large affinity range are overlaid to this geometry. Based on this overlay, a common pharmacophore is generated, which also takes the affinity values into account. The tool Forge (http://www.cresset-group.com/forge/) builds QSAR models based on the overlay of ligands employing pharmacophore field points. The output shows exactly which parts around the common scaffold should be decorated in order to increase/decrease affinity. This information is also very useful later, for docking mode validation.

3.3 Ligand Docking and Final Model Selection

The prerequisite for ligand docking is that high-quality homology models are available and the bioactive conformation of the ligands has been generated. Now the overlaid ligands have to be docked into the model regions with high-energy waters and the ligand-derived pharmacophore has to match with the cavity residues.

3.3.1 Rigid Ligand Docking

Docking programs are designed to generate a large ensemble of ligand conformations and to place them into the binding site of a protein followed by an energy scoring of the complex. In the present case the bioactive conformation has already been generated, and therefore rigid docking of the ligands into the top scoring homology models has to be performed. Most docking programs, such as Gold, allow the setting that the ligand has to be docked in its input conformation. Ideally, several different docking poses for the rigid ligand in each homology model are generated. If the binding cleft is narrow or if sidechains obstruct the putative binding site it is advisable to allow side-chain flexibility during docking. In this case, the procedure allows the binding site residues to adopt various accessible rotamers and the binding site adapts to the ligand to a certain degree (*see* **Note 7**).

3.3.2 Docking Pose Assessment

Two parameters are important for the assessment of the ligand pose in a homology model. First, the ligand interactions must agree with experimental findings. In the best case single-point mutagenesis results are available and certain interactions are found to be crucial. All poses that do not show these interactions can be deleted. And second, the pharmacophore that has been derived by the overlay of the ligands has to be compatible with the docking pose. If there are only few pose/model combinations left, this check can be done manually by overlaying the superimposed bunch of ligands on the docking pose. Clashes of potent ligands with the receptor are a clear sign for a problem of the pose/model combination. This step needs a thorough knowledge of the SAR of a compound class, because only if the existing SAR can be satisfyingly explained, the model and the pose are suitable to make predictions for ligand modifications.

3.3.3 Postprocessing and Stability Assessment

The best way to check the reliability of a docking pose is to perform molecular dynamics (MD) simulations and observe if the ligand changes its orientation or if it even moves out of the binding site. MD simulations should be done for at least 100 ns in triplicate (with different random seeds) but these calculations are computationally very expensive. Therefore, only the best poses should undergo simulation as final means to assess the stability of different docking poses. The setup of an MD simulation goes beyond the scope of this chapter; however, few modeling suites such as Maestro allow the setup through an intuitive graphics interface and MD simulations can also be run by non-experts. For the analysis the

RMSD of the ligand positions is the most important parameter to assess the stability of the binding pose. It is best to generate a 2D–RMSD plot, which allows the identification of the most common binding pose during a simulation. Once the stability of a binding mode is confirmed, it is advisable to take a step back and check if the SAR of the compound class is still in line with the binding mode after the MD. Once this is ensured, the identified snapshot should be energy minimized and can be used for prospective modeling.

4 Notes

Obviously, there are many ways to modify the suggested protocol. However, the largest issues arise, when no close template for homology modeling is available or if only very flexible ligands are available. In the following, several issues are described and workarounds are proposed.

1. Only remotely homologous templates available (sequence identity <20%): The combination of multiple templates of low homology vastly increases the conformational space covered by the homology models. Therefore, the generation of a large number of models is recommended. The difficulty is to identify the most useful models to proceed to the next steps. In this case model selection by ligand docking can be employed. All models that are not able to accommodate highly active ligands should be rejected. The largest variability in the generic binding site comes from different EL2 geometries, and therefore the docking should be done without EL2.
2. The template selection process at GPCRdb also allows the definition of the regions of a GPCR which should be used for comparison. Extremely useful are the predefined sets such as the TM-regions and the class-specific generic ligand binding pockets.
3. Other binding sites: Without experimental confirmation (mutagenesis), the ligand binding pocket is a priori unknown. In many cases, the generic retinal binding site is the correct one, but if the whole modeling procedure does not yield satisfying ligand poses, one should try docking to other pockets. In recent X-ray structures pockets in the membrane interface or in the intracellular region are found. As a first guess these locations can be checked for binding crevices and should be used for subsequent docking runs.
4. Activation states: For most GPCRs X-rays for only one activation state are available. If the closest template is in the wrong state, the addition of additional templates in the correct

activation state ensures the conformational sampling of the correct activation state.

5. Ligand flexibility: If the ligands of interest are very flexible and no unique bioactive conformation can be generated, the use of ligands from the literature with enhanced rigidity as starting points may be useful.
6. Automated homology modeling: There are web resources where GPCR models with a good quality can be built (GPCR-SSFE [44] and GPCR-ModSim). These procedures should also be employed because they are based on different modeling approaches. The ligand placement still has to be done afterward.
7. Involvement of water in ligand binding: In many GPCR structures water is involved in ligand binding [45]. To account for this, highly stable water molecules can be retained during ligand docking.

References

1. Palczewski K, Kumasaka T, Hori T, Behnke CA, Motoshima H, Fox BA, Trong IL, Teller DC, Okada T, Stenkamp RE, Yamamoto M, Miyano M (2000) Crystal structure of rhodopsin: a G protein-coupled receptor. Science 289 (5480):739–745. https://doi.org/10.1126/science.289.5480.739
2. Kimura SR, Tebben AJ, Langley DR (2008) Expanding GPCR homology model binding sites via a balloon potential: a molecular dynamics refinement approach. Proteins: Struct Funct Bioinform 71(4):1919–1929. https://doi.org/10.1002/prot.21906
3. Cherezov V, Rosenbaum DM, Hanson MA, Rasmussen SGF, Thian FS, Kobilka TS, Choi HJ, Kuhn P, Weis WI, Kobilka BK, Stevens RC (2007) High-resolution crystal structure of an engineered human beta2-adrenergic G protein coupled receptor. Science 318 (5854):1258–1265
4. Rasmussen SG, Choi HJ, Rosenbaum DM, Kobilka TS, Thian FS, Edwards PC, Burghammer M, Ratnala VR, Sanishvili R, Fischetti RF, Schertler GF, Weis WI, Kobilka BK (2007) Crystal structure of the human beta2 adrenergic G-protein-coupled receptor. Nature 450(7168):383–387. Doi: nature06325 [pii];10.1038/nature06325 [doi]
5. Michino M, Abola E, participants GD, Brooks CL, Dixon JS, Moult J, Stevens RC (2009) Community-wide assessment of GPCR structure modelling and ligand docking: GPCR dock 2008. Nat Rev Drug Discov 8(6):455–463. doi:http://www.nature.com/nrd/journal/v8/n6/suppinfo/nrd2877_S1.html
6. Kufareva I, Katritch V, Stevens Raymond C, Abagyan R (2014) Advances in GPCR modeling evaluated by the GPCR dock 2013 assessment: meeting new challenges. Structure 22 (8):1120–1139. https://doi.org/10.1016/j.str.2014.06.012
7. Kufareva I, Rueda M, Katritch V, Stevens Raymond C, Abagyan R (2011) Status of GPCR modeling and docking as reflected by community-wide GPCR dock 2010 assessment. Structure 19(8):1108–1126. https://doi.org/10.1016/j.str.2011.05.012
8. Costanzi S, Skorski M, Deplano A, Habermehl B, Mendoza M, Wang K, Biederman M, Dawson J, Gao J (2016) Homology modeling of a class A GPCR in the inactive conformation: a quantitative analysis of the correlation between model/template sequence identity and model accuracy. J Mol Graph Model 70:140–152. https://doi.org/10.1016/j.jmgm.2016.10.004
9. Gleeson MP (2008) Generation of a set of simple, interpretable ADMET rules of thumb. J Med Chem 51(4):817–834. https://doi.org/10.1021/jm701122q
10. Costanzi S, Tikhonova IG, Harden TK, Jacobson KA (2009) Ligand and structure-based methodologies for the prediction of the activity of G protein-coupled receptor ligands. J Comput Aided Mol Des 23(11):747–754. https://doi.org/10.1007/s10822-008-9218-3

11. Levoin N, Calmels T, Poupardin-Olivier O, Labeeuw O, Danvy D, Robert P, Berrebi-Bertrand I, Ganellin CR, Schunack W, Stark H, Capet M (2008) Refined docking as a valuable tool for lead optimization: application to histamine H3 receptor antagonists. Arch Pharm 341(10):610–623. https://doi.org/10.1002/ardp.200800042
12. Berman HM, Westbrook J, Feng Z, Gilliland G, Bhat TN, Weissig H, Shindyalov IN, Bourne PE (2000) The protein data Bank. Nucleic Acids Res 28(1):235–242
13. Tautermann CS, Pautsch A (2011) The Implication of the First Agonist Bound Activated GPCR X-ray Structure on GPCR in Silico Modeling. ACS Med Chem Lett 2(6):414-418. https://doi.org/10.1021/ml100247s
14. Dosa PI, Amin EA (2016) Tactical approaches to interconverting GPCR agonists and antagonists. J Med Chem 59(3):810–840. https://doi.org/10.1021/acs.jmedchem.5b00982
15. Köppen H (2009) Virtual screening - what does it give us? Curr Opin Drug Discov Dev 12(3):397–407
16. Beuming T, Sherman W (2012) Current assessment of docking into GPCR crystal structures and homology models: successes, challenges, and guidelines. J Chem Inf Model 52 (12):3263–3277. https://doi.org/10.1021/ci300411b
17. Bortolato A, Tehan BG, Bodnarchuk MS, Essex JW, Mason JS (2013) Water network perturbation in ligand binding: adenosine A2A antagonists as a case study. J Chem Inf Model 53(7):1700–1713. https://doi.org/10.1021/ci4001458
18. Storer RI, Brennan PE, Brown AD, Bungay PJ, Conlon KM, Corbett MS, DePianta RP, Fish PV, Heifetz A, Ho DKH, Jessiman AS, McMurray G, de Oliveira CAF, Roberts LR, Root JA, Shanmugasundaram V, Shapiro MJ, Skerten M, Westbrook D, Wheeler S, Whitlock GA, Wright J (2014) Multiparameter optimization in CNS drug discovery: Design of Pyrimido[4,5-d]azepines as potent 5-Hydroxytryptamine 2C (5-HT2C) receptor agonists with exquisite functional selectivity over 5-HT2A and 5-HT2B receptors. J Med Chem 57(12):5258–5269. https://doi.org/10.1021/jm5003292
19. Heifetz A, Storer RI, McMurray G, James T, Morao I, Aldeghi M, Bodkin MJ, Biggin PC (2016) Application of an integrated GPCR SAR-modeling platform to explain the activation selectivity of human 5-HT2C over 5-HT2B. ACS Chem Biol 11(5):1372–1382. https://doi.org/10.1021/acschembio.5b01045
20. Heifetz A, Morris GB, Biggin PC, Barker O, Fryatt T, Bentley J, Hallett D, Manikowski D, Pal S, Reifegerste R, Slack M, Law R (2012) Study of human orexin-1 and -2 G-protein-coupled receptors with novel and published antagonists by modeling, molecular dynamics simulations, and site-directed mutagenesis. Biochemistry 51(15):3178–3197. https://doi.org/10.1021/bi300136h
21. Heifetz A, Barker O, Morris GB, Law RJ, Slack M, Biggin PC (2013) Toward an understanding of agonist binding to human Orexin-1 and Orexin-2 receptors with G-protein-coupled receptor modeling and site-directed mutagenesis. Biochemistry 52(46):8246–8260. https://doi.org/10.1021/bi401119m
22. Venkatakrishnan AJ, Deupi X, Lebon G, Tate CG, Schertler GF, Babu MM (2013) Molecular signatures of G-protein-coupled receptors. Nature 494(7436):185–194
23. Tautermann CS (2011) The use of G-protein coupled receptor models in lead optimization. Future Med Chem 3(6):709–721. https://doi.org/10.4155/fmc.11.24
24. Heifetz A, James T, Morao I, Bodkin MJ, Biggin PC (2016) Guiding lead optimization with GPCR structure modeling and molecular dynamics. Curr Opin Pharmacol 30:14–21. https://doi.org/10.1016/j.coph.2016.06.004
25. Kowalsman N, Niv MY (2014) GPCR & company: databases and servers for GPCRs and interacting partners. Adv Exp Med Biol 796:185–204. https://doi.org/10.1007/978-94-7-7423-0_9
26. UniProt Consortium (2014) UniProt: A hub for protein information. Nucleic Acids Res 43 (D1):D204–D212. https://doi.org/10.1093/nar/gku989
27. Munk C, Isberg V, Mordalski S, Harpsøe K, Rataj K, Hauser AS, Kolb P, Bojarski AJ, Vriend G, Gloriam DE (2016) GPCRdb: the G protein-coupled receptor database – an introduction. Br J Pharmacol 173 (14):2195–2207. https://doi.org/10.1111/bph.13509
28. Southan C, Sharman JL, Benson HE, Faccenda E, Pawson AJ, Alexander Stephen PH, Buneman OP, Davenport AP, McGrath JC, Peters JA, Spedding M, Catterall WA, Fabbro D, Davies JA (2015) The IUPHAR/BPS guide to PHARMACOLOGY in 2016: towards curated quantitative interactions between 1300 protein targets and 6000 ligands. Nucleic Acids Res 44(D1):D1054–D1068. https://doi.org/10.1093/nar/gkv1037

29. Bento AP, Gaulton A, Hersey A, Bellis LJ, Chambers J, Davies M, Krüger FA, Light Y, Mak L, McGlinchey S, Nowotka M, Papadatos G, Santos R, Overington JP (2014) The ChEMBL bioactivity database: an update. Nucleic Acids Res 42(Database issue): D1083–D1090. https://doi.org/10.1093/nar/gkt1031
30. Esguerra M, Siretskiy A, Bello X, Sallander J, Gutiérrez-de-Terán H (2016) GPCR-ModSim: a comprehensive web based solution for modeling G-protein coupled receptors. Nucleic Acids Res 44(Web Server issue): W455–W462. https://doi.org/10.1093/nar/gkw403
31. Isberg V, de Graaf C, Bortolato A, Cherezov V, Katritch V, Marshall FH, Mordalski S, Pin J-P, Stevens RC, Vriend G, Gloriam DE (2015) Generic GPCR residue numbers – aligning topology maps while minding the gaps. Trends Pharmacol Sci 36(1):22–31. https://doi.org/10.1016/j.tips.2014.11.001
32. Eswar N, Webb B, Marti-Renom MA, Madhusudhan MS, Eramian D, Shen M-y, Pieper U, Sali A (2001) Comparative protein structure modeling using MODELLER. In: Current protocols in protein science. John Wiley & Sons, Inc., Hoboken, NJ. https://doi.org/10.1002/0471140864.ps0209s50
33. Breiten B, Lockett MR, Sherman W, Fujita S, Al-Sayah M, Lange H, Bowers CM, Heroux A, Krilov G, Whitesides GM (2013) Water networks contribute to enthalpy/entropy compensation in protein–ligand binding. J Am Chem Soc 135(41):15579–15584. https://doi.org/10.1021/ja4075776
34. Truchon J-F, Pettitt BM, Labute P (2014) A cavity corrected 3D-RISM functional for accurate solvation free energies. J Chem Theory Comput 10(3):934–941. https://doi.org/10.1021/ct4009359
35. Friesner RA, Banks JL, Murphy RB, Halgren TA, Klicic JJ, Mainz DT, Repasky MP, Knoll EH, Shelley M, Perry JK, Shaw DE, Francis P, Shenkin PS (2004) Glide: a new approach for rapid, accurate docking and scoring. 1. Method and assessment of docking accuracy. J Med Chem 47(7):1739–1749. https://doi.org/10.1021/jm0306430
36. Jones G, Willett P, Glen RC, Leach AR, Taylor R (1997) Development and validation of a genetic algorithm for flexible docking1. J Mol Biol 267(3):727–748. https://doi.org/10.1006/jmbi.1996.0897
37. Morris GM, Huey R, Lindstrom W, Sanner MF, Belew RK, Goodsell DS, Olson AJ (2009) AutoDock4 and AutoDockTools4: automated docking with selective receptor flexibility. J Comput Chem 30(16):2785–2791. https://doi.org/10.1002/jcc.21256
38. Adeniyi A, Ajibade P (2013) Comparing the suitability of autodock, gold and glide for the docking and predicting the possible targets of Ru(II)-based complexes as anticancer agents. Molecules 18(4):3760
39. Chung SY, Subbiah S (1996) A structural explanation for the twilight zone of protein sequence homology. Structure 4 (10):1123–1127. https://doi.org/10.1016/S0969-2126(96)00119-0
40. Kneissl B, Leonhardt B, Hildebrandt A, Tautermann CS (2009) Revisiting automated G-protein coupled receptor modeling: the benefit of additional template structures for a Neurokinin-1 receptor model. J Med Chem 52 (10):3166–3173. https://doi.org/10.1021/jm8014487
41. Chun E, Thompson AA, Liu W, Roth CB, Griffith MT, Katritch V, Kunken J, Xu F, Cherezov V, Hanson MA, Stevens RC (2012) Fusion partner Toolchest for the stabilization and crystallization of G protein-coupled receptors. Structure(London, England:1993) 20 (6):967–976. https://doi.org/10.1016/j.str.2012.04.010
42. Tautermann CS, Seeliger D, Kriegl JM (2015) What can we learn from molecular dynamics simulations for GPCR drug design? Comput Struct Biotechnol J 13:111–121. https://doi.org/10.1016/j.csbj.2014.12.002
43. Weisel M, Proschak E, Schneider G (2007) PocketPicker: analysis of ligand binding-sites with shape descriptors. Chem Cent J 1(1):7. https://doi.org/10.1186/1752-153x-1-7
44. Worth CL, Kreuchwig A, Kleinau G, Krause G (2011) GPCR-SSFE: a comprehensive database of G-protein-coupled receptor template predictions and homology models. BMC Bioinform 12(1):185. https://doi.org/10.1186/1471-2105-12-185
45. Congreve M, Dias JM, Marshall FH (2014) Chapter one - structure-based drug design for G protein-coupled receptors. In: Lawton G, Witty DR (eds) Progress in medicinal chemistry, vol 53. Elsevier, Amsterdam, pp 1–63. https://doi.org/10.1016/B978-0-444-63380-4.00001-9

Chapter 6

GPCRs: What Can We Learn from Molecular Dynamics Simulations?

Naushad Velgy, George Hedger, and Philip C. Biggin

Abstract

Advances in the structural biology of G-protein Coupled Receptors have resulted in a significant step forward in our understanding of how this important class of drug targets function at the molecular level. However, it has also become apparent that they are very dynamic molecules, and moreover, that the underlying dynamics is crucial in shaping the response to different ligands. Molecular dynamics simulations can provide unique insight into the dynamic properties of GPCRs in a way that is complementary to many experimental approaches. In this chapter, we describe progress in three distinct areas that are particularly difficult to study with other techniques: atomic level investigation of the conformational changes that occur when moving between the various states that GPCRs can exist in, the pathways that ligands adopt during binding/unbinding events and finally, the influence of lipids on the conformational dynamics of GPCRs.

Key words Simulation, Ligand binding, Computational, Lipid, Metadynamics, Enhanced sampling

1 Introduction

Molecular dynamics (MD) simulations provide an ideal tool to explore the dynamical aspects of proteins that may otherwise be difficult to obtain via experimental methods [1, 2]. They enable an atomistic interpretation of kinetic and thermodynamic properties of biomolecules while allowing direct control of the parameters of interest [3].

As a protein family, GPCRs have benefitted tremendously in recent years from the application of MD simulation approaches. It is estimated that approximately 40% of all drugs on the market target GPCRs or GPCR-mediated processes [4, 5] and that annual revenues for GPCR drugs are around $30 billion [6]. MD simulations can be used to explore a range of dynamic properties including receptor flexibility, ligand binding modes, binding kinetics, mechanism of action, and many others [3, 7, 8].

In recent years, there has been excellent progress in obtaining high-resolution structural information for GPCRs (*see* **Note 1**).

Alexander Heifetz (ed.), *Computational Methods for GPCR Drug Discovery*, Methods in Molecular Biology, vol. 1705,
https://doi.org/10.1007/978-1-4939-7465-8_6,

However, the information from such studies has some limitations. As well as it being unlikely that every one of the more than 800 human GPCRs [9, 10] will be solved to a high-resolution by X-ray crystallography, the structures that do exist are often modified to help crystallization and/or diffraction and usually reflect a single-conformation state of the protein. Furthermore, they are often in an environment that does not reflect the natural surrounding lipid, an aspect that is receiving increasing attention [11].

MD simulations can provide additional information that can help address these issues. Even prior to any crystal structures being available for use in homology modeling, several research groups focused on acquiring a theoretical understanding of receptor structure by combining modeling and molecular dynamics simulations: Strahs and Weinstein identified different receptor microenvironments and concerted motions in the core of the opioid receptors [12]; Scheer and colleagues investigated the role of Arg143 in constitutively active homology models of the α_{1B}–adrenergic receptor [13]. This study was also one of the earliest reports on computational modeling of polar networks; Czaplewski and colleagues reported on interactions between a model of the human vasopressin V2 receptor and 2 agonists ([arginine^{8}]vasopressin and [D-arginine^{8}]vasopressin) [14]; and Sansom and Weinstein provided a detailed look into the role of prolines in helical hinges, further elucidating the role of helical kinks in the activation of GPCRs ([15], and references therein).

In 2000, Palczewski and colleagues successfully crystallized bovine rhodopsin (and solved the structure to a resolution of 2.8 Å) [16]. There was a subsequent growth in GPCR MD articles published since then, as this work paved the way for a myriad of studies, both experimental and computational, that attempted to characterize various aspects of rhodopsin function, including intramolecular signal transduction [17], the effect of bilayers [18, 19], and the modulation by cholesterol [20] among other aspects.

The crystallization of bovine rhodopsin also resulted in better homology models of other rhodopsin-like GPCRs and consequently more studies on these were published. Seeber and colleagues used the coordinates of the bovine rhodopsin to create homology models of the 5-HT_{1A} receptor and subsequently studied ligand-induced dynamics which may be related to how ligand binding relays information to the G-protein binding site [21]. Huang and colleagues used molecular dynamics to gain insight into the relationship between the binding mechanism of SDF-1α and signal transduction of the CXCR4 receptor [22] and Zhang and colleagues built homology models of the μ opioid receptor based on the bovine rhodopsin structure and identified critical residues involved in the binding of the opioid antagonist naltrexone [23]. The latter study was also one of the first to report MD simulations on an opioid receptor-membrane complex, with

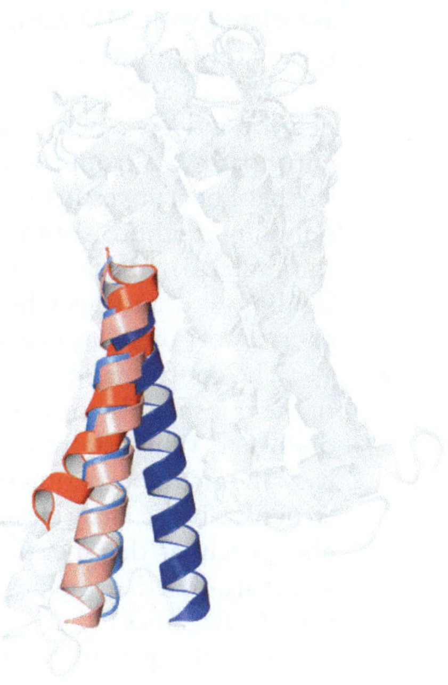

Fig. 1 The predominant movement in GPCR activity is in TM6 (*coloured*). In the inactive state (*dark blue*, PDB: 4N6H [25]), the receptor is unable to bind a G protein. TM6 moves away from the rest of the protein when bound to arrestin (*light blue*, PDB: 4ZWJ [26]), to G protein mimetic nanobodies (*peach*, PDB: 5C1M [27]) and even further when bound to a G protein (*red*, PDB: 3SN6 [28])

1,2-dimiristoyl-SN-glycero-3-phosphocholine (DMPC) serving as the lipid membrane (*see* later and **Note 2**). An earlier article reported simulations of the κ opioid receptor in complex with the endogenous ligand, dynorphin, and other benzomorphan ligands, in a dipalmitoylphosphatidylcholine (DPPC) bilayer [24].

In order to bind an intracellular effector (such as a G protein or an arrestin), GPCRs undergo structural changes. These changes are most prominent in the intracellular end of transmembrane helix 6 (TM6; *see* Fig. 1). Other transmembrane helices, particularly TM3 and TM7, also undergo conformational changes in different activity states, though these are not as accentuated as for TM6 [29–33]. Exploring these changes, and in particular the activation mechanism of GPCRs in the presence of various ligands, has been the focus of many studies using both conventional long-time scale MD and methods that employ advanced sampling tricks. These studies have shown the importance of capturing intermediate conformations that otherwise might be inaccessible to X-ray crystallography, and how ligands play a crucial role in modulating the free-energy surface of receptors. Importantly, these studies have also highlighted the advances made in both software and hardware associated with MD simulations. The first part of this chapter will review this progress.

A lot of attention has also been devoted to the use of MD to develop a deeper understanding of drug binding kinetics. Ligand entry pathways from the bulk solution into the canonical binding site have been investigated by both conventional MD and enhanced sampling methods. Both approaches provide a unique insight into the molecular mechanism by which ligands access their canonical binding site and also how this can be modulated. We give an overview of these aspects in Section 2.2.

The conformational plasticity reported in the intracellular ends of the transmembrane helices has great implications for GPCR drug discovery. While static structures, such as the ones obtained from crystallography experiments, have proven highly effective at aiding structure-based drug design of new experimental ligands [34, 35] and potential therapeutic compounds [36], the full potential of structure-based drug design requires a deeper understanding of GPCR dynamics.

Several segments of the GPCR topology are highly flexible, even within a particular conformational state [37]. As such, drug design studies carried out on a single structure may be restrictive. In such cases, the ability to study the effects of ligands in multiple conformations of the same receptor, or to better explore the plasticity of the binding site of one particular conformation, is desirable in high-throughput drug discovery.

One route to improve the effectiveness of such studies is to add the information resulting from MD simulations into drug discovery protocols. In addition to docking studies performed in static receptors, using software packages such as GOLD [38], Autodock-VINA [39] (henceforth VINA), and Glide [40], among others (refer to [41] for a more complete list), it is possible to use MD to analyze the dynamic behavior of binding pockets and explore rare conformations, giving researchers the ability to design drugs that more specifically target the receptor [42].

Another area of recent interest has been in the interaction between GPCRs and lipids and indeed this is a complicating factor for thermostability. For example, Gater and colleagues analyzed the unfolding of the β_2AR in the presence or absence of cholesterol, noting that higher melting temperatures (T_m) are seen with increasing concentrations of cholesterol, saturating at 1 mol% [43]. However, the precise nature of protein-cholesterol interactions may be difficult to ascertain using experimental methods alone. MD simulations have been used recently to explore the molecular mechanisms behind cholesterol modulation of receptors and we review this area further in the last part of this chapter.

2 Methods

2.1 From Long-Time MD to Enhanced Sampling

With the most recent advances in processing power, it has become possible to carry out long-time scale (routinely over 100 ns) MD simulations of GPCRs embedded in lipid bilayers. A variety of MD software packages now offer efficient parallelisability providing excellent performance [44]. The latest versions of popular MD packages, such as GROMACS [45] and AMBER16 [46], are equipped to efficiently distribute MD calculations to several CPU cores, as well as enabling usage of GPU cores for calculations, taking advantage of the performance boost obtained from integrating CUDA into the MD code [47]. Novel software packages, such as Desmond [48], have previously been reported to achieve performances of ~470 ns/day on commodity clusters (for a system of ~23,000 atoms on 1024 cores) [49]. In addition to parallelization, performance can be boosted by utilizing specialized hardware as is the case with Anton; a supercomputer designed for the purpose of accelerating MD simulations [50] that can achieve performance of up to 10 μs/day [49].

The advent of such performance boosts has led many researchers to simulate systems for longer time scales. Using Anton, Rosebaum and colleagues reported the first long-time scale all-atom MD simulation of the β_2AR in a lipid bilayer. The researchers used both experimental and computational methods to develop an irreversible agonist for the β_2AR, making full use of performance boosting techniques to simulate ~30 μs of receptor/ligand activity. Furthermore, Rosebaum and colleagues concluded that, in the absence of either a G protein or a mimetic nanobody, the receptor active state spontaneously destabilizes, transitioning to inactive states [51].

Using a similar approach, and making full use of crystals structures of the β_2AR in the active state, Dror and colleagues studied the mechanisms by which GPCRs transition from inactive to active states [52]. The researchers did so by simulating the spontaneous deactivation of active structures and analyzing motion of key features throughout the trajectories. In short, 92 simulations were performed, totalling approximately 656 μs, of which 76 were based on the coordinates from the active β_2AR in complex with the G protein mimetic nanobody Nb80 (PDB: 3P0G [53]), five were based on the coordinates from the active β_2AR in the complex with a G protein (PDB: 3SN6 [28]), and 2 were based on the coordinates from the inactive structure (PDB: 2RH1 [54]). In this study, Dror and colleagues highlighted a key structural region that connects the canonical binding site to the G-protein binding site, termed the "connector region" (Fig. 2). Consisting of 2 hydrophobic residues (Ile$^{3.40}$ and Phe$^{6.44}$ that are part of the key structural PIF motif [55]), this region is one of several key motifs (*see*

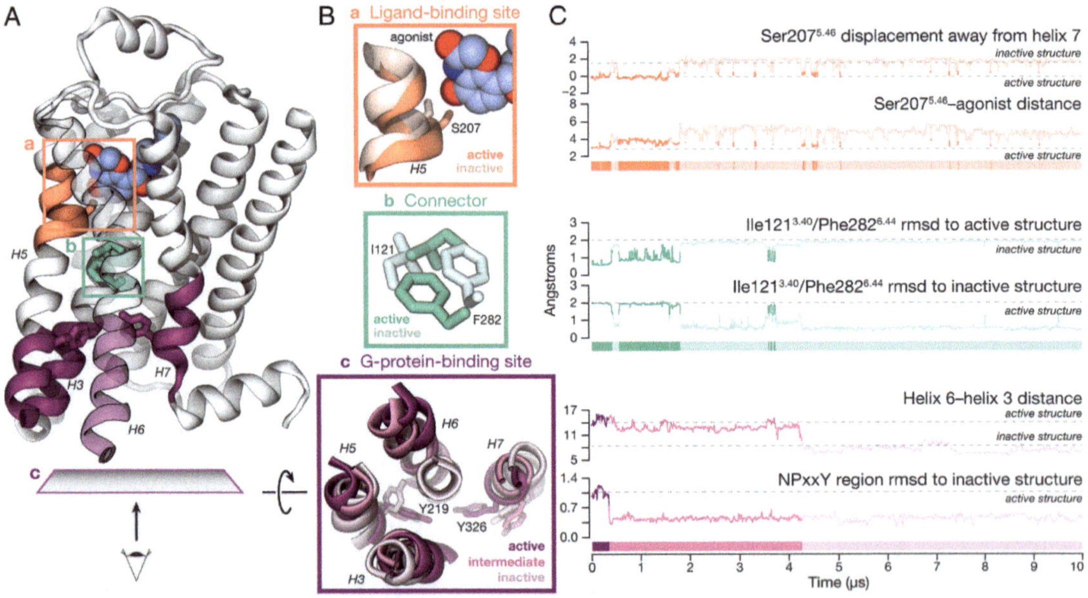

Fig. 2 Examining the conformational state of β_2AR during simulation (adapted from [52]). (**a**) Overview of the allosteric network between the canonical binding site (*orange*), the connector region (*green*), and the G protein binding site (*purple*). The receptor is represented by *silver cartoon*; ligand atoms are shown as spheres; key residues in each region are shown as sticks (hydrogens excluded). (**b**) Structure of the aforementioned regions in active and inactive conformations. (**c**) Metrics used during simulation to determine activity state of receptor

Note 3) that is thought to allosterically couple ligand binding to the G-protein binding region. The conformation of the G protein binding site is thought to be the key determinant of the connector region's conformation, as the inactive G protein binding site restricts the connector to its inactive conformation [52].

Another conclusion from this study was that the activation of GPCRs may well begin from spontaneous activation of the G protein binding site, as it transitions to an intermediate state, and that the presence of an agonist likely shifts the equilibrium of the ligand binding site to a more active conformation [52].

Kohlhoff et al. [29], in a long-time scale MD study, provided more support to confirm the hypothesis put forth by Dror and colleagues. Using the Google Exacycle system [56] they successfully simulated a total of 2.15 ms of β_2AR dynamics, using both PDB entries 2RH1 [54] and 3P0G [53] as starting points, aggregated the results using Markov state models, and measured activation of the receptor in the presence of an agonist and an inverse agonist using many of the same metrics discussed by Dror and colleagues. In addition to the root mean square deviation (RMSD) of the connector region, the RMSD of the NPxxY motif [57], and the distance between TM3 and TM6, the researchers expanded the definition of an active conformation of the ligand binding site to include $S^{5.43}$ and $F^{5.47}$. Using these metrics, the study showed two interesting results; the receptor exhibits different behavior based on the activity of the agonist, and mutual information networks differ based on the ligand present [29].

In the absence of any ligand and in the presence of an inverse agonist, the receptor remained in an inactive conformation throughout 150 μs of aggregated simulation time. However, in the presence of a full agonist, the receptor is capable of spontaneously transitioning to an active-like state (if only for a short time; approximately 2.5 μs). Additionally, based on mutual information networks, the researchers noted that agonists strengthen connections between the canonical binding site and the G protein binding site, whereas inverse agonists discourage these connections, limiting them to the intracellular G protein binding site [29].

Using a similar strategy, Schneider and colleagues performed long-time scale MD simulations on the active state of the μ opioid receptor (PDB entry 5C1M [27]) to study the differences in mutual information networks generated in the presence of a full agonist and in the presence of a biased agonist. With the aid of performance boosts from Anton, the researchers performed a total of 50 μs of simulation time, the bulk of which was dedicated to the binding pathway of the biased agonist (discussed in more detail below).

Using mutual information networks, Schneider and colleagues [58] showed that the full agonist morphine produces a much larger mutual information network compared to that produced by the biased agonist TRV-130, particularly in key areas of interest, such as the sodium allosteric site. Their results suggest that biased ligands allosterically communicate with a smaller set of residues, making it possible to design experiments to study the nature of the interactions, as well as new drugs with increased therapeutic use that exploit these contacts.

Long-time scale studies, such as the ones discussed above, have profound implications for the design of better, more efficient drugs. Unfortunately, they come at huge computational resource cost. However, it is worth noting that there are alternatives to running long-time MD, while enabling the exploration of the same free energy surface. One of the main drawbacks of conventional (or classical) MD is that simulations can only reliably explore low-energy events, such as breaking or forming salt bridges, or perhaps the equilibration of a ligand pose in a binding pocket. High-energy events, such as a GPCR transitioning from an inactive to an active conformation, are rarely sampled using conventional MD techniques [59], unless one has the dedicated resources for long simulations as described above. These high-energy events also occur at time scales of milliseconds or more, making conventional MD unsuitable as a conformational exploration tool [60].

Many enhanced sampling techniques to explore the free energy landscape of complexes have been developed including metadynamics [61, 62], targeted MD [63–65], steered MD [65, 66], biased MD [65, 67, 68], accelerated MD (aMD) [69, 70], and dual-boost aMD [69, 71] among others. Some of these tools have been directly implemented in the aforementioned MD packages, as is the case with aMD and dual boost aMD being implemented in AMBER (since version 12), while others have been developed as plugins, as is the case with metadynamics and PLUMED [72]. Using such techniques and novel protocols enables us to explore similar dynamic properties with reduced computational resources.

The guiding principle behind enhanced sampling tools, such as metadynamics, is to use one or more collective variables (CVs) to describe the evolution of the system over time [59] (*see* **Note 4**). Upon choosing collective variables (some examples given below), a history-dependent bias is added to the original force-field parameters, discouraging the system from visiting states that have already been explored [73]. Subsequently, a free-energy surface can be calculated using a re-weighing algorithm [74] as a function of system variables [73].

These principles can be used to find metastable conformations of receptors that might otherwise be inaccessible experimentally, which is of particular importance given that the presence of certain ligands might stabilize such conformations. The first reported case of the use of biased MD to explore the activation pathway of GPCRs was reported in 2010 [75]. In later studies, using a variant of metadynamics, well-tempered metadynamics [76], Provasi and colleagues showed how ligands with different physiological properties (agonists, neutral antagonists, and inverse agonists) modulate the free energy landscape of the receptor [73].

After ascertaining binding poses for ligands in the β_2AR, either by positioning them according to an experimentally determined structure or by calculating their pose via docking, Provasi and colleagues used the Cα RMSD as a basis for computing their CVs, simulating the systems until convergence (300 ns). Then, using the distance between $R^{3.50}$ and $E^{6.30}$ (the "ionic lock"; known to break upon activation of the receptor [77]), the rotameric angle of $W^{6.48}$ (the "toggle switch" [78], or sometimes "micro-switch," thought to influence receptor activation [57]), and the displacement of TM6 (showing the biggest change upon receptor activation; *see* Fig. 1), the researchers used the previously mentioned re-weighing algorithm to estimate the free-energy of the system.

The results of this study, using well-tempered metadynamics as an approach to sample the free-energy surface, provide quantitative descriptions for the dynamics of receptors. In the absence of ligands and in the presence of neutral antagonists, receptors explore both inactive and active-like intermediate conformations. Inverse agonists appear to destabilize active-like intermediate states, substantially increasing the energy required for receptors to transition to said states; and agonists appear to stabilize active-like states, both disrupting the ionic lock and influencing the rotameric angle of $W^{6.48}$. Results from further studies into the rotameric angle of this critical residue, also by implementing metadynamics tools, postulate that rotation of the sidechain of $W^{6.48}$ results in disruption of the hydrophobic core within GPCRs, weakening connections between TM3 and TM6, and allowing waters to inundate the receptor core [79]. It is worth noting that the partial agonists and very weak partial agonists appear to be less capable of disrupting the ionic lock compared to full agonists [73].

Zia et al. also used an accelerated MD approach [80]. Using the crystal structure of the A_{2A} adenosine receptor (PDB: 4EIY [81]), they designed and implemented a strategy using steered MD to identify regions where structural waters might be located, and to subsequently identify the role of each region. Water molecules have been implicated in many roles within GPCRs, including strengthening interactions between helices [82], affecting ligand affinity and binding kinetics [80, 83] and may act as "low-energy molecular switches" that stabilize active and inactive conformations of GPCRs [27]. While multiple software packages have been developed to predict their location within receptors (such as WaterDock [84], GCMC [85], and grid cell theory (GCT) [86, 87] among others [88–90]), the results don't often match and in certain situations convergence may be hard to achieve [80]. Zia et al. designed a protocol to force desolvation of particular areas of interest based on a specifically designed collective variable (CV). In this case, a bias is implemented that actively repulses water oxygen atoms via fictitious charges. In addition, the behavior of water molecules in the receptor was analyzed in the presence of a diversity of ligands, the positions of which were obtained from other crystal structures. The resulting *apo* simulation was compared to a 100 ns simulation of the *apo*-receptor using conventional MD methods.

One of the key outcomes from this study was how implementing novel strategies, aided by enhanced sampling methods, can help reduce simulation time required to find the same results one would find using conventional MD. Compared to the 100 ns required by conventional MD, the researchers reported simulations times as low as 1.2 ns per receptor [80].

Taken together, these studies show how the recent innovations in hardware architecture and software programming have enabled the sampling of longer-timescale events, ultimately allowing researchers to have a better, more detailed understanding of important processes that affect GPCRs (and, indeed, many other larger systems).

2.2 Ligand Binding Pathways

MD simulations can be used to study several aspects of ligand-protein interactions. Ligand binding modes and the influence of ligands on protein dynamics are some of the properties that can easily be studied using MD simulations. One aspect of protein-ligand interactions that is not very easily probed by experimental techniques is what are the precise pathways a ligand takes to gain access to its fully bound state? The growing interest in kinetic properties of drugs means that this question is becoming increasingly important, especially as the kinetics of drug binding are intimately connected to therapeutic effects [91]. For example, it has been suggested that slow unbinding may enhance therapeutic effects [92–96].

Unfortunately, little is known about how the ligand moves from bulk solvent into the canonical binding site (*see* **Note 5**). One of the earliest studies used all-atom MD simulations of the β_1- and β_2-adrenoceptor (β_1AR and β_2AR, based on PDB entries 2VT4 [97] and 2RH1 [54], respectively), each in complex with a variety of ligands (Fig. 3a). The authors showed that there is a dominant binding pathway for ligands in the adrenoceptors characterized by 2 energetic barriers that hinder ligand binding from the bulk solution. These barriers reflect a dewetting of the ligand as it enters the vestibule region [91] (Fig. 3b).

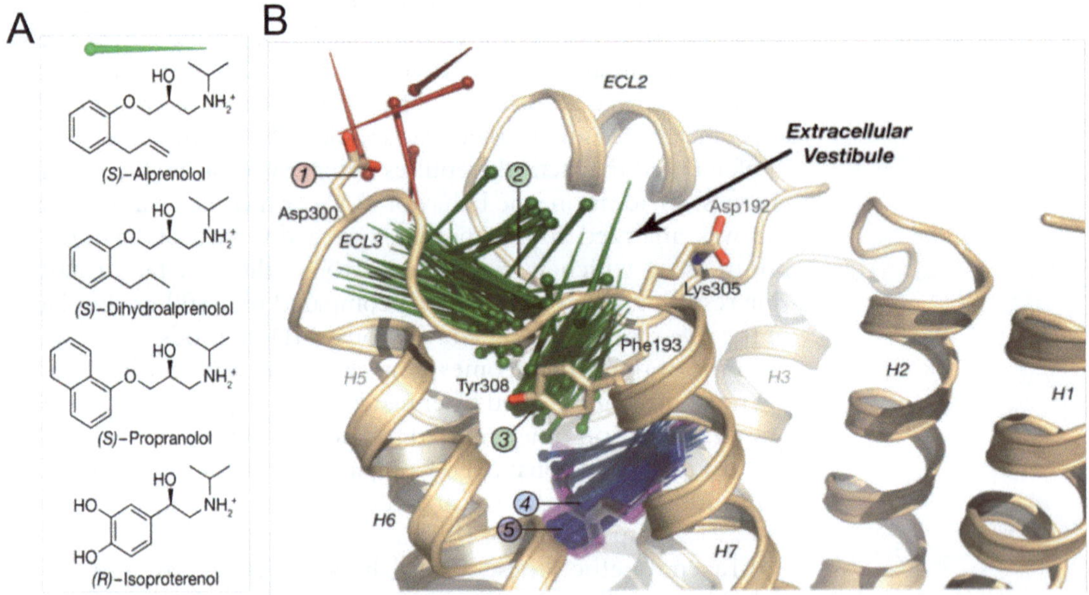

Fig. 3 Identification of the critical "vestibule region" within the β_2AR. (**a**) Ligands used by Dror and colleagues [91]. (**b**) Ligand entry pathway from bulk solution. Receptor shown as *tan ribbon* and ligands shown as pins, where the aromatic pharmacophore is represented by the round end and the positively charged nitrogen atom is represented by the pin point (as per (**a**)). Ligands are *colored* based on the RMSD from the crystal structure (*red*: in bulk solvent; *green*: in vestibule region; *blue*: in canonical binding site). Reproduced from [91] with permission

In this work, 21 of 82 unbiased all-atom MD simulations resulted in spontaneous binding events, a total of 232 μs of simulation time [91]. A similar protocol was applied in a recent study of the μ opioid receptor, where researchers randomly inserted 10 Oliceridine (TRV-130) molecules in the extracellular bulk and studied the binding pathway to the canonical active site. Analysis of approximately 50 μs of simulation revealed a similar binding pathway where TRV-130 ligands first visited the vestibule region before entering the canonical binding site [58]. Contrary to the approach by Dror and colleagues, the researchers reduced the computational time by implementing exclusion criteria: all simulations in which all ligands were bound to the lipid bilayer after a certain amount of simulation time were discontinued.

Both studies are examples of how classical MD can help elucidate ligand binding into the canonical binding site, and may be used to discover new binding sites for small molecules and modulators (also see the GPCR-lipid interactions section below). However, simulating 50 μs or more is computationally very expensive and time consuming and not generally available to most research groups. It is therefore always useful to develop new protocols to decrease the computational time and increase performance.

One way this can be achieved involves "shepherding" of ligands in classical MD simulations. An example of this, termed Supervised MD (SuMD), is capable of reducing the total simulation time from the microsecond timescale to the nanosecond timescale [98]. In short, the distance between the center of mass (CoM) of the ligand and the residues constituting the binding site is monitored over a predefined timestep. If the distance decreases, the simulation is continued; if the distance increases, the simulation is restarted from the previous checkpoint (*see* Fig. 4).

This method was subsequently used to analyze the binding pathway of several ligands to the human A_{2A} adenosine receptor: ZM241385 (PDB: 3EML [99]); T4G, (PDB: 3UZA [100]); T4E (PDB: 3UZC [100]); and Caffeine (PDB: 3RFM [101]). The authors report that the binding pathway of ZM241385 reproduces the crystallographic pose in less than 60 ns, the T4G simulations reproduce the crystallographic pose in less than 65 ns, and the T4E system reproduces the crystallographic pose in less than 110 ns. The authors reported that caffeine also enters the canonical site but do not report how long the binding event takes [98].

In a later study, the SuMD protocol was applied to the identification and recognition of possible allosteric pathways of LUF6000, an A_3 adenosine receptor (A_3AR) positive allosteric modulator [102]. This study used SuMD to identify 2 distinct mechanisms by which LUF6000 mediates agonist-receptor interactions, particularly interactions between the A_3 adenosine receptor and adenosine. Briefly, adenosine was supervised into the binding pocket of the A_3AR (taking just over 20 ns to reach this configuration) and

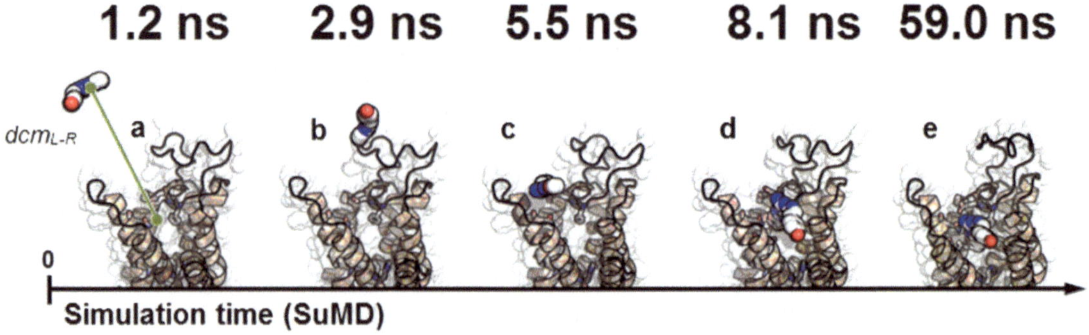

Fig. 4 Schematic representation of SuMD. Notice the decrease in time taken compared to previously reported β_2AR ligand entry times. The ligand, ZM241385, is shown as spheres, and A_{2A}AR is shown as *tan ribbons*. Reproduced from [98] with permission

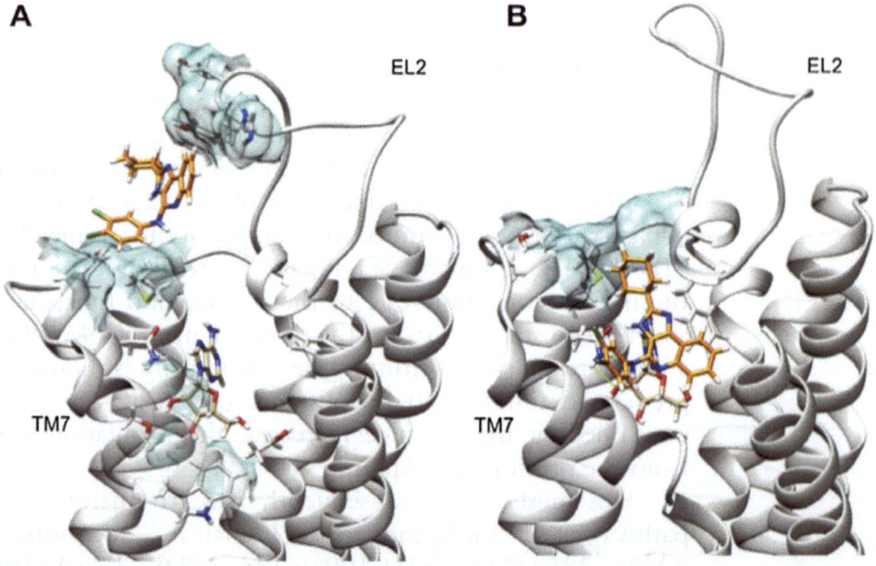

Fig. 5 Possible allosteric mechanisms of LUF6000. (**a**) Conformation changes in EL2 induced by LUF6000 (*orange sticks*) resulting in more favorable interactions between adenosine (*tan sticks*) and A_3AR (*silver ribbons*). (**b**) LUF6000 working as an "Orthosteric pocket cap," shielding adenosine from bulk solvent. Reproduced with permission from Deganutti et al. [102]

the subsequent adenosine-A_3AR complex was used to observe how LUF6000 enters the receptor and how this translates to possible allosteric modulation mechanisms. The authors report that LUF6000 either (a) alters the conformation state of the second extracellular loop, thus resulting in more energetically favorable interactions between adenosine and residues deeper in the canonical binding site (Fig. 5a), or (b) acts as an "orthosteric pocket cap," reducing adenosine exposure to bulk solvent (Fig. 5b) [102].

However, the idea of using distance constraints to drive a small molecule to a binding pocket is not new. Combined with some previously discussed enhanced sampling techniques, the distance between the CoM of a ligand and the CoM of the accepted binding pocket can be used as a CV to calculate the free energy of a receptor-ligand complex, as the ligand moves from the bulk solution to the canonical binding pocket. Using this distance as a primary CV and the distance between the CoM of the canonical pocket and the CoM of heavy atoms of the extracellular loop 2 (ECL2) as a secondary CV, Provasi and colleagues used well-tempered metadynamics to explore the free energy of the δ opioid receptor antagonist Naltrexone as it moved from the solution to the accepted opioid binding site [103]. Based on their results, the researcher calculated an equilibrium constant (K_{eq}) of 80±13 nM, a value close to the experimentally determined K_i values from the radiolabeled ligand binding assays.

2.3 GPCR-Lipid Interactions

In addition to modulation by small organic molecules, G proteins, peptide ligands, and ions, lipids can also influence the function GPCRs. Cholesterol, and a number of other key lipid species, have become the focus of extensive in vitro and in silico studies and have been shown to modulate the stability, oligomerization and ligand binding activity of GPCRs [104, 105].

Early attempts to identify potential cholesterol sites in GPCRs used long time scale classical MD simulations of adenosine-bound A_{2A}AR (PDB:2YDO [106]) embedded in a POPC:cholesterol (7:3) bilayer to identify three binding sites (*see* **Note 6**). One site was located in the intracellular end of the receptor, linking TM3 and TM6, while two sites were identified in the extracellular end of the receptor, one linking TM1 and TM7, and one linking TM2 and TM3 [107]. This third cholesterol site showed good agreement with a subsequently determined X-ray crystallographic structure of the same receptor (PDB: 4EIY [81]).

Long time scale MD can also be used to elucidate the mechanism by which certain lipids may modulate receptor structure and function. One such example came from simulations of the β_2AR embedded in either simple zwitterionic or mixed zwitterionic: anionic lipid bilayers [108]. Employing over 0.25 ms of simulation the authors found that embedding the apo receptor in these two bilayer environments had distinct effects on the stability of the active states and on the critical salt bridge between $Arg^{3.50}$ and $Glu^{6.30}$, the "ionic lock" [108].

Their discoveries were twofold. Starting from different crystal structures (PDB: 3SN6 [28] with G protein removed, and PDB: 3P0G [53] with nanobody removed), Neale and colleagues found that the structures derived from the nanobody-stabilized crystal were statistically more likely to deactivate based on the TM3-TM6 distance (as measured by the distance between $Arg^{3.50}$

and Leu$^{6.34}$) and ionic lock formation (defined as the distance between N_η-O_ϵ of Arg$^{3.50}$ and Glu$^{6.30}$, respectively). The authors estimate an active state half-life of 2.9±0.3 μs for the G protein-derived structure, and a significantly shorter half-life of 0.6±0.2 μs for the nanobody-derived structure. This could suggest that the conformational state required to bind to a nanobody is distinct to that required to bind the canonical G protein [108].

The second major discovery pertained to the insertion of lipids into the G protein binding site. The reported simulations suggest that phospholipids are capable of competing with Glu$^{6.40}$ for the interaction with Arg$^{3.50}$, effectively preventing the formation of the ionic lock, and thus conferring additional stability to the active states. The simulations also suggest that this effect is more pronounced in the presence of anionic lipids, which more strongly compete with Glu$^{6.40}$, resulting in a threefold increase in the half-life of active states. Critically, the simulations suggested that lipid binding in the intracellular end of the receptor might sterically hinder G protein binding, leading the researchers to propose a novel mechanism explaining the modulation of GPCRs by phospholipids (Fig. 6).

The conclusion that the active state of β_2AR is preferentially stabilized by anionic lipids was later confirmed in electron paramagnetic resonance (EPR)-based experiments employing spin-labeled β_2AR embedded in nanodiscs to monitor TM6 movements in the presence of a range of different lipid species [109]. Taken together, these two studies provide a key example of how MD can be applied predictively, in complement to experiment. Another such example of how MD can shed some light into processes, otherwise difficult to capture experimentally, came from investigations into the all-atom allosteric mechanism of cholesterol (and analogues) modulation of the β_2AR. Manna et al. [110] performed simulations of the β_2AR (totalling ca. 100 μs) embedded in a 1,2-dioleoyl-sn-glycero-3-phosphocholine (DOPC) bilayer of varying cholesterol content (ranging from 0 to 40% cholesterol). The authors found that the receptor exhibited a decrease in conformational flexibility, adopting fewer intermediate conformations (conformations between active and inactive states) over the simulated time course when the concentration of cholesterol was between 10 and 40%. This effect, measured by the distance between key atoms in the ligand binding site (C_α atoms of Asp$^{3.32}$ and Ser$^{5.46}$) and the G protein binding site (C_α atoms of Arg$^{3.50}$ and Glu$^{6.30}$), was seen regardless of starting receptor conformation. Critically, in the active state cholesterol was found at the interface between TM5 and TM6, restricting the movement of the latter helix toward its inactive conformation and thus preventing spontaneous deactivation [110].

As discussed in the sections above, capturing rare events can be achieved by very long time scale conventional MD simulations or

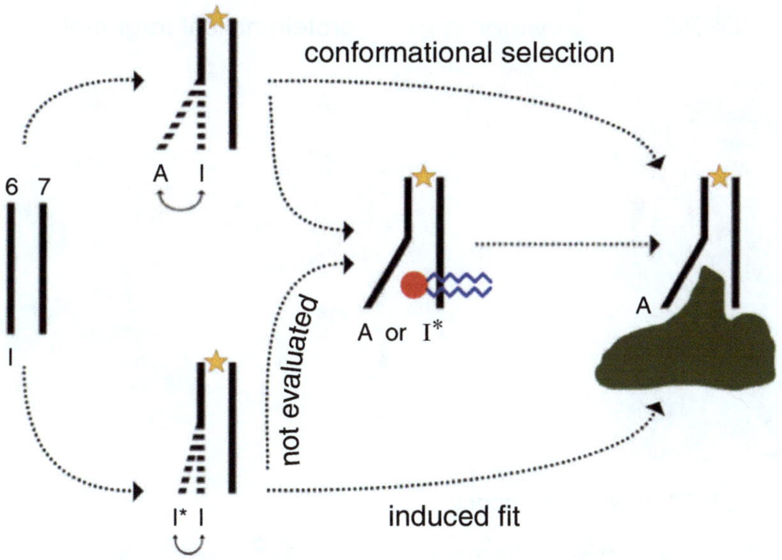

Fig. 6 Schematic of the proposed "foot-in-door" mechanism to explain receptor modulation by phospholipids. Only helices 6 and 7 are shown for simplicity. The top pathway depicts the classical view of GPCR activity; an agonist (represented by a *star*) binds to the receptor in the canonical active site, resulting in TM6 moving from an inactive (I) to an active (A) conformation, then binding to the G protein (or another partner protein, such as a nanobody, shown as a *dark green blob*). The bottom pathway depicts the receptor undergoing conformations that are hard to capture via crystallography (I*), to then be partnered with an intracellular signaller and undergoing further conformational changes to reach the active state (A). The work by Neale and colleagues is depicted and summarized by the central pathway, upon ligand binding, where the structure stabilized by a phospholipid (*red circle with blue tails*) acts as a precursor to intracellular binding. It is worth noting that this effect might also be possible without the presence of an extracellular ligand. Reproduced from [108] with permission

by exploiting advanced sampling strategies. Another route to accessing longer time and length scale events is to employ coarse-grained (CG) simulations. By mapping several non-hydrogen atoms (typically 4) into a single bead, the properties of which correspond to the properties of the mapped atoms (Fig. 7), the number of particles in the system is reduced and the computations involved at each step of the simulation are simplified. This approach thus allows access to longer time and length scales. The ability to simulate large timescales is of crucial importance, as most biologically interesting phenomena, such as higher-order protein complexes, protein folding, and signal transduction, are beyond the capabilities of atomistic MD simulations, despite the advances previously mentioned. Details of the design principles and particle interactions using the MARTINI force field have previously been reviewed [112].

Periole and colleagues employed the CG MARTINI model to investigate visual rhodopsin association in a range of bilayers with varying hydrophobic thickness [111]. They reported on systems of

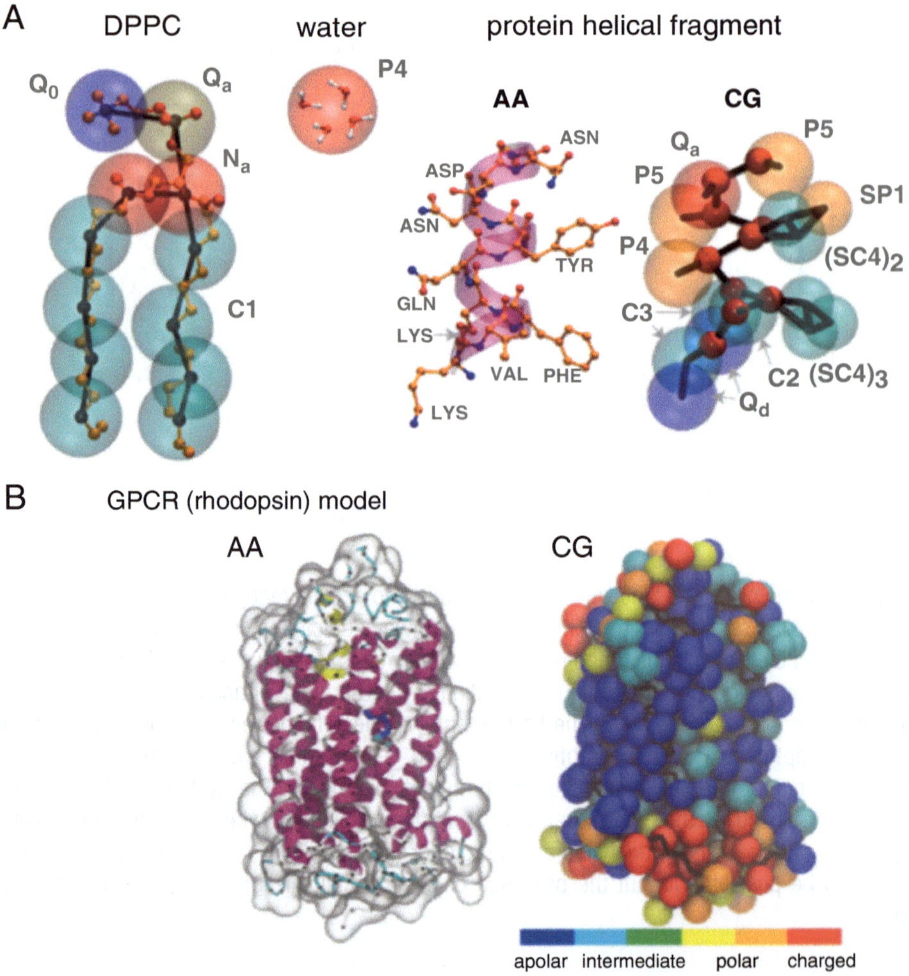

Fig. 7 Schematic of the MARTINI mapping protocol. (**a**) Examples of lipid (in this case, DPPC), water, and protein secondary structure (in this case, an alpha helix) mapping. In all cases, the atomistic representation is shown as a ball-and-stick model, while the coarse-grained representation is shown as van der Waals with underlying bonds as *black sticks*. In the helical fragment, the *red* spheres represent the backbone. Bead types are shown in transparent *gray* next to the corresponding bead. (**b**) Rhodopsin model showing the full atomistic structure as ribbons (left), and the corresponding CG representation. The bead color represents the property of the atoms incorporated into it. Adapted from [111] with permission

up to 16 rhodopsin receptors embedded in 1600 lipid molecules, employing 8 μs of simulation. The sizes of the systems were approximately 200 Å × 200 Å. Their results showed that spontaneous oligomerization of receptors depends on hydrophobic mismatch (the difference between the length of the hydrophobic part of a membrane protein and the equilibrium hydrophobic bilayer thickness [113]), in agreement with previously published atomic force microscopy (AFM) [114] and FRET [115] data. This work was the first of its kind reported, and presented exciting new prospects for

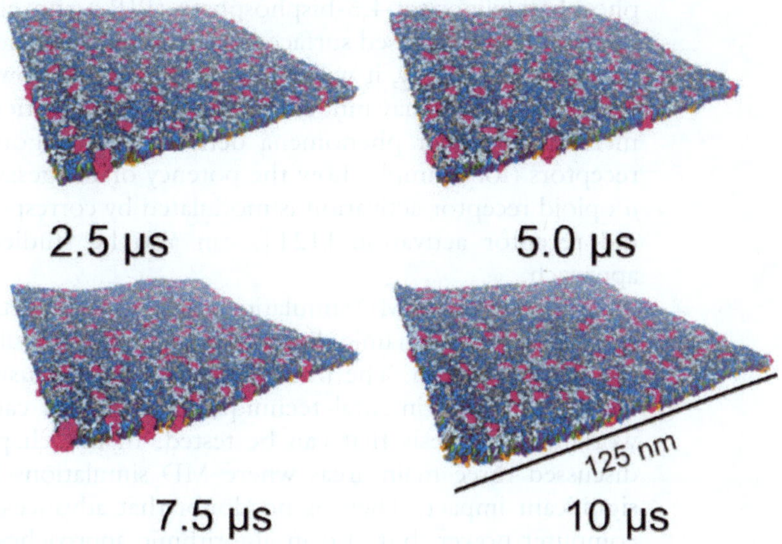

Fig. 8 Oligomerization of $S1P_1$ receptors (PDB: 3V2W [120]) within a complex lipid bilayer over 10 μs of CG simulation. The lipid composition mimics that of a mammalian plasma membrane. The receptors are shown in *pink* (144 in total). Lipid color code: *blue*—POPC; *purple*—POPE; *dark gray*—Sphingomyelin; *light blue*—monosialodihexosylganglioside (GM3); *green*—cholesterol; *light grey*—phosphatidylserine (PS); and *orange*—phosphatidylinositol-4,5-bisphosphate (PIP_2) (Based on [119] with permission)

understanding oligomerization of receptors based on bilayer properties [111, 116]. Using a coarse-grained representation facilitates computing properties for systems of such size and timescale.

The spontaneous oligomerization of rhodopsin, as demonstrated by Periole and colleagues, has since been demonstrated for the β_2AR [117], opioid receptors [118], and a range of other GPCRs [118]. Koldsø and Sansom recently reported large simulations of 144 sphingosine 1 ($S1P_1$)-membranes [119] that mimic the lipidome of a mammalian plasma membrane (Fig. 8). The authors reported on $S1P_1$ receptors embedded in 59,616 lipid molecules; a ninefold increase in the number of receptors and a 37-fold increase in the number of lipids compared to the rhodopsin study published by Periole and colleagues. Their systems measured approximately 1250 Å × 1250 Å, an approximate 40-fold increase in surface area simulated [119]. Simulation of such lengths and timescales are only made feasible by CGMD and, in this case, access to quite large compute resources.

Of particular interest, they found that, in contrast to the previous studies of GPCR oligomerization in simple lipid mixtures [116], the $S1P_1$ remained predominantly monomeric over 10 μs of CGMD simulation (Fig. 8) [119]. This suggests differences in receptor oligomerization patterns when including more complex, biologically relevant membrane compositions [116]. The authors also identified lipid binding sites for the regulatory lipids phosphatidylinositol-4,5-bisphosphate (PIP_2) and cholesterol on the membrane-exposed surface of the $S1P_1$. In addition to increasing lipid complexity, it will be interesting to see how inclusion of multiple proteins may influence their lateral interactions and oligomerization. Other phenomena derived from oligomerization of receptors (for example, how the potency of analgesia provided by μ opioid receptor activation is modulated by corresponding δ opioid receptor activation [121]) can also be studied using this approach.

In recent years, MD simulations have provided extremely useful insight into the dynamic, ligand binding, and modulatory aspects of GPCR function. The information obtained is usually complementary to experimental techniques and indeed can help refine working hypothesis that can be tested. In this chapter, we have discussed three main areas where MD simulations have made a significant impact. There is no doubt that advances, not only in computer power, but also in algorithmic approaches will help to provide even more important insight into the function and modulation of GPCRs. A better of understanding of such properties at the molecular level can surely only increase our chances of developing better drugs in the future.

3 Notes

1. The GPCR database (http://gpcrdb.org) is an up-to-date online resource that provides excellent tools for GPCR analysis. It contains alignment tools, representation tools (such as snake diagrams for example), protein-ligand interaction diagrams generators, mutation data, and much more [122].
2. Despite how complex physiological lipid bilayers are simulating them can become burdensome particularly since receptor activity is often mediated by lipids (*see* ***GPCR-lipid interactions***). In most cases, and depending on the hypothesis being investigated, researchers opt for pure bilayers (typically containing POPC or DPPC) rather than more complex ones. Recently, however, more articles have been published with authors using mixed POPC/cholesterol bilayers, or even more complex lipid rafts (for examples of such bilayers, consult [119, 123]) in order to simulate more native-like environments.
3. GPCRs, particularly those belonging to Class A, share many conserved residues and structural motifs. The most frequently mentioned motifs include:
   (a) The DRY motif, located in TM3;
   (b) The CWxP motif, located in TM6, where W is $Trp^{6.48}$, believed to be a micro-switch in receptor activation;
   (c) The NPxxY motif, of which $Asn^{7.49}$ has been shown to be part of an allosteric network connected to a core sodium

ion in certain GPCRs (see the ultra-high-resolution structure of the human δ opioid receptor, PDB 4N6H), and $Tyr^{7.53}$, a highly flexible residue that is thought to be involved in G protein binding and;

(d) The PIF motif, connecting TMs 3, 5, and 6.

Some of these motifs are used as metrics to predict the activity state of receptors (see Fig. 8, where the RMSD of both NPxxY and IF of the PIF motif are used as indicators of receptor deactivation).

4. The choice of which CV to use of non-trivial, as it very much depends on what aspect of the receptor is being studied. One strategy previously employed was to use dihedral angles of key residues, such as $Trp^{6.48}$, as CVs. The dihedral angle of $Phe^{6.44}$ can also be used, as the position of this residue is closely related to that of $Trp^{6.48}$. One could also use the RMSD of the motif as a whole as a CV. The best strategy to decide on a CV is to first understand the question from a modeling perspective (e.g., what residues are involved in this process), how these can be exploited in an MD scenario (e.g., what are the most representative changes you want to explore), and then decide on a metadynamics/accelerated MD tool to explore those changes.
5. As previously mentioned, software packages such as VINA and GOLD are used to dock ligands into binding pockets. Docking provides a good approximation of how ligands are oriented in the binding pocket. However, one of the main drawbacks of this method is that the predicted binding affinities calculated from the various tools are not always compatible (i.e., the highest scoring binding pose from VINA won't always match the highest scoring binding pose from GOLD). With little or no knowledge of binding site, researchers might be more inclined to allow ligands to diffuse naturally into the binding pocket, rather than using docking programs. The tradeoff is, as mentioned, computational time.
6. Mixed bilayers are now more and more common in the literature (*see* also **Note 2**). Building mixed bilayers usually involves 2 steps:

(a) Building a pure bilayer by either using an equilibrated patch or by using coarse-grained self-assembly. VMD contains atomistic topologies for equilibrated patches of pure bilayers and the MemProtMD database (http://sbcb.bioch.ox.ac.uk/memprotmd/) contains structures embedded in a pure DPPC bilayer in both atomistic and coarse-grained representations. It is possible to also use coarse-grained self-assembly methods to embed GPCRs into other pure bilayers, and subsequently revert those to atomistic representations (using scripts such as CG2AT [124] or Backward [125]).

(b) Then, pure bilayers can be converted to mixed bilayers using software packages such as Exchange Lipids [119]. This script randomly replaces a certain percentage of lipids from the pure bilayer, and allows users to build more complex bilayers.

It is worth noting that both the steps can be performed using the web-based CHARMM-GUI [126], a tool that also provides input files for GROMACS and other MD software. Another tool that is now more commonly used is the *insane* tool, which allows users to generate complex bilayers, as discussed above, and also allows users to generate custom periodic boundary conditions [127]. Typically, GPCRs are simulated using cubic periodic boundary conditions.

Acknowledgments

NV funded by the MRC/EPSRC and Pfizer via the Systems Approaches to Biomedical Sciences Doctoral Training Centre (EP/G037280/1). Research in the laboratory of PCB is supported by the BBSRC and MRC. G.H. is funded by a fellowship from the MRC.

References

1. Adcock SA, Mccammon JA (2006) Molecular dynamics: survey of methods for simulating the activity of proteins. Chem Rev 106:1589–1615
2. Karplus M, McCammon JA (2002) Molecular dynamics simulations of biomolecules. Nat Struct Biol 9:646–652
3. McRobb FM, Negri A, Beuming T, Sherman W (2016) Molecular dynamics techniques for modeling G protein-coupled receptors. Curr Opin Pharm 30:69–75. https://doi.org/10.1016/j.coph.2016.07.001
4. Filmore D (2004) It's a GPCR world. J Modern D Discov 7:24–27
5. Salon JA, Lodowski DT, Palczewski K (2011) The significance of G protein-coupled receptor crystallography for drug discovery. Pharm Rev 63:901–937. https://doi.org/10.1124/pr.110.003350.901
6. Martins SMA, Trabuco JRG, Monteiro GA, Prazeres DM (2012) GPCR screening and drug discovery: challenges and latest trends. Eur Pharm Rev 17
7. Ciancetta A, Sabbadin D, Federico S, Spalluto G, Moro S (2015) Advances in computational techniques to study GPCR–ligand recognition. Trends Pharm Sci 36 (12):878–890. https://doi.org/10.1016/j.tips.2015.08.006
8. Johnston JM, Filizola M (2011) Showcasing modern molecular dynamics simulations of membrane proteins through G protein-coupled receptors. Curr Opin Struc Biol 21:552–558. https://doi.org/10.1016/j.sbi.2011.06.008
9. Fredriksson R, Lagerström MC, Lundin L-G, Schiöth HB (2003) The G-protein-coupled receptors in the human genome form five main families. Phylogenetic analysis, paralogon groups, and fingerprints. Mol Endocrinol 63:1256–1272. https://doi.org/10.1124/mol.63.6.1256
10. Hill SJ (2006) G-protein-coupled receptors: past, present and future. Brit J Pharm 147: S27–S37. https://doi.org/10.1038/sj.bjp.0706455

11. Xiang J, Chun E, Liu C, Jing L, Al-Sahouri Z, Zhu L, Liu W (2016) Successful strategies to determine high-resolution structures of GPCRs. Trends Pharm Sci (in press). https://doi.org/10.1016/j.tips.2016.09.009.
12. Strahs D, Weinstein H (1997) Comparative modeling and molecular dynamics studies of the delta, kappa and mu opioid receptors. Protein Eng 10(9):1019–1038
13. Scheer A, Fanelli F, Costa T, Benedetti PGD, Cotecchia S (1996) Constitutively active mutants of the a1B-adrenergic receptor: role of highly conserved polar amino acids in receptor activation. EMBO J 15:3566–3578
14. Czaplewski C, Kazmierkiewicz R, Ciarkowski J (1998) Molecular modeling of the human vasopressin V2 receptor/agonist complex. J Comput Aided Mol Des 12:275–287
15. Sansom MSP, Weinstein H (2000) Hinges, swivels and switches: the role of prolines in signalling via transmembrane alpha helices. TIPS 21:445–451
16. Palczewski K, Kumasaka T, Hori T, Behnke CA, Motoshima H, Fox BA, Le Trong I, Teller DC, Okada T, Stenkamp RE, Yamamoto M, Miyano M (2000) Crystal structure of rhodopsin: a G protein-coupled receptor. Science 289:739–745
17. Rohrig UF, Guidoni L, Rothlisberger U (2002) Early steps of the intramolecular signal transduction in rhodopsin explored by molecular dynamics simulations. Biochemistry 41:10799–10809
18. Crozier PS, Stevens MJ, Forrest LR, Woolf TB (2003) Molecular dynamics simulation of dark-adapted rhodopsin in an explicit membrane bilayer: coupling between local retinal and larger scale conformational change. J Mol Biol 333:493–514. https://doi.org/10.1016/j.jmb.2003.08.045
19. Huber T, Botelho AV, Beyer K, Brown MF (2004) Membrane model for the G-protein-coupled receptor rhodopsin : hydrophobic interface and dynamical structure. Biophys J 86:2078–2100. https://doi.org/10.1016/S0006-3495(04)74268-X
20. Pitman MC, Grossfield A, Suits F, Feller SE (2005) Role of cholesterol and polyunsaturated chains in lipid - protein interactions: molecular dynamics simulation of rhodopsin in a realistic membrane environment. J Am Chem Soc 127:4576–4577
21. Seeber M, Benedetti PGD, Fanelli F (2003) Molecular dynamics simulations of the ligand-induced chemical information transfer in the 5-HT_{1A} receptor. J Chem Inf Comp Sci 43:1520–1531
22. Huang X, Shen J, Cui M, Shen L, Luo X, Ling K, Pei G, Jiang H, Chen K (2003) Molecular dynamics simulations on SDF-1α: binding with CXCR4 receptor. Biophys J 84:171–184. https://doi.org/10.1016/S0006-3495(03)74840-1
23. Zhang Y, Sham YY, Rajamani R, Gao J, Portoguese PS (2005) Homology modeling and molecular dynamics simulations of the mu opioid receptor in a membrane – aqueous system. Chem Bio Chem 6:853–859. https://doi.org/10.1002/cbic.200400207
24. Iadanza M, Holtje M, Ronsisvalle G, Holtje H-D (2002) k-Opioid receptor model in a phospholipid bilayer: molecular dynamics simulation. J Med Chem 45:4838–4846
25. Fenalti G, Giguere PM, Katritch V, Huang X-P, Thompson AA, Cherezov V, Roth BL, Stevens RC (2014) Molecular control of [dgr]-opioid receptor signalling. Nature 506 (7487):191–196. https://doi.org/10.1038/nature12944
26. Kang Y, Zhou XE, Gao X, He Y, Liu W, Ishchenko A, Barty A, White TA, Yefanov O, Han GW, Xu Q, Waal PWD, Ke J, Tan MHE, Zhang C, Moeller A, West GM, Pascal BD, Eps NV, Caro LN, Vishnivetskiy SA, Lee RJ (2015) Crystal structure of rhodopsin bound to arrestin by femtosecond X-ray laser. Nature 523:561–567. https://doi.org/10.1038/nature14656
27. Huang W, Manglik A, Venkatakrishnan AJ, Laeremans T, Feinberg EN, Sanborn AL, Kato HE, Livingston KE, Thorsen TS, Kling RC, Granier S, Gmeiner P, Husbands SM, Traynor JR, Weis WI, Steyaert J, Dror RO, Kobilka BK (2015) Structural insights into μ-opioid receptor activation. Nature 524 (7565):315–321. https://doi.org/10.1038/nature14886
28. Rasmussen SGF, DeVree BT, Zou Y, Kruse AC, Chung KY, Kobilka TS, Thian FS, Chae PS, Pardon E, Calinski D, Mathiesen JM, Shah STA, Lyons JA, Caffrey M, Gellman SH, Steyaert J, Skiniotis G, Weis WI, Sunahara RK, Kobilka BK (2011) Crystal structure of the β2 adrenergic receptor-Gs protein complex. Nature 477(7366):549–555
29. Kohlhoff KJ, Shukla D, Lawrenz M, Bowman GR, Konerding DE, Belov D, Altman RB, Pande VS (2014) Cloud-based simulations on Google Exacycle reveal ligand modulation of GPCR activation pathways. Nat Chem 6 (1):15–21. https://doi.org/10.1038/nchem.1821
30. Prioleau C, Visiers I, Ebersole BJ, Weinstein H, Sealfon SC (2002) Conserved helix 7 tyrosine acts as a multistate conformational switch

in the 5HT2C receptor. Identification of a novel "locked-on" phenotype and double revertant mutations. J Biol Chem 277:36577–36584. https://doi.org/10.1074/jbc.M206223200

31. Sealfon SC, Chi L, Ebersole BJ, Rodic V, Zhang D, Ballesteros J, Weinstein H (1995) Related contribution of specific heix 2 and 7 residues to conformational activation of the serotonin 5-HT_{2A} receptor. J Bio Chem 270:16683–16688. https://doi.org/10.1074/jbc.270.28.16683
32. Tiburu EK, Bowman AL, Struppe JO, Janero DR, Avraham HK, Makriyannis A (2009) Solid-state NMR and molecular dynamics characterization of cannabinoid receptor-1 (CB1) helix 7 conformational plasticity in model membranes. Biochim Biophys Acta 1788:1159–1167. https://doi.org/10.1016/j.bbamem.2009.02.002
33. Venkatakrishnan AJ, Deupi X, Lebon G, Tate CG, Schertler GF, Babu MM (2013) Molecular signatures of G-protein-coupled receptors. Nature 494(7436):185–194
34. Negri A et al (2013) Discovery of a novel selective kappa-opioid receptor agonist using crystal structure-based virtual screening. J Chem Inf Model 53(3):521–526. https://doi.org/10.1021/ci400019t
35. Weiss DR et al (2013) Conformation guides molecular efficacy in docking screens of activated β-2 adrenergic G protein coupled receptor. ACS Chem Biol 8(5):1018–1026. https://doi.org/10.1021/cb400103f
36. Manglik A et al (2016) Structure-based discovery of opioid analgesics with reduced side effects. Nature 537(7619):185–190. https://doi.org/10.1038/nature19112
37. Lebon G et al (2012) Agonist-bound structures of G protein-coupled receptors. Curr Opin Struct Biol 22(4):482–490. https://doi.org/10.1016/j.sbi.2012.03.007
38. Verdonk ML et al (2003) Improved protein-ligand docking using GOLD. Proteins 52 (4):609–623. https://doi.org/10.1002/prot.10465
39. Trott O, Olson AJ (2010) AutoDock Vina: improving the speed and accuracy of docking with a new scoring function, efficient optimization, and multithreading. J Comput Chem 31(2):455–461. https://doi.org/10.1002/jcc.21334.
40. Friesner RA et al (2006) Extra precision glide: docking and scoring incorporating a model of hydrophobic enclosure for protein–ligand complexes. J Med Chem 49 (21):6177–6196. https://doi.org/10.1021/jm051256o
41. Chen YC (2015) Beware of docking! Trends Pharmacol Sci 36(2):78–95. https://doi.org/10.1016/j.tips.2014.12.001
42. Amaro RE et al (2008) An improved relaxed complex scheme for receptor flexibility in computer-aided drug design. J Comput Aided Mol Des 22(9):693–705. https://doi.org/10.1007/s10822-007-9159-2
43. Gater DL, Saurel O, Iordanov I, Liu W, Cherezov V, Milon A (2014) Two classes of cholesterol binding sites for the β2AR revealed by thermostability and NMR. Biophys J 107:2305–2312. https://doi.org/10.1016/j.bpj.2014.10.011
44. Harvey MJ, De Fabritiis G (2012) High-throughput molecular dynamics: the powerful new tool for drug discovery. Drug Discov Today 17:1059–1062. https://doi.org/10.1016/j.drudis.2012.03.017
45. Abraham MJ, Murtola T, Schulz R, Páll S, Smith JC, Hess B, Lindahl E (2015) GROMACS: high performance molecular simulations through multi-level parallelism from laptops to supercomputers. SoftwareX 1–2:19–25. https://doi.org/10.1016/j.softx.2015.06.001
46. Case DA, Betz RM, Cerutti DS, Cheatham TE III, Darden TA, Duke RE, Giese TJ, Gohlke H, Goetz AW, Homeyer N, Izadi S, Janowski P, Kaus J, Kovalenko A, Lee TS, LeGrand S, Li P, Lin C, Luchko T, Luo R, Madej BD, Mermelstein D, Merz KM, Monard G, Nguyen H, Nguyen HT, Omelyan I, Onufriev A, Roe DR, Roitberg A, Sagui C, Simmerling CL, Botello-Smith WM, Swails J, Walker RC, Wang J, Wolf RM, Wu X, Xiao L, Kollman PA (2016) AMBER 2016. University of California, San Francisco
47. Liu W, Schmidt B, Voss G, Müller-Wittig W (2008) Accelerating molecular dynamics simulations using graphics processing units with CUDA. Compu Phys Comm 179:634–641. https://doi.org/10.1016/j.cpc.2008.05.008
48. Bowers KJ, Chow E, Xu H, Dror RO, Eastwood MP, Gregersen BA, Klepeis JL, Kolossvary I, Moraes MA, Sacerdoti FD, Salmon JK, Shan Y, Shaw DE (2006) Scalable algorithms for molecular dynamics simulations on commodity clusters. In: Proceedings of the 2006 ACMI/IEEE conference on supercomputing. ACM Press, New york
49. Klepeis JL, Lindorff-Larsen K, Dror RO, Shaw DE (2009) Long-timescale molecular dynamics simulations of protein structure and function. Curr Opin Struct Biol 19 (2):120–127

50. Shaw DE et al (2008) Anton, a special-purpose machine for molecular dynamics simulation. Comm ACM 51(7):91–97. https://doi.org/10.1145/1364782.1364802
51. Rosenbaum DM, Zhang C, Lyons JA, Holl R, Aragao D, Arlow DH, Rasmussen SGF, Choi H-j, Devree BT, Sunahara RK, Chae PS, Gellman SH, Dror RO, Shaw DE, Weis WI, Caffrey M, Gmeiner P, Kobilka BK (2011) Structure and function of an irreversible agonist-β2 adrenoceptor complex. Nature 469:236–240. https://doi.org/10.1038/nature09665
52. Dror RO, Arlow DH, Maragakis P, Mildorf TJ, Pan AC, Xu H, Borhani DW, Shaw DE (2011) Activation mechanism of the β2-adrenergic receptor. Proc Natl Acad Sci U S A 108(46):18684–18689
53. Rasmussen SGF, Choi H-J, Fung JJ, Pardon E, Casarosa P, Chae PS, Devree BT, Rosenbaum DM, Thian FS, Kobilka TS, Schnapp A, Konetzki I, Sunahara RK, Gellman SH, Pautsch A, Steyaert J, Weis WI, Kobilka BK (2011) Structure of a nanobody-stabilized active state of the β(2) adrenoceptor. Nature 469:175–180. https://doi.org/10.1038/nature09648
54. Cherezov V, Rosenbaum DM, Hanson MA, Rasmussen SGF, Thian FS, Kobilka TS, Choi H-J, Kuhn P, Weis WI, Kobilka BK, Stevens RC (2007) High-resolution crystal structure of an engineered human β2-adrenergic G protein–coupled receptor. Science 318 (5854):1258–1265
55. Marti-Solano M, Sanz F, Pastor M, Selent J (2014) A dynamic view of molecular switch behavior at serotonin receptors: implications for functional selectivity. PLoS One 9: e109312. https://doi.org/10.1371/journal.pone.0109312
56. Hellerstein JL, Kohlhoff KJ, Konerding DE (2012) Science in the cloud: accelerating discovery in the 21st century. IEEE Internet Comput 16(4):64–68. https://doi.org/10.1109/MIC.2012.87
57. Nygaard R, Frimurer TM, Holst B, Rosenkilde MM, Schwartz TW (2009) Ligand binding and micro-switches in 7TM receptor structures. Trends Pharm Sci 30:249–259. https://doi.org/10.1016/j.tips.2009.02.006
58. Schneider S, Provasi D, Filizola M (2016) How Oliceridine (TRV-130) binds and stabilizes a μ-opioid receptor conformational state that selectively triggers G protein-signaling pathways. Biochemistry 55:6456–6466. https://doi.org/10.1021/acs.biochem.6b00948
59. Rodríguez-Espigares I, Kaczor AA, Selent J (2016) In silico exploration of the conformational universe of GPCRs. Mol Inform 35:227–237. https://doi.org/10.1002/minf.201600012
60. Pierce LCT, Salomon-Ferrer R, De Oliveira C AF, JA MC, Walker RC (2012) Routine access to millisecond time scale events with accelerated molecular dynamics. J Chem Theory Comput 8:2997–3002
61. Bussi G, Laio A, Parrinello M (2006) Equilibrium free energies from nonequilibrium metadynamics. Phys Rev Lett 96(9):090601
62. Leone V, Marinelli F, Carloni P, Parrinello M (2010) Targeting biomolecular flexibility with metadynamics. Curr Opin Struc Biol 20:148–154. https://doi.org/10.1016/j.sbi.2010.01.011
63. Schlitter J, Engels M, Krüger P, Jacoby E, Wollmer A (1993) Targeted molecular dynamics simulation of conformational change-application to the T ↔ R transition in insulin. Mol Sim 10:291–308. https://doi.org/10.1080/08927029308022170
64. Schlitter J, Engels M, Krueger P (1994) Targeted molecular dynamics: a new approach for searching pathways of conformational transitions. J Mol Graph 1994:84–89
65. Huang H, Ozkirimli E, Post CB (2009) Comparison of three perturbation molecular dynamics methods for modeling conformational transitions. J Chem Theor Comput 5:1304–1314. https://doi.org/10.1021/ct9000153
66. Grubmuller H, Heymann B, Tavan P (1996) Ligand binding: molecular mechanics calculation of the streptavidin-biotin rupture force. Science 271:997–999
67. Marchi M, Ballone P (1999) Adiabatic bias molecular dynamics : a method to navigate the conformational space of complex molecular systems. J Chem Phys 110:3697–3702. https://doi.org/10.1063/1.478259
68. Paci E, Karplus M (1999) Forced unfolding of fibronectin type 3 modules: an analysis by biased molecular dynamics simulations. J Mol Bio 288:441–459
69. Hamelberg D, De Oliveira CAF, McCammon JA (2007) Sampling of slow diffusive conformational transitions with accelerated molecular dynamics. J Chem Phys 127:155102. https://doi.org/10.1063/1.2789432
70. Markwick PRL, McCammon JA (2011) Studying functional dynamics in bio-molecules using accelerated molecular dynamics. Phys Chem Chem Phys 13:20053–20065. https://doi.org/10.1039/c1cp22100k

71. Miao Y, Nichols SE, Gasper PM, Metzger VT, McCammon JA (2013) Activation and dynamic network of the M2 muscarinic receptor. Proc Natl Acad Sci U S A 110 (27):10982–10987. https://doi.org/10.1073/pnas.1309755110
72. Bonomi M, Branduardi D, Bussi G, Camilloni C, Provasi D, Raiteri P, Donadio D, Marinelli F, Pietrucci F, Broglia RA, Parrinello M (2009) PLUMED: a portable plugin for free-energy calculations with molecular dynamics. Comput Phys Comm 180 (10):1961–1972. https://doi.org/10.1016/j.cpc.2009.05.011
73. Provasi D, Artacho MC, Negri A, Mobarec JC, Filizola M (2011) Ligand-induced modulation of the free-energy landscape of G protein-coupled receptors explored by adaptive biasing techniques. PLoS Comput Biol 7: e1002193. https://doi.org/10.1371/journal.pcbi.1002193
74. Bonomi M, Barducci A, Parrinello M (2009) Reconstructing the equilibrium boltzmann distribution from well-tempered metadynamics. J Comput Chem 30:1615–1621. https://doi.org/10.1002/jcc
75. Provasi D, Filizola M (2010) Putative active states of a prototypic g-protein-coupled receptor from biased molecular dynamics. Biophys J 98:2347–2355. https://doi.org/10.1016/j.bpj.2010.01.047
76. Barducci A, Bussi G, Parrinello M (2008) Well-tempered metadynamics: a smoothly converging and tunable free-energy method. Phys Rev Lett 100(2):020603
77. Ballesteros JA, Jensen AD, Liapakis G, Rasmussen SGF, Shi L, Gether U, Javitch JA (2001) Activation of the β2 -adrenergic receptor involves disruption of an ionic lock between the cytoplasmic ends of transmembrane segments 3 and 6. J Biol Chem 276:29171–29177. https://doi.org/10.1074/jbc.M103747200
78. Shi L, Liapakis G, Xu R, Guarnieri F, Ballesteros JA, Javitch JA (2002) β2 adrenergic receptor activation: modulation of the proline kink in transmembrane 6 by a rotamer toggle switch. J Biol Chem 277:40989–40996. https://doi.org/10.1074/jbc.M206801200
79. Li J, Jonsson AL, Beuming T, Shelley JC, Voth GA (2013) Ligand-dependent activation and deactivation of the human adenosine $A2_A$ receptor. J Am Chem Soc 135:8749–8759
80. Zia SR, Gaspari R, Decherchi S, Rocchia W (2016) Probing hydration patterns in class-a GPCRs via biased MD: the A_{2A} receptor. J Chem Theor Comput 12:6049–6061. https://doi.org/10.1021/acs.jctc.6b00475
81. Liu W, Chun E, Thompson AA, Chubukov P, Xu F, Katritch V, Han GW, Roth CB, Heitman LH, Ijzerman AP, Cherezov V, Stevens RC (2012) Structural basis for allosteric regulation of GPCRs by sodium ions. Science 337 (6091):232–236
82. Angel TE, Chance MR, Palczewski K (2009) Conserved waters mediate structural and functional activation of family a (rhodopsin-like) G protein-coupled receptors. Proc Natl Acad Sci U S A 106:8555–8560
83. Kaszuba K, Róg T, Bryl K, Vattulainen I, Karttunen M (2010) Molecular dynamics simulations reveal fundamental role of water as factor determining affinity of binding of β-blocker nebivolol to β-adrenergic receptor. J Phys Chem B 114:8374–8386. https://doi.org/10.1021/jp909971f
84. Ross GA, Morris GM, Biggin PC (2012) Rapid and accurate prediction and scoring of water molecules in protein binding sites. PLoS One 7(3):e32036
85. Ross GA, Bodnarchuk MS, Essex JW (2015) Water sites, networks, and free energies with grand canonical Monte Carlo. J Am Chem Soc 137(47):14930–14943. https://doi.org/10.1021/jacs.5b07940
86. Gerogiokas G, Southey MWY, Mazanetz MP, Hefeitz A, Bodkin M, Law RJ, Michel J (2015) Evaluation of water displacement energetics in protein binding sites with grid cell theory. Phys Chem Chem Phys 17:8416–8426. https://doi.org/10.1039/C4CP05572A
87. Gerogiokas G, Southey MWY, Mazanetz MP, Heifetz A, Bodkin M, Law RJ, Henchman RH, Michel J (2016) Assessment of hydration thermodynamics at protein interfaces with grid cell theory. J Phys Chem B 120:10442–10452. https://doi.org/10.1021/acs.jpcb.6b07993
88. Michel J, Tirado-Rives J, Jorgensen WL (2009) Prediction of the water content in protein binding sites. J Phys Chem 113:13337–13346. https://doi.org/10.1021/jp9047456
89. Sciabola S, Stanton RV, Mills JE, Flocco MM, Baroni M, Cruciani G, Perruccio F, Mason JS (2010) High-throughput virtual screening of proteins using GRID molecular interaction fields. J Chem Inf Model 50:155–169
90. Uehara S, Tanaka S (2016) AutoDock-GIST: incorporating thermodynamics of active-site water into scoring function for accurate protein-ligand docking. Molecules 21:

E1604. https://doi.org/10.3390/molecules21111604

91. Dror RO, Pan AC, Arlow DH, Borhani DW, Maragakis P, Shan Y, Xu H, Shaw DE (2011) Pathway and mechanism of drug binding to G-protein-coupled receptors. Proc Natl Acad Sci U S A 108(32):13118–13123
92. Guo D, Hillger JM, Ijzerman AP, Heitman LH (2014) Drug-target residence time—a case for G protein-coupled receptors. Med Res Rev 34(4):856–892. https://doi.org/10.1002/med.21307
93. Lu H, Tonge PJ (2010) Drug-target residence time: critical information for lead optimization. Curr Opin Chem Biol 14:467–474. https://doi.org/10.1016/j.cbpa.2010.06.176
94. Swinney DC (2008) Applications of binding kinetics to drug discovery: translation of binding mechanisms to clinically differentiated therapeutic responses. Pharmaceut Med 22:23–34. https://doi.org/10.1007/BF03256679
95. Swinney DC (2009) The role of binding kinetics in therapeutically useful drug action. Curr Opin Drug Discov Devel 12:31–39
96. Swinney DC, Haubrich BA, Van Liefde I, Vauquelin G (2015) The role of binding kinetics in GPCR drug discovery. Curr Top Med Chem 15:2504–2522
97. Warne T, Serrano-Vega MJ, Baker JG, Moukhametzianov R, Edwards PC, Henderson R, Leslie AGW, Tate CG, Schertler GFX (2008) Structure of a β1-adrenergic G-protein-coupled receptor. Nature 454:486–491. https://doi.org/10.1038/nature07101
98. Sabbadin D, Moro S (2014) Supervised molecular dynamics (SuMD) as a helpful tool to depict GPCR-ligand recognition pathway in a nanosecond time scale. J Chem Inf Model 54:372–376. https://doi.org/10.1021/ci400766b
99. Jaakola VP, Griffith MT, Hanson MA, Cherezov V, Chien EY, Lane JR, Ijzerman AP, Stevens RC (2008) The 2.6 angstrom crystal structure of a human A2A adenosine receptor bound to an antagonist. Science 322(5905):1211–1217. https://doi.org/10.1126/science.1164772
100. Congreve M, Andrews SP, Dore AS, Hollenstein K, Hurrell E, Langmead CJ, Mason JS, Ng IW, Tehan B, Zhukov A, Weir M, Marshall FH (2012) Discovery of 1,2,4-triazine derivatives as adenosine A_{2A} antagonists using structure based drug design. J Med Chem 55:1898–1903
101. Doré AS, Robertson N, Errey JC, Ng I, Hollenstein K, Tehan B, Hurrell E, Bennett K, Congreve M, Magnani F, Tate CG, Weir M, Marshall FH (2011) Structure of the adenosine a 2A receptor in complex with ZM241385 and the xanthines XAC and caffeine. Structure 19:1283–1293. https://doi.org/10.1016/j.str.2011.06.014
102. Deganutti G, Cuzzolin A, Ciancetta A, Moro S (2015) Understanding allosteric interactions in G protein-coupled receptors using supervised molecular dynamics: a prototype study analysing the human A3 adenosine receptor positive allosteric modulator. Bioorg Med Chem 23:4065–4071. https://doi.org/10.1016/j.bmc.2015.03.039
103. Provasi D, Bortolato A, Filizola M (2009) Exploring molecular mechanisms of ligand recognition by opioid receptors with metadynamics. Biochemistry 48:10020–10029. https://doi.org/10.1021/bi901494n
104. Dickson CJ, Hornak V, Velez-Vega C, McKay DJJ, Reilly J, Sandham DA, Shaw D, Fairhurst RA, Charlton SJ, Sykes DA, Pearlstein RA, Duca JS (2016) Uncoupling the structure-activity relationships of β2 adrenergic receptor ligands from membrane binding. J Med Chem 59:5780–5789. https://doi.org/10.1021/acs.jmedchem.6b00358
105. Hedger G, Sansom MSP (2016) Lipid interaction sites on channels, transporters and receptors: recent insights from molecular dynamics simulations. Biochim Biophys Acta 1858(10):2390–2400. https://doi.org/10.1016/j.bbamem.2016.02.037
106. Lebon G, Warne T, Edwards PC, Bennett K, Langmead CJ, Leslie AGW, Tate CG (2011) Agonist-bound adenosine A2A receptor structures reveal common features of GPCR activation. Nature 474:521–525. https://doi.org/10.1038/nature10136
107. Lee JY, Lyman E (2012) Predictions for cholesterol interaction sites on the A2A adenosine receptor. J Am Chem Soc 134:16512–16515
108. Neale C, Herce HD, Pomès R, García AE (2015) Can specific protein-lipid interactions stabilize an active state of the Beta 2 adrenergic receptor? Biophys J 109:1652–1662. https://doi.org/10.1016/j.bpj.2015.08.028
109. Dawaliby R, Trubbia C, Delporte C, Masureel M, Van Antwerpen P, Kobilka BK, Govaerts C (2016) Allosteric regulation of G protein–coupled receptor activity by phospholipids. Nature Chem Biol 12:35–41. https://doi.org/10.1038/nchembio.1960

110. Manna M, Niemelä M, Tynkkynen J, Javanainen M, Kulig W, Müller DJ, Rog T, Vattulainen I (2016) Mechanism of allosteric regulation of β2-adrenergic receptor by cholesterol. eLife 5:e18432. https://doi.org/10.7554/eLife.18432
111. Periole X, Huber T, Marrink S-j, Sakmar TP (2007) G Protein-coupled receptors self-assemble in dynamics simulations of model bilayers. J Am Chem Soc 129:10126–10132. https://doi.org/10.1021/ja0706246
112. Marrink SJ, Tieleman DP (2013) Perspective on the Martini model. Chem Soc Rev 42 (16):6801–6822. https://doi.org/10.1039/C3CS60093A
113. Killian JA (1998) Hydrophobic mismatch between proteins and lipids in membranes. Biochim Biophys Acta 1376:401–415. https://doi.org/10.1016/S0304-4157(98)00017-3
114. Casuso I, Khao J, Chami M, Paul-Gilloteaux P, Husain M, Duneau J-P, Stahlberg H, Sturgis JN, Scheuring S (2012) Characterization of the motion of membrane proteins using high-speed atomic force microscopy. Nature Nanotech 7:525–529. https://doi.org/10.1038/nnano.2012.109
115. Botelho AV, Huber T, Sakmar TP, Brown MF (2006) Curvature and hydrophobic forces drive oligomerization and modulate activity of rhodopsin in membranes. Biophys J 91:4464–4477. https://doi.org/10.1529/biophysj.106.082776
116. Periole X (2016) Interplay of G protein-coupled receptors with the membrane: insights from supra-atomic coarse grain molecular dynamics simulations. Chem Rev 117:156–185. https://doi.org/10.1021/acs.chemrev.6b00344
117. Prasanna X et al (2014) Cholesterol modulates the dimer interface of the β_2-adrenergic receptor via cholesterol occupancy sites. Biophys J 106(6):1290–1300. https://doi.org/10.1016/j.bpj.2014.02.002
118. Provasi D et al (2015) Preferred supramolecular organization and dimer interfaces of opioid receptors from simulated self-association. PLoS Comp Biol 11(3):1–21. https://doi.org/10.1371/journal.pcbi.1004148
119. Koldsø H, Sansom MSP (2015) Organization and dynamics of receptor proteins in a plasma membrane. J Am Chem Soc 137:14694–14704. https://doi.org/10.1021/jacs.5b08048
120. Reinhart G, Desale H, Clemons B, Cahalan SM, Schuerer SC (2012) Crystal structure of a lipid G protein – coupled receptor. Science 335:851–856
121. Jiang Q, Mosberg HI, Porreca F (1990) Modulation of the potency and efficacy of μ-mediated antinociception by delta agonists in the mouse. J Pharmacol Exp Ther 254:683–689
122. Isberg V, Mordalski S, Munk C, Rataj K, Harpsøe K, Hauser AS, Vroling B, Bojarski AJ, Vriend G, Gloriam DE (2016) GPCRdb: an information system for G protein-coupled receptors. Nucleic Acids Res 44(D1):D356–D364. https://doi.org/10.1093/nar/gkv1178
123. Bartuzi D, Kaczor AA, Matosiuk D (2015) Activation and allosteric modulation of human μ opioid receptor in molecular dynamics. J Chem Inf Model 55(11):2421–2434. https://doi.org/10.1021/acs.jcim.5b00280
124. Stansfeld PJ, Sansom MSP (2011) From coarse grained to atomistic: a serial multiscale approach to membrane protein simulations. J Chem Theory Comput 7(4):1157–1166. https://doi.org/10.1021/ct100569y
125. Wassenaar TA, Pluhackova K, Böckmann RA, Marrink SJ, Tieleman DP (2014) Going backward: a flexible geometric approach to reverse transformation from coarse grained to atomistic models. J Chem Theor Comput 10 (2):676–690. https://doi.org/10.1021/ct400617g
126. Lee J, Cheng X, Swails JM, Yeom MS, Eastman PK, Lemkul JA, Wei S, Buckner J, Jeong JC, Qi Y, Jo S, Pande VS, Case DA, Brooks CL, MacKerell AD, Klauda JB, Im W (2016) CHARMM-GUI input generator for NAMD, GROMACS, AMBER, OpenMM, and CHARMM/OpenMM simulations using the CHARMM36 additive force field. J Chem Theor Comput 12(1):405–413. https://doi.org/10.1021/acs.jctc.5b00935
127. Wassenaar TA, Ingólfsson HI, Böckmann RA, Tieleman DP, Marrink SJ (2015) Computational lipidomics with insane: a versatile tool for generating custom membranes for molecular simulations. J Chem Theor Comput 11 (5):2144–2155. https://doi.org/10.1021/acs.jctc.5b00209

Chapter 7

Methods of Exploring Protein–Ligand Interactions to Guide Medicinal Chemistry Efforts

Paul Labute

Abstract

We present a number of techniques to analyze protein–ligand interactions in the context of medicinal chemistry: crystal Contract Preferences, Electrostatic Maps and pharmacophore screening using Hückel Theory. Contact Preferences is a statistical technique to predict hydrophobic and hydrophilic geometry in receptor active sites. Electrostatic Maps use the Poisson-Boltzmann Equation to model solvation effects and are particularly useful for predicting hydrophobic regions. Pharmacophore annotation with Hückel Theory provides finer detail of hydrogen bonding interactions, including CH..O interactions. Applications to AblK:Gleevec and CDK2 virtual screening are presented.

Key words Molecular interactions, Contact statistics, Electrostatic maps, Pharmacophore annotation

1 Introduction

The combination of the widespread availability of X-ray crystal structure coordinates of macromolecular structures (e.g., as found in the Protein Data Bank [1] and affordable 3D computer graphics hardware) has made it possible for scientists to routinely visualize such structures in their efforts to design better inhibitors and drugs.

While visually inspecting atomic coordinates and bonding patterns can be revealing, it is often not easy to perceive structures such as the shape of a receptor's active site. A molecular surface is drawn about the receptor (e.g., a Connolly surface [2] can provide such information in a readily discernable manner). Similarly, a molecular surface constructed from an iso-contour of the van der Waals energy between a receptor and a hypothetical "probe" atom can highlight sterically accessible regions in an active site. Such surfaces are often useful when contemplating modifications to a ligand in order to improve activity or selectivity. Receptor properties such as atom type, electrostatic energy, and hydrogen bond information can be used to color code the molecular surfaces in an effort to

Alexander Heifetz (ed.), *Computational Methods for GPCR Drug Discovery*, Methods in Molecular Biology, vol. 1705,
https://doi.org/10.1007/978-1-4939-7465-8_7,

further assist with the interpretation of the macromolecular structure information.

In medicinal chemistry applications, crystal structures of protein–ligand complexes are used to understand important interactions, such as hydrogen bonds, halogen bonds, ionic and hydrophobic interactions, in an effort to guide ligand modification for improved potency, selectivity, or physical properties. Hydrogen bonding interactions tend to be directional, localized and can be modulated by ligand substituents, especially the weaker CH...O interactions. Hydrophobic and ionic interactions tend to be more diffuse and can be affected by solvation/cavitation effects, which present challenges when attempting to specify a pharmacophore for a particular active site.

In this chapter, we will examine a number of techniques of determining and analyzing non-bonded interactions between proteins and ligands with the intended application of guiding ligand design. This guidance takes the form of understanding existing interactions, predicting potential new or modulated interactions and the specification of pharmacophores for virtual screening.

2 Contact Preferences

The x-ray crystal structures in the Protein Databank and the Cambridge Structural Database [3] are a good source of non-bonded contact information from which statistics can be derived. Methodologies such as X-Site [4] and SuperStar [5] are knowledge-based techniques that use statistical distributions derived from crystallographic data to describe or predict non-bonded contacts between ligands and proteins. The main idea is to describe (statistically) the geometry of non-bonded interactions and use the statistical descriptions to predict the likelihood of interactions in a specific protein active site. Knowledge-based methods are appealing because they rely on experimental data and not molecular mechanics, precise hydrogen coordinates, or other approximations; however, they do rely on sufficient experimental data and appropriate a priori classification of interacting atom types (e.g., sp^2 donor vs. sp^3 acceptor).

We will describe one example of a statistical method, *Contact Preferences* [6], for determining geometric preferences for polar and hydrophobic atoms given the 3D coordinates of a protein receptor. Fundamentally, Contact Preference maps are contours of probability densities indicating a percentage likelihood of a non-bonded contact between a protein receptor and a particular ligand atom type; in other words, the likelihood that a given contact geometry would be observed in the crystallographic databases. Interactions in high-probability regions (e.g., donor and acceptor pair at ideal geometry) are considered good and interactions in low probability regions (e.g., hydrophobic atoms covering a hydrogen bond position) are considered non-optimal.

We are interested in the probability density

$$\Pr(l, x|s)$$

the probability of observing a ligand atom of type l at position x in contact with a structure s consisting of a collection of atoms with types $\{t_i\}$. We approximate.

$$\Pr(l, x \mid s) = \Pr(x \mid l, s)\Pr(l \mid s)$$

$$\approx c \max_{i} \Pr(x \mid l, t_i)\Pr(l \mid t_{i.})$$

where c is a normalization constant and have assumed that we can compose the distribution over the entire structure from the individual receptor atoms.

The Pr $(l \mid t_i)$ term (the probability that a ligand type l is in contact with a receptor type, t_i) is estimated straightforwardly using relative frequencies of atomic contacts in a training set; for example, the relative frequency that the (l, t) pair is observed within 4.5 Å over sidechain-sidechain contacts in the PDB.

Two vectors **u** and **v** (*see* Fig. 1) are used to define a coordinate system about each receptor atom located at Cartesian coordinates, y, depending on hybridization and heavy atom coordination number. In some cases **v** is taken to be 0 (e.g., 3-coordinated sp^2 atoms). For an interacting atom located at x, we define the spherical-type coordinates

$$r = |x - y| \quad a = \frac{180}{\pi} cos^{-1}\left[u \cdot \frac{x - y}{r}\right] \quad p = \frac{180}{\pi} sin^{-1}\left[v \cdot \frac{x - y}{r}\right]$$

where r is the distance between l and t_i, a is the angle relative to **u** in the plane normal to v, and p is angle relative to **u** in the **uv** plane. Assuming independence of the individual coordinates we approximate

$$\Pr(x \mid l, t_i) \approx \Pr(r \mid l, t_i) \Pr(a \mid l, t_i) \Pr(p \mid l, t_i)$$

and estimate the individual one-dimensional probability densities from a collection of sidechain-sidechain (l, t) contacts found in the PDB.

For each ligand receptor atom type pair, histograms are collected over the PDB and these histograms are fitted with analytical distributions. For example, Fig. 2 shows the histograms and fitted distributions for the receptor atom type "oQ1," an sp^2 oxygen with one heavy atom neighbor. One can clearly see the directional preference for the in-plane hydrogen bonding interactions at lone pair directions ($a = 60°$) and the out-of-plane hydrophobic interactions ($p > 45°$).

Figure 3 depicts the 95% probability contour levels for a collection of receptor atom types created from the analytical probability

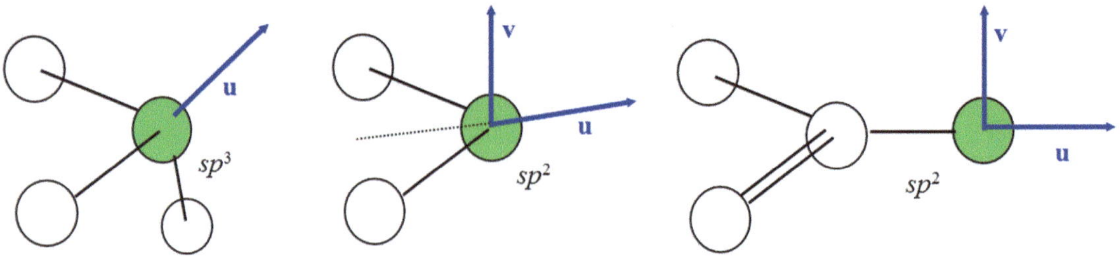

Fig. 1 Two vectors **u** and **v** centered on a receptor atom (*green*) define a local coordinate system; the directions of these vectors depend on the receptor atom's hybridization and heavy coordination number

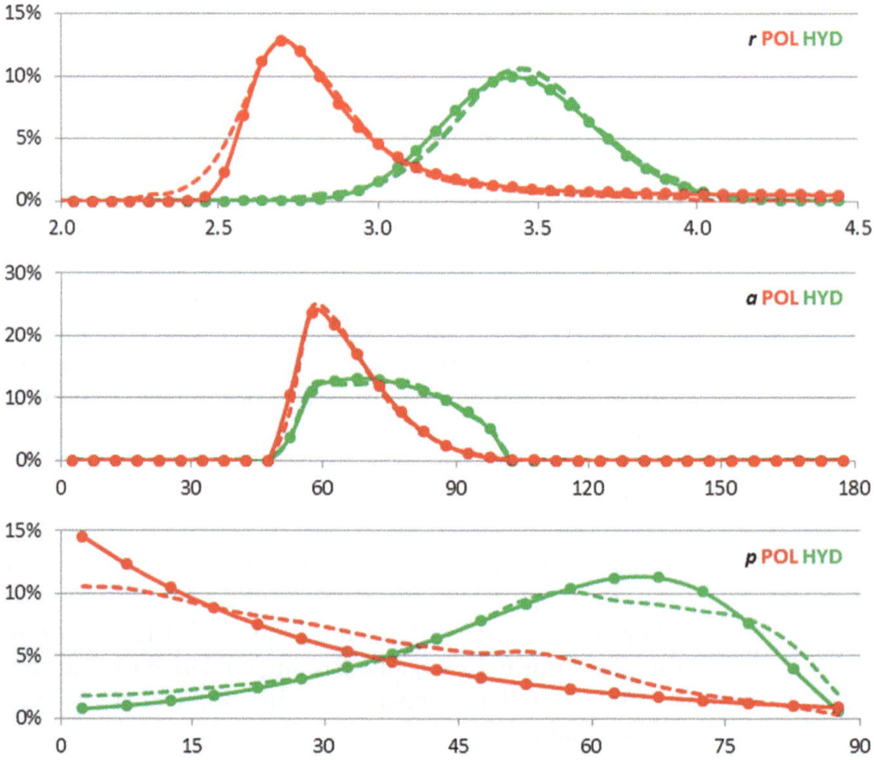

Fig. 2 Distributions of non-bonded contact coordinates between ligand atom types polar (POL in red) and hydrophobic (HYD in *green*) and a receptor atom of type "oQ1", sp^2 oxygen with one heavy neighbor. The *dashed lines* are the histograms collection from sidechain-sidechain contacts in the PDB and the *solid lines* are analytical fitted distributions. The top distribution is for the inter-atom distance, the middle the in-plane angle, and the bottom is the out-of-plane angle (*see* the text)

distributions fitted to PDB statistics. Receptor atom types are denoted by their element (lower case for sp^2) and heavy atom neighbor count. Ligand atom types are either polar (POL) or hydrophobe (HYD). Hydrogens are displayed for reference; however, these were not used in the statistical fit, which is why, for example, the primary amine type "NQ1" does not show directionality with respect to the neighboring hydrogens.

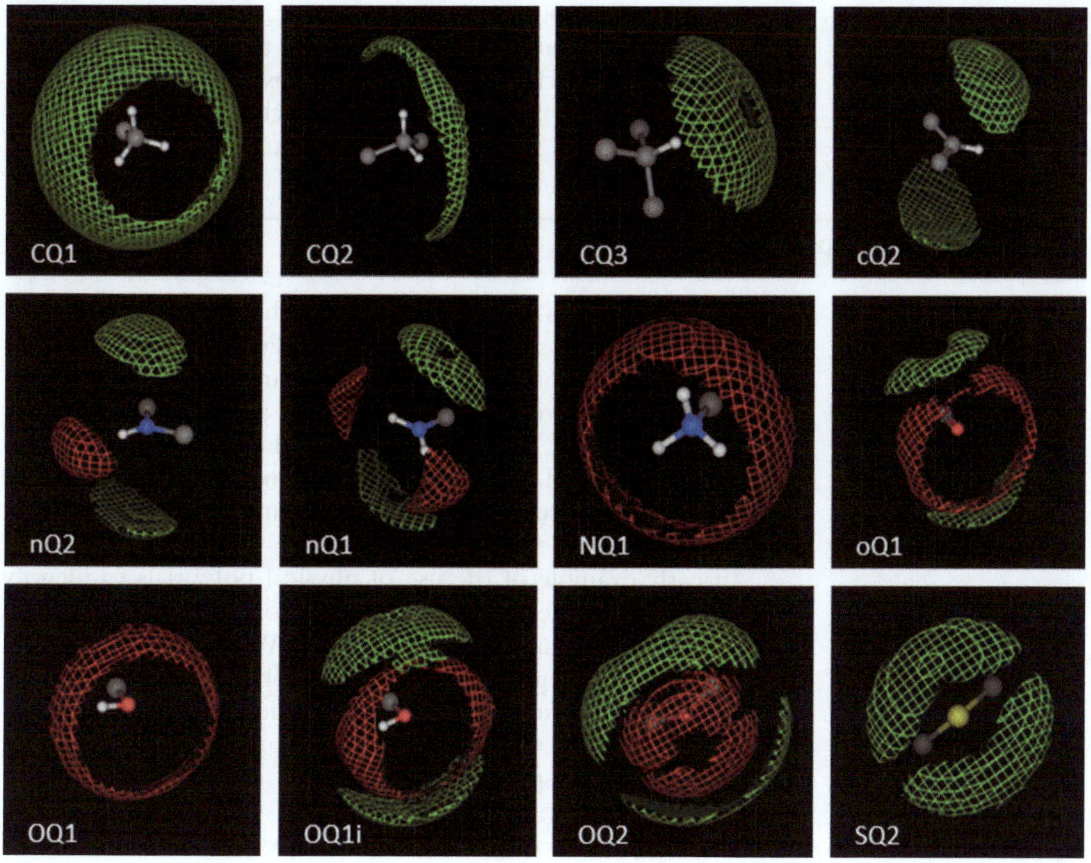

Fig. 3 95% probability contour levels for interactions with a receptor atom type and ligand atom types POL (*red*) and HYD (*green*)

There are a number of uses for such probability contours. Composite contour level plots in receptor active sites can indicate regions of high probability of polar or hydrophobic contacts which can be useful in structure-based design. The directional preferences are also useful for annotating ligand atoms' hypothetical contact partners in pharmacophore 3D search techniques. Also, the statistical distributions can be used to assess the geometry quality of hydrogen bonds in graphical interaction displays.

However, there are notable drawbacks to statistical techniques. First, the method is sensitive to the a priori atom typing which causes a loss of detail of chemical context; for example, phenol oxygens (type "OQ1i") are generally good hydrogen bond acceptors except for weakening due to substituent effects in ligands. The statistics are gathered for *all* phenol oxygens which conflate the chemical contexts leading to potentially misleading contact probabilities. Second, because hydrogen atoms are generally not visible in the PDB there is loss of distinction between donor and acceptor partners, for example, with hydroxyl oxygens. Third, each receptor atom is treated independently; consequently, compositing the maps in a receptor active site neglects correlation and longer range electrostatic effects. Fourth, the probability maps reflect, in some sense, only the frequency of occurrence of contacts in crystal structures and not necessarily a strength or energy value.

3 Electrostatic Maps

Molecular mechanics potentials are one method of addressing some of the deficiencies of statistical approaches. Forcefield terms such as Coulomb or Reaction Field electrostatics and Lennard-Jones van der Waals energies are physically motivated and rely on only a few parameters per atom. In principle, distinct chemical contexts can be better described and interaction correlation effects can be better treated (in contrast to the independent-atom assumption of statistical approaches).

An early example of the use of physical potentials to identify favorable binding sites in a receptor is Goodford's GRID [7, 8] in which iso-energetic surfaces of the van der Waals, electrostatic and hydrogen bond energy between a small probe atom (or fragment) and a macromolecular structure are used to identify favorable binding sites. The surfaces are displayed on a graphics terminal and the binding regions in space are identified by suitably adjusting the energy values used to produce the contour surfaces.

Originally, a Coulomb potential was used; however, this gas-phase model overemphasized the electrostatics interactions. Moreover, hydrophobic interactions were difficult to quantify due to the lack of an implicit solvent model. Subsequently, an implicit solvent model was incorporated based upon spatial differences in the dielectric of solvent versus solute. Notwithstanding the fact that the implicit solvent model took only the macromolecular structure and a particular location of the probe into account, the GRID methodology has found utility in efforts to understand ligand binding [9–11]

A more sophisticated treatment of macromolecular electrostatics uses the Poisson–Boltzmann Equation (PBE) which is a partial differential equation used in the study of electrolytes. Methods to solve the PBE have been the subject of much research [12–14] because of their ability to model implicit solvent and buffer effects. Typically, a macromolecular structure is used to define a region of low dielectric (e.g., 1) in a certain volume and the remaining points of space are assigned a high dielectric (e.g., 80). The solution to the PBE has largely been used to predict pK_a values for ionizable groups or to color code molecular surfaces. Such color coding can be revealing since it presents a self-consistent view of the electrostatic field near a solute with solvent screening taken into account.

It is tempting to use the PBE electrostatic field in GRID-type methodology; however, the application of a single solution is flawed because the electrostatic field depends on the low dielectric cavity which depends on all particles, including the probe. Either a prohibitively long calculation would be required (one PBE solution for every lattice value calculated) or the cavitation effects of the probe would have to be neglected.

An alternative, called Electrostatic Maps [15], uses solutions of the PBE to augment the GRID-type methodology. The starting point is the Poisson equation which is a fundamental equation of physics that relates the equilibrium electrostatic potential φ at a point in space and a charge density ρ at that point:

$$\nabla \cdot \nabla \varphi + \rho = 0$$

where $\nabla \cdot \nabla = \partial^2/\partial x^2 + \partial^2/\partial y^2 + \partial^2/\partial z^2$ is the Laplacian operator. Given a charge density, ρ, one can calculate the electrostatic field φ by solving the partial differential equation. In the special case of a collection of point charges, φ is the familiar Coulomb potential.

Suppose that a charge density (from a protein or small molecule) ρ is interacting with a number of mobile ions (e.g., from salts) each carrying a formal charge, q_I. At thermodynamic equilibrium these mobile ions will be distributed about ρ according to a charge density which we will denote by ρ_I. The electrostatic potential φ from the combined charge densities must satisfy the Poisson equation

$$\nabla \cdot \nabla \varphi + \rho_I + \rho = 0$$

For multiple types of mobile ions, the foregoing equation is easily extended by replacing the lone ρ_I with a sum of similar densities (one for each type). For the purposes of exposition, only one mobile ion type will be considered.

If the mobile ions are spatially restricted (e.g., the ions cannot be inside a protein and must remain in solvent), then such a restriction can be specified with a spatial potential u_I that has high energy in the forbidden regions of space and low energy in the allowed regions of space. Taken with the electrostatic potential φ the potential energy of a mobile ion at a particular point in space will be, at equilibrium, $q_I\varphi + u_I\, q_I$. We now assume that the ion location density, ρ_I, follows a Boltzmann distribution and so the ion charge density is its formal charge multiplied by the probability it will be located at a particular point in space

$$\rho_I(x) = q_I c\, e^{-[q_I\varphi(x) + u_I(x)]/kT}$$

where c is normalization constant, k is Boltzmann's constant, and T is the temperature of the system. If φ and u_I are zero sufficiently far from ρ then at such distances the exponential term tends to one which means that c may be taken as the bulk concentration, C_I, of the mobile ions (e.g., in mol/L). The combination of the previous two equations results in a Poisson-Boltzmann type equation, namely that

$$\nabla\cdot\nabla\varphi + q_I C_I\, e^{-[q_I\varphi(x)+u_I(x)]/kT} + \rho = 0$$

which specifies not only the resulting electrostatic potential, but also the ionic charge densities both of which result from an original charge density ρ.

The Electrostatic Maps methodology consists of the display of the equilibrium potential energy iso-surfaces of mobile "oxygen" and "hydrogen" particles carrying partial charges q_O and q_H, respectively, and subject to spatial Lennard-Jones van der Waals potentials u_O and u_H, respectively. Upon (numerical) solution of the above PBE, we have that

$$\nabla\cdot\nabla\varphi + Cq_O e^{-[q_O\varphi+u_O]/kT} + 2Cq_H e^{-[q_H\varphi+u_H]/kT} + \rho = 0$$

where C is the bulk concentration of water and ρ is the charge distribution of a macromolecule under consideration. The potential function $q_O\varphi + u_O$ is the potential energy landscape of the implicit "oxygen" particle and $q_H\varphi + u_H$ is the potential energy landscape of the implicit "hydrogen" particle. These two potentials represent the free energy minimizing energy landscapes of the neutralizing, or screening, solvent particles. In some sense these represent the energies required to displace solvent at particular points in space under the assumption that the remaining points in space are ideally (in a mean field sense) solvated or neutralized. An additional "hydrophobic" potential, $-(q_O + q_H)\varphi + u_C$, is defined (where u_C is a Lennard-Jones potential for carbon) which will have negative values in neutral regions (absence of "O" or "H" particles).

Fundamentally, Electrostatic Maps are plots of van der Waals and electrostatic energies of implicit particles and, as such, they are similar in spirit to the GRID methodology with the mobile particles playing the role of the "probes." The difference, however, is that a single, nonlinear, self-consistent calculation is performed in which each implicit mobile particle feels the effects of the other implicit mobile particles in order to better model screening effects. This nonlinearity and self-consistency ultimately manifests itself in a more localized distribution of implicit particle densities. Approximate, or even exact $\varepsilon = 80$ solvation models can fail to adequately screen chemical groups leading to electrostatic domination of a region because there is a nonlinear screening effect in the buffered solution that cannot be modeled by linear ε dielectric models.

A simple example will serve to illustrate this point. Figure 4 shows the contour plots of the $q_O\varphi + u_O$ and $q_H\varphi + u_H$ functions for two different solvent models. On the left, a linear screening dielectric of $\varepsilon = 80$ outside of a small molecule was used in the PBE without mobile particles. One can see the large extent of the blue (preference for positive) contour created by the carboxylate group.

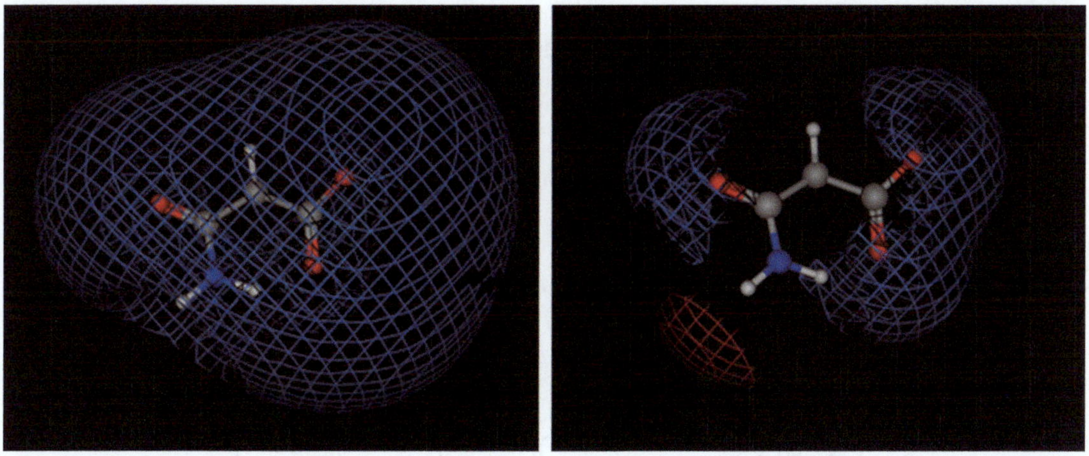

Fig. 4 (*Left*) Electrostatic and van der Waals energy contour plots of an anionic small molecule calculated by solving the Poisson equation in a polar medium ($\varepsilon = 80$). Positive preference is indicated in *blue* and negative preference is indicated in *red*. Notice that the carboxylate dominates the entire region and no red contour is visible. (*Right*) Electrostatic with van der Waals contour plots of the same small molecule calculated by solving the nonlinear PBE with "O" and "H" mobile screening particles (*see* the text). Notice that the screening effects are much stronger and that more localized positive and negative preferences are visible

On the right are the contours of the proposed Electrostatic Map calculation in which the contours are much more localized; so much so, that a preference for negative (red) appears next to the hydrogen in the amide group. Charged groups in aqueous environments are neutralized by the "oxygen" mobile counter-ions (modeled in Electrostatic Maps with the "oxygen" and "hydrogen" particles) that provide nonlinear screening effects and must be modeled using nonlinear Boltzmann term of the PBE. By using the mobile self-consistent Boltzmann particles of the PBE, the carboxylate group will be sufficiently screened so that groups beyond its first solvation shell will not feel an unrealistically strong electrostatic field.

We shall now use the AblK:Gleevec complex to show how Electrostatic Maps can be used to rationalize the activity trends in a ligand series. The Bcr-Abl chimeric protein has been implicated in the development of chronic myeloid leukemia (CML) and acute lymphoblastic leukemia [16, 17] Gleevec (Imatinib) is a small molecule inhibitor of the dephosphorylated (inactive) form of Bcr-Abl kinase currently being used clinically as a treatment for CML. Patients can develop resistance to Gleevec if Bcr-Abl expression is enhanced or if point mutations occur within the Abl kinase domain [17] and consequently there is interest in further developing next-generation Bcr-Abl kinase inhibitors.

It has been shown [18] that the Gleevec piperazine ring fits in the solvent exposed opening of the pocket and forms hydrogen bonds to Thr315, Ile360, Asp381, and Met318 (*see* Fig. 5). One

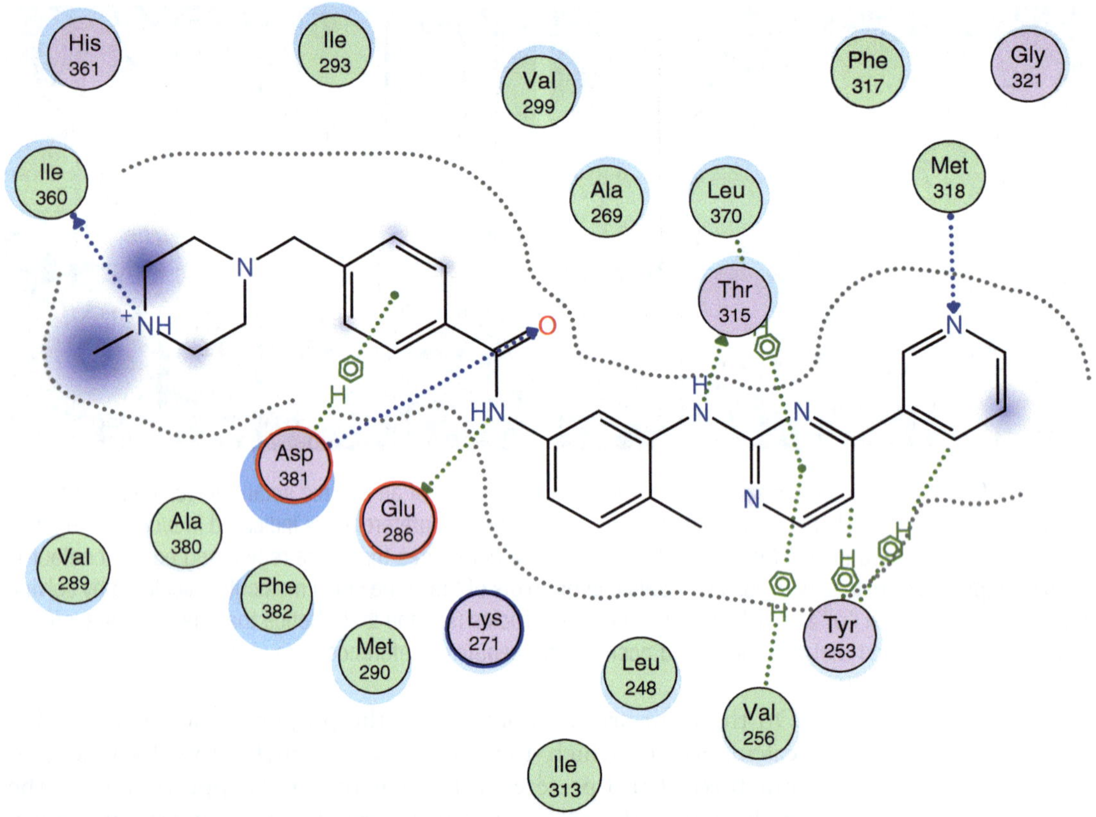

Fig. 5 A diagram of the interactions of Gleevec with Abl kinase in PDB:1IEP. Residues are represented by *discs* with polar residues in *pink* (acidic residues with a red contour and basic residues with a *blue contour*) and hydrophobic residues in *green*. *Dotted arrows* indicate hydrogen bonding to sidechain (*green*) and backbone (*blue*) atoms respectively. Blue "clouds" on ligand atoms indicate the solvent-exposed surface area of ligand atoms (larger means more exposure). Light-blue "halos" around residues indicate the degree of surface area contact with ligand atoms (larger means more contact). The dotted contour reflects steric room for methyl substitution.

would expect that an Electrostatic Map calculated from AblK alone to reproduce these interactions and possibly suggest regions of possible optimization.

The AblK protein was prepared for electrostatic analysis as follows. His361 was neutralized according to its immediate environment so that it could form hydrogen bonds to the backbone nitrogen of Asp363 and the backbone oxygen of Ala380. The remaining residues were assigned standard ionization states and protons were added in extended conformation. Partial charges were then calculated using the MMFF94 forcefield [19] and the electrostatic map was calculated from the receptor structure alone near the active site. Figure 6 shows the results of the calculation with the positive (blue) density plotted at −2.5 kcal/mol, the negative (red) density at −2.0 kcal/mol, and the hydrophobic (green) density at −3.0 kcal/mol. The calculated Electrostatic

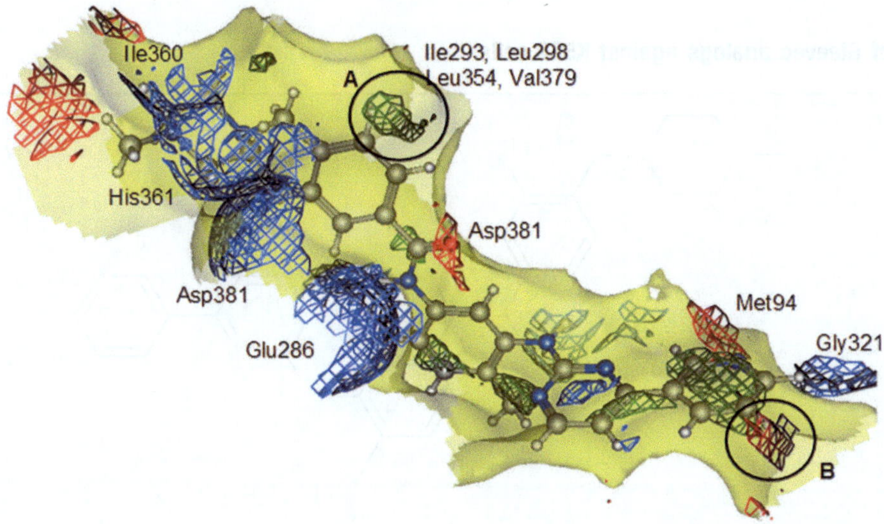

Fig. 6 The Electrostatic Map for the active site of AblK:Gleevec (1IEP) calculated from receptor atoms only; positive preference is indicated in *blue*, negative in *red*, and neutral in *green*. The *yellow* interaction surface shows the steric boundary of the pocket. *A* and *B* denote map densities not filled by Gleevec (*see* the text)

Map (*see* Fig. 6) agrees well with clear positive and negative charge densities corresponding to all of the important hydrogen bonding interactions of Gleevec. The predicted hydrophobic regions overlap well with both the methyl group attached to the phenyl linker as well as the pyridine ring fragment indicating strong favorable hydrophobic interactions.

There are two significant regions in the AblK Electrostatic Map that are not filled by Gleevec and therefore are potential regions of ligand optimization. The unfilled hydrophobic density (Fig. 6 region A) suggests that a small hydrophobic group such as a halogen, CH_3, or CF_3 would have favorable interactions with AblK and could be a location for optimization. This is in fact the case; Asaki et al. [18] reported biological activity data for a series of 3-substituted benzamide derivatives as Bcr-AblK inhibitors reproduced in Table 1; these compounds differ from Gleevec only by the phenyl ring substituent in the hydrophobic subpocket (created by Ile293, Leu298, Leu354 and Val379). The IC_{50} data of Table 1 shows that small hydrophobic substituents improve activity, in particular, the 3-trifluromethylated benzamide. The same 3-trifluoromethyl moiety is also present in NS-187 [20], a recently proposed Gleevec analog that not only binds more strongly to AblK than Gleevec, but maintains potency in the presence of many known point mutations. It is interesting that the gain in potency of approximately -2.1 kcal/mol $= -kT \ln (IC_{50}(CF_3)/IC_{50}(\text{Gleevec})$ of the CF_3 substituent is in rough agreement with the -3 kcal/mol hydrophobic contour value of the Electrostatic Map. This is most likely due to the fact that the Electrostatic Map (free) energies estimate the solvation effects that largely determine differences in affinity in this highly homologous series.

Table 1
Activity of Gleevec analogs against K562 cells

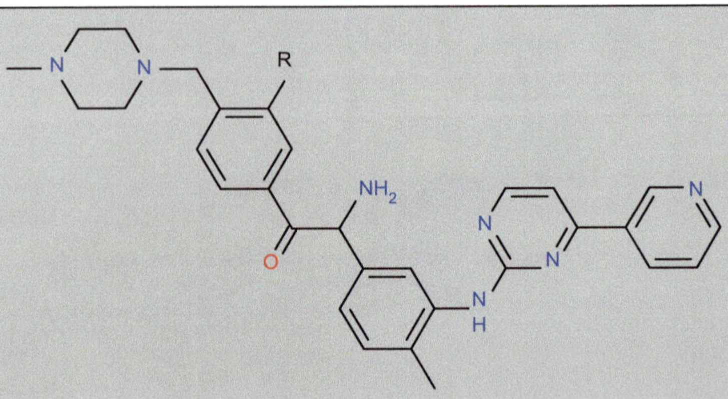

Compound	R	IC_{50} (nM)
Gleevec	H	183
1	F	63
2	Cl	10
3	Br	7
4	I	10
5	CF_3	5

The second unfilled region of the Electrostatic Map for AblK (Fig. 6 region B) suggests a preference for negative charge. This negative density (created by the hydroxyl of Tyr253) lies about 1.5 Å from the *meta* position of the pyridine ring at about 45° out-of-plane. Consequently, it is a difficult feature to reach. Asaki et al. [18] synthesized pyrimidine derivatives instead of the pyridine ring to compensate for the increased hydrophobicity caused by trifluoromethyl group addition to Gleevec and to avoid interfering with the Tyr253 (which stabilizes the inactive form of AblK) with bulky substitutions. Their results show no significant improvement in activity against Bcr-Abl K562 cell lines (estimated $\Delta\Delta G$ of 0.13 kcal/mol). This experimental conclusion could possibly have been elucidated by the analysis of the Electrostatic Map around the pyridine moiety because the negative (red) density is relatively far from the contemplated pyrimidine nitrogen. A contour level of −1 kcal/mol begins to show overlap with the contemplated nitrogen; however, since the hydroxyl and the nitrogen would be far from optimal hydrogen bonding geometry it is reasonable to conclude that an interaction would be weak, even if mildly favorable.

The Electrostatic Maps methodology has a number of advantages. The method is physically motivated and relies on molecular mechanics parameters (partial charges and van der Waals parameters). It is fast to calculate, requiring a few seconds on common computer hardware. Correlation and solvation effects are taken into account in a self-consistent way (in contrast to statistical composites). The method is particularly useful for identifying important hydrophobic (electrostatically neutral) regions in an active site (e.g., for pharmacophore specification); this is in contrast to purely geometric approaches such as Contact Preferences which neglect longer range interactions. More sophisticated approaches such as 3D–RISM [21] have been the subject of much research and can deal with higher order correlations and cavitation effects. These techniques are much slower to calculate but can identify "non-standard" hydrophobic regions; for example, hydrophobic regions in the immediate vicinity of polar atoms created by the shape of an active site.

4 Pharmacophore Screening

The virtual screening of 3D conformation databases using a pharmacophore as a query is a widely used technique in computer-aided drug discovery, with hundreds of published accounts of its application since its advent in the late 1970s [22, 23] and the first published account of a successful virtual screen in 1992 [24]. Typically, a pharmacophore query (possibly derived from a receptor structure) specifies a collection of molecular features (hydrogen bond donor/acceptor, cation, anion, hydrophobe, etc.) intended to capture the essentials of ligand binding interactions to a receptor, along with inter-feature distance constraints. The query is applied to a database of small-molecule conformations to search for arrangements of features in 3D that match the pharmacophore query with respect to both feature type and distance constraints.

Commonly, features of ligands are assigned by substructure searching methods from a fixed template or rules collection. For example, a typical hydrogen bond donor heavy atom can be encoded with the SMARTS pattern "[#7,#8;!H0]" meaning "nitrogen or oxygen with at least one hydrogen." Assigning hydrogen bond acceptor features is not as straightforward—electron withdrawal and hyper-conjugation must be taken into account since, among others, conjugated ether oxygen and nitro oxygen are not good hydrogen bond acceptors [25]. The problem becomes particularly acute when one attempts to accurately encode CH hydrogen bond donors (e.g., in kinase inhibitors [26]). Electron withdrawal and substituent effects are very difficult to consistently encode in a (small) collection of substructure patterns.

An alternative to substructure patterns for assigning hydrogen bond donor and acceptor features is to use Hückel Theory quantum calculation to provide a consistent treatment of electronic effects. A small organic molecule is specified as a collection of atoms, each with an atomic number, Z, a formal charge, Q, and a collection of bonds $\{(i,j)\}$ between the atoms. A self-consistent modified Hückel Theory calculation is performed on the chemical graph (without coordinates) using a well-known method [27]; the calculation produces a partial charge (or electron occupancy) per atom and a fractional π bond order per bond.

The partial charges and fractional bond orders are used to determine a hydrogen bond "donor strength," ε_{don}, and "acceptor strength," ε_{acc}, per atom using the methods described by Gerber [28] and are the key component of the non-bonded terms of the MAB forcefield [27]. In this forcefield, the energy of a hydrogen bond at ideal geometry is the product of the donor and acceptor strengths, $E_{\text{HB}} = \varepsilon_{\text{don}} \times \varepsilon_{\text{acc}}$; for example, an ideal water-water hydrogen bond has an energy of 1.56 kcal/mol. In the development of the strength coefficients, Gerber demonstrated their clear relationship with free energies of hydration.

It seems natural to use strength coefficient thresholds to define donors and acceptors in small molecules; in other words, a large enough ε_{acc} value would indicate an acceptor and a large enough ε_{don} value would indicate a donor. Using $\varepsilon_{\text{acc}} > 0.83$ (a hydrogen bond energy with water greater than 1 kcal/mol) we find that indicated oxygen atoms (in red) in the following structures

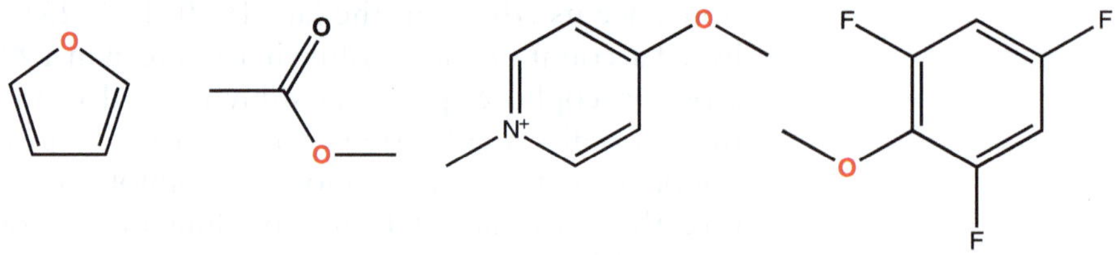

are not deemed acceptors, along with other similar cases that normally would have to be treated as special case exceptions in rule-based systems, while the carbon atoms in isonitrile and carbon monoxide were deemed acceptors (e.g., in metal ligation). Conventional alcohols, phenols, alphatic amines, etc. are properly annotated. Generally, there is very good agreement with hand curated acceptor rules. Using the same principle for the ε_{don} cutoff, we find very good agreement with hand curated rules. Alcohols, phenols, conjugated amines, conjugated thiols were deemed donors, while neutral aliphatic amines and aliphatic thiols were not deemed donors unless sufficient electron withdrawing groups were present (e.g., FCH_2SH)

Unfortunately, the strength threshold scheme does not produce satisfactory separation for CH donors: there does not appear to be any single a priori cutoff value that satisfactorily separates CH donors from non-CH donors, especially in substituted heterocycles. This is due to the fact that CH hydrogen bonds tend to be weak and have a rather continuous strength distribution. In fact a similar issue exists for borderline acceptors; for example, the following structures

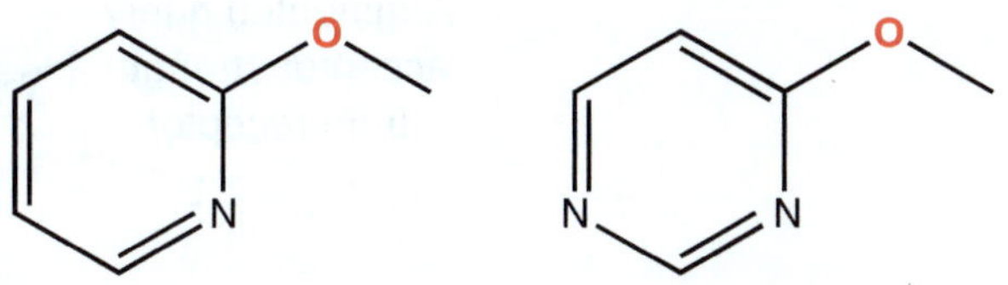

straddle the $\varepsilon_{acc} > 0.83$ cutoff line, which in any case is somewhat arbitrary. These borderline examples highlight that what is required is to model the *interaction* between the small molecule and a hypothetical receptor in the pharmacophore query itself rather than an isolated small-molecule feature without context. In other words, the pharmacophore query should contain (hypothesized) information about the receptor's hydrogen bonding partner atom. For example, a strong acceptor in the receptor could match strong or weak donors in the ligand while a weak acceptor in the receptor should only match strong donors in the ligand.

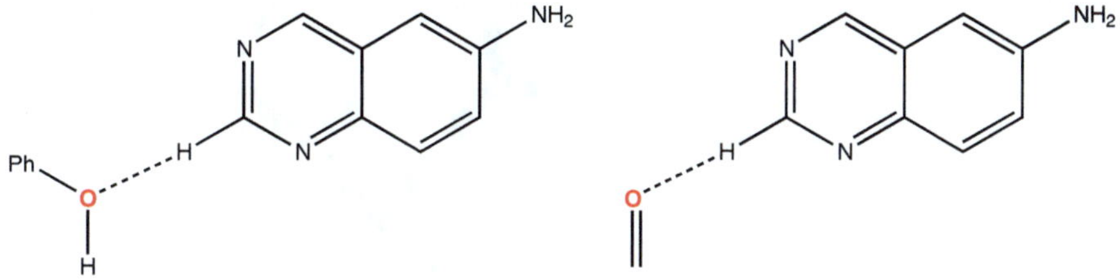

A CH donor interacting with a Tyr OH (above) is an example of a weak acceptor—weak donor interaction with $\varepsilon_{don} = 0.57$ $\varepsilon_{acc} = 0.87$ and $E_{HB} = 0.50$ kcal/mol, whereas a backbone carbonyl oxygen produces a strong acceptor—weak donor interaction with $\varepsilon_{don} = 0.57$ $\varepsilon_{acc} = 2.00$, and $E_{HB} = 1.14$ kcal/mol. Clearly, the threshold comparison should take place on the hydrogen bond energy of the interaction and not the individual ligand strengths. In other words, a CH is a donor if it interacts with a suitably strong acceptor, otherwise it is not.

To make use of this concept, a pharmacophore query is augmented with a strength parameter on each hydrogen bond feature. In this way, the query becomes a kind of receptor model that contains inter-feature distance constraints, feature types, and a strength coefficient intended to model the nature of the receptor's corresponding hydrogen bond partner.

Figure 7 shows an augmented pharmacophore query; the augmented query (magenta sphere) contains information, either known or hypothesized, about the strength of the receptor's acceptor: $\varepsilon_{acc} = 2.1$ (a typical carbonyl oxygen value). The donor strength of the ligand is calculated as the search proceeds. If the geometric and feature type constraints are met, then the strength product is formed $0.8 \times 2.1 = 1.7$ kcal/mol. If this interaction energy is greater than a user-supplied threshold, say 1 kcal/mol, the structure is emitted as a hit, otherwise it is rejected.

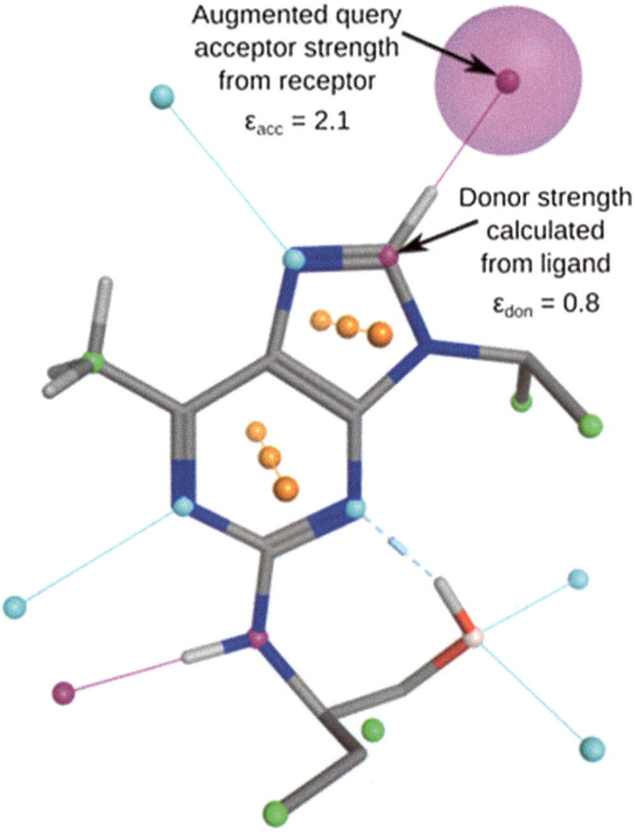

Fig. 7 The augmented pharmacophore query for CDK2 hinge binders with CH donors; the magenta sphere is a query feature with an encoded hydrogen bond acceptor strength of 2.1 of a hypothesized receptor atom. The small spheres represent potential pharmacophore features calculated from the ligand

CH donor interactions are often seen in ligands binding to the GluPheLeu hinge in kinases. Hinge binders form tridentate bond interactions with the backbone of the hinge binding in a similar fashion to ATP. We obtained the crystal coordinates of CDK2 in complex with pyrazolo[4,3-d]pyrimidine (PDB:3PJ8). The crystal structure was prepared with Protonate3D [29] and the coordinates refined. A pharmacophore query was constructed in an attempt to replace the pyrazolopyrimidine scaffold, while maintaining interactions with the hinge so that the substituent presentation vectors would be preserved. The query contained four features: two aromatic centroids, a donor, and an acceptor feature. The strength coefficient of the donor feature was set to 1.91 to model the acceptor strength of the carbonyl oxygen of the backbone of Glu81 in CDK2. Figure 8 (left) depicts the cocrystallized staring ligand in CDK2 along with the pharmacophore query.

We searched a 3D fragment database using the modified pharmacophore search (including the Hückel strengths) and the result was 66 hits that satisfied the pharmacophore query and the

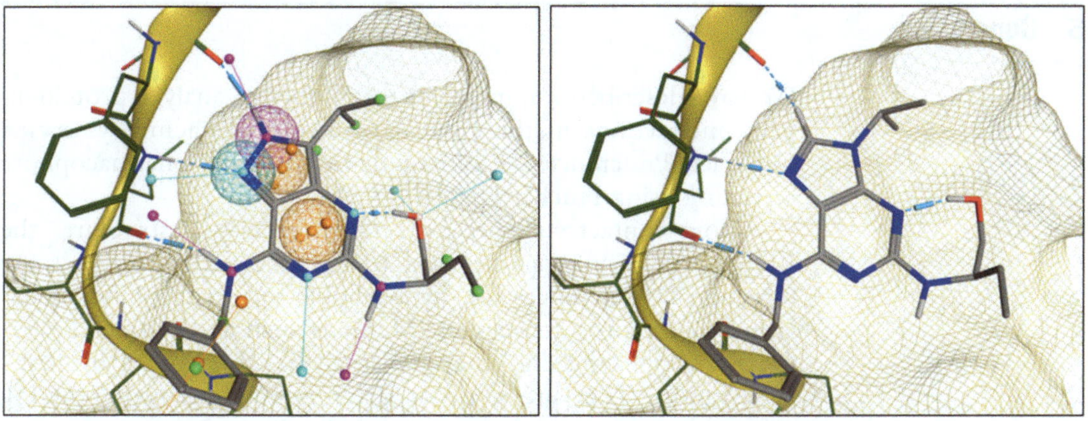

Fig. 8 *Left*: pyrazolo[4,3-d]pyrimidine cocrystallized with CDK2 (PDB:3PJ8) along with the four feature pharmacophore queries used in a scaffold replacement experiment. *Right*: top ranking CH donor imidazopyrimidine scaffold with attached substituents in a calculated binding mode with CDK2

substituent presentation vectors. Among the hits were conventional hydrogen bonding scaffolds (the parenthesized quantities are the Hückel hydrogen bond energies to the hinge, in kcal/mol):

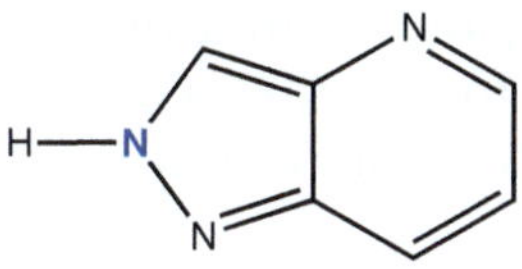

pyrazolopyridine
(6,67)

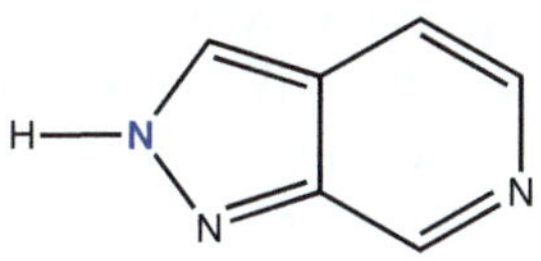

pyrazolopyridine
(6.69)

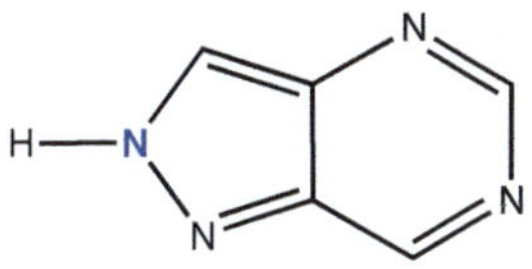

pyrazolopyrimidine
(6.88)

Note that the search recovered the original pyrazolo-pyrimidine scaffold. The search also retrieved scaffolds with CH donors that interacted with the hinge.

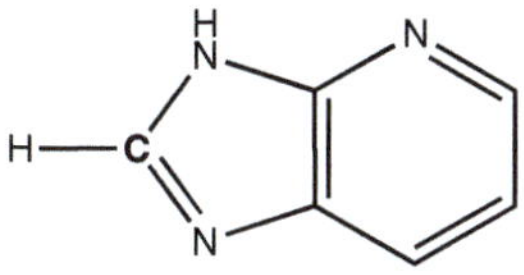

imidazopyridine
(1.25)

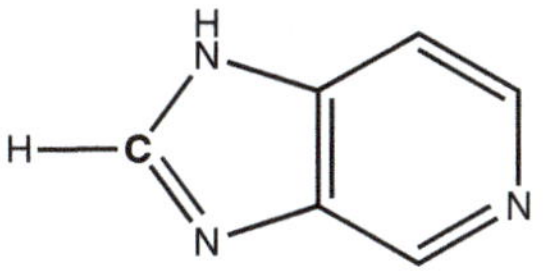

imidazopyridin e
(1.21)

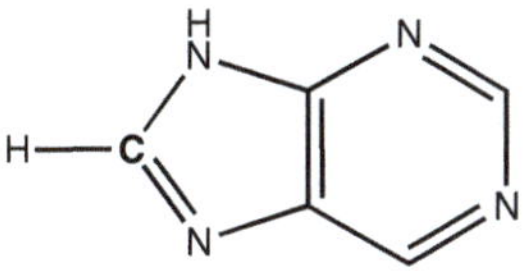

imidazopyrimidine
(1.34)

Note that the imidazopyrimidine framework, the top ranking CH scaffold, results in the compound seliciclib (purvalanol class of molecules); *see* Fig. 8 (right). A similar search using conventional pharmacophore typing and the identical pharmacophore query resulted in only 35 hits, none of which had the CH donor motif.

5 Conclusion

We have described a number of techniques to analyze protein–ligand interactions in the context of medicinal chemistry: crystal Contract Preferences, Electrostatic Maps, and pharmacophore screening using Hückel Theory.

Atom contact statistics extracted from crystal structures the PDB or the CSD can be used to describe and/or predict the directional preferences of hydrophobic or hydrophilic ligand atoms in receptor active sites. Methods such as Contract Preferences are appealing since they are largely based on an experimental observation. They are good at determining hydrogen bonding geometric preferences but poorer at identifying important hydrophobic regions in an active site. This failing is largely due to the neglect of long-range forces and correlation effects.

Methods based upon the Poisson-Boltzmann Equation such as Electrostatic Maps can capture long-range forces and solvation effects. These methods require assignment of hydrogen coordinates to crystal structures and atomic partial charges which capture local chemical context in ligands. The details of polar interactions are generally in broad agreement with statistical methods; however, important hydrophobic regions in an active site are more readily identified. An application to the AblK:Gleevec complex demonstrated the utility of the Electrostatic Maps method for structure-based design.

Fine detail of hydrogen bonding interactions, including the weaker CH donor interactions, requires a more sophisticated treatment. We presented a method based upon Hückel Theory to assign hydrogen bond strengths to protein and ligand atoms in the context of 3D pharmacophore search. An application to CDK2 was presented showing a replacement scaffold including a CH..O interaction could be determined with pharmacophore search.

References

1. Berstein FC, Koetzle TF, Williams GJB, Meyer EF Jr, Brice MD, Rodgers JR, Kennard O, Shimanouchi T, Tasumi M (1977) The protein data bank: a computer-based archival file for macromolecular structures. J Mol Biol 112:535–542
2. Connolly ML (1983) Solvent-accessible surfaces of proteins and nucleic acids. Science 221:709–713
3. Groom CR, Bruno IJ, Lightfoot MP, Ward SC (2016) The Cambridge structural database. Acta Cryst B72:171–179
4. Laskowski RA, Thornton JM, Humblet C, Singh J (1998) X-SITE: use of empirically derived atomic packing preferences to identify Favorable interaction regions in the binding sites of proteins. J Mol Biol 259:175–201
5. Nissink JWM, Verdonk ML, Klebe G (2000) Simple knowledge-based descriptors to predict protein-ligand interactions. Methodology and validation. J Comput Aided Mol Des 14:787–803
6. Labute P (2001) Contact preference maps. Molecular operating environment version 2001.01, Chemical Computing Group Inc., 1010 Sherbrooke St. W. #910, Montreal, QC, Canada H3A 2R7
7. Goodford PJ (1985) A computational procedure for determining energetically favorable

binding sites on biologically important macromolecules. J Med Chem 28:849–856

8. Wade RC, Clark KJ, Goodford PJ (1993) Further development of hydrogen bond functions for use in determining energetically Favorable binding sites on molecules of known structure. 1. Ligand probe groups and the ability to form two hydrogen bonds. J Med Chem 36:140–147
9. Pastor M, Cruciani G, Clementi S (1997) Smart region definition: a new way to improve the predictive ability and interpretability of three-dimensional quantitative structure-activity relationships. J Med Chem 40:1455–1464
10. Melani F, Gratteri P, Adamo M, Bonaccini C (2001) FILO (field interaction ligand optimization): a simplex strategy for searching the optimal ligand interaction field in drug design. J Comput Aided Mol Des 15:57–66
11. Crivori P, Zamora I, Speed B, Orrenius C, Poggesi I (2004) Model based on GRID-derived descriptors for estimating CYP34A enzyme stability of potential drug candidates. J Comput Aided Mol Des 18:155–166
12. Gilson M, Sharp KA, Honig B (1987) Calculating electrostatic interactions in biomolecules: method and error assessment. J Comput Chem 9:327–335
13. Sharp KA, Honig B (1990) Calculating Total electrostatic energies with the non-linear Poisson–Boltzmann equation. J Phys Chem 94:7684–7692
14. Grant JA, Pickup BT, Nicholls A (2001) A smooth permittivity function for Poisson-Boltzmann solvation methods. J Comput Chem 22:608–640
15. Labute P (2006) Electrostatic maps. Molecular operating environment version 2006.08, Chemical Computing Group Inc., 1010 Sherbrooke St. W. #910, Montreal, QC, Canada H3A 2R7
16. Clark SS, McLaughlin J, Timmons M, Pendergast AM, Ben-Neriah Y, Dow LW, Crist W, Rovera G, Smith SD, Witte ON (1988) Expression of a distinctive BCR-ABL oncogene in Ph1-positive acute lymphocytic Leukemia (ALL). Science 239:775–777
17. Cortes JE, Talpaz M, Beran M, O'Brien SM, Rios MB, Stass M, Kantarjian HM (1995) Philadelphia chromosome-negative chronic Myelogenous Leukemia with rearrangement of the breakpoint cluster region: long-term follow-up results. Cancer 75:464–470
18. Asaki T, Sugiyama Y, Hamamoto T, Higashioka M, Umehara M, Naito H, Niwa T (2006) Design and synthesis of 3-substituted Benzamide derivatives as bcr-abl kinase inhibitors. Bioorg Med Chem Lett 16:1421–1425
19. Halgren TA (1995) The Merck molecular force field. J Comput Chem 20:720–774
20. Kimura S, Nalto H, Segawa H, Kuroda J, Yuasa T, Sato K, Yokota A, Kamitsuji Y, Kawata E, Ashihara E, Nakaya Y, Naruoka H, Wakayama T, Nasu K, Asaki T, Niwa T, Hirabayashi K, Maekawa T (2005) NS-187, a potent and selective dual Bcr-Abl/Lyn tyrosine kinase inhibitor is a novel agent for Imatinib-resistant Leukemia. Blood 106:3948–3954
21. Kovalenko A, Hirata F (1999) Self-consistent description of a metal-water Interface by the Kohn-sham density functional theory and the three dimensional reference interaction site model. J Chem Phys 110:10095–10112
22. Gund P (1979) Pharmacophoric pattern searching and receptor mapping. Ann. Rep Med Chem 14:299–308
23. Marshall GR, Barry CD, Bosshard HE, Dammkoehler RA, Dunn DA (1979) The conformational parameter in drug design: the active analog approach. In: Olson EC, Christoffersen RE (eds) Computer-assisted drug design. American Chemical Society, Columbus, OH, pp 205–226
24. Martin YC (1992) 3D Database searching in drug design. J Med Chem 9:1649–1964
25. Böhm H-J, Brode S, Hesse U, Klebe G (1996) Oxygen and nitrogen in competitive situations: which is the hydrogen-bond acceptor? Chem Eur J 2:1509–1513
26. Pierce AC, Sandretto KL, Bemis GW (2002) Kinase inhibitors and the case for CH...O hydrogen bonds in protein-ligand binding. Proteins 49:567–576
27. Gerber PR, Müller K (1995) MAB, a generally applicable molecular force field for structure modelling in medicinal chemistry. J Comput Aided Mol Des 9:251–268
28. Gerber PR (1998) Charge distribution from a simple molecular orbital type calculation and non-bonding interaction terms in the force field MAB. J Comput Aided Mol Des 12:37–51
29. Labute P (2008) Protonate 3D: assignment of ionization states and hydrogen coordinates to macromolecular structures. Protein Struct Funct Bioinform 75:187–205

Chapter 8

Exploring GPCR-Ligand Interactions with the Fragment Molecular Orbital (FMO) Method

Ewa I. Chudyk, Laurie Sarrat, Matteo Aldeghi, Dmitri G. Fedorov, Mike J. Bodkin, Tim James, Michelle Southey, Roger Robinson, Inaki Morao, and Alexander Heifetz

Abstract

The understanding of binding interactions between any protein and a small molecule plays a key role in the rationalization of affinity and selectivity. It is essential for an efficient structure-based drug design (SBDD) process. FMO enables ab initio approaches to be applied to systems that conventional quantum-mechanical (QM) methods would find challenging. The key advantage of the Fragment Molecular Orbital Method (FMO) is that it can reveal atomistic details about the individual contributions and chemical nature of each residue and water molecule toward ligand binding which would otherwise be difficult to detect without using QM methods. In this chapter, we demonstrate the typical use of FMO to analyze 19 crystal structures of β1 and β2 adrenergic receptors with their corresponding agonists and antagonists.

Key words GPCR G-protein-coupled receptors, Chemical interactions, Pair-interaction energy, Drugs, Receptor, Modeling, QM, Quantum Mechanics, FMO, Fragment Molecular Orbitals method, CADD, Computer-Aided Drug Design, SBDD, Structure Based Drug Design, GAMESS, General Atomic and Molecular Electronic Structure System, PIEDA, Pair Interaction Energies Decomposition Analysis

1 Introduction

1. G-protein-coupled receptor (GPCR)–ligand interactions are fundamental to almost all processes occurring in living organisms, and as such it is perhaps unsurprising that they are the targets of about 40% of all prescribed drugs [1–3]. What is surprising is that these drugs only target around 50 of the 800 known GPCRs [4]. Thus, there is huge potential with respect to the number of targets for new therapies to be designed against [5].
2. Further progress in drug discovery for GPCRs is highly dependent upon gaining an in-depth understanding of the

Alexander Heifetz (ed.), *Computational Methods for GPCR Drug Discovery*, Methods in Molecular Biology, vol. 1705, https://doi.org/10.1007/978-1-4939-7465-8_8, © Springer Science+Business Media LLC 2018

structure-activity relationships (SAR) and the individual interactions between the receptor and a small molecule (drug candidate) [4, 6–8]. The efficiency and cost-effectiveness of the drug-discovery process can be accelerated by the availability of structural data regarding the target protein, and by the reliability of the computational tools for data exploration [6–8]. However, even with the crystal structure "visual inspection" and force field-based molecular mechanics (MM) calculations often used for the rationalization of ligand-protein potency cannot always explain the full complexity of molecular interactions [9].

3. There is increasing evidence [9–12] that there are a number of non-intuitive interactions such as CH/π [13, 14], halogen/π [15], cation/π [16], non-classical H-bonds [17], and others that play important roles in protein-ligand binding that are not sufficiently parameterized in the most popular force fields (FF) [11]. The application of quantum mechanical (QM) methods has been employed to improve the reliability of the exploration of protein-ligand interactions [18, 19]. Historically, the application of high level ab initio quantum mechanical calculations was limited to molecular systems consisting of a small number of atoms, usually low molecular weight organic molecules. Recent and continuing advances in computer science have enabled the method to be applied to much larger biological molecules such as kinases [20].
4. The fragment molecular orbital (FMO) method [14, 19, 21] offers a considerable mark-up in computational speed over traditional QM methods [22]. One of the key advantages of the FMO approach is that the output from these calculations includes a list of the interactions made between the ligand and the receptor along with a chemically intuitive breakdown of these interactions [20]. Such information is essential for medicinal chemists to be able to adopt a rational approach in the modification of lead compounds in order to enhance favorable interactions. It works by partitioning a large system into small fragments (Fig. 1). For example, in proteins, each residue can be represented as a fragment. Similarly, the ligand can be represented by single or multiple fragments as necessary. By performing QM calculations on the fragments, one is able to attain a high level of accuracy with previously untenable efficiency.
5. The pair interaction energy (PIE) between any two fragments calculated by FMO is the sum of four energy terms: electrostatic, exchange-repulsion, charge transfer, and dispersion, provided by pair interaction energy decomposition analysis (PIEDA) [23]—*see* Fig. 1. The electrostatic and charge transfer terms are predominant in salt-bridge, hydrogen bond, and polar interactions, while the dispersion term generally

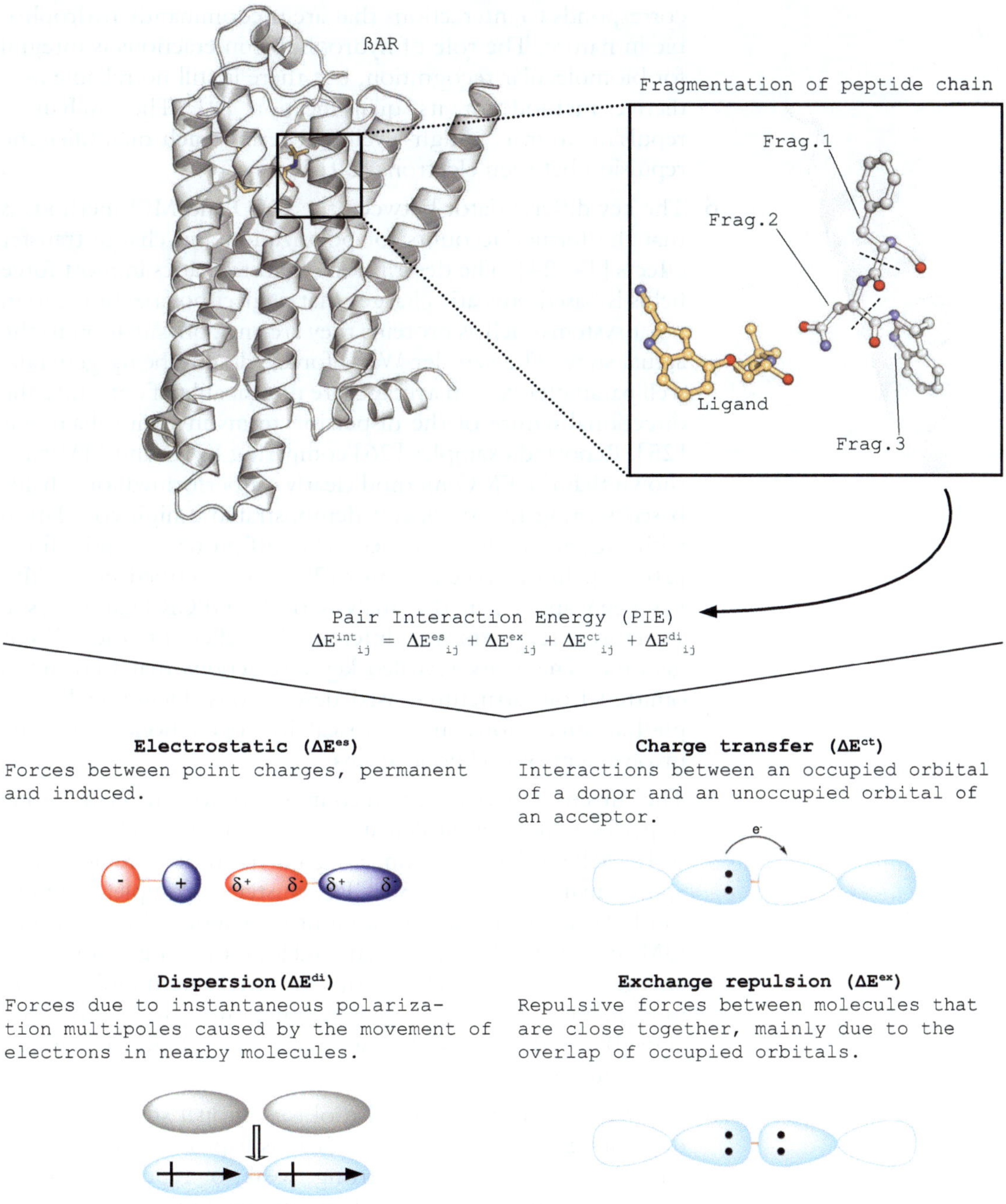

Fig. 1 Workflow for FMO calculations and details on each of PIE terms that are computed [20]. The electrostatic component arises from the Coulomb interaction between polarized charge distributions of fragments. The exchange repulsion term is derived from the interaction between fragments situated in close proximity and is always repulsive; it is due to the Pauli repulsion and is related to the overlap of the two occupied orbitals. The charge transfer term arises from the interaction between the occupied orbitals of a donor and the unoccupied orbitals of an acceptor. The dispersion arises as the interaction between instantaneous dipole moments of two fragments, it is hydrophobic (non-polar) in nature and is obtained in PIEDA from the correlation energy of electrons

corresponds to interactions that are predominantly hydrophobic in nature. The role of hydrophobic interactions is integral for biomolecular recognition, but there is still no reliable predictive method for its quantification [9]. The exchange-repulsion term is a high level QM term which quantifies the repulsion between electrons [20].

6. The key differentiator between the FMO and MM methods is that the former accounts for polarization and charge transfer effects [14, 24]. The description of electrostatics in most force fields is based on static charges that neglect polarization and in polar systems such as proteins they are an approximation to the actual state. The van-der-Waals forces, despite being generally well parameterized on average, are not capable of detecting the directional nature of the dispersion terms involving halogens [25]. Reported examples [26] comparing FMO and MM have shown that the FMO method clearly outperformed force field-based scoring functions and demonstrated a high correlation with experimentally measured values of protein-ligand affinity [26, 27]. In our recent report [27], we described how FMO has been applied in the analysis of 18 GPCR-ligand crystal structures representing different branches of the GPCR genome. This work revealed key interactions that were often omitted from structure-based descriptions, including hydrophobic interactions, non-classical hydrogen bonds, and the involvement of backbone atoms.

7. The current state of the art in computer processing enables one to perform high-level calculations relatively routinely. A typical FMO-MP2 calculation on a ligand-receptor complex takes approximately 4 h on a 36 CPU cluster to complete, significantly faster than weeks to a month (or more) for traditional QM approaches that have been used for estimating binding free energies. Recent developments in FMO methodology and implementation with density-functional tight binding (DFTB) method [28] further reduce the high-throughput capabilities of the method.

8. FMO can be a highly useful tool for rational structure-based drug design (SBDD) [14, 29, 30], as it provides an accurate and comprehensive list of strong, weak, or repulsive interactions between the ligand and its surrounding residues. Such information is highly instructive in rational SBDD in terms of modifications, scaffold replacement (scaffold hoping) linking (specifically in case of fragment-based drug discovery), or extension of chemical moieties to form stronger or new interactions with the protein or alternatively to remove repulsions. FMO can also be applied in analysis of ligand-water-protein networks, to distinguish between energetically favorable and unfavorable water molecules to enable the design of ligands

that can interact or displace certain waters. As previously demonstrated [27], significant correlation between protein-ligand affinity and FMO energy terms [26] indicates that they can be efficiently used as descriptors in QSAR modeling to predict the binding affinities of new molecules. FMO has been successfully applied in the discovery of novel Hsp90 inhibitors [29] and in many of our confidential drug discovery programs.

9. From our experience, the application of the FMO method in hit-to-lead and lead optimization stages of drug discovery is a highly valuable approach for the design, evaluation, and filtering of targets for synthesis which significantly decreases the effort and cost of chemical synthesis (for more details *see* Chapter 19 of this book).

2 Methods

1. *FMO* [22, 31] is a code implemented in General Atomic and Molecular Electronic Structure System (GAMESS). In FMO calculations, a large biological system is partitioned into fragments (Fig. 1) [14, 19]. Each residue is characterized as a fragment, and the interaction energies reported herein correspond to actual amino acid residues as opposed to residue fragments. The ligand can also be represented as one fragment or can be fragmented; some ligands can be very large and dividing a ligand into several fragments has the benefit of both reducing the computational cost and providing a more detailed analysis. The detailed description of the FMO strategy and methodology can be found in the published reports [14, 19, 23], including a detailed mathematical formulation that is beyond the scope of this manuscript.

2. The FMO calculation consists of the following steps: (a) Fragmentation (i.e., assigning atoms in a system to a fragment); (b) Fragment self-consistent field (SCF) calculations in the embedding polarizable potential, so that fragments mutually polarize each other in a self-consistent fashion whereby intra-fragment charge transfer and other quantum effects are accounted for; (c) Fragment pair SCF calculations, bringing in inter-fragment charge transfer; (d) Total property (energy, gradient, etc.) evaluation. By performing QM computations on fragments one can achieve high efficiency, often resulting from linear scaling and computational speed. The FMO method has been parallelized for PC clusters [22]. By default we use the MP2 method (Second Order Møller-Plesset perturbation theory [32]) with the 6-31G* basis set. This basis set is most commonly used and is often considered the best compromise

between speed and accuracy [26]. Residues and water molecules within a radius of ≤4.5 Å around the ligand atoms were included in the FMO calculations, since previous work demonstrated [33] that including these atoms significantly increases the speed of the calculation without compromising the results. FMO can be implemented with a polarizable continuum solvent model (PCM), which ameliorates the effect of charged residues around the system of interest [34].

3. The pair interaction energy (PIE, ΔE^{int}) between fragments i and j is a sum of four terms: electrostatics (ΔE^{es}), exchange-repulsion (ΔE^{ex}), charge transfer (ΔE^{ct}), and dispersion (ΔE^{dis}) (Eq. 1) as described in Fig. 1.

$$\Delta E_{ij}^{\mathrm{int}} = \Delta E_{ij}^{\mathrm{es}} + \Delta E_{ij}^{\mathrm{ex}} + \Delta E_{ij}^{\mathrm{ct}} + \Delta E_{ij}^{\mathrm{di}} \tag{1}$$

4. The PIE is not a difference between energies of the protein-ligand complex and the sum of the "free" protein and ligand, but rather represents the "strength" of the interaction between the ligand and protein residues in the complex. The Δ denotes the differences in total QM energy of a fragment pair ij and two individual fragments i and j, both computed in the receptor-ligand complex. In the present work, fragment i is the ligand and the other n fragments are receptor residues and water molecules. In the equations below there is no self-interaction (the sums exclude $j = i$). The total PIE calculated by the FMO method describes the stability of the receptor-ligand complex. This stability correlates to, but is not the same as, the binding energy [26]. The difference lies in the polarization factors—the ligand is polarized by the protein and vice versa [22]. Based on previous reports [14], we consider any interaction with an absolute PIE greater than or equal to 3.0 kcal/mol to be significant.

5. The difference between $\mathrm{PIE}^{\mathrm{lig1}}$ and $\mathrm{PIE}^{\mathrm{lig2}}$ ($\Delta\Delta E_{lig1,lig2}^{\mathrm{int}}$), and the corresponding component energy terms, can be calculated using Eq. 2. This equation allows for a detailed comparison of the interactions and their composition which potentially gives rise to the difference in observed activities between any two ligands.

$$\Delta\Delta E_{lig1,lig2}^{\mathrm{int}} = \Delta\Delta E_{lig1,lig2}^{\mathrm{es}} + \Delta\Delta E_{lig1,lig2}^{\mathrm{ex}} + \Delta\Delta E_{lig1,lig2}^{\mathrm{ct}} + \Delta\Delta E_{lig1,lig2}^{\mathrm{di}} \tag{2}$$

6. *Structure preparation* is performed using MOE (Chemical Computing Group). Hydrogen atoms are added to crystal structures at physiological pH (7.4) with the Protonate3D tool. A constrained minimization procedure, where each atom was allowed to deviate up to 0.5 Å from its original position in crystal structure, was applied in order to optimize the positions of each atom and remove potential steric clashes. The semi-empirical AMBER10:EHT forced field was used to model the proteins interatomic potential. Finally, residues within a radius of ≤4.5 Å around the ligand atoms were selected for inclusion in the FMO calculations. The position of each TM's amino acid residue was identified by its unique sequence number as well as by its generic number proposed by Ballesteros and Weinstein [35, 36] in superscript. Ballesteros and Weinstein numbering scheme allows comparison between topologically equivalent residues in different GPCR structures.

3 Notes

1. To illustrate the typical results obtained through FMO (*see* Subheading 2, **step 1**), we performed FMO calculations on 19 β_1 and β_2 adrenergic receptor-ligand crystal structures (Table 1) in order to identify conserved interactions and to investigate their chemical nature. Structures containing agonist and antagonist ligands were included to explore distinct interaction patterns. We focused on the interaction patterns of agonists versus antagonists, and on the observation that agonists appear to interact preferentially and in a conserved fashion with residue $S^{5.46}$, while antagonists with residue $W^{6.48}$.
2. β_1 and β_2 adrenergic receptors (βARs) belong to the superfamily of GPCRs [7] and are expressed largely in the tissues of the cardiac, vascular and respiratory systems [37, 38]. More specifically, β_1AR is the predominant subtype in the normal myocardium, representing 75–80% of total βAR density, followed by β_2AR, which comprises approximately 15–18% of the total cardiomyocyte βARs [39]. β_1AR is also located in the kidney [7]. Indeed, the β_1AR subtype is involved in physiological processes including the heart beat and blood pressure regulation [7, 38], whereas, activation of β_2AR notably results in the dilatation of smooth muscles of the lungs, uterus and blood vessels [7, 38].
3. Catecholamines, the endogenous agonists of βARs, activate βARs and stimulate the sympathetic nervous system required for the regulation of the unconscious actions of the body. The most well-known catecholamines include epinephrine (adrenaline), norepinephrine (noradrenaline) [40, 41], and dopamine

Table 1
Overview of βAR complexes

Receptor	Species	Ligand	Ligand function	Year	Res (A)	PDB-ID
β1	Turkey	Carmoterol	fAGO	2011	2.60	2Y02 [49]
β1	Turkey	Isoprenaline	fAGO	2011	2.85	2Y03 [49]
β1	Turkey	Dobutamine	pAGO	2011	2.50	2Y00 [49]
β1	Turkey	Salbutamol	pAGO	2011	3.05	2Y04 [49]
β1	Turkey	Cyanopindolol	ANT	2014	2.10	4BVN [48]
β1	Turkey	Arylpiperazine 20	ANT	2013	2.70	3ZPR [51]
β1	Turkey	Arylpiperazine 19	ANT	2013	2.80	3ZPQ [51]
β1	Turkey	Bucindolol	ANT	2012	3.20	4AMI [58]
β1	Turkey	Carvedilol	iAGO	2012	2.30	4AMJ [58]
β1	Turkey	Carazolol	iAGO	2011	3.00	2YCW [59]
β2	Human	BI-167107	fAGO	2013	2.79	4LDE [50]
β2	Human	HBI	fAGO	2013	3.10	4LDL [50]
β2	Human	Adrenaline (epinephrine)	fAGO	2013	3.20	4LDO [50]
β2	Human	FAUC37	cAGO	2014	3.30	4QKX [53]
β2	Human	Alprenolol	ANT	2010	3.16	3NYA [52]
β2	Human	Carazolol	iAGO	2007	2.40	2RH1 [60]
β2	Human	Timolol	iAGO	2008	2.80	3D4S [44]
β2	Human	ICI-118,551	iAGO	2010	2.84	3NY8 [52]
β2	Human	VS hit (Kolb)	iAGO	2010	2.84	3NY9 [52]

Abbreviations: fAGO corresponds to full agonist, AGO to agonist, cAGO to covalent agonist, pAGO to partial agonist, ANT to antagonist and iAGO to inverse agonist

[42]. Adrenaline is a neurotransmitter which is involved in flight-or-flight response by increasing blood flow to muscles, cardiac output, pupil dilation, and blood sugar [43]. Noradrenaline and dopamine are neurotransmitters that widely modulate brain circuits and behaviors [42]. Noradrenaline is involved in arousal, attention, memory, and stress, whereas dopamine is implicated in learning, reward, attention, and movement control [42]. Catecholaminergic dysfunctions are associated with cognitive, emotional, and motor disorders [42] and can initiate cardiac, vascular, respiratory pathologies.

4. Antagonists and inverse agonists of βARs [44], so called β-blockers, are employed to treat diseases such as hypertension and heart failure [45], as well as to increase cardiac frequency and the force of cardiac contraction. In contrast, agonists of

β_2AR, known as β_2-sympathomimetics, are most frequently used for the treatment of respiratory problems caused by asthma or chronic obstructive pulmonary disease (COPD) [46]. Moreover, activation of β_1ARs increases heart rate and therefore cardiac output [47]. It is known that βARs can access a broad spectrum of functional states [41], which can be modulated through the binding of small molecules. Understanding receptor-ligand interactions at the atomic scale can therefore provide insights into the molecular determinants of agonists versus antagonist behavior, and contribute toward the design of the next generation of small-molecule modulators of βARs.

5. *Analysis and comparison of β_1AR-cyanopindolol and β_1AR-carmoterol complexes*—Herein, we present an application of the FMO method for the analysis of the β_1AR-cyanopindolol crystal structure (PDB access code 4BVN [48]). FMO highlighted 14 significant interactions (Fig. 2a) between the ligand and the following residues: D121$^{3.32}$, V122$^{3.33}$, F201^{ECL2}, Y207$^{5.38}$, A208$^{5.39}$, S211$^{5.42}$, S212$^{5.43}$, W303$^{6.48}$, F306$^{6.51}$, F307$^{6.52}$, N310$^{6.55}$, N329$^{7.39}$, Y333$^{7.43}$ and HOH3024. The majority of these interactions are consistent with previous reports [48]. However, FMO also reveals some polar interactions that have been previously reported in the literature, which include the residues A208$^{5.39}$ and Y333$^{7.43}$ and the water molecule HOH3024.

   In the case of carmoterol (PDB access code 2Y02 [49]) the FMO calculations reveal 11 significant interactions (Fig. 2b). Similar interactions were reported in the literature with the following residues: D121$^{3.32}$, V122$^{3.33}$, F201^{ECL2}, S215$^{5.46}$, F307$^{6.52}$, N310$^{6.55}$, V326$^{7.36}$, N329$^{7.39}$, and Y333$^{7.43}$. Further unreported interactions with S212$^{5.43}$ and HOH2010 were also identified.

   We analyzed the differences in the interaction profiles between the two complexes. FMO reveals ten mutual interactions (Fig. 2c), where six interactions were stronger with the agonist carmoterol, with residues F201^{ECL2}, S212$^{5.43}$, I213$^{5.44}$, S215$^{5.46}$, N310$^{6.55}$ and V326$^{7.35}$. On the other hand, four interactions were stronger with the antagonist cyanopindolol: with residues T203^{ECL2}, A208$^{5.39}$, I209$^{5.40}$, and S211$^{5.42}$.

6. *βAR complexes reveal conserved ligand-protein interactions*—By analyzing all 19 ligand-receptor *βAR* complexes we were also able to demonstrate general trends in ligand binding for both β_1AR and β_2AR receptors (Fig. 3). For example, residues D$^{3.32}$, V$^{3.33}$, F201/193^{ECL2}, F$^{6.51}$, F$^{6.52}$, N$^{6.55}$, N7$^{.39}$, and Y$^{7.43}$ make a considerable contribution to receptor-ligand binding and are quite conserved (>70%, Fig. 3). Residues in the other positions are less frequently involved and are more specific for particular ligands.

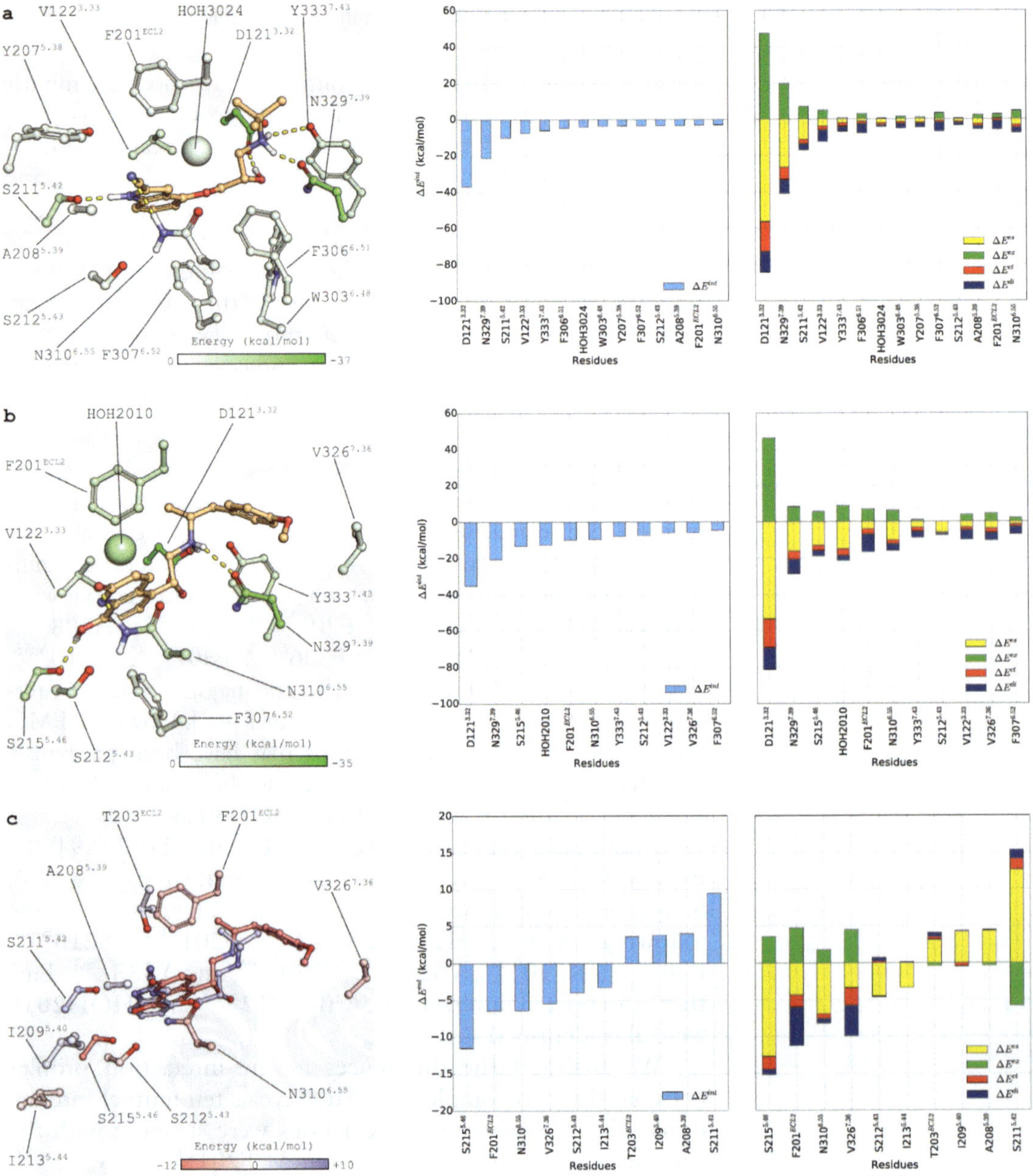

Fig. 2 Comparison of interactions between an antagonist and an agonist binding to β_1AR. (**a**) Turkey β_1AR in complex with antagonist cyanopindolol (PDB entry 4BVN). (**b**) Turkey β_1AR complex with agonist carmoterol (PDB entry 2Y02). The carbon atoms of the ligands are shown in light orange and for the receptor are colored according to PIE values calculated by FMO. Nitrogen atoms are shown in *blue*, oxygen in *red*, sulfur in *yellow*, and chlorine in *light green*. The classical hydrogen bonds formed between the receptor and the ligand are marked as *yellow dashed line*. The bar charts on the left show the sorted PIE for the residues interacting with energies larger than −3 kcal/mol, and chart on the right describe the PIEDA of these interactions. PIE terms: electrostatics, dispersion, charge-transfer and exchange repulsion are colored coded *yellow*, *blue*, *red* and *green* respectively. (**c**) Difference of shared interactions between cyanopindolol and carmoterol-β_1AR complexes. In this case, cyanopindolol is shown in *light blue* and carmoterol in *salmon pink*, with the residues interacting more strongly with cyanopindolol shown on a *white* to *light blue* spectrum, and residues interacting more strongly with carmoterol on a *white* to *salmon* spectrum, where white means equal interaction energy for both ligands. Accordingly, on the bar chart on the right, negative $\Delta\Delta E$ values represent a stronger interaction with carmoterol, while negative $\Delta\Delta E$ values a stronger interaction with cyanopindolol

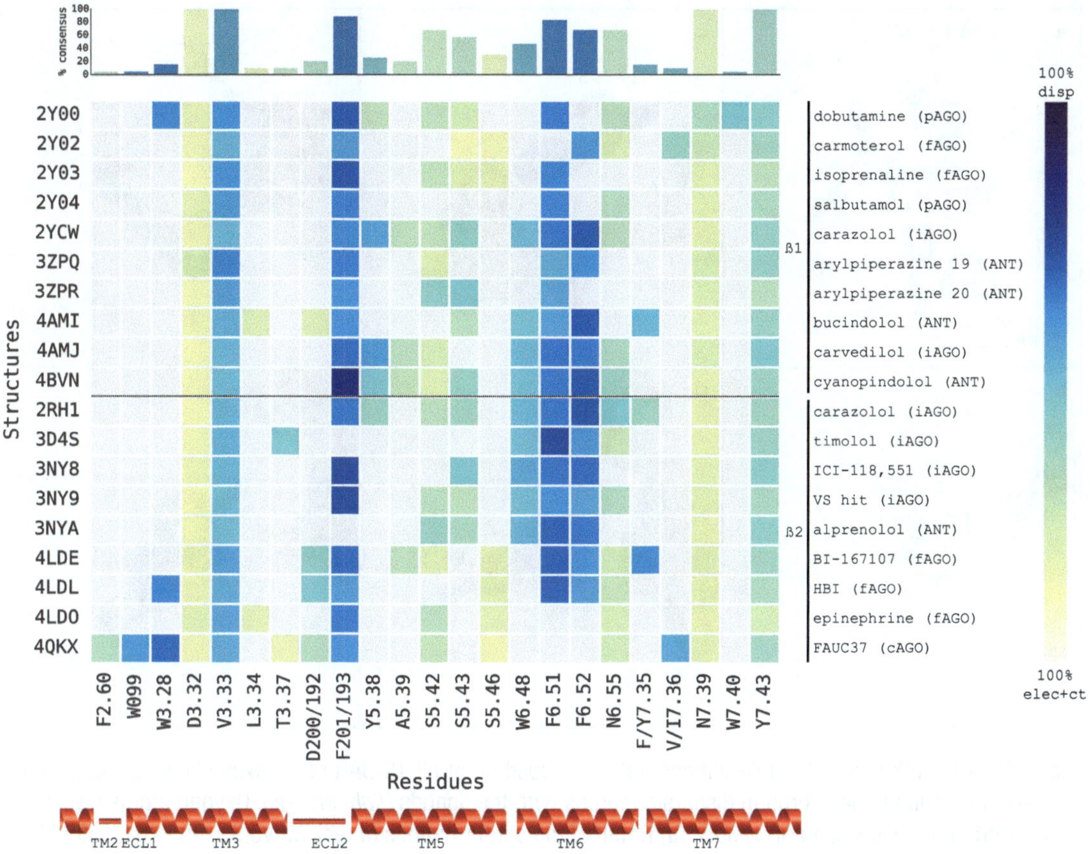

Fig. 3 Overview of all significant interactions for each βAR complex. Each row represents a structure, for a total of 19 rows, where the PDB-IDs are shown on the left of each row. The name and action of the ligands are shown on the right side of each row. Columns represent the residues interacting with the ligands, identified through their Ballesteros–Weinstein numbers. In the matrix, the presence of a contact between the ligand and the residue is illustrated as a colored cell, and the absence of a contact is illustrated as *gray* cell. Cells representing a contact are colored according to their PIEDA signature: from *dark blue* (100% dispersion) to *yellow* (100% electrostatic and charge-transfer). A mixed contribution (e.g., 50% dispersion, and 50% electrostatic and charge-transfer) therefore results in a cell being colored in *green* to *light blue*, as indicated by the spectrum bar on the right. The percentage of consensus for any ligand-residue interaction is shown at the top of the figure as a histogram, with each bar color-coded according to average PIEDA signature, following the same scheme as for the individual residue-ligand interactions (*blue* to *yellow*)

7. Residues $V^{3.33}$, $F201/193^{ECL2}$, $F^{6.51}$, $F^{6.52}$, and $Y^{7.43}$ form interactions with predominantly hydrophobic in nature (Fig. 4). While electrostatic and polar interactions are widely reported in the literature [50–53], hydrophobic interactions are often neglected, these interactions are difficult to identify through visual inspection alone. Key hydrophobic residues in TM3, ECL2, TM6, and TM7 form a consensus core of the βAR ligand binding pocket, and retain a very similar

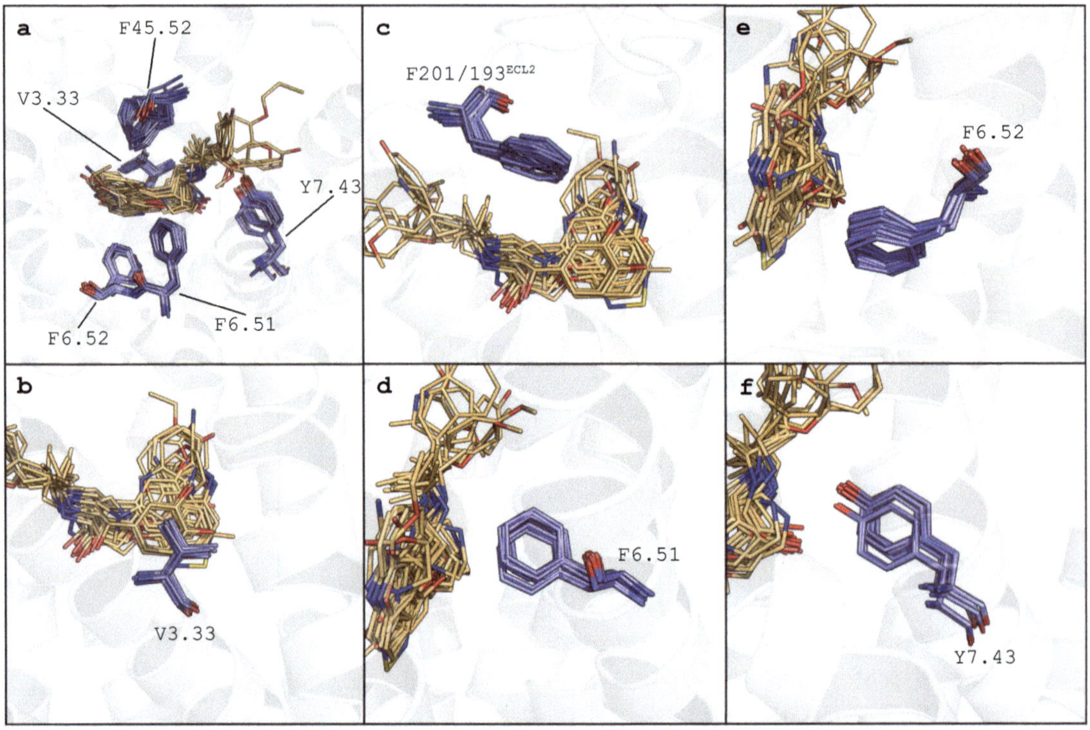

Fig. 4 Conserved hydrophobic interactions. (**a**) Superposition of all 19 βAR complexes showing the residues (*purple*) interacting largely through dispersion forces with the ligands (*light orange*). Oxygen atoms are shown in *red*, nitrogen in *blue*, sulfur in *yellow*. (**b**) Detail of the conformations of residue V3.33, (**c**) F201/193^{ECL2}, (**d**) F6.51, (**e**) F6.52, and (**f**) Y7.43

conformation in all the structures analyzed (Fig. 4). Published mutations in these positions frequently decrease ligand affinity [54].

8. *Agonist* vs. *antagonist*—Next, we focused on exploring the differences in interaction patterns between agonists and antagonists. We therefore separated and compared the FMO results calculated for antagonists (Fig. 5b) and agonists (Fig. 5c). FMO reveals that residues: $D^{3.32}$, $V^{3.33}$, F201/193^{ECL2}, $S^{5.42}$, $F^{6.51}$, $F^{6.52}$, $N^{6.55}$, $N^{7.39}$, $Y^{7.43}$ are important for binding of both agonists and antagonists. The most striking difference in the interaction profile between agonists and antagonists is the presence of a strong polar interaction with $S^{5.46}$ (present in 6 out of 8 agonists, Fig. 5a), This interaction is not observed in the case of the antagonists. In addition to this, the presence of a hydrophobic interaction with $W^{6.48}$ for 9 out 11 antagonists (Fig. 5a) is not observed in the agonist cases.

9. Visual inspection of the overlaid structures (Fig. 6a) reveals two distinct conformers in which $S^{5.46}$ and $W^{6.48}$ (Fig. 6b) exist in the active versus inactive forms of βARs. The “active”

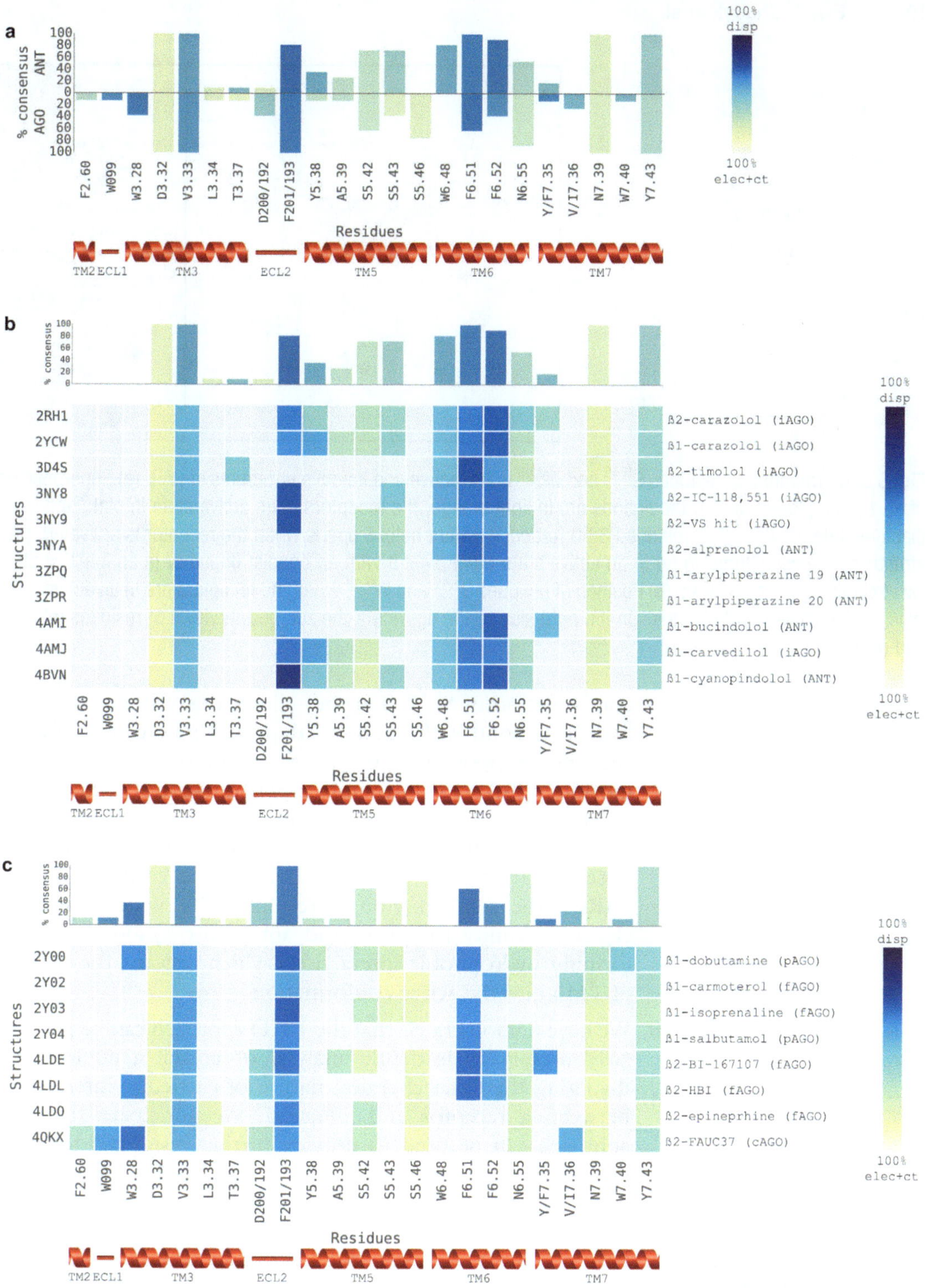

Fig. 5 (**a**) Histograms showing the consensus of interactions across the complexes containing agonists (AGO) and antagonists (ANT). The height of each bar represents the percentage of structures in which a strong (larger than 3 kcal/mol) interaction is present, while its color summarizes the chemical nature of the interaction (*yellow* for mainly electrostatic and *blue* for mainly hydrophobic). (**b**) Overview heat map for βARs in complex with an antagonist and (**c**) with an agonist

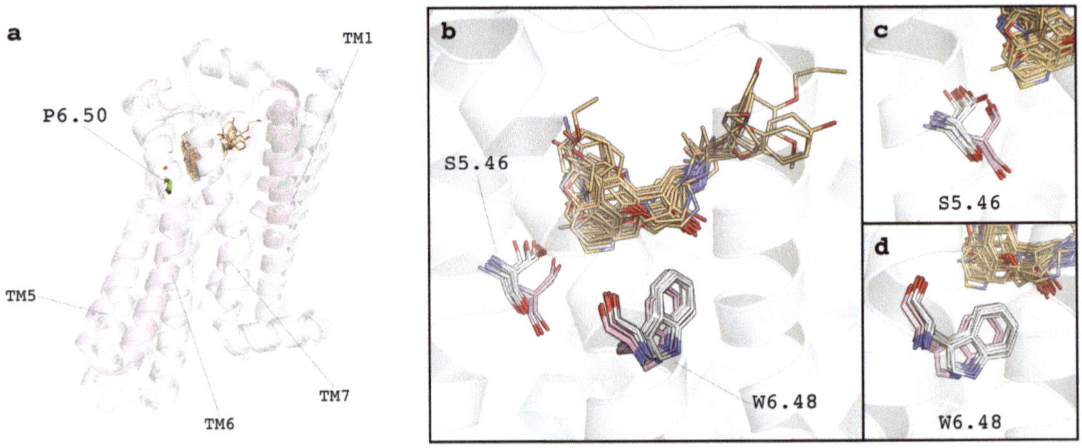

Fig. 6 Conformations of residues $S^{5.46}$ and $W^{6.48}$ in the active and inactive structures of βAR. (**a**) Superposition of all 19 complexes, where the ligands are in light orange, the receptors in an active state in pink, and in an inactive state in white. The residue P6.50 (*green*) is where helix 6 opens when the receptor is activated. βAR structures are considered to be in an active state (*light pink*) if with an agonist bound or in an inactive state (*white*) if antagonist is bound. (**b**) Position of residues $S^{5.46}$ and $W^{6.48}$ when the receptors are in an active (*light pink*) and inactive (*white*) states. On the right side of the figure are the zoomed-in views of residues (**c**) $S^{5.46}$ and (**d**) $W^{6.48}$

conformation of $S^{5.46}$ (Fig. 6c) allows this residue to form a hydrogen bond with agonists. Reported mutagenesis studies [55, 56] support this observation and suggest that TM6 motion in β_2AR [57] depends largely on agonists engaging in polar interactions with $S203^{5.42}$ and $S207^{5.46}$ which stabilize a 2 Å inward shift of the extracellular part of TM5. One key observation in the analysis of the inactive conformation of βARs is the conformation of residue $W^{6.48}$ which resides in closer proximity to the bound antagonist ligands (Fig. 6d) forming hydrophobic interactions which prevent the motion of TM6 into the active conformation.

10. We have demonstrated that the FMO approach can be particularly useful for in-depth analysis of crystal-structures and divulging the exact chemical nature of particular interactions between a receptor and a ligand. We anticipate that this approach can be used to provide further insights into many protein-ligand interactions, with a view to rationally design novel therapeutic compounds.

Acknowledgments

A.H. and R.R. would like to acknowledge the support of EU H2020 CompBioMed project (http://www.compbiomed.eu/675451) and the BBSRC Flexible Interchanger Programme project (BB/P004245/1).

References

1. Rask-Andersen M, Masuram S, Schioth HB (2014) The druggable genome: evaluation of drug targets in clinical trials suggests major shifts in molecular class and indication. Annu Rev Pharmacol Toxicol 54:9–26
2. Wise A, Gearing K, Rees S (2002) Target validation of G-protein coupled receptors. Drug Discov Today 7:235–246
3. Overington JP, Al-Lazikani B, Hopkins AL (2006) How many drug targets are there? Nat Rev Drug Discov 5:993–996
4. Heifetz A, Schertler GF, Seifert R, Tate CG, Sexton PM, Gurevich VV, Fourmy D, Cherezov V, Marshall FH, Storer RI, Moraes I, Tikhonova IG, Tautermann CS, Hunt P, Ceska T, Hodgson S, Bodkin MJ, Singh S, Law RJ, Biggin PC (2015) GPCR structure, function, drug discovery and crystallography: report from Academia-Industry International Conference (UK Royal Society) Chicheley Hall, 1–2 September 2014. Naunyn Schmiedebergs Arch Pharmacol 388 (8):883–903
5. Dohlman HG (2015) Thematic minireview series: new directions in G protein-coupled receptor pharmacology. J Biol Chem 290:19469–19470
6. Tautermann CS (2014) GPCR structures in drug design, emerging opportunities with new structures. Bioorg Med Chem Lett 24:4073–4079
7. Shonberg J, Kling RC, Gmeiner P, Lober S (2015) GPCR crystal structures: medicinal chemistry in the pocket. Bioorg Med Chem 23:3880–3906
8. Jazayeri A, Dias JM, Marshall FH (2015) From G protein-coupled receptor structure resolution to rational drug design. J Biol Chem 290:19489–19495
9. Bissantz C, Kuhn B, Stahl M (2010) A medicinal chemist's guide to molecular interactions. J Med Chem 53:5061–5084
10. Tong Y, Mei Y, Li YL, Ji CG, Zhang JZ (2010) Electrostatic polarization makes a substantial contribution to the free energy of avidin-biotin binding. J Am Chem Soc 132:5137–5142
11. Raha K, Peters MB, Wang B, Yu N, Wollacott AM, Westerhoff LM, Merz KM Jr (2007) The role of quantum mechanics in structure-based drug design. Drug Discov Today 12:725–731
12. Beratan DN, Liu C, Migliore A, Polizzi NF, Skourtis SS, Zhang P, Zhang Y (2015) Charge transfer in dynamical biosystems, or the treachery of (static) images. Acc Chem Res 48:474–481
13. Ozawa T, Okazaki K, Kitaura K (2011) CH/pi hydrogen bonds play a role in ligand recognition and equilibrium between active and inactive states of the beta2 adrenergic receptor: an ab initio fragment molecular orbital (FMO) study. Bioorg Med Chem 19:5231–5237
14. Fedorov DG, Nagata T, Kitaura K (2012) Exploring chemistry with the fragment molecular orbital method. Phys Chem Chem Phys 14:7562–7577
15. Lu Y-X, Zou J-W, Wang Y-H, Yu Q-S (2007) Substituent effects on noncovalent halogen/π interactions: theoretical study. Int J Quantum Chem 107:1479–1486
16. Gallivan JP, Dougherty DA (1999) Cation-pi interactions in structural biology. Proc Natl Acad Sci U S A 96:9459–9464
17. Johnston RC, Cheong PH (2013) C-H...O non-classical hydrogen bonding in the stereomechanics of organic transformations: theory and recognition. Org Biomol Chem 11:5057–5064
18. Yu N, Li X, Cui G, Hayik SA, Merz KM 2nd (2006) Critical assessment of quantum mechanics based energy restraints in protein crystal structure refinement. Protein Sci 15:2773–2784
19. Fedorov DG, Kitaura K (2007) Extending the power of quantum chemistry to large systems with the fragment molecular orbital method. J Phys Chem A 111:6904–6914
20. Phipps MJ, Fox T, Tautermann CS, Skylaris CK (2015) Energy decomposition analysis approaches and their evaluation on prototypical protein-drug interaction patterns. Chem Soc Rev 44:3177–3211
21. Kitaura K, Ikeo E, Asada T, Nakano T, Uebayasi M (1999) Fragment molecular orbital method: an approximate computational method for large molecules. Chem Phys Lett 313:701–706
22. Alexeev Y, Mazanetz MP, Ichihara O, Fedorov DG (2012) GAMESS as a free quantum-mechanical platform for drug research. Curr Top Med Chem 12:2013–2033
23. Fedorov DG, Kitaura K (2007) Pair interaction energy decomposition analysis. J Comput Chem 28:222–237
24. Fedorov DG, Kitaura K (2012) Energy decomposition analysis in solution based on the fragment molecular orbital method. J Phys Chem A 116:704–719
25. El Kerdawy A, Murray JS, Politzer P, Bleiziffer P, Hesselmann A, Gorling A, Clark T (2013) Directional noncovalent interactions:

repulsion and dispersion. J Chem Theory Comput 9:2264–2275

26. Mazanetz MP, Ichihara O, Law RJ, Whittaker M (2011) Prediction of cyclin-dependent kinase 2 inhibitor potency using the fragment molecular orbital method. J Cheminform 3:2
27. Heifetz A, Chudyk E, Gleave L, Aldeghi M, Cherezov V, Fedorov DG, Biggin PC, Bodkin M (2015) The fragment molecular orbital method reveals new insight into the chemical nature of GPCR-ligand interactions. J Chem Inf Model 56(1):159–172
28. Morao I, Fedorov DG, Robinson R, Southey M, Townsend-Nicholson A, Bodkin MJ, Heifetz A (2017) Rapid and accurate assessment of GPCR-ligand interactions using the fragment molecular orbital-based density-functional tight-binding method. J Comput Chem 38(23):1987–1990
29. Barker JJ, Barker O, Courtney SM, Gardiner M, Hesterkamp T, Ichihara O, Mather O, Montalbetti CA, Muller A, Varasi M, Whittaker M, Yarnold CJ (2010) Discovery of a novel Hsp90 inhibitor by fragment linking. ChemMedChem 5:1697–1700
30. Sawada T, Fedorov DG, Kitaura K (2010) Binding of influenza A virus hemagglutinin to the sialoside receptor is not controlled by the homotropic allosteric effect. J Phys Chem B 114:15700–15705
31. Fedorov DG, Kitaura K (2004) The importance of three-body terms in the fragment molecular orbital method. J Chem Phys 120:6832–6840
32. Fedorov DG, Kitaura K (2004) Second order Moller-Plesset perturbation theory based upon the fragment molecular orbital method. J Chem Phys 121:2483–2490
33. Heifetz A, Chudyk EI, Gleave L, Aldeghi M, Cherezov V, Fedorov DG, Biggin PC, Bodkin MJ (2016) The fragment molecular orbital method reveals new insight into the chemical nature of GPCR-ligand interactions. J Chem Inf Model 56:159–172
34. Li H, Fedorov DG, Nagata T, Kitaura K, Jensen JH, Gordon MS (2010) Energy gradients in combined fragment molecular orbital and polarizable continuum model (FMO/PCM) calculation. J Comput Chem 31:778–790
35. Ballesteros JA, Weinstein H (1995) Integrated methods for the construction of three-dimensional models and computational probing of structure-function relations in G protein-coupled receptors. Methods Neurosci 25:366–428
36. Prioleau C, Visiers I, Ebersole BJ, Weinstein H, Sealfon SC (2002) Conserved helix 7 tyrosine acts as a multistate conformational switch in the 5HT2C receptor. Identification of a novel "locked-on" phenotype and double revertant mutations. J Biol Chem 277:36577–36584
37. Van Iterson EH, Karpen SR, Baker SE, Wheatley CM, Morgan WJ, Snyder EM (2015) Impaired cardiac and peripheral hemodynamic responses to inhaled beta2-agonist in cystic fibrosis. Respir Res 16:103
38. Bylund DB, Eikenberg DC, Hieble JP, Langer SZ, Lefkowitz RJ, Minneman KP, Molinoff PB, Ruffolo RR Jr, Trendelenburg U (1994) International union of pharmacology nomenclature of adrenoceptors. Pharmacol Rev 46:121–136
39. Ferron AJ, Jacobsen BB, Sant'Ana PG, de Campos DH, de Tomasi LC, Luvizotto Rde A, Cicogna AC, Leopoldo AS, Lima-Leopoldo AP (2015) Cardiac dysfunction induced by obesity is not related to beta-adrenergic system impairment at the receptor-signalling pathway. PLoS One 10:e0138605
40. Evans BA, Sato M, Sarwar M, Hutchinson DS, Summers RJ (2010) Ligand-directed signalling at beta-adrenoceptors. Br J Pharmacol 159:1022–1038
41. Rosenbaum DM, Rasmussen SG, Kobilka BK (2009) The structure and function of G-protein-coupled receptors. Nature 459:356–363
42. Nomura S, Bouhadana M, Morel C, Faure P, Cauli B, Lambolez B, Hepp R (2014) Noradrenalin and dopamine receptors both control cAMP-PKA signaling throughout the cerebral cortex. Front Cell Neurosci 8:247
43. Rhoades R, Bell DR (2009) Medical physiology : principles for clinical medicine. Lippincott Williams & Wilkins, Philadelphia
44. Hanson MA, Cherezov V, Griffith MT, Roth CB, Jaakola VP, Chien EY, Velasquez J, Kuhn P, Stevens RC (2008) A specific cholesterol binding site is established by the 2.8 A structure of the human beta2-adrenergic receptor. Structure 16:897–905
45. Prichard BN, Cruickshank JM, Graham BR (2001) Beta-adrenergic blocking drugs in the treatment of hypertension. Blood Press 10:366–386
46. Waldeck B (2002) β-Adrenoceptor agonists and asthma—100 years of development. Eur J Pharmacol 445:1–12
47. Wong GW, Laugerotte A, Wright JM (2015) Blood pressure lowering efficacy of dual alpha and beta blockers for primary hypertension. Cochrane Database Syst Rev 8:Cd007449
48. Miller-Gallacher JL, Nehme R, Warne T, Edwards PC, Schertler GF, Leslie AG, Tate CG (2014) The 2.1 A resolution structure of

cyanopindolol-bound beta1-adrenoceptor identifies an intramembrane Na+ ion that stabilises the ligand-free receptor. PLoS One 9: e92727

49. Warne T, Moukhametzianov R, Baker JG, Nehme R, Edwards PC, Leslie AG, Schertler GF, Tate CG (2011) The structural basis for agonist and partial agonist action on a beta(1)-adrenergic receptor. Nature 469:241–244
50. Ring AM, Manglik A, Kruse AC, Enos MD, Weis WI, Garcia KC, Kobilka BK (2013) Adrenaline-activated structure of beta2-adrenoceptor stabilized by an engineered nanobody. Nature 502:575–579
51. Christopher JA, Brown J, Dore AS, Errey JC, Koglin M, Marshall FH, Myszka DG, Rich RL, Tate CG, Tehan B, Warne T, Congreve M (2013) Biophysical fragment screening of the beta1-adrenergic receptor: identification of high affinity arylpiperazine leads using structure-based drug design. J Med Chem 56:3446–3455
52. Wacker D, Fenalti G, Brown MA, Katritch V, Abagyan R, Cherezov V, Stevens RC (2010) Conserved binding mode of human beta2 adrenergic receptor inverse agonists and antagonist revealed by X-ray crystallography. J Am Chem Soc 132:11443–11445
53. Weichert D, Kruse AC, Manglik A, Hiller C, Zhang C, Hubner H, Kobilka BK, Gmeiner P (2014) Covalent agonists for studying G protein-coupled receptor activation. Proc Natl Acad Sci U S A 111:10744–10748
54. Dore AS, Robertson N, Errey JC, Ng I, Hollenstein K, Tehan B, Hurrell E, Bennett K, Congreve M, Magnani F, Tate CG, Weir M, Marshall FH (2011) Structure of the adenosine A(2A) receptor in complex with ZM241385 and the xanthines XAC and caffeine. Structure 19:1283–1293
55. Strader CD, Sigal IS, Register RB, Candelore MR, Rands E, Dixon RA (1987) Identification of residues required for ligand binding to the beta-adrenergic receptor. Proc Natl Acad Sci U S A 84:4384–4388
56. Liapakis G, Ballesteros JA, Papachristou S, Chan WC, Chen X, Javitch JA (2000) The forgotten serine. A critical role for Ser-2035.42 in ligand binding to and activation of the beta 2-adrenergic receptor. J Biol Chem 275:37779–37788
57. Rasmussen SG, Choi HJ, Fung JJ, Pardon E, Casarosa P, Chae PS, Devree BT, Rosenbaum DM, Thian FS, Kobilka TS, Schnapp A, Konetzki I, Sunahara RK, Gellman SH, Pautsch A, Steyaert J, Weis WI, Kobilka BK (2011) Structure of a nanobody-stabilized active state of the beta(2) adrenoceptor. Nature 469:175–180
58. Warne T, Edwards PC, Leslie AG, Tate CG (2012) Crystal structures of a stabilized beta1-adrenoceptor bound to the biased agonists bucindolol and carvedilol. Structure 20:841–849
59. Moukhametzianov R, Warne T, Edwards PC, Serrano-Vega MJ, Leslie AG, Tate CG, Schertler GF (2011) Two distinct conformations of helix 6 observed in antagonist-bound structures of a beta1-adrenergic receptor. Proc Natl Acad Sci U S A 108:8228–8232
60. Cherezov V, Rosenbaum DM, Hanson MA, Rasmussen SG, Thian FS, Kobilka TS, Choi HJ, Kuhn P, Weis WI, Kobilka BK, Stevens RC (2007) High-resolution crystal structure of an engineered human beta2-adrenergic G protein-coupled receptor. Science 318:1258–1265

Chapter 9

Molecular Basis of Ligand Dissociation from G Protein-Coupled Receptors and Predicting Residence Time

Dong Guo and Adriaan P. IJzerman

Abstract

G protein-coupled receptors (GPCRs) are integral membrane proteins and represent the largest class of drug targets. During the past decades progress in structural biology has enabled the crystallographic elucidation of the architecture of these important macromolecules. It also provided atomic-level visualization of ligand-receptor interactions, dramatically boosting the impact of structure-based approaches in drug discovery. However, knowledge obtained through crystallography is limited to static structural information. Less information is available showing how a ligand associates with or dissociates from a given receptor, whose importance is in fact increasingly recognized by the drug research community. Owing to recent advances in computer power and algorithms, molecular dynamics stimulations have become feasible that help in analyzing the kinetics of the ligand binding process. Here, we review what is currently known about the dynamics of GPCRs in the context of ligand association and dissociation, as determined through both crystallography and computer simulations. We particularly focus on the molecular basis of ligand dissociation from GPCRs and provide case studies that predict ligand dissociation pathways and residence time.

Key words Binding kinetics, Molecular dynamics simulations, G protein-coupled receptor, Dissociation rate, Dissociation pathway

1 Introduction

G protein-coupled receptors (GPCRs) constitute an important family of integral membrane proteins. These receptors are able to respond to a divergent array of molecules outside the cell, ranging from small ions and photons to large glycoproteins, and to subsequently trigger a variety of intracellular signaling cascades. To date, about 800 members of the GPCR family have been identified [1]. They are of significant interest in pharmaceutical research owing to their involvement in a plethora of important physiological and pathophysiological processes [2]. A status quo analysis of major protein families as drug targets showed that over one-third of drugs act on GPCRs, hence representing the largest class of drug targets [3].

Alexander Heifetz (ed.), *Computational Methods for GPCR Drug Discovery*, Methods in Molecular Biology, vol. 1705,
https://doi.org/10.1007/978-1-4939-7465-8_9, © Springer Science+Business Media LLC 2018

In the past decades dramatic progress toward the discovery of effective therapeutics for GPCRs has been witnessed. This is particularly boosted by recent breakthroughs in GPCR structural biology. The solved crystal structures are templates for homology modelling of related GPCRs, and allow comparison of active and inactive forms of receptors to help understand the mechanistic details of activation [4]. The atomic-resolution structures also enable direct visualization of the binding mode of a ligand to its cognate orthosteric or allosteric binding pockets [5, 6], hence enhancing structure-based drug design (SBDD) approaches. The knowledge obtained by understanding the structural basis of ligand-receptor interactions has critical implications for rational drug design. However, such information remains limited. First, the number of available structures is far lagging behind in a comprehensive coverage of the entire GPCR superfamily. Second, the available structures often refer to static and heavily engineered conformational states, which is in stark contrast with the inherently dynamic nature of the ligand-GPCR interactions in their natural lipid environment.

In recent years, there has been a growing interest in linking structures to binding kinetics, i.e., structure-kinetics relationship studies (SKRs), next to the classical structure-affinity relationship studies (SARs) [7, 8]. Kinetic measurements enable an experimental determination of both molecular recognition (k_{on}) and complex stability (k_{off}) of a ligand-receptor interaction process. Its importance in drug discovery is now appreciated and is receiving increasing attention from the drug research community [9]. In particular, drug-target residence time (RT), the time a target is occupied by a ligand ($RT = 1/k_{off}$), is suggested to predict in vivo pharmacological activity better than binding affinity per se, as extensively discussed in several recent reviews [10, 11]. Furthermore, kinetic characterization of ligand-receptor interactions provides additional information to aid compound advancement in the hit-to-lead campaign, particularly useful in triaging a series of compounds that are otherwise biologically or chemically similar [12].

Experimental approaches that enable medium- to high-throughput screening are available, which greatly enhance kinetics-directed drug design and development [13]. Next to the development of experimental approaches for kinetic investigation, atomic-level molecular dynamics (MD) simulations have become substantially more powerful in predicting ligand binding/unbinding trajectories and the associated kinetics [14]. Several studies were reported by colleagues from the D. E. Shaw Research Institute, in which unbiased molecular dynamics simulations, thanks to increased computer power and advanced algorithms, were carried out to capture the full process of a ligand associating to a given GPCR [15]. As an example, Dror and colleagues reported that several beta-blockers and a beta-receptor agonist all traverse the same well-defined, dominant pathway—without any artificial guiding or biased forces—as they bind to the β_1- or β_2-adrenergic receptor [16]. The compound first interacts with a vestibule on each receptor's extracellular surface. Such metastate appears to represent the largest energetic barrier to binding, which includes substantial dehydration that takes place as the drug associates with the vestibule. After the initial contact the compound further enters into the binding pocket assuming the crystallographic ligand binding pose. Notably, in this second step less energy was needed although further entry into the binding pocket requires the receptor to deform and the drug to travel through a narrow passage. The two-step binding mode appears to hold for other GPCRs as well. For instance, ligands for the muscarinic M_3 receptor also "pause" at a similar alternative binding site at the extracellular vestibule *en route* to their binding pockets [17, 18].

The currently accessible time scale for all-atom MD simulations of proteins is more than a millisecond in length [15]. Such time scale guarantees reliable prediction of pathways and kinetics of the association process, as exemplified by the abovementioned studies on the β-adrenergic receptors [16]. In comparison, the ligand unbinding process from a given target occurs on timescales that remain inaccessible to direct all-atom MD simulation—as the residence time of molecules can be in the order of seconds, minutes, or even hours [19]. Methods that allow enhanced sampling are therefore being developed and optimized to shorten the long time scale of MD simulations into an accessible range [20]. This can be done by either simulating a system along a set of predefined collective variables (metadynamics), applying external forces to the system (adiabatic, targeted, steered, and accelerated MD), or simulating multiple parallel replicas at varying temperatures (replica exchange MD) [21].

In the following sections, we will illustrate the breadth of MD simulations as a tool to understand ligand unbinding trajectories and to predict residence time through several case studies.

2 Case Study 1: The Dissociation of Tiotropium from the Muscarinic Receptor and Its Residence Time

Tiotropium is a long-acting muscarinic antagonist and a first-line bronchodilator for the treatment of chronic obstructive pulmonary disease (COPD). Kinetic investigation of this compound revealed its long receptor residence time (i.e., over 24 h) on the muscarinic M_3 receptor, which was closely linked to its long duration of action [22, 23]. Recent elucidation of the M_3 receptor structure (from rat) in complex with tiotropium [Protein Data Bank (PDB) ID: 4DAJ] provided a clear image for the molecular mechanism underling its

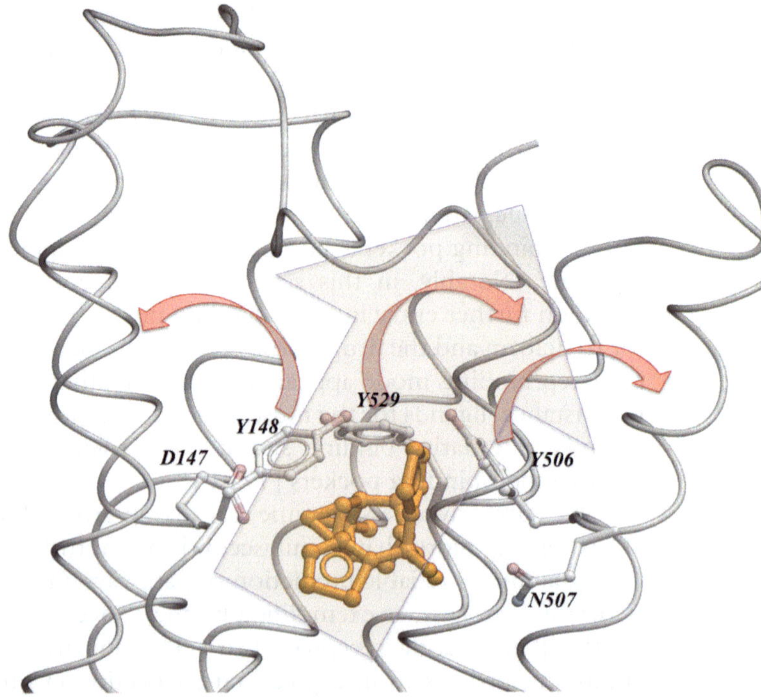

Fig. 1 The structure of tiotropium in the complex with the rat M_3 receptor and an illustrative dissociation process of the ligand from its binding pocket. This figure was generated with ICM Browser v3.8 (Molsoft) from PDB code: 4DAJ [17]. Tiotropium (*orange*) binds within the pocket. Three tyrosine residues, $Y148^{3.33}$, $Y506^{6.51}$, and $Y529^{7.39}$, together prevent the ligand from moving out of the receptor. $N507^{6.52}$ interacts with the carbonyl and hydroxyl groups through H-bonds, while the ligand's typical quaternary ammonium group interacts with $D147^{3.32}$. The movement of the three residues (*red arrows*) clears a path for tiotropium's dissociation from the orthosteric site to the extracellular vestibule and finally from the receptor

long RT profile [17]. As presented in Fig. 1, tiotropium binds deeply in the receptor core and is covered by an aromatic "lid" comprising three conserved tyrosines—$Y148^{3.33}$, $Y506^{6.51}$, and $Y529^{7.39}$ [Residue superscripts refer to the Ballesteros-Weinstein numbering [24]]. This aromatic "lid" nearly occludes the ligand from the solvent thus preventing it from being "wetted" by water molecules and then "washed" off the binding pocket. In addition, $N507^{6.52}$ interacts with the carbonyl and hydroxyl groups of tiotropium through H-bonds, while the ligand's typical quaternary amine interacts with $D147^{3.32}$.

The elucidation of the cocrystal structure of tiotropium in the rat M_3 receptor also facilitated molecular dynamics simulations to characterize the pathway by which tiotropium binds to and dissociates from the M_3 receptor [17]. Naturally, the egress of a ligand from a receptor can be considered the reverse process of the ligand *en route* to its binding pocket. Indeed, the dissociation of tiotropium consists of two steps. First, the compound leaves the orthosteric ligand binding pocket and then "pauses" at an extracellular vestibule—the same intermediate binding site as the ligand associating to the pocket. After forming the loosely connected metastate, tiotropium finally dissociates from the receptor. The RT of tiotropium is greatly determined by the first step, which is rate limiting for the ligand dissociation process. The first step formed a large energy barrier that involved the movement of the extracellular loop 2 (ECL2), which disrupts the hydrophobic cluster (Y148$^{3.33}$, Y506$^{6.51}$, and Y529$^{7.39}$) around the charged ligand head group (Fig. 1). Such motion clears a path for tiotropium's dissociation from the orthosteric site to the extracellular vestibule. In comparison, the increased mobility of ECL2 in another muscarinic receptor subtype (i.e., the M_2 receptor), as observed in the simulations, appears to facilitate tiotropium's traversal of the largest energetic barrier on the dissociation pathway. This finding is in line with tiotropium's different residence times on the M_3 and M_2 receptors, a phenomenon thought to provide clinically important "kinetic selectivity" of this drug for M_3 receptors despite tiotropium's similar equilibrium binding affinities for both subtypes [25].

A similar finding was observed in the microsecond MD simulations study by Tautermann and colleagues [26]. They confirmed that loop flexibility indeed had a strong effect on the dissociation rate of tiotropium from the human M_3 receptor. As evidence, mutations that give rise to increased flexibility in the upper part of the exit channel lead to enhanced dissociation rates. The structurally important residues are from the aromatic "lid" topping tiotropium, i.e., Y149$^{3.33}$, Y507$^{6.51}$, and Y530$^{7.39}$—equivalent to the above-mentioned hydrophobic cluster, i.e., Y148$^{3.33}$, Y506$^{6.51}$, and Y529$^{7.39}$ in the rat M_3 receptor. Mutation of these residues to alanine significantly accelerated the dissociation by up to two orders of magnitude. Such observations further corroborate the role of the aromatic cluster as a mechanical barrier that keeps tiotropium in the binding pocket. The MD simulations also highlighted the role of the double hydrogen bonded interaction of N508$^{6.52}$ with tiotropium, which hinders the compound from moving into the exit channel by reducing the frequency of tyrosine-lid opening movements. In accordance, the dissociation of tiotropium was accelerated by more than one order of magnitude upon mutation of N508$^{6.52}$ to alanine. Apparently, the interaction with N508$^{6.52}$ also plays a critical role in the development of slowly dissociating muscarinic receptor ligands.

3 Case Study 2: The Dissociation of ZM241385 from the Adenosine Receptor and Its Residence Time

The multi-step dissociation process appears to hold for other GPCRs as well. In recent research from our laboratory, we discovered that the egress of an adenosine A_{2A} receptor antagonist from the receptor also followed a similar dissociation pathway, consecutively interacting with topographically distinct regions of the receptor [27]. In this study, we applied temperature-accelerated molecular dynamics (TAMD) simulations to the crystal structure of the $A_{2A}R$ (PDB ID: 4EIY) [6] to probe the structural hotspots that have potential interactions with the crystallographic ligand, ZM241385, along its dissociation pathway from the binding pocket. In total, 16 residues were identified from the MD simulations. These residues are located in the upper part of the receptor, in either the transmembrane helices or the extracellular loops. The residues were then mutated to alanine and examined in both equilibrium and kinetic binding assays. As a result, the binding affinity of ZM241385 was only marginally influenced, except for E169A and Y271A, which were known to be involved in direct interaction with the $A_{2A}R$ ligands [28, 29]. In contrast, ZM241385's residence time was dramatically altered. Notably, most of the selected residues would have gone unnoticed in a more classical site-directed mutagenesis study with a primary emphasis on loss- or gain-of-affinity mutations. Here, with the combination of biochemical and computational studies we discovered two topographically different clusters in the $A_{2A}R$ crystal structure, one formed by $E169^{ECL2}$, $T256^{6.58}$, and $H264^{7.29}$, the other by $I66^{2.63}$, $S67^{2.64}$, and $L267^{7.32}$. Residues in the first cluster are located at the intersection of the binding cavity and the extracellular loops. Mutating these residues into alanine significantly accelerated ZM241385's egress from these mutants (less than 5 min vs. 84 min from the wild-type receptor). The MD simulations of ZM241385's egress from the $A_{2A}R$ further support the experimental observations. In the 4EIY crystal structure from which the simulations were initiated, the residues whose mutation accelerates dissociation form a "triad" interacting with ZM241385 through hydrogen bonding together with a structural water molecule (W2517) (Fig. 2). The breaking of the interaction between $H264^{7.29}$, $E169^{ECL2}$, and the ligand—a motion loosening the hydrogen bond network formed by the "triad" and enlarging the opening of the binding pocket—preceded dissociation. Mutating residues in this cluster loosens the hydrogen bond network, thus facilitating further movement of ZM241385 toward the extracellular space. The important role of residues in the triad was further confirmed in a follow-up study to improve the understanding of the molecular mechanism of ligands dissociating from the adenosine A_{2A} receptor [30]. In this study the A_{2A} receptor was cocrystalized with a series of ligands related to and including ZM241385. These ligands had been reported to have high affinities for the human $A_{2A}R$ but with divergent dissociation rate constants [31]. The elucidation of high-resolution X-ray structures of these ligands in the complex with the $A_{2A}R$ highlighted differences in the interactions between the ligands and the E169–H264 salt bridge, which may contribute to

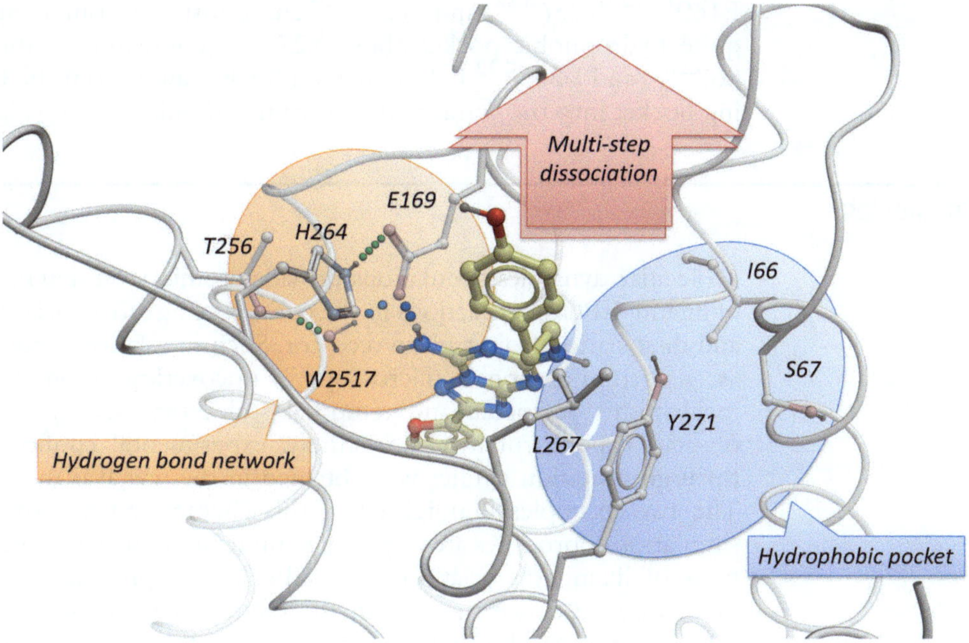

Fig. 2 The egress of ZM241385 from the human adenosine A_{2A} receptor. This figure was generated with ICM Browser v3.8 (Molsoft) from PDB code: 4EIY [6]. The ligand appears to follow a multistep pathway, first breaking the hydrogen bond network formed by the triad of $E169^{ECL2}$, $T256^{6.58}$, and $H264^{7.29}$ and transiently contacting the quite hydrophobic pocket above $Y271^{7.36}$ consisting of $I66^{2.63}$, $S67^{2.64}$, and $L267^{7.32}$ before moving further away from the binding pocket into the extracellular domain and bulk solvent

the variation in dissociation kinetics. The residence time of the ligands correlated with the energy required to break the salt bridge as estimated in silico using a metadynamics protocol. Long residence time ligands appear to stabilize the E169–H264 ionic interaction, while fast off-rate derivatives were generally predicted to destabilize this salt bridge.

An alternative binding intermediate along the dissociation pathway in the MD simulations involved ZM241385 interacting with a hydrophobic pocket formed by $I66^{2.63}$, $S67^{2.64}$, and $L267^{7.32}$. This forces ZM241385 to assume a pose similar to the one observed in another $A_{2A}R$/ZM241385 crystal structure (PDB ID: 3PWH) (Dore et al., 2011), where the antagonist's phenol group projected into the aforementioned domain ($I66^{2.63}$, $S67^{2.64}$, and $L267^{7.32}$). Mutation of these residues into much smaller alanine reduces steric hindrance and increases the ligand's freedom of rotation. As a result, ZM241385 displayed significantly increased RTs at these three mutant receptors.

Taken together, both biochemical and computational results provide a molecular description of the dissociation of ZM241385 from the $A_{2A}R$. The ligand appears to follow a multistep pathway, first breaking the hydrogen bond network formed by the triad of

E169^{ECL2}, T256$^{6.58}$, and H264$^{7.29}$ and transiently contacting the quite hydrophobic pocket above Y271$^{7.36}$ consisting of I66$^{2.63}$, S67$^{2.64}$, and L267$^{7.32}$ before moving further away from the binding pocket into the extracellular domain and bulk solvent (Fig. 2).

4 Conclusion

Molecular dynamics simulations provide atomic-level descriptions of ligand-GPCR interactions in the context of ligand association to and dissociation from a given receptor. Such simulations may serve as a "computational microscope," uncovering biomolecular mechanisms at spatial and temporal scales that are difficult to observe experimentally. Interestingly, experimental confirmation through mutation studies was obtained in the few available cases. The two examples highlighted in this chapter indicate that the simulation approaches are powerful in understanding the molecular basis of ligand dissociation from GPCRs and provide clues for predicting residence time. With the rapidly evolving state of the art for atomic-resolution biomolecular simulations, in combination with the growing body of structural information, we expect that MD techniques will be increasingly applied to enhance kinetically informed structure-based drug design and lead optimization for GPCRs.

Acknowledgments

This research received support from the Innovative Medicines Initiative Joint Undertaking under K4DD (www.k4dd.eu), grant agreement no. 115366, resources of which are composed of financial contribution from the European Union's Seventh Framework Programme (FP7/2007-2013) and European Federation of Pharmaceutical Industries and Associations (EFPIA) companies' in-kind contribution. For more information, see www.imi.europa.eu. This study was further supported by National Natural Science Foundation of China (no. 81603170 to D.G.) and Natural Science Foundation of Jiangsu Province (no. BK20160234 to D.G.).

References

1. Bjarnadottir TK, Gloriam DE, Hellstrand SH, Kristiansson H, Fredriksson R, Schioth HB (2006) Comprehensive repertoire and phylogenetic analysis of the G protein-coupled receptors in human and mouse. Genomics 88:263–273
2. De Amici M, Dallanoce C, Holzgrabe U, Trankle C, Mohr K (2010) Allosteric ligands for G protein-coupled receptors: a novel strategy with attractive therapeutic opportunities. Med Res Rev 30:463–549
3. Santos R, Ursu O, Gaulton A, Bento AP, Donadi RS, Bologa CG, Karlsson A, Al-Lazikani B, Hersey A, Oprea TI, Overington JP (2017) A comprehensive map of

molecular drug targets. Nat Rev Drug Discov 16(1):19–34

4. Cooke RM, Brown AJH, Marshall FH, Mason JS (2015) Structures of G protein-coupled receptors reveal new opportunities for drug discovery. Drug Discov Today 20:1355–1364
5. Tan Q, Zhu Y, Li J, Chen Z, Han GW, Kufareva I, Li T, Ma L, Fenalti G, Li J, Zhang W, Xie X, Yang H, Jiang H, Cherezov V, Liu H, Stevens RC, Zhao Q, Wu B (2013) Structure of the CCR5 chemokine receptor-HIV entry inhibitor maraviroc complex. Science 341:1387–1390
6. Liu W, Chun E, Thompson AA, Chubukov P, Xu F, Katritch V, Han GW, Roth CB, Heitman LH, IJzerman AP, Cherezov V, Stevens RC (2012) Structural basis for allosteric regulation of GPCRs by sodium ions. Science 337:232–236
7. Andres M, Buil MA, Calbet M, Casado O, Castro J, Eastwood PR, Eichhorn P, Ferrer M, Forns P, Moreno I, Petit S, Roberts RS (2014) Structure-activity relationships (SAR) and structure-kinetic relationships (SKR) of pyrrolopiperidinone acetic acids as CRTh2 antagonists. Bioorg Med Chem Lett 24:5111–5117
8. Andersson K, Karlsson R, Lofas S, Franklin G, Hamalainen MD (2006) Label-free kinetic binding data as a decisive element in drug discovery. Expert Opin Drug Dis 1:439–446
9. Guo D, Heitman LH, IJzerman AP (2015) The role of target binding kinetics in drug discovery. ChemMedChem 10:1793–1796
10. Copeland RA (2016) The drug-target residence time model: a 10-year retrospective. Nat Rev Drug Discov 15:87–95
11. Swinney DC, Haubrich BA, Liefde IV, Vauquelin G (2015) The role of binding kinetics in GPCR drug discovery. Curr Top Med Chem 15:2504–2522
12. Nunez S, Venhorst J, Kruse CG (2012) Target-drug interactions: first principles and their application to drug discovery. Drug Discov Today 17:10–22
13. Hoffmann C, Castro M, Rinken A, Leurs R, Hill SJ, Vischer HF (2015) Ligand residence time at G-protein-coupled receptors-why we should take our time to study it. Mol Pharmacol 88:552–560
14. Latorraca NR, Venkatakrishnan AJ, Dror RO (2017) GPCR dynamics: structures in motion. Chem Rev 117(1):139–155
15. Dror RO, Dirks RM, Grossman JP, Xu H, Shaw DE (2012) Biomolecular simulation: a computational microscope for molecular biology. Annu Rev Biophys 41:429–452
16. Dror RO, Pan AC, Arlow DH, Borhani DW, Maragakis P, Shan Y, Xu H, Shaw DE (2011) Pathway and mechanism of drug binding to G-protein-coupled receptors. Proc Natl Acad Sci U S A 108:13118–13123
17. Kruse AC, Hu J, Pan AC, Arlow DH, Rosenbaum DM, Rosemond E, Green HF, Liu T, Chae PS, Dror RO, Shaw DE, Weis WI, Wess J, Kobilka BK (2012) Structure and dynamics of the M_3 muscarinic acetylcholine receptor. Nature 482:552–556
18. Kappel K, Miao Y, McCammon JA (2015) Accelerated molecular dynamics simulations of ligand binding to a muscarinic G-protein-coupled receptor. Q Rev Biophys 48:479–487
19. Mollica L, Decherchi S, Zia SR, Gaspari R, Cavalli A, Rocchia W (2015) Kinetics of protein-ligand unbinding via smoothed potential molecular dynamics simulations. Sci Rep 5:11539
20. Zuckerman DM (2011) Equilibrium sampling in biomolecular simulations. Annu Rev Biophys 40:41–62
21. McRobb FM, Negri A, Beuming T, Sherman W (2016) Molecular dynamics techniques for modeling G protein-coupled receptors. Curr Opin Pharmacol 30:69–75
22. Casarosa P, Bouyssou T, Germeyer S, Schnapp A, Gantner F, Pieper M (2009) Preclinical evaluation of long-acting muscarinic antagonists: comparison of tiotropium and investigational drugs. J Pharmacol Exp Ther 330:660–668
23. Dowling MR, Charlton SJ (2006) Quantifying the association and dissociation rates of unlabelled antagonists at the muscarinic M_3 receptor. Br J Pharmacol 148:927–937
24. Ballesteros J, Weinstein H (1995) Integrated methods for the construction of three-dimensional models and computational probing of structure-function relations in G protein-coupled receptors. In: Methods Neurosci, vol 25. Elsevier, Amsterdam, pp 366–428. doi:citeulike-article-id:7694060. https://doi.org/10.1016/s1043-9471(05)80049-7
25. Disse B, Speck GA, Rominger KL, Witek TJ Jr, Hammer R (1999) Tiotropium (Spiriva): mechanistical considerations and clinical profile in obstructive lung disease. Life Sci 64:457–464
26. Tautermann CS, Kiechle T, Seeliger D, Diehl S, Wex E, Banholzer R, Gantner F, Pieper MP, Casarosa P (2013) Molecular basis for the long duration of action and kinetic selectivity of tiotropium for the muscarinic m3 receptor. J Med Chem 56:8746–8756

27. Guo D, Pan AC, Dror RO, Mocking T, Liu R, Heitman LH, Shaw DE, IJzerman AP (2016) Molecular basis of ligand dissociation from the adenosine A_{2A} receptor. Mol Pharmacol 89:485–491
28. Kim J, Jiang Q, Glashofer M, Yehle S, Wess J, Jacobson KA (1996) Glutamate residues in the second extracellular loop of the human A_{2a} adenosine receptor are required for ligand recognition. Mol Pharmacol 49:683–691
29. Kim J, Wess J, van Rhee AM, Schoneberg T, Jacobson KA (1995) Site-directed mutagenesis identifies residues involved in ligand recognition in the human A_{2a} adenosine receptor. J Biol Chem 270:13987–13997
30. Segala E, Guo D, Cheng RK, Bortolato A, Deflorian F, Dore AS, Errey JC, Heitman LH, IJzerman AP, Marshall FH, Cooke RM (2016) Controlling the dissociation of ligands from the adenosine A_{2A} receptor through modulation of salt bridge strength. J Med Chem 59:6470–6479
31. Guo D, Xia L, van Veldhoven JP, Hazeu M, Mocking T, Brussee J, IJzerman AP, Heitman LH (2014) Binding kinetics of ZM241385 derivatives at the human adenosine A_{2A} receptor. ChemMedChem 9:752–761

Chapter 10

Methodologies for the Examination of Water in GPCRs

Andrea Bortolato, Benjamin G. Tehan, Robert T. Smith, and Jonathan S. Mason

Abstract

The following chapter examines some of the current "state-of-the-art" tools for predicting, scoring, and examining explicit water molecules in proteins and protein/ligand complexes, highlighting some of the ways information can be readily examined in a manner that is useful in a drug discovery process.

Key words Water, WaterFLAP, Molecular dynamics, WaterMap, Water energetics, Water perturbation

1 Introduction

1.1 Waters and Drug Discovery

Water molecules and their networks are now realized to have crucial functions for both protein function and ligand binding. In terms of protein function they affect protein plasticity, allosterism, protein-protein interactions and the mediation of ligand binding, the focus of this chapter [1–3]. In terms of ligand binding, waters play a key role in potency, with displacement of "unhappy" (relative to bulk solvent) waters from lipophilic regions a key driver, but also in the modulation of selectivity and kinetics. It has been found important to take into account the perturbation of the remaining water network as well as the displacement of waters. With the richness of new GPCR structures the importance of waters has been shown computationally, enabled and supported by X-ray structural information [4–6]. Water network energetics can explain trends in off-rate kinetics, with for example trapped "unhappy" waters predicted to occur in members of a series of adenosine A_{2A} antagonists with fast off-rates [7]. Waters are a key component of molecular dynamics (MD) simulations, which enable a physics-based method to predict binding affinities etc. using FEP (Free Energy Perturbation), a very exciting method that with the FEP+ software (Schrödinger LLC) on GPUs can now be done routinely in a timely fashion. MD and water dynamics have been used for kinetic off-rate prediction,

Alexander Heifetz (ed.), *Computational Methods for GPCR Drug Discovery*, Methods in Molecular Biology, vol. 1705, https://doi.org/10.1007/978-1-4939-7465-8_10,

which has been a computational chemistry challenge [8]. The role of waters as the critical third dimension alongside the protein and ligand is now clear, and that they are important for all aspects of computational chemistry. This spans from enhancing the docking process, where the energetic perturbation of the remaining waters as well as the displaced ones can be visualized and used to evaluate known or new idea ligands, through the design of potency, selectivity, and kinetics, including FEP calculations, finally a reality in the drug designer's toolbox. A powerful druggability assessment is enabled by an analysis of (pseudo)apo water networks in conjunction with GRID lipophilic hotspots. We use three complementary methods WaterFLAP [9, 10] in conjunction with MD, WaterMap [11–13], and WaterFLAP [9, 10] alone, discussed in this chapter. All use explicit waters, unlike some other approaches, that are the key to evaluating network energies and perturbations thereof. The use of MD in the first WaterFLAP MD method and WaterMap approach provide a powerful physics-based method to evaluate full water networks. Furthermore, the WaterMap method delineates the entropic and enthalpic components from an all-atom explicit solvent MD simulation followed by a statistical thermodynamic analysis of water clusters (hydration sites). WaterFLAP alone provides a fast empirical method that enables routine use for all complexes and docked idea structures with a relatively robust energetic evaluation, and works well for seeding MD simulations with full networks including regions with trapped waters in the ligand complex.

2 Materials

The protocols described in this chapter have been implemented and tested using the software packages and scripting languages listed below. It must be noted that the scoring and functions shown below have been calibrated from our analysis of numerous reference structures [4] and our own in-house experience of GPCR structures. All of the code used for these analyses is shown in the "methods" section and utilizes open-source languages such as python, and wherever possible open source programs, such as PyMol [14] & GROMACS [15].

2.1 WaterFLAP MD Method

1. WaterFLAP is used for the initial placement of the water molecules.
2. Optimization of the water network utilizes GROMACS.
3. Ligand topology is generated using the GAFF force field [16].
4. GROMACS used for pseudo apo and receptor/ligand complex water refinement.

5. PyMol is used for protein alignment.
6. Optimized water network rescored in WaterFLAP.
7. Final classification of water is based upon reference analyses [4].

2.2 WaterMap Method

1. WaterFLAP used for initial placement of water molecules.
2. WaterMap used for placement and analyses of waters.
3. Final classification of water is based upon reference analyses [4].

2.3 WaterFLAP Method

1. WaterFLAP (flapvs) used to generate GRIDS for protein.
2. WaterFLAP (flapwater) used to generate water network for pseudo apo protein.
3. WateFLAP (flapdock) used to generate protein/ligand complex water network.
4. Final classification of water is based upon reference analyses [4].

3 Methods

3.1 WaterFLAP MD Method

3.1.1 Coupling WaterFLAP with Molecular Dynamics

WaterFLAP allows the prediction of the location of water molecules and the evaluation of their free energy [5, 7]. It is based on GRID [10] a software to probe a protein binding site using a range of different functional groups, including water, to identify areas of attraction (hotspots). The water probe is used to detect favorable locations for water molecules in a ligand-protein complex. The energy of the waters is then evaluated combing different probes: OH2 to evaluate the hydrophilic character of the pocket, CRY (a combination of DRY and the carbon sp^2 C1 = probe) to evaluate its hydrophobic/apolar components. Positional entropy of waters is estimated evaluating the energy landscape around its location. Trapped (low entropy) water molecules are located in a deep and narrow energy well, while bulk-like (high entropy) waters correspond to shallow energy basins. Since GRID is based on a united-atom force field, we introduced a short molecular dynamics step to generate an all-atoms system including a hydrogen bond network useful to understand the water network role in the stability of ligand docked poses.

The final protocol called HepWaterFlap.py is an easy-to-use python script requiring as input a protein and a docked ligand. It consists of three steps:

1. Calculation of the water network using WaterFLAP (apo or protein ligand complex).
2. Short optimization of the water network using a short MD in GROMACS.
3. Refinement and rescoring of the water network using WaterFLAP.

HepWaterFlap.py import **os** for the execution of the WaterFLAP/GROMACS and **Pymol** for the protein alignment since the MD step will translate the system to another frame of reference. **OptionParser** is used to manage the input flags.

```
#!/bin/python

import os
import time

import __main__

__main__.pymol_argv = ['pymol', '-qc']  # Pymol: quiet and no GUI
import pymol

pymol.finish_launching()

from optparse import OptionParser

###############################################################
```

The main program is shown below, the steps are:

1. Read the input files.
2. Creation of an unique working directory.
3. Creation of the starting water network.
4. Equilibration using MD in GROMACS.
5. Alignment of the equilibrated system to the original frame of reference in Pymol.
6. Rescoring of the water network.
7. Preparation of the output.

```
def main():
    # read the input files and flags
    protein, ligand, cpu, mode = readinput()
    print time.strftime("%c")
    # create working directory whit unique name
    # This directory is called with the protein and ligand name and an unique number
    # this unique number will allow to run the same ligand protein several times without over writing the results
    proteinbasename = os.path.splitext(protein)[0]
    ligandbasename = os.path.splitext(ligand)[0]
```

```
    n = 1
    while os.path.isdir(proteinbasename + "_" + ligandbasename + "_" + str(n)) or os.path.exists(
                                    proteinbasename + "_" + ligandbasename + "_" + str(n) + ".tar.gz"):
        n += 1
    os.system('mkdir ' + proteinbasename + "_" + ligandbasename + "_" + str(n))
    workdir = proteinbasename + "_" + ligandbasename + "_" + str(n)
    fixbenpdb(protein)
    # copy the files to the working directory and go there
    os.system("mv protein.pdb " + workdir + "/protein.pdb")
    os.system("cp " + ligand + " " + workdir + "/ligand.mol2")
    print " working directory: " + workdir
    os.chdir(workdir)
    # create starting network
    print "creating starting water network"
    createwaternetwork(mode)
    # run md
    print "equilibrating water network"
    # we check for the output, in case we try again (max 12 times)
    t = 0
    while t < 12 and not os.path.isfile('emFinal.gro'):
        fixchains()
        if mode == "complex":
            prepareligand()
            createproteintop()
        if mode == "apo":
            createproteintopAPO()
        solvatebox()
        runmin(cpu, mode)
        runmd(cpu)
        t += 1
    # we align the translated system after MD to the original frame of reference
    alignpdb(mode)
    # rescore waters
    print "rescoring water network"
    rescorewaters(mode)
    # create output
    createoutput(mode, ligandbasename, workdir)


if __name__ == "__main__":
    main()
```

The following code uses **OptionParser** to read the input files. You need a protein prepared in Maestro (Schrödinger) including hydrogens and a ligand as mol2 with partial charges. You need to provide the number of CPUs to use for the MD step and if you want to calculate the apo water network or in the presence of the ligand.

```
def readinput():
    usage = "usage: %prog -p protein.pdb -l ligand.mol2 -c CPUnumber -m apo/complex\nhelp: %prog -h"
    parser = OptionParser(usage,
                    version="%prog 1.0, March 2015\nAndrea Bortolato\nandrea.bortolato@heptares.com\nHeptares Therapeutics - All rights reserved")
    parser.add_option("-p", "--protein", dest="protein",
                help="protein pdb alone with H prepared by Maestro")
    parser.add_option("-l", "--ligand", dest="ligand",
                help="ligand as mol2 with H")
    parser.add_option("-c", "--cpu", dest="cpu",
                help="cpu: number of CPUs")
    parser.add_option("-m", "--mode", dest="mode",
                help="mode: apo or complex")
    (options, args) = parser.parse_args()
    if len(args) != 0:
      parser.error("incorrect number of arguments")
    if str(options.protein) == 'None' or str(options.ligand) == 'None' or str(options.cpu) == 'None' or str(
        options.mode) == 'None':
      if str(options.protein) == 'None':
        parser.error("-p protein.pdb missing in the input")
      if str(options.bound) == 'None':
        parser.error("-l ligand.mol2 missing in the input")
      if str(options.cpu) == 'None':
        parser.error("-c CPU Number missing in the input")
      if str(options.mode) == 'None':
        parser.error("-m apo/complex missing in the input")
    if str(options.mode) != 'apo' and str(options.mode) != 'complex':
      parser.error("-m select mode to use: choose apo or complex")
    return str(options.protein), str(options.ligand), str(options.cpu), str(options.mode)
```

The next steps create the water network using WaterFLAP. Please note the path to WaterFLAP executable is hardcoded for Heptares' cluster and it will need to be changed. The version of WaterFLAP used is from January 2015, but it should be possible to use newer versions (please discuss directly with Molecular Discovery).

```
def createwaternetwork(mode):
    # create a pdb with only the ATOM lines for WaterFLAP
    os.system("grep ATOM protein.pdb | grep -v '          H' > proteinNoH.pdb")
    # if the mode is apo, generate the apo network and use the ligand only to define the binding site around the ligand
    if mode == "apo":
        os.system(
            "/apps/WaterFLAP/20150122/flapwater -w -i proteinNoH.pdb -o flapWaters.pdb -gl ligand.mol2 -gr 8 -ms 3 -se -8 -fe -1 -p CRY -cp 0 -sm 0 -rf 1 -O0 -iw 50 -it 5 -wn")
    # if the mode is complex, generate the water network around the ligand
    elif mode == "complex":
        os.system('/apps/WaterFLAP/20150122/flapwater -lp ligand.mol2 proteinNoH.pdb complex4waterFlap.pdb ')
        os.system(
            "/apps/WaterFLAP/20150122/flapwater -w -i complex4waterFlap.pdb -o flapWaters.pdb -gl ligand.mol2 -gr 8 -ms 3 -se -8 -fe -1 -p CRY -cp 0 -sm 0 -rf 1 -O0 -iw 50 -it 5 -wn")
    flap = open('newWaters.pdb', 'w')
    # write the pdb in a format usable by GROMACS and remove low density waters
    with open('flapWaters_H2O.pdb', 'r') as w:
        for i in w:
            if len(i) > 40:
                if i.split()[0] == 'HETATM':
                    if float(i[54:60]) <= -8.0:  # filter low density waters
                        flap.writelines('ATOM  ' + i[6:11] + '  O   SOL  ' + i[23:54] + '  1.00  0.00           O\n')
    flap.close()
```

Possible errors in the pdb are corrected:

- The histidine name is changed to be compatible with GROMACS based on the hydrogens present to define the histidine tautomeric state or protonation state.
- Chains are added to take into account the possibility of missing residues.

```
# fix waters and chain name if there is a gap
def fixchains():
    # fix his
    his = {}
    with open('protein.pdb', 'r') as p:
        for i in p:
            if i.split()[0] == 'ATOM':
                res = int(i[22:27])
                if i[13:20] == 'ND1 HIS' or i[13:20] == 'NE2 HIS':
                    if i.split()[-1] == 'N1+':
                        his[res] = 'HIP'
                elif i[13:20] == 'HD1 HIS' and res not in his:
                    his[res] = 'HID'
                elif i[13:20] == 'HE2 HIS' and res not in his:
                    his[res] = 'HIE'
                    # else:
                    # his[res] = 'HIS'
    ou = open('ProteinAmber0.pdb', 'w')
    chains = ['A', 'B', 'C', 'D', 'E', 'F', 'G', 'H']
    n = 0
    resbefore = 0
    with open('protein.pdb', 'r') as p:
        for i in p:
            if len(i) > 30:
                if i.split()[0] == 'ATOM':
                    if i.split()[-1] != 'H':
                        res = int(i[22:27])
                        if resbefore == 0:
                            resbefore = res
                        if res == resbefore or res == resbefore + 1:
                            if i[17:20] == 'HIS':
                                ou.writelines(i[:17] + his[res] + ' ' + chains[n] + i[22:])
                            else:
                                ou.writelines(i[:21] + chains[n] + i[22:])
                            resbefore = res
                        else:  # new chain
                            n += 1
                            if i[17:20] == 'HIS':
                                ou.writelines('TER\n' + i[:17] + his[res] + ' ' + chains[n] + i[22:])
                            else:
                                ou.writelines('TER\n' + i[:21] + chains[n] + i[22:])
                            resbefore = 0
    ou.close()
    os.system('cat ProteinAmber0.pdb newWaters.pdb > ProteinAmber.pdb')
```

Different functions are used for the apo or the complex to prepare the topology for the MD in GROMACS.

3.1.2 Protein/Ligand Complex

The ligand topology is generated using the GAFF force field exploiting the acpype.py script [16] (https://github.com/t-/acpype). This script requires the net charge of the ligand in the input to calculate the partial charges correctly. The net charge is automatically calculated summing atoms partial charges.

```
# prepare ligand topology
def prepareligand():
  # create standard names
  mol2standard = open('Ligand.mol2', 'w')
  atoms = False
  totalcharge = 0.0
  with open('ligand.mol2', 'r') as mol2:
    for i in mol2:
      if i == '@<TRIPOS>BOND\n':
        atoms = False
      if atoms:
        mol2standard.writelines(i[:59] + 'LIG' + i[62:])
        # get total charge summing partial charges in the mol2
        totalcharge += float(i.split()[-1])
      else:
        mol2standard.writelines(i)
      if i == '@<TRIPOS>ATOM\n':
        atoms = True
  mol2standard.close()
  # ligand topology
  print 'preparing ligand topology'
  os.system("acpype.py -i Ligand.mol2 -c bcc -n " + str(int(round(totalcharge))))
```

The protein-ligand complex topology including the predicted WaterFLAP waters is finally created in the following function:

```
def createproteintop():
  print 'creating protein-waters topology'
  os.system("pdb2gmx -f ProteinAmber.pdb -ff amber99sb-ildn -water spc -ignh -o Protein2.pdb -p Protein.top")
  # Merge Protein2.pdb + updated Ligand_NEW.pdb -> Complex.pdb
  os.system('grep ATOM Protein2.pdb | grep -v HOH > Protein2only.pdb')
  os.system('grep HOH Protein2.pdb  > Protein2wateronly.pdb')
  os.system('cat Protein2only.pdb Ligand.acpype/Ligand_NEW.pdb Protein2wateronly.pdb > Complex.pdb')
```

```
# Edit Protein.top -> Complex.top
os.system("cp Ligand.acpype/Ligand_GMX.itp Ligand.itp")
os.system("cp Protein.top Complex.top")
os.system("cat Complex.top | sed \'/forcefield\.itp\"/a\#include \"Ligand.itp\"\' >Complex2.top")
top = open('Complex2.top', 'r')
topl = top.readlines()
topfix = open('Complex3.top', 'w')
totalwater = 0
ligandadded = False
for i in topl:
  if i[:3] == "SOL":
    if not ligandadded:
      topfix.writelines('Ligand   1\n') # fix the ligand position in the topology
      ligandadded = True
    totalwater += int(i.split()[1])
  else:
    topfix.writelines(i)
topfix.writelines("SOL    " + str(totalwater) + "\n")
top.close()
topfix.close()
os.system('mv Complex3.top Complex.top')
```

3.1.3 Pseudo Apo

The code required for the apo protein is instead simple:

```
# protein topology
def createproteintopAPO():
  print 'creating protein-waters topology'
  os.system("pdb2gmx -f ProteinAmber.pdb -ff amber99sb-ildn -water spc -ignh -o Protein2.pdb -p Protein.top")
  os.system("cp Protein.top Complex.top")
  os.system("cp Protein2.pdb Complex.pdb")
```

Once the topology is ready, a triclinic box is added and the system is solvated. Ions are added to create a neutral system before minimization with constraints on the protein-ligand atoms:

```
def solvatebox():
    os.system('editconf -bt triclinic -f Complex.pdb -o ComplexBox.gro -d 1.0')
    os.system('genbox -cp ComplexBox.gro -cs spc216.gro -o Complex_b4ion.gro -p Complex.top')


def runmin(cpu, mode):
    # create minimization file
    os.system('cp Complex.top Complex.top_bkup')
    minimization = open('em0.mdp', 'w')
    em = ['define              = -DPOSRES\n', 'integrator          = steep\n',
        'nsteps              = 1000\n', 'constraints         = none\n', 'emtol              = 1.0\n',
        'emstep              = 0.01 ; used with steep\n', 'nstcomm             = 1\n',
        'coulombtype         = PME\n', 'ns_type             = grid\n', 'rlist              = 1.0\n',
        'rcoulomb            = 1.0\n', 'rvdw                = 1.0\n', 'Tcoupl             = no\n',
        'Pcoupl              = no\n', 'gen_vel             = no\n',
        'nstxout             = 0 ; write coords every # step\n', 'cutoff-scheme = Verlet\n']
    for i in em:
        minimization.writelines(i)
    minimization.close()
    # Run minimizaton
    print "initial minimization"
    os.system('grompp -f em0.mdp -c Complex_b4ion.gro -p Complex.top -o em.tpr -maxwarn 10')
    # check that the number of ions is correct
    if mode == "complex":
        os.system('echo 15| genion -s em.tpr -o ComplexIons.gro -neutral -p Complex.top -conc 0.001')
    elif mode == "apo":
        os.system('echo 13| genion -s em.tpr -o ComplexIons.gro -neutral -p Complex.top -conc 0.001')
    if not os.path.isfile('ComplexIons.gro'):
        ('editconf -f Complex.pdb -o ComplexIons.gro')
    # create posre for the ligand
    os.system('echo 0 | genrestr -f Ligand.acpype/Ligand_GMX.gro -o Ligandposre')
    itplig = open('Ligand.itp', 'a')
    itplig.writelines('\n; Include Position restraint file\n#ifdef POSRES\n#include "Ligandposre.itp"\n#endif\n')
    itplig.close()
    os.system('grompp -f em0.mdp -c ComplexIons.gro -p Complex.top -o em.tpr -maxwarn 10')
    os.system('mdrun -v -ntomp ' + cpu + ' -deffnm em -pin auto')
```

The water hydrogen bonding network is then optimized during a 20 ps MD simulation with positional restrains on the ligand-protein atoms:

```
def runmd(cpu):
    # Create md.mdp file
    mdfile = open('md.mdp', 'w')
    md = ['integrator              = md\n',
          'define                  = -DPOSRES\n',
          'nsteps                  = 10000; 20ps\n',
          'dt                      = 0.002\n',
          'constraints             = all-bonds\n',
          'ns_type                 = grid\n',
          'rlist                   = 1.1\n',
          'rcoulomb                = 1.1\n',
          'rvdw                    = 1.1\n',
          'vdwtype                 = Cut-off\n',
          'rvdw-switch             = 0.9\n',
          'coulombtype             = PME\n',
          'Tcoupl                  = v-rescale\n',
          'tau_t                   = 0.1 0.1\n',
          'tc-grps                 = protein non-protein\n',
          'ref_t                   = 300 300\n',
          'Pcoupl                  = Berendsen\n',
          'Pcoupltype              = isotropic\n',
          'tau_p                   = 0.5\n',
          'compressibility         = 4.5e-5\n',
          'ref_p                   = 1.0\n',
          'gen_vel                 = yes ;;;\n',
          'nstxout                 = 500 ; write coords every # step\n',
          'lincs-iter              = 2\n',
          'DispCorr                = EnerPres\n',
          'optimize_fft            = yes\n',
          'refcoord-scaling        = com\n',
          'cutoff-scheme           = Verlet']
    for i in md:
        mdfile.writelines(i)
    mdfile.close()
    # Run a short simulation
    print "short MD"
    os.system('grompp -f md.mdp -c em.gro -p Complex.top -o md.tpr -maxwarn 10')
    os.system('mdrun -v -ntomp ' + cpu + ' -deffnm md -pin auto')
    print "run final minimization"
    os.system('grompp -f em0.mdp -c md.gro -p Complex.top -o emFinal.tpr -maxwarn 10')
    os.system('mdrun -v -ntomp ' + cpu + ' -deffnm emFinal -pin auto')
```

The resulting MD output is aligned to the original frame of reference using pymol:

```
def alignpdb(mode):
    if mode == 'complex':
        ndx = open('ndx_input.txt', 'w')
        ndx.writelines('0 & !a H* & ! 18\nq\n')
        ndx.close()
        os.system('make_ndx -f emFinal.gro < ndx_input.txt')
        os.system('echo 22 | trjconv -f emFinal.gro -s emFinal.tpr -o emFinal.pdb -pbc whole -n index')
        os.system('echo 0 | trjconv -f emFinal.gro -s emFinal.tpr -o emFinalH.pdb -pbc whole -n index')
    elif mode == 'apo':
        ndx = open('ndx_input.txt', 'w')
        ndx.writelines('0 & !a H* & ! 16\nq\n')
        ndx.close()
        os.system('make_ndx -f emFinal.gro < ndx_input.txt')
        os.system('echo 18 | trjconv -f emFinal.gro -s emFinal.tpr -o emFinal.pdb -pbc whole -n index')
        os.system('echo 0 | trjconv -f emFinal.gro -s emFinal.tpr -o emFinalH.pdb -pbc whole -n index')
    # waters for waterflap:
    pymol.cmd.load('protein.pdb')  # reference protein
    pymol.cmd.load('emFinal.pdb')  # final system
    pymol.cmd.select('finalprot', 'emFinal and not resn SOL')
    pymol.cmd.align('finalprot', 'protein')
    if mode == 'complex':
        pymol.cmd.extract('water', 'resn LIG around 8 and resn SOL')
    elif mode == 'apo':
        pymol.cmd.load('ligand.mol2')  # reference ligand
        pymol.cmd.extract('water', 'ligand around 8 and resn SOL')
    pymol.cmd.save('FinalMDwaters.pdb', 'water')
    pymol.cmd.reinitialize()
    # waters from MD with H
    pymol.cmd.load('protein.pdb')  # reference protein
    pymol.cmd.load('emFinalH.pdb')  # final system
    pymol.cmd.select('finalprot', 'emFinalH and not resn SOL')
    pymol.cmd.align('finalprot', 'protein')
    if mode == 'complex':
        pymol.cmd.select('waterO', 'resn LIG around 8 and resn SOL and element O')
    elif mode == 'apo':
        pymol.cmd.load('ligand.mol2')  # reference ligand
        pymol.cmd.select('waterO', 'ligand around 8 and resn SOL and element O')
    pymol.cmd.select('waterH', 'neighbor waterO')
    pymol.cmd.extract('water', 'waterO or waterH')
    pymol.cmd.save('FinalMDwatersH.pdb', 'water')
    os.system("grep ATOM FinalMDwatersH.pdb > FinalMDwatersHfixed.pdb")
```

The optimized water network is then refined and rescored in WaterFLAP:

```
def rescorewaters(mode):
    wfwaters = open('FinalMDwatersFixed.pdb', 'w')
    waternetwork = open('FinalMDwaters.pdb', 'r')
    waternetworklines = waternetwork.readlines()
    for w in waternetworklines:
        fixedline = w.replace("ATOM  ", "HETATM").replace("OW  SOL", "OH2 HOH")
        wfwaters.writelines(fixedline)
    wfwaters.close()
    if mode == "complex":
        os.system('grep ATOM complex4waterFlap.pdb > complex4waterFlapProtein.pdb ')
        os.system('grep HETATM complex4waterFlap.pdb > complex4waterFlapLigand.pdb ')
        os.system(
            "cat complex4waterFlapProtein.pdb complex4waterFlapLigand.pdb FinalMDwatersFixed.pdb > system4rescoring.pdb")
    if mode == "apo":
        os.system("cat proteinNoH.pdb FinalMDwatersFixed.pdb > system4rescoring.pdb")
    # evaluate OH2
    os.system(
        "/apps/WaterFLAP/20150122/flapwater -r -i system4rescoring.pdb -o refine-FinalOutput.pdb -gl ligand.mol2 -gr 10 -fe 9 -p CRY -cp 0 -rf 1 -it 3 -wn")
    # CRY score without the network
    os.system(
        "/apps/WaterFLAP/20150122/flapwater -r -i refine-FinalOutput.pdb -o CRY-FinalOutput.pdb -gl ligand.mol2  -gr 10 -fe 9 -p CRY -cp 0 -rf 0 -ws")
    # evaluate entropy
    os.system(
        "/apps/WaterFLAP/20150122/flapwater -r -i refine-FinalOutput.pdb -o OH2ENT-FinalOutput.pdb -gl ligand.mol2  -gr 10 -fe 9 -p ENTR -cp 0 -rf 0 -wn")
```

The final output is then generated. This includes a pdb with the optimized MD water network and another pdb with these waters after refinement from WaterFLAP. This pdb is coded to include additional information:

- The waters are classified using the element from unhappy (oxygen: red; sulfur: yellow) to bulk-like (carbon: gray) and happy (nitrogen: blue). The rules for the different classes are shown in the function **dahliawaterscore** shown below.
- The total ΔG in kcal/mol is multiplied by ten (to remove the decimal) and included after the element.

- The B factor column includes the OH2 probe energy (kcal/mol) and the occupancy column the CRY probe energy (kcal/mol).

```
def createoutput(mode, ligandbasename, workdir):
    cry = open("CRY-FinalOutput_H2O.pdb", "r")
    crylines = cry.readlines()
    oh2ent = open("OH2ENT-FinalOutput_H2O.pdb", "r")
    oh2entlines = oh2ent.readlines()
    # waterflap output
    finalresult = open('WF' + mode + ligandbasename + '.pdb', 'w')
    # md ouput with H and energy from waterflap
    # finalresultwaters = open('HOH' + mode + ligandbasename + '.pdb', 'w')
    wline = 0
    n = 1
    # nmd = 0
    for i in crylines:
        if i[:6] == 'HETATM':
            CRY = float(i[60:66])
            OH2 = float(oh2entlines[wline][54:60])
            ENT = float(oh2entlines[wline].split()[-1])
            element, DG = dahliawaterscore(CRY, OH2, ENT, mode)
            if DG < 0:
                if -10 * DG > 100:
                    dgcode = "99N"
                else:
                    dgcode = str(int(round(-10 * DG))).zfill(2) + 'N'
            elif DG >= 0:
                if 10 * DG > 100:
                    dgcode = "99P"
                else:
                    dgcode = str(int(round(10 * DG))).zfill(2) + 'P'
            finalresult.writelines('ATOM' + str(n).rjust(7) + '  ' + element + dgcode + 'HOH  ' + str(n)
.rjust(4) + i[26:54] + ('%.2f' % OH2).rjust(6) + ('%.2f' % CRY).rjust(6) + '\nTER\n')
            n += 1
        wline += 1
    cry.close()
    oh2ent.close()
    finalresult.close()
    os.system('mv WF' + mode + ligandbasename + '.pdb ../')
    # os.system('mv HOH' + mode + ligandbasename + '.pdb ../')
    os.system('cp  FinalMDwatersHfixed.pdb ../HOH' + mode + ligandbasename + '.pdb')
    os.chdir('..')
    print "WateFlap waters written to: WF" + mode + ligandbasename + ".pdb"
    print "MD waters written to: HOH" + mode + ligandbasename + ".pdb"
```

```
    print "you can remove now the working directory: " + workdir


def dahliawaterscore(CRY, OH2, ENT, mode):
    if CRY > 0.0 and mode == "complex":
        DG = OH2 + 14 + ENT + 1.5
        # DG = OH2 + 5
    elif CRY <= 0.0 and mode == "complex":
        DG = OH2 + 14 + ENT + 1.5 - 1.5 * CRY
        # DG = OH2 + 5 - CRY
    elif CRY > 0.0 and mode == "apo":
        DG = OH2 + 14 + ENT + 3.0
        # DG = OH2 + 5
    elif CRY <= 0.0 and mode == "apo":
        DG = OH2 + 14 + ENT + 3.0 - 1.5 * CRY
        # DG = OH2 + 5 - CRY
    if ENT <= -3.0:
        element = "C"
    elif DG >= 3.5:
        element = "O"
    elif DG < 3.5 and DG >= 2:
        element = "S"
    elif DG < 2 and DG >= -2:
        element = "C"
    elif DG < -2:
        element = "N"
    return element, DG
```

3.2 WaterMap-Based Method

3.2.1 Background

WaterMap [9–11] is based on the inhomogeneous solvation theory of Lazaridis [17] where enthalpy is taken directly from non-bonded interactions and entropy is computed from a local expansion of spatial and orientational correlation functions. WaterMap calculations involve running converging MD simulations with explicit water molecules, the resultant trajectories are then analyzed to cluster hydration sites (a region of space where water molecules aggregate). Entropy and enthalpy are computed for each hydration site and energy terms are computed relative to bulk solvent. The results are presented graphically for easy visualization of hydration sites, with corresponding tables to enable in-depth analysis of the specific enthalpic or entropic contributions to any specified hydration site. The technique has been applied to numerous protein ligand complexes with many successes [4].

3.2.2 Running WaterMap

The default parameters in the WaterMap setup work well in most cases, and in most pseudo apo simulations these will suffice. Unfortunately in many GPCR structures we often find numerous water-mediated interactions from the ligand to the protein in addition to many ligand occluded regions that the grand canonical Monte Carlo (GCMC) water placement method can sometimes struggle to hydrate in a manner that is consistent with our in-house crystal structures, *see* Fig. 1. Although in the most recent update to the Schrodinger force field, OPLS3, the number of waters more closely resembles that seen in our crystal structures.

It should be noted that crystal structures are snapshots of the protein at temperatures close to absolute zero, and thus may not be real representations of what is actually happening. Unfortunately, this is all we have to work with and thus must be pragmatic in our approach and perhaps not absolutely theoretically correct.

To overcome the issues associated with the GCMC placement of waters, we have utilized the water placement within WaterFLAP [9, 10] to place the waters around the ligand. The water placements around the ligand were generated using the function "createwaternetwork" discussed previously in the "3.1.0 WaterFLAP MD

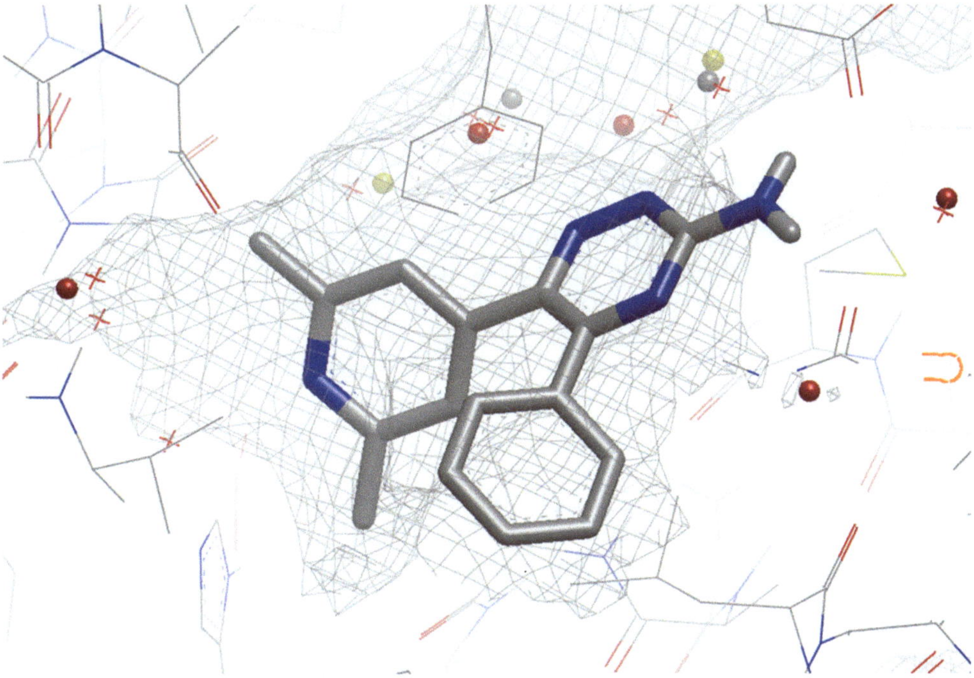

Fig. 1 The round spheres show the waters predicted by the Grand Canonical Monte Carlo (GCMC) method using the OPLS2005 force field within WaterMap, the red crosses are the location of the crystallographic waters seen within our structure

method" section. These water placements are then added to the starting structure and a standard protein preparation process, from Schrodinger, is run to incorporate the waters around the ligand into the protein/ligand/water complex.

The WaterMap process is then run via the command line after altering the input or command files. This is done by writing out the command files, or XXX.msj & XXX.maegz files, instead of running them from the GUI, and editing them so that the "solvate ligand" step within the XXX.msj file reads "should_skip = true."

XXX.msj

```
.
.
.
solvate_pocket {
 backend = {
   buffer = 10.0
   ligand_proximity = 10.0
   protein_proximity = 5.0
   proximity_resolution = 0.5
 }
 ligand_file = ?
 num_output = 1
 should_skip = true
}
.
.
.
```

At the bottom of the XXX.msj file is the command line job submission that can be run after altering the file.

command example:

```
# $SCHRODINGER/watermap -JOBNAME 3PWH_WM -HOST compchem-node3
-cpu 32 -m 3PWH_WM.msj 3PWH_WM-in.maegz
```

3.2.3 Processing WaterMap

To ensure comparable results are obtained for this method as have been obtained previously, the WaterMap results are analyzed and the csv file of the energies, from "Export to CSV..." in "WaterMap – Examine Results" (*see* Fig. 2), in combination with the placement of the waters, "XXX-watermap.pdb", are combined.

The following script with a simplified colouring scheme to what had been discussed earlier in the function "dahliawaterscore," in the previous section, is then applied.

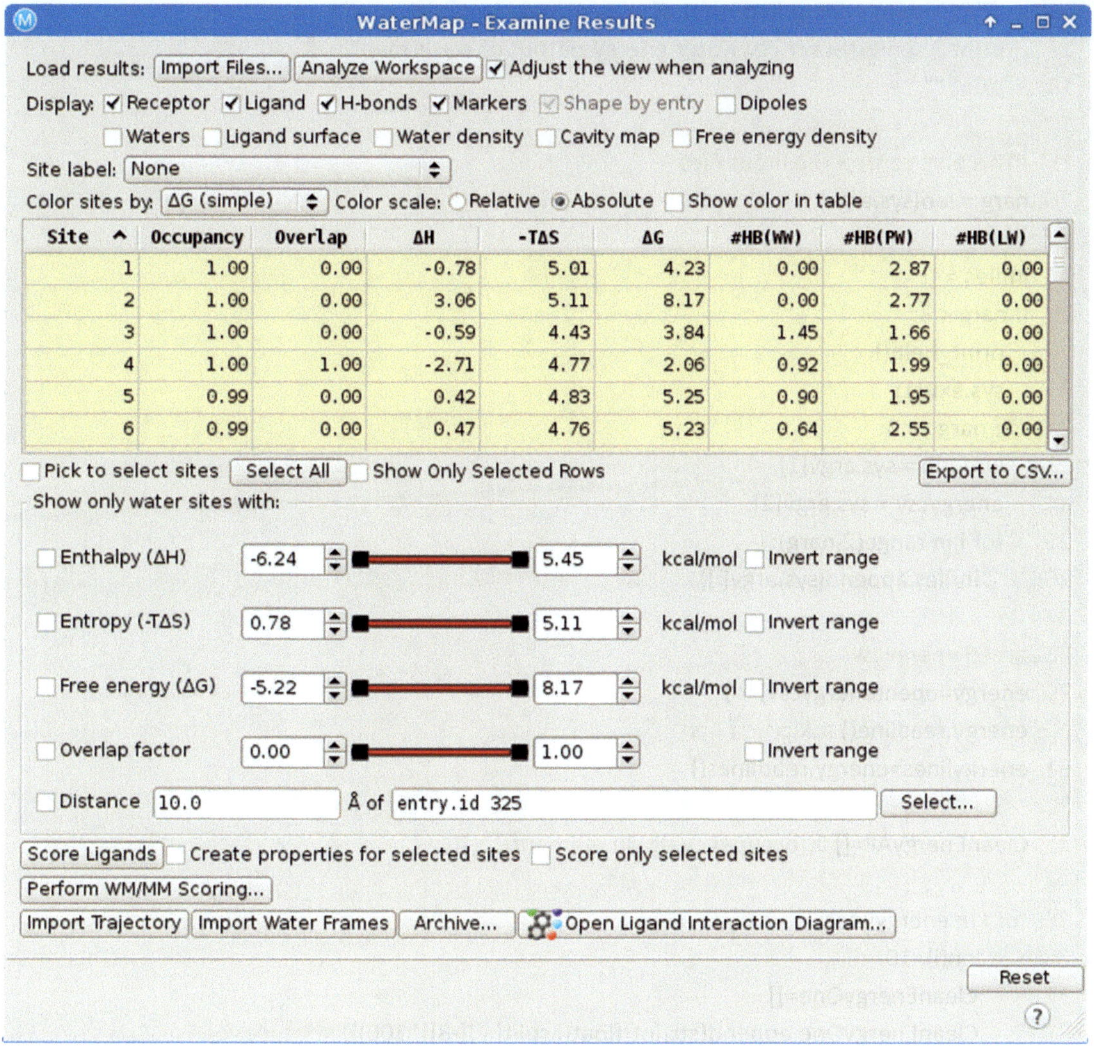

Fig. 2 Tab within maestro for examining the results from a WaterMap calculation, the "Export to CSV..." button is located on this tab

```
#!/usr/bin/python

import os,sys

def print_help():
    print ""
    print " prepareWM.py v0.1"
    print ""
    print " python prepareWM.py <wm.pdb> <energy.csv>"
    print ""
```

```python
    print " <wm.pdb> pdb output of watermap waters"
    print " <energy.csv> csv water energy output of watermap"
    print ""

#This part control the input files
narg = len(sys.argv)

infiles = []
if narg < 3:
    print_help()
    sys.exit(1)
elif narg >= 3:
    wmpdb = sys.argv[1]
    energycsv = sys.argv[2]
    for i in range(2,narg):
        infiles.append(sys.argv[i])

#read energycsv
energy=open(energycsv, 'r')
energy.readline() #skip first line
energylines=energy.readlines()

CleanEnergyAll=[] #[[occupancy,dg,dh,element], ...]

for i in energylines:
    if len(i)>10:
        CleanEnergyOne=[]
        CleanEnergyOne.append(str(int(float(i.split(',')[-8])*100))) #occupancy
        dg=i.split(',')[-4]
        CleanEnergyOne.append(dg) #dg
        CleanEnergyOne.append(i.split(',')[-6]) #dh
        if float(dg)>3.5: CleanEnergyOne.append('O') #element
        elif float(dg)>2.0: CleanEnergyOne.append('S')
        elif float(dg)<-1.0: CleanEnergyOne.append('N')
        else: CleanEnergyOne.append('C')
        CleanEnergyAll.append(CleanEnergyOne)


#open output
output=open('Prepared_'+wmpdb,'w')
```

```
#read wm.pdb
wm=open(wmpdb,
wmline=wm.readline()

n=0
while wmline:
   #write starting pdb
   if wmline[:6]!='HETATM'
   else:
      output.writelines(wmline[:12]+str(CleanEnergyAll[n][3]+CleanEnergyAll[n][0]).ljust(5)+wmline[17:54]+('%.2f'
 )).rjust(6)+'\nTER\n'
      n+=1
   wmline=wm.readline()
```

The resultant pdb file has the waters that are classified using the element from unhappy (oxygen: red; sulphur: yellow) to bulk-like (carbon: gray) and happy (nitrogen: blue).

3.3 WaterFLAP-Based Method

3.3.1 Protein GRID Generation

Typically, we generate the GRID [10] interaction fields for each protein we dock ligands to, in order to identify hotspots and understand the energetics of the water molecules and ligand within the receptor site. To generate the GRID interaction fields, we use WaterFLAP [9], and execute it using the following command line option:

(Path to WaterFLAP)/flapvs -d TEMP -gg 0.75 -pp 4 C1= C3 H O -gr 6 -cpu 8 -gl ligand.mol2 protein.pdb

The key command line options we employ are:

-d which defines the directory for FLAP to use (TEMP in the above case).

-gg which defines the spacing between the GRID points.

-pp which defines the number of probes (4 in the above case), followed by the probes we wish to run (C1=, C3, H, O in the above case).

-gr which defines the distance from the ligand to explore using GRID.

-cpu which defines how many CPU cores to use for generating the GRID (maximum of 8).

-gl which defines the ligand around which we want to centre the GRID.

An alternative command line for running GRID, without a ligand is the following:

(Path to WaterFLAP)/flapvs -d TEMP –gg 0.75 -g 5 0 66 12 12 12 -pp 4 C1= C3 H O -cpu 8 protein.pdb

Where the key command line options we employ that are alternative to the -gr and -gl options above are:

-g which defines the centre of the GRID cube to explore, in the format x y z lx ly lz (where lx, ly, lz is the respective side lengths of the cube).

Due to the fact we are running GRID around a cube, which we have defined, we do not require a ligand to be present in the receptor site.

The key outputs from these commands are the .xplor files, which either give a shape for the pocket (in the case of the H probe), or give an energy for the attractive force felt between the GRID probe and the protein. In the event a position within the protein is unfavorable for the probe a 0 is recorded in the file.

These GRIDs are calculated and used internally by WaterFLAP during the apo and complex water network generation described below.

3.3.2 APO Water Network Generation

Prior to running WaterFLAP on a protein, we first take it through the Protein Preparation Wizard within Maestro (by Schrodinger). After the preparation, we save the protein as protein.pdb, and the ligand as ligand.mol2.

A new automated flag has been introduced by Molecular Discovery, which captures many of the key options we had previously set within our Python scripts, making them redundant. For the more recent versions of WaterFLAP (released during 2016), to generate an apo water network, we run the following command line:

(Path to WaterFLAP)/flapwater -w-auto -i protein.pdb -o WAT_-PRED_OCT.pdb -gl ligand.mol2 -gr 6 -cpu 8 >apowaterflap. log 2>apowaterflap.log2

The key command line options we employ are:

-w-auto which runs the automated water protocol, with the default settings.

-i which defines the protein file.

-o which defines where we want the output to be saved.

-gl which defines the ligand that was present within the pocket.

-gr which defines the distance from the ligand to explore using GRID (and the resulting waters).

-cpu which defines how many CPU cores to use for generating the water network (maximum of 8).

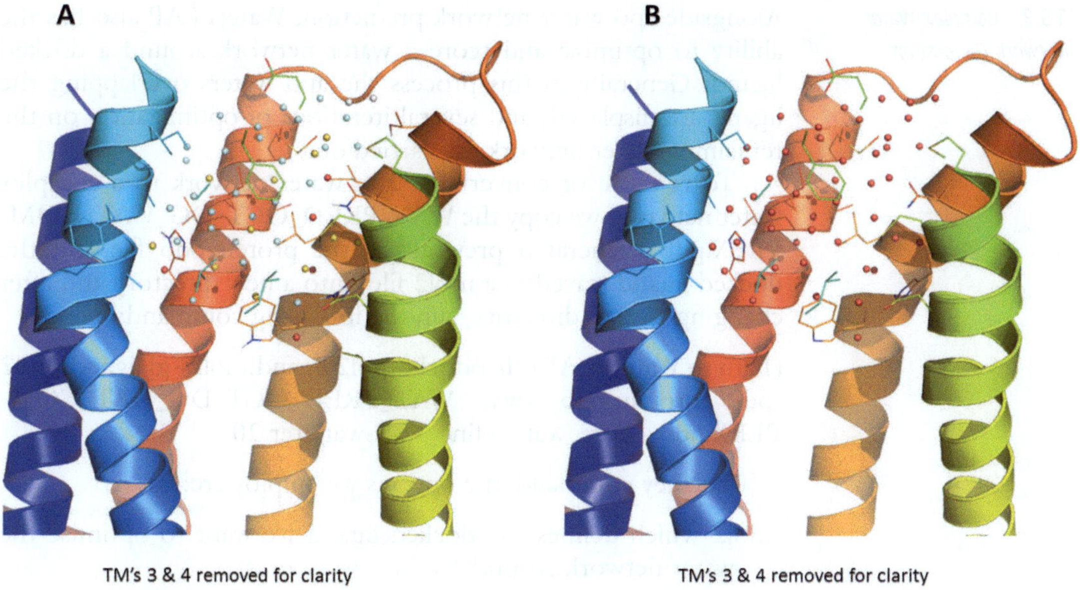

Fig. 3 WaterFLAP pseudo apo water network predictions with (**a**) showing the protein with the predicted waters from the WAT_PRED_OCT_DG_WAT_H2O_ele.pdb file. The waters are colored *gray* (depicted as iron) for bulk waters, *red* for high-energy waters (depicted as oxygen), *yellow* for mid energy waters (depicted as sulfur), and *blue* for low-energy waters (depicted as nitrogen). The energy of the waters is given in the b-factor column within the pdb. And (**b**) showing the WAT_PRED_OCT_DG_WAT_COMPLEX.pdb file which is utilized in the generation of the protein/ligand complex water network

As given in the command line above, we typically fill the pocket defined within 6 Å from the ligand with water.

The output from the above command is a series of files, with a separate pdb file for each refinement step as high-energy waters are iteratively removed. Two key output files from WaterFLAP are kept, the WAT_PRED_OCT_DG_WAT_H2O_ele.pdb file and the WAT_PRED_OCT_DG_WAT_COMPLEX.pdb.

The WAT_PRED_OCT_DG_WAT_H2O_ele.pdb file (Fig. 3a) contains the waters that WaterFLAP has predicted would be present within the receptor pocket, with different elements for the different classes of waters present. The classes of waters that can be present are: bulk waters (by default depicted as iron), high-energy waters (by default depicted as oxygen), mid energy waters (by default depicted as sulfur), and low-energy waters (by default depicted as nitrogen). The energy of the waters is given in the b-factor column within the pdb.

The WAT_PRED_OCT_DG_WAT_COMPLEX.pdb file (Fig. 3b) contains the original protein, with waters filling the pocket which contained the ligand, and is retained for optimizing the apo water network when an analogue of the reference ligand is docked, in order to give a complex water network.

3.3.3 Complex Water Network Generation

Alongside apo water network prediction, WaterFLAP also has the ability to optimize and score a water network around a docked ligand. Generally in this process the apo waters overlapping the ligand are displaced, and several iterations of optimization on the remaining water network are carried out.

To re-score or convert the apo water network to a complex water network, we copy the WAT_PRED_OCT_DG_WAT_COMPLEX.pdb (generated previously), the protein.pdb file and the docked ligand (saved as a mol2 file) into a new directory and after changing to that directory, run the following command:

(Path to WaterFLAP)/flapdock -mol2 ligand.mol2 -gl ligand.mol2 -pdb protein.pdb -wat WAT_PRED_OCT_DG_WAT_COMPLEX.pdb -score_wat -refine_wat -wat_iter 20

The key command line options we employ are:

-mol2 which defines the docked ligand we want to optimise the water network around.

-gl which defines the docked ligand we want to optimise the water network around.

-pdb which defines the protein pdb we have docked ligand.mol2 into.

-wat which defines the apo water network with the receptor pdb file (the WAT_PRED_OCT_DG_WAT_COMPLEX.pdb generated during the apo prediction).

-score_wat which informs WaterFLAP that we want to score the waters, post refinement.

-refine_wat which informs WaterFLAP that we want to refine the waters present within the –wat pdb file.

-wat_iter which defines the number of iterations WaterFLAP employs when optimising the water network around the docked ligand.

As with the apo water network generation, this generates a number of files for output, but the key file we use is WATERFLAP_Delta_DG_DG_WAT_H2O_ele.pdb.

The WATERFLAP_Delta_DG_DG_WAT_H2O_ele.pdb file (Fig. 4) returned contains waters in their perturbed positions relative to the input apo state from WAT_PRED_OCT_DG_WAT_COMPLEX.pdb. The element type in the output file gives an indication as to whether the water molecule has been stabilized by the ligands presence in the receptor, or destabilized. In this file, you can have waters heavily stabilized by the ligand (depicted as nitrogen atoms), somewhat destabilized by the ligand (depicted as sulfur atoms), destabilized by the ligand (depicted as oxygen atoms), or unaffected by the ligand (depicted as iron atoms).

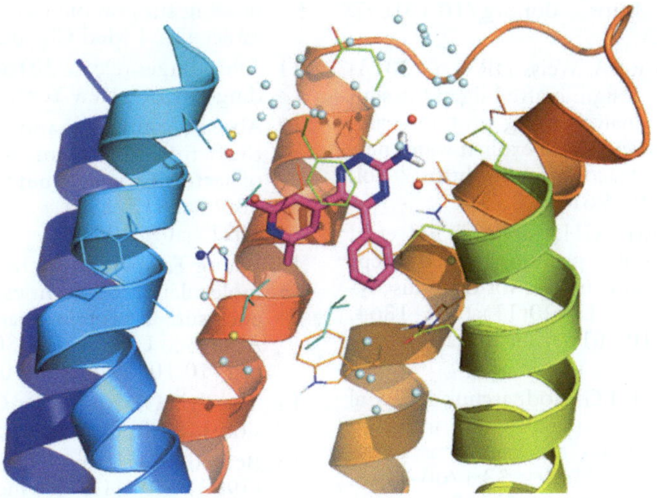

Fig. 4 WaterFLAP perturbed complex water network prediction from the WATERFLAP_Delta_DG_DG_WAT_H2O_ele.pdb file showing waters perturbed from the apo state (Fig. 3a) under the influence of the ligand. Waters shown: *blue* (nitrogen) are stabilized, *gray* (iron) are unaffected, *yellow* (sulfur) partially destabilized, and *red* (oxygen) significantly destabilized

3.3.4 Final Output

The output is then combined into a multi molecule SD file, which contains the docked ligand, key interactions between the docked ligand and the protein, a minimized form of the ligand and the output from WaterFLAP. This combined multi molecule file allows us to review whether the docked ligand is in a relatively high energy conformation, if there are a significant number of unfavorable interactions, and whether the ligand results in a more energetically favorable water network.

To retrieve the key interactions, we use the poseviewer_interactions.py script supplied within a default Schrodinger install, and this can be run using the following.

$SCHRODINGER/run poseviewer_interactions.py protein.pdb ligand.pdb

The output from this command is a file called protein_pv_interactions.txt, which contains details of interactions between the residues and the ligand atoms.

References

1. Ball P (2008) Water as an active constituent in cell biology. Chem Rev 108(1):74–108. https://doi.org/10.1021/cr068037a
2. Chaplin M (2006) Do we underestimate the importance of water in cell biology? Nat Rev Mol Cell Biol 7(11):861–866. https://doi.org/10.1038/nrm2021
3. Ball P (2008) Water as a biomolecule. ChemPhysChem 9(18):2677–2685. https://doi.org/10.1002/cphc.200800515
4. Mason JS, Bortolato A, Congreve M et al (2012) New insights from structural biology into the druggability of G protein-coupled receptors. Trends Pharmacol Sci 33

(5):249–260. https://doi.org/10.1016/j.Tips.2012.02.005

5. Mason JS, Bortolato A, Weiss DR et al (2013) High end GPCR design: crafted ligand design and druggability analysis using protein structure, lipophilic hotspots and explicit water networks. Silico Pharmacol 1(1):23. https://doi.org/10.1186/2193-9616-1-23
6. Cooke RM, Brown AJH, Marshall FH et al (2015) Structures of G protein-coupled receptors reveal new opportunities for drug discovery. Drug Discov Today 20(11):1355–1364. https://doi.org/10.1016/j.Drudis.2015.08.003
7. Bortolato A, Tehan BG, Bodnarchuk MS et al (2013) Water network perturbation in ligand binding: adenosine A_{2A} antagonists as a case study. J Chem Inf Model 53:1700–1713. https://doi.org/10.1021/ci4001458
8. Bortolato A, Deflorian F, Weiss DR, Mason JS (2015) Decoding the role of water dynamics in ligand-protein unbinding: CRF1R as a test case. J Chem Inf Model 55(9):1857–1866. https://doi.org/10.1021/acs.Jcim.5b00440
9. FLAP 2.2, Molecular Discovery Ltd. https://www.moldiscovery.com/
10. Goodford PJ (1985) A computational procedure for determining energetically favorable binding sites on biologically important macromolecules. J Med Chem 28(7):849–857
11. Schrödinger release 2016-4: WaterMap. Schrödinger, LLC, New York, NY, 2016
12. Abel R, Young T, Farid R et al (2008) Role of the active-site solvent in the thermodynamics of factor Xa ligand binding. J Am Chem Soc 130(9):2817–2831. https://doi.org/10.1021/ja0771033
13. Young T, Abel R, Kim B et al (2007) Motifs for molecular recognition exploiting hydrophobic enclosure in protein–ligand binding. Proc Natl Acad Sci U S A 104:808–813. https://doi.org/10.1073/pnas.0610202104
14. The PyMOL Molecular Graphics System, Version 1.7.4.0, Open-Source
15. Berendsen HJ, van der Spoel D, van Drunen R (1995) GROMACS: a message-passing parallel molecular dynamics implementation. Comput Phys Commun 91(1–3):43–56
16. Sousa da Silva AW, Vranken WF (2012) ACPYPE - AnteChamber PYthon parser interfacE. BMC Res Notes 5(1):367
17. Lazaridis T (1998) Inhomogeneous fluid approach to solvation thermodynamics. 1. Theory. J Phys Chem B 102:3531–3541

Chapter 11

Methods for Virtual Screening of GPCR Targets: Approaches and Challenges

Jason B. Cross

Abstract

Virtual screening (VS) has become an integral part of the drug discovery process and is a valuable tool for finding novel chemical starting points for GPCR targets. Ligand-based VS makes use of biochemical data for known, active compounds and has been applied successfully to many diverse GPCRs. Recent progress in GPCR X-ray crystallography has made it possible to incorporate detailed structural information into the VS process. This chapter outlines the latest VS techniques along with examples that highlight successful applications of these methods. Best practices for increasing the likelihood of VS success, as well as ongoing challenges, are also discussed.

Key words Virtual screening, G protein-coupled receptor, Molecular docking, Data fusion, Data mining, Shape search, Pharmacophore search, Homology modeling, Fingerprint similarity, Machine learning

1 Introduction

The G protein-coupled receptors (GPCRs) represent a pharmaceutically important class of proteins, accounting for approximately 40% of marketed drugs [1]. Situated in the cellular membrane, these receptors detect chemical signals from outside the cell and translate these into internal cellular responses, making them ideal pharmaceutical targets for a variety of therapeutic indications, including central nervous system (CNS) disorders and pain, inflammation and immune response, cancer, and endocrine dysfunction. As such, there is considerable and ongoing interest in targeting GPCRs with small-molecule therapeutics, as evidenced by the continued efforts to bring these drugs to the market [2]. The identification of tractable chemical starting points for medicinal chemistry optimization is a crucial requirement to progress any GPCR drug discovery program.

Virtual screening (VS) is an increasingly common knowledge-based approach [3] for separating active molecules from inactive

Alexander Heifetz (ed.), *Computational Methods for GPCR Drug Discovery*, Methods in Molecular Biology, vol. 1705,
https://doi.org/10.1007/978-1-4939-7465-8_11,

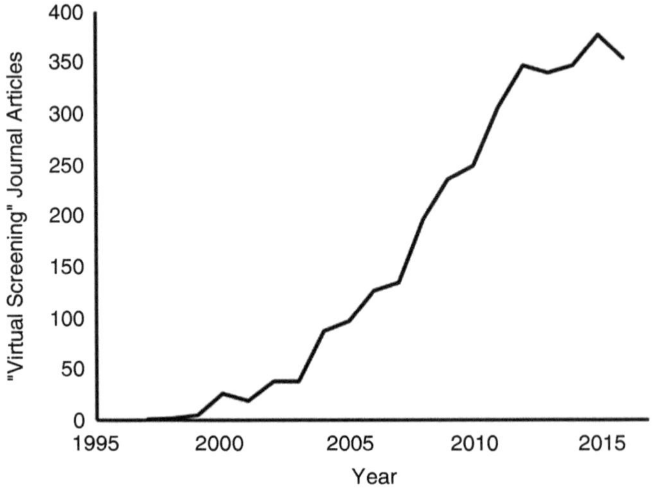

Fig. 1 Results of Google Scholar search for "virtual screening" in journal article titles, broken down by year (search conducted 1/4/2017)

ones in a chemical library (Fig. 1). In a practical sense, VS aims to reduce the number of compounds to be experimentally screened from millions down to a library of hundreds or thousands, which is enriched with actives. This smaller library can be prosecuted using low or medium throughput screening methods, saving time and resources, and satisfying the primary objective of any screening effort, which is the identification of novel chemical starting points. Due to these advantages, VS has become an integral part of the overall screening strategy at many pharmaceutical and biotechnology companies, as well as academic labs, either as a lone screening technique or as one component in an effort that encompasses other screening methods, such as high-throughput screening (HTS) and fragment-based approaches [4, 5].

A general overview of a VS workflow is shown in Fig. 2. Although there are a wide variety of algorithms and techniques available to perform the actual in silico screening, the overall workflow is essentially the same. This starts with the collection of active compound information used to inform model development, library curation, and method validation (data mining). Protein sequence and structural data can be included in this category as well, since bioinformatics analyses and protein models are critical pieces of information that feed into a structure-based virtual screening (SBVS) workflow. Generation of a curated compound library to search against is also a prerequisite for VS and may take many forms, from SMILES strings to conformationally expanded databases, depending on the search method employed. The search methodology generally falls into one of two major groups: ligand-based or structure-based. Ligand-based methods require knowledge of previously discovered active compounds to build search models, such

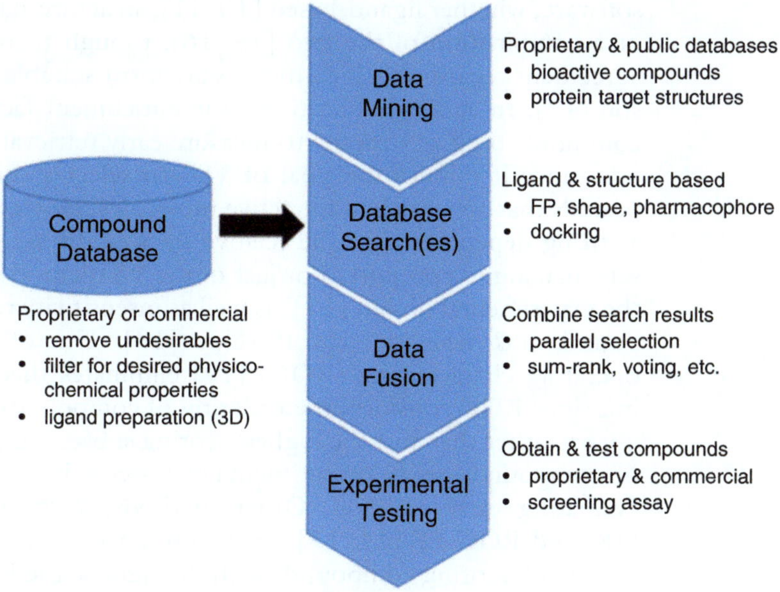

Fig. 2 Streamlined VS workflow, including collection of input data (data mining and compound databases), ligand and/or structure-based database searching, combination of output from different search methods and compound selection (data fusion), and procurement of compounds for experimental testing

as pharmacophore or shape queries. Structure-based methods, which generally involve molecular docking and require either an X-ray structure of the protein target or a high-quality homology model, can also make excellent use of ligand activity data, but have the advantage of not requiring this type of data. In cases where more than one search is conducted using either a single method or multiple techniques, which has become increasingly common, a scheme to incorporate the output from these methods into a final "hit list" is required (data fusion). Once this list of compounds is selected computationally, it is good practice to visually inspect the output molecules to ensure their suitability for medicinal chemistry follow-up before experimental testing. All of these topics will be addressed in the following sections.

While the ultimate success of a VS will be dependent on the identification of hits amenable to medicinal chemistry optimization, it can be difficult to estimate the likelihood of success a priori. The success of retrospective screens can be neatly quantified, since the number of actives and decoys are known and can be controlled as part of the virtual experiment. Public benchmark datasets, such as the Directory of Useful Decoys (DUD) [6] for SBVS, the Maximum Unbiased Validation (MUV) set [7] for ligand-based virtual screening (LBVS), as well as recently curated GPCR-focused datasets [8, 9], have proven valuable in standardizing method comparisons. Many articles have been published that attempt to measure the relative success rates of different VS methods and software, whether ligand-based [10, 11], structure-based [12–14], or a combination of the two [15, 16], though there can be disagreement regarding which metrics are most suitable for comparison [17]. In earlier publications, the enrichment factor (EF) was commonly used as a means to measure early retrieval of VS methods, since it mimics the goal of VS—to select a subset of compounds that are enriched for active molecules. However, EF shows a strong dependency on the relative sizes of the active and decoy sets, making it a property not just of the VS method used but also the experimental design [17], as well as insensitivity to active compound rank near the cutoff value [18]. Use of the receiver operating characteristic (ROC) [19] eliminates this dependency, but since ROC considers the ordering of compounds in the entire library rather than just the highest scoring subset (i.e., the short list of compounds to be experimentally tested), it is not useful in evaluating early retrieval. Other methods, such as BED-ROC [18] and ROC EF [14, 17], seek to balance the focus of the EF on the top scoring compounds with the rigor of the ROC methodology, and are higher quality metrics for evaluating retrospective VS. An alternate method, Robust Initial Enhancement (RIE) [20], addresses some issues with EF, but does not have all the advantages of ROC. Unfortunately, prospective screens cannot be quantified in exactly the same way as retrospective screens, since the proportion of actives and inactives is unknown. However, if there are sufficient known actives to run a small retrospective screen using the same methodology intended for the prospective screen, it is possible to obtain an estimate of the likelihood of VS success.

2 Materials

The methods and protocols outlined in this chapter can be implemented on a variety of hardware and operating system platforms, including *nix (Linux and MacOS being the most common) and Windows. However, certain compute-intensive applications are restricted to specific high-performance architectures and platforms. Many of the programs listed are free of charge or are available at reduced cost to academics, and open source options have been included, where available, in addition to commercial products.

2.1 Computer Hardware

The type of computer hardware recommended strongly depends on the nature of the computations involved. For example, computationally expensive calculations, such as molecular dynamics simulations, are best handled by a CPU cluster or GPU workstation, while many ligand-based methods and data fusion techniques may require only a high-performance laptop computer. For those without access to high-performance hardware, the availability of cloud computing resources (e.g., Amazon Web Services, Microsoft Azure, Google Cloud Platform) is becoming a viable option.

2.2 Computer Software

1. There are several commercial products focused on facilitating data mining, including GSK BIO's GoStar database, Jubilant Life Science's ChemBioBase, and Clarivate Analytics' Integrity database. Publicly available resources include ChEMBL [21] (including GPCR SarFari), PubChem [22], and BindingDB [23].
2. Most modern molecular modeling suites have workflows for molecular database preparation. This includes MOE [24], Schrödinger LigPrep [25], and OpenEye OMEGA [26]. Pipelining software, such as Pipeline Pilot [27] and KNIME [28], along with standalone tools are also available.
3. Fingerprint methods are available in comprehensive molecular modeling packages (e.g., MOE [24] and Schrödinger CANVAS [29]) as well as pipelining software and toolkits (e.g., RDKit [30]).
4. A variety of pharmacophore-based methods are widely available, including CATALYST [31], MOE [24], Schrödinger PHASE [32, 33], and LigandScout [34].
5. Shape-based comparison methods include ROCS [35] from OpenEye, ShapeScreen [36] from Schrödinger, and PARAFIT [37]. Molecular field and surface-based software include the tools from Cresset [38] and Surflex-Sim [39].
6. There are many open source options available that contain machine learning code, including RDKit [30] and R [40].
7. Molecular docking tools are widely available. Some of the most popular programs include DOCK [41], AutoDock [42], GOLD [43], FlexX [44], Surflex [45], Glide [46, 47], and FRED [48]. Induced-fit docking and interaction fingerprint analysis are also available as part of many comprehensive software packages.
8. There are several high-quality public tools for homology modeling available, including SWISS-MODEL [49] and I-TASSER [50, 51]. Commercial homology modeling software is also widely available, including MODELLER [52], Schrödinger PRIME [53], and MOE [24].
9. Many options are available for running protein simulations. AMBER [54] and NAMD [55] are common academic choices, while Desmond [56] is an example of a more recent entry in the commercial sphere.
10. Several methods for calculating water placement and energetics are widely available, including GRID [57, 58], SZMAP [59], Schrödinger WaterMap [60–62], and 3D-RISM [63].
11. Data fusion can be performed by using standard office productivity software, such as spreadsheets, chemically aware database management systems in molecular modeling suites, or automated using pipelining tools, such as Pipeline Pilot [27] or KNIME [28].

3 Methods

3.1 Data Mining

1. The success of any VS campaign, as well as the selection of techniques utilized, is strongly dependent on the quantity, quality, and type of data available to computational chemist at its outset. Regardless of the availability of protein structural information, whether from structural biology or protein modeling, curation of a dataset of known active compounds for the target of interest, as well as other closely related targets, is a key first step in enabling VS. This data can be used directly as input for ligand-based approaches, as a validation set for structure-based methods, and as a platform for the enumeration of enriched virtual libraries.
2. Sources of this data depend on the resources available to the computational chemist. Those in large pharmaceutical companies may have access to a wealth of proprietary data from previous HTS campaigns and earlier drug discovery projects. While researchers in smaller companies or academia may not have this luxury, public databases are available for compiling comprehensive databases of bioactive compounds.
3. Privileged structures, or scaffolds that show activity on two or more receptors yet can be rendered selective using specific substitution patterns, are powerful tools for drug discovery. Evans et al. [64] introduced this concept for a benzodiazepine scaffold that had previously been optimized for CCK_1 receptor antagonism, but could be tuned for activity against CCK_2 receptor with certain substitutions. The identification and application of these privileged structures can be quite useful within GPCR families. A sampling of these scaffolds, as well as their substructures (obtained via methods such as RECAP [65]), can provide starting points not only for 2D searches, but also for virtual library enumeration in advance of VS.

3.2 Compound Database Design and Preparation

1. Compounds in a VS database may come from several sources, but regardless of the source similar considerations and issues arise during database construction and preparation. While pharmaceutical and biotech companies will likely have a proprietary, physical screening library that can serve as a starting point for building a compound database, there is often a desire to expand the number of compounds beyond those available within the company. Likewise, small companies and academics may not have access to a physical library, but still require a compound database for VS. Online resources such as ZINC [66] (zinc.docking.org) can fill this gap by providing an aggregated database of compounds that are readily available for purchase from commercial vendors. Alternatively, a database

of "virtual compounds" that are based on available chemical building blocks and established chemistry can be computationally enumerated; however, this requires commitment of additional time and resources for the synthesis of VS hits.

2. The quality of hits delivered by VS is necessarily dependent on the quality of the compound database being screened; hence, great care should be exercised in the construction and curation of this resource [67]. Filtering out unwanted chemical structures based on specific rule sets is an effective way to improve database quality in a predictable and consistent manner. Filters such as Rapid Elimination of Swill (REOS) [68] can efficiently remove the most undesirable compounds from a database using criteria like physical property and reactive group filters. Removal of compounds that have a high likelihood of assay interference, or Pan-Assay INterference compounds (PAINS) [69] can further improve compound database quality by reducing the number of false positives. Compound aggregate formation is another common source of false positives and although these compounds are very challenging to identify computationally, making their removal from a compound database difficult; addition of detergent to assay protocols is often sufficient to eliminate them from further consideration [70, 71].

3. Additional "drug-like" [72] or "lead-like" [73] physicochemical property filters can also be applied to improve database quality (Table 1). The "Rule of 5" [74], which applies to orally administered drugs, is a useful tool for assessing drug-likeness, but is more appropriately applied during lead optimization than compound database design. Since hits are generally expected to add size and lipophilicity during the optimization process [75, 76], a "lead-like" compound database with more stringent limits on molecular weight and clogP is more appropriate for screening. Relaxation or modification of these rules

Table 1
Properties of orally administered "drug-like" molecules (Rule of 5) [74], "lead-like" molecules [73], and molecules able to cross the blood-brain barrier [79]

Rule of 5	Lead-like	CNS MPO (most desired)
• MW $\leq$ 500 Da	• MW $\leq$ 350 Da	• MW $\leq$ 360 Da
• clog P $\leq$ 5	• clog P $\leq$ 3	• clog P $\leq$ 3
• HBD $\leq$ 5	• Affinity ~0.1 μM	• clog D $\leq$ 2
• HBA (N + O) $\leq$ 10		• 40 < TPSA $\leq$ 90
		• HDB $\leq$ 0.5
		• pK_a $\leq$ 8

may be required if a different route of administration is targeted (e.g., IV or nasal administration), in order to keep from unduly limiting the chemical space available for VS. Furthermore, consideration of the properties required for a drug or lead to access-specific compartments within the body can also play a key role during compound database construction. For example, compounds that need to cross the blood-brain barrier tend to have a more constrained set of physical properties than other drugs [77–79], which can be reflected in the compound database. However, it may not be practical to apply these "drug-like", "lead-like", or compartment-specific filters at the database construction stage, since there is often a need to use the compound database for a variety of projects that may have different needs. In this case, the construction of a "master" compound database with relaxed physicochemical properties, maximizing chemical space coverage, is appropriate. This "master" database can be broken out into an appropriate sub-library before VS (saving computational resources) or more stringent filters can be applied to the VS hits post-screen.

4. With the contents of the compound database assembled, it is necessary to prepare it for use by the software tools that will be engaged for the VS search process. Commonly, this involves taking raw compound structure files, often from multiple proprietary, commercial, and public sources, and converting them into a format, such as canonical SMILES, that is conducive to duplicate removal. At this point compound property filters, described in the previous section, can be applied. Generation of 3D coordinates, saturated ring conformations, and stereoisomers (often with a limit on the number of stereocenters treated) is followed by enumeration of tautomers and protonation states (confined to a limited range centered on pH 7). An expanded conformer library can then be created for use in methods such as pharmacophore and shape searching. Most comprehensive molecular modeling software suites come with this type of ligand preparation workflow in place, though it may be necessary to modify default settings to obtain the desired results.

3.3 Ligand-Based Methods

1. LBVS has shown immense value in identifying novel chemical matter, including as a means for scaffold hopping [80]. This technique requires knowledge of compounds that have the desired bioactivity, since these serve directly as search queries or are used to build models that serve this purpose. Many techniques fall into this category, including fingerprint (FP) methods, pharmacophores, shape searching, molecular fields and surface-based methods, and machine learning. Each of these methods will be explored in this section.

2. FP-based methods are one of the most common techniques for performing LBVS due to their simplicity and speed. There are many excellent review papers that describe the different types of FPs and the reader is directed to those for further information [81, 82]. In essence, FPs encode the structure of ligands in a 1D bit string, greatly simplifying either 2D or 3D ligand information and enabling rapid comparison with a compound database. Among the major FP types (path-based, circular, and structural keys), circular FPs, such as ECFP4 and ECFP6, tend to perform best when ranking compounds with diverse structures and are recommended for VS [83]. MACCS structural keys have also shown an ability to distinguish between target- and family-selective compounds [84].

3. Pharmacophore methods have long been one of the preferred techniques for GPCR VS due to their ability to encode complex 3D interactions into a simplified set of chemical features and "scaffold hop" into unexplored chemical space [85]. An excellent overview of recent developments in this field is available [86]. Pharmacophore models can also be derived from protein structural information, obviating the need for detailed ligand activity data as long as a pharmacologically-relevant binding site is evident, as in the SNOOKER methodology [87]. Regardless of the source of data for building pharmacophore models, the selection of features to include in the search model can be difficult; select too many features and very few hits are found, and include too few features and the model becomes indiscriminate. The use of a pilot retrospective VS to validate the model and methodology is recommended if enough compound activity data is available. In many cases, pharmacophore screening is applied as part of an overall VS workflow involving additional techniques (*see* Subheading 3.5).

4. Inclusion of additional 3D ligand-based methods for VS has become widespread in pharmaceutical drug discovery. Shape matching methods use a 3D representation of an active ligand as a query and compare the molecular shape and chemical features to a compound database. Due to the successful application of shape matching to a wide variety of target classes, including biogenic amines [88], there has been significant effort in trying to further advance the performance of these methods by improving the quality of the input queries [89], as well as the proper weighting of chemical information [90]. Molecular field [38] and surface-based [39] methods are also viable strategies to include 3D information in VS queries and searches.

5. LBVS has also seen the use of machine learning becomes more common in recent years [91, 92]. There are many machine learning methods in regular use for VS, including decision trees

(DT), support vector machines (SVM), naive Bayesian, artificial neural networks (ANN), and trend vectors, to name a few. No one method has become dominant, since success when applied to VS appears to be specific to each experiment [93]. Due to their speed, chemogenomics approaches, or very large-scale VS across multiple GPCRs, are possible using machine learning.

3.4 Structure-Based Methods

1. Although the X-ray crystal structure of bovine rhodopsin was solved in 2000 [94], it was not until structures of the human β2-adrenoceptor were released in 2007 [95] that SBVS of GPCRs became a more productive path for lead identification. The reason for this delay in truly successful application of SBVS was due to the presence of key structural differences between the rhodopsin structure and other Class A GPCRs, particularly in the second extracellular loop (ECL2), leading to models with a distorted orthosteric binding site. Since that time, many additional X-ray crystal structures that cover an ever-widening region of the GPCR phylogenetic tree have been solved, making it possible to perform SBVS directly on the target of interest or by using a high-quality homology model. In addition, the activated form of β2-adrenoceptor was solved [96], giving insights into shifts in transmembrane helix 6 (TM6) and rearrangements in TM5 and TM7 that are associated with activation. To date, there are more than 100 GPCR X-ray structures in the Protein Data Bank [97], including those from pharmaceutically important groups such as chemokine receptors [98–101], biogenic amine binding receptors [95, 102–105], and opioid receptors [106–109], among others. Though fewer in number, there are now several structures that cover Class B [110], C [111], and F [112] GPCRs. This recent wealth of structural data has enabled the application of structure-based methods to GPCR VS.
2. Molecular docking is the primary engine driving structure-based drug discovery, but additional methods, such as molecular dynamics (MD) simulations, solvation analysis, and advanced scoring methodologies, are finding increased application [113]. When docking to GPCR X-ray structures or homology models, for which there are numerous successful applications to SBDD in the literature [114, 115], the availability of mutagenesis data or ligand pharmacophore models can be extremely helpful in improving docked pose accuracy [116]. One of the primary issues that still remains for molecular docking is the quality and predictability of scoring functions. While modern scoring functions are generally adequate for selecting low-energy ligand poses, there is still room for improvement in affinity prediction and ligand ranking, which

is at the heart of any SBVS operation [117]. Use of alternative scoring strategies such as interaction fingerprints (IFPs) can help in this regard. In addition, the GPCR Dock assessments have served as a valuable benchmark for the modeling community [118–120], identifying some best practices as well as lingering deficiencies in methodology. Although not used as the primary docking engine due to computational cost, the results of some advanced techniques are now being incorporated into model development [121] and compound selection [122]. A sample SBVS workflow is presented in Fig. 3, and of

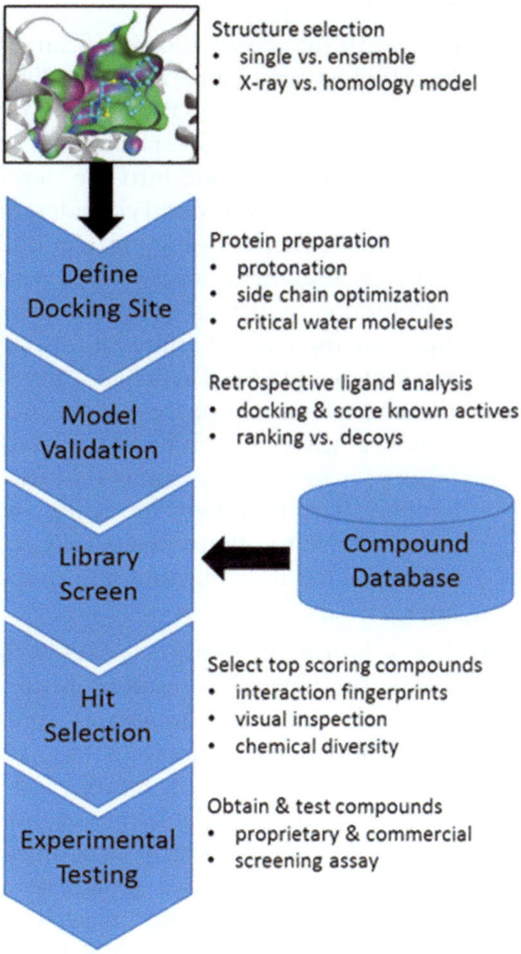

Fig. 3 Generalized SBVS workflow. This process begins with model selection, which can include X-ray structures or homology models, single models or ensembles from MD snapshots. Docking site definition delimits the search space within the protein and involves optimization of residue position and protonation, as well as location of critical water molecules. A retrospective VS can be run to validate the model, if sufficient data regarding active compounds is available. Screening of the compound database via molecular docking is followed by a hit selection procedure, often involving ranking by docking score, IFPs, visual inspection, and consideration of compound diversity. Selected compounds are then obtained and tested experimentally

these steps protein structure selection and preparation along with hit selection strategies have a large influence on success.

3. Protein-ligand IFPs are a simple way to encode 3D information in a 1D format, making it easy to identify whether critical interactions are maintained across a large number of docking poses [123]. It was also found, based on a retrospective VS analysis of five GPCR targets, that detailed IFPs encoding atom-atom information perform better than simple amino acid-based methods that encode residue-ligand contacts [124]. With enough training data, IFPs have successfully distinguished partial/full agonists from antagonist/inverse agonists.
4. Even with the recent and significant advances in GPCR structural biology, many pharmaceutically important targets do not have X-ray structures available. To fulfill the desire to use structure to inform compound design as well as VS, a great deal of effort has gone into the development of high-quality homology models. Careful consideration should be given to all the steps of the homology model building process, including sequence alignment, location and nature of helical kinks, biasing binding cavities using ligand information, and loop building, since there is still limited information to guide this process for GPCRs [125]. Particular care should be taken in the sequence alignment phase and the incorporation of mutagenesis data at this stage can be helpful in orienting helical residues. Indeed, many VS campaigns that used bovine rhodopsin as a template for homology modeling, with its structural differences relative to other Class A GPCRs, likely owe their success to the use of known ligands and mutation data to guide binding site shape.
5. The solution of additional GPCR X-ray structures beyond bovine rhodopsin has resulted in the availability of additional homology modeling templates spread across more of the GPCR phylogenetic tree. In fact, many targets now have a closely related X-ray structure, and even though there are many templates available, good homology models can be built from a single nearby structure. However, there are still many regions with poor structural coverage; in these cases, multi-template homology modeling can produce superior results [126]. Prospective VS are now being regularly run using these improved homology models as well. Several studies have also directly compared VS hit-rates using X-ray structures and homology models side by side. Interestingly, these studies show a range of results, from a CXCR4 X-ray structure giving better enrichment than a homology model [127], to near identical results between the two protein structure sources for D_3 receptor [128]. In some circumstances, even though both

protein structure sources yielded acceptable enrichments, the overlap between hit compounds is minimal [129]. This suggests that use of X-ray and homology model structures may be complementary, or at least able to sample different protein conformations leading to non-overlapping hits.

6. As computer hardware and algorithms have improved, the application of MD simulations to GPCR structure has become more commonplace. The use of MD snapshots as structures for VS is no longer cost prohibitive and prospective studies have been shown to find quality hits [121]. Retrospective VS studies of multiple GPCRs have also suggested that MD snapshots can actually outperform X-ray structures and homology models, possibly due to the additional conformational sampling and reduced bias toward a single chemotype [130]. Protein dynamics play a critical role in GPCR activation as well [131], and given the insights provided by active state structures [96] the use of MD simulations will likely continue to play a key role in the refinement of these models and their application to VS [132].
7. Careful consideration of water molecules in GPCR structures has become more common, both as the techniques used to do so gain traction with the wider computational chemistry community and as higher resolution GPCR structures with visible water networks become increasingly available. Since there are relatively few GPCR X-ray structures of sufficiently high resolution to unambiguously assign full water networks, placement of water molecules often comes from either protein simulations or other solvation methods.

3.5 Data Fusion

1. Data fusion procedures link data sources to improve the overall quality of data points, and have shown utility in drug repositioning [133] as well as polypharmacology and safety profile analysis [134]. Any of the VS methods described in previous sections can be used in isolation; however, combining these techniques can lead to better overall enrichment and a wider diversity of hit structures. In the VS context, there are two primary ways data fusion is implemented: sequentially and in parallel (*see* Fig. 4). Additionally, there are many alternative ways of combining methods, especially when hybrid approaches and machine learning are incorporated [135, 136].
2. Sequential data fusion is commonly used to increase the throughput of a VS protocol by searching a compound database using a computationally inexpensive technique (e.g., FP similarity) and progressing the best hits as the input for more computationally intensive techniques (e.g., docking). This is particularly useful when computer hardware resources are limited or when downstream VS methods truly become a

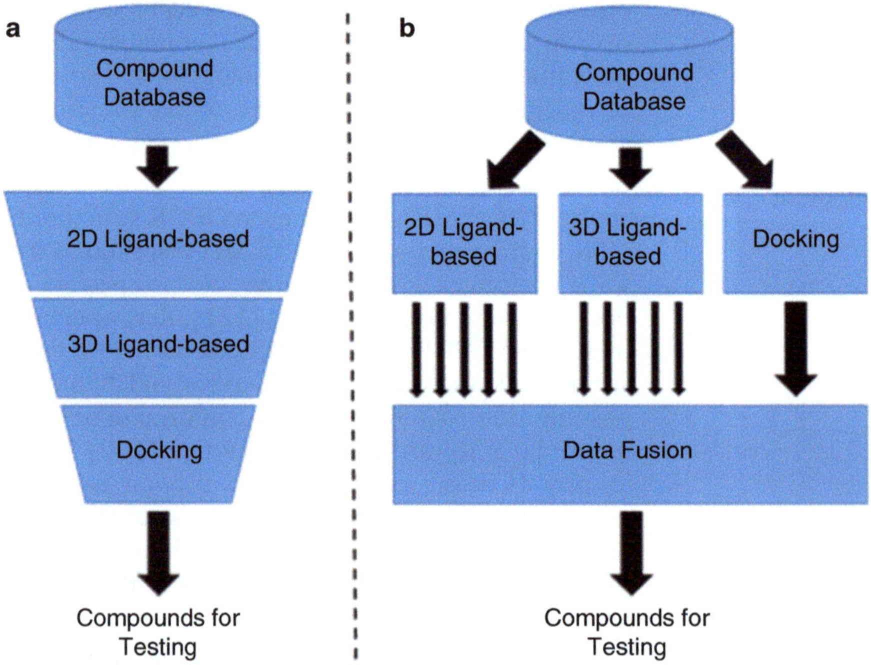

Fig. 4 Generalized data fusion workflows. (**a**) Sequential data fusion starts with computationally inexpensive methods, such as 2D similarity search, and progresses the best scoring compounds to more computationally costly methods, such as docking. Several methods may be threaded together in this manner to reduce overall resource cost and improve enrichment. (**b**) Parallel data fusion takes the output of several independent VS techniques, or multiple queries from a single technique, and combines them using fusion algorithms to improve enrichment and diversity. These two paradigms can also be combined into hybrid methods or modified using machine learning

bottleneck (i.e., induced-fit docking). In theory, any number of VS techniques can be placed in sequence, though regular practice often uses only two or three methods in a given screen. Pipelining tools can be invaluable in designing these workflows and in standardizing best practices across an institution or for a given target class. Though often successful, and sometimes necessary due to limits on computational resources, sequential workflows can suffer from lack of diversity in hits since false negatives are lost at each stage and subsequent methods have no way to retrieve them, even if they would have produced high scores.

3. The dramatic increase in computing power in recent years has made it practical to consider running several VS techniques, scoring strategies, or queries in parallel rather than sequentially, and then using data fusion methods for the selection of a combined hit list of compounds for experimental testing. There are a number of standard methods for combining VS data in a way that increases the possibility of improving hit enrichment (e.g., MAX, SUM, EUC), but an in-depth analysis

of these is beyond the scope of this chapter (see Willett [137] and Plewczynski et al. [93]). In VS, these methods have been used very effectively for combining the results of 2D similarity searches [138, 139] due to the relatively low computational cost of performing these searches against multiple query molecules. Similar concepts using docking, rather than similarity searches, are also possible. Consensus scoring uses the same set of docked poses, but re-evaluates them with different scoring functions and combines them into an aggregate score or ranking [140]. Consensus docking takes this a step further, either incorporating docking with a single program against an ensemble of protein structures [141] or docking with multiple programs against a single-protein structure [142].

4. Combinations of VS techniques have also been shown to improve enrichment [143]. Studies combining ligand and structure-based methods have shown improved enrichment [144, 145], and in several of these cases [146, 147] parallel selection of the top ranked hits from each technique or query produced the best results. Since different VS techniques identify hit compounds with different structures, parallel selection offers an advantage by selecting a more diverse set of compounds than some other data fusion methods.
5. Once hit selection is complete, there is typically a final step before procurement of compounds for testing—visual inspection. This step involves examining the structure of each compound in the hit list and, preferably, its relationship to the model or query that was used in its selection. This ensures that no unwanted chemical structures escaped the filtering process and that no errors occurred during tautomerization/protonation. If done in conjunction with medicinal chemistry colleagues, visual inspection can be a constructive way to promote ownership of the results by the chemistry team, since compounds tested experimentally will have already passed their review.

4 Notes

4.1 Data Mining

1. Additional examples of the privileged structure concept for GPCRs and other targets have been described previously [148].
2. There are a number of promiscuous scaffolds that interact with members of the chemokine receptor family, such as CCR2/CCR5 aryl sulfonamide antagonists [149], and CCR2/CCR9 antagonists such as PF-4178903 [150] and cencriviroc [151]; since these compounds interact with multiple family members they may qualify as privileged structures.

3. A recent example of a successful prospective VS using substructure-based searching involved A_{2A} receptor antagonists and resulted in eight hits found out of 36 compounds tested [152].

4.2 Ligand-Based Methods

1. In addition to finding lead matter for medicinal chemistry optimization, pharmacophore-based VS has been used to find molecules that allosterically stabilize the R* state of rhodopsin, based on a peptide template rather than small-molecule dataset [153].
2. Though not a VS per se, Low et al. [154] used molecular fields to find two new scaffolds for CCK_2 receptor, demonstrating the applicability of these methods to GPCRs.
3. Retrospective analysis for a set of biogenic amine receptor targets using an SVM built with 3D pharmacophore FPs was able to successfully scaffold hop to new core structures [155].
4. Bock and Gough [156] used a chemogenomics approach to identify putative ligands for 55 orphan GPCRs from a library of ~35,000 compounds, using only the residue properties of the primary sequence and 2D ligand feature connectivity as input. Jacob et al. [157] followed a similar approach to demonstrate the ability of an SVM trained on GPCR-ligand interaction pairs to find ligands of orphan GPCRs.

4.3 Structure-Based Methods

1. There are now a growing number of success stories based on the SBVS of GPCRs [158, 159]. One recent publication outlined a retrospective and prospective VS of A_{2A} receptor [160]. In the end, 9 of 20 predicted agonists turned out to bind the receptor, which is an excellent result, but the hits lacked the desired ability to activate A_{2A} receptor. The authors' conclusion was that it is still difficult to accurately model functional states, though compound database composition may also play an important role in success or failure.
2. Weiss et al. [161] were able to identify two novel β2-adrenoceptor binders with a similar signaling profile to that of the co-crystallized ligand, illustrating that it is possible to find compounds with specific functional activity. However, when this structure was used to build an active form D_2 receptor homology model, VS resulted in only a few marginal hits, once again demonstrating the difficulty of accurately describing active states without X-ray structures.
3. A VS of rhodopsin R* focused on the intracellular loop region rather than the orthosteric site, aimed at interfering with transducin binding, was able to identify molecules that modulated signal transduction [162], showing that targeting allosteric or protein-protein interaction sites is a viable strategy given enough structural information.

4. Sato et al. [163] used IFPs to evaluate models used in retrospective SBVS and were able to use the protocol to predict important interactions in 5-HT_{2A} receptor. There are now several published examples of IFPs being successfully applied to prospective SBVS of GPCRs [123, 164, 165]. In one study, the use of IFPs was able to identify novel antagonists/inverse agonists of H_1 receptor and antagonists of β2-adrenoceptor [166].
5. Differentiation between partial/full agonists and antagonist/inverse agonists was demonstrated in retrospective studies of β2-adrenoceptor [167, 168] where the incorporation of IFPs was critical for the identification of agonists and partial agonists.
6. Heifetz et al. [169] demonstrated the power of using sugar-based library hits to refine a MCH_1 receptor model, which was a key step in enabling a subsequent VS that had a 14% hit-rate and identified 10 novel antagonist chemotypes.
7. Information on known ligands and mutational data was a key factor in the success of retrospective analyses of $mGlu_5$ receptor for the identification of negative allosteric modulators (NAMs) [170] and of CB_2 receptor for the detection of antagonists [171], as well as prospective studies against A_{2A} receptor (37 of 80 compounds with $K_i < 10\ \mu M$) [172], MCH_1 receptor (6 of 129 compounds with $K_i = 7\text{–}20\ \mu M$) [173], and H_4 receptor (28 of 120 compounds with $IC_{50} < {\sim}4\ \mu M$) [174].
8. A number of retrospective VS analyses demonstrate the effectiveness of using high-quality homology models [175–177].
9. Even using the human β2-adrenoceptor as a template, McRobb et al. [178] found that only 6 of 9 biogenic amine binding receptor models gave reasonable enrichment rates. Binding site residue and ECL2 placement were cited as explanations for the failed models.
10. An A_{2A} receptor model based on turkey β1-adrenoceptor template yielded a 9% hit-rate [179], while a TA_1 receptor model based on β2-adrenoceptor produced three hits in the μM range out of 42 compounds tested [180].
11. Although not used directly for VS, WaterMap analysis [60–62] was applied to the A_{2A} receptor and successfully explained the SAR of a triazolylpurine series [181]. Mason et al. [182] used calculated water molecules from WaterMap [60–62], SZMAP [59], and GRID [57, 58], as part of an assessment of druggability for GPCRs, which is a useful prelude to a VS campaign. Recently, it was shown that the inclusion of specific water molecules in the VS of A_{2A} receptor improved enrichment relative to a non-solvated model, though careful optimization of the hydrogen positions was required [183]. It was also not

necessary for water molecules to come from X-ray structures to improve enrichment, since those from MD snapshots worked just as well.

4.4 Data Fusion

1. Tömöri et al. [184] describe a recent large-scale VS success against PDE5 that used sequential data fusion, starting with FP similarity on 5 million compounds followed by docking of ~2000 compounds, resulting in 48 hits with >55% inhibition or $IC_{50} < 1\ \mu M$.
2. Hert et al. [185] showed that combining multiple queries and different similarity metrics for the same query molecule is a successful strategy for improving recall in retrospective studies.
3. Baber et al. [186] studied the use of multiple ligand-based methods against four GPCRs and found that enrichment was improved the most by using sum rank and logistic regression data fusion. It was concluded that the improved performance was primarily due to increased sampling as well as the scoring functions being in approximate agreement regarding ranking of actives.

4.5 Practical Integration of Virtual Screening Methods—Allosteric Modulators

1. The identification of allosteric modulators of metabotropic glutamate receptors, which are important pharmaceutical targets for a variety of CNS disorders, provides an excellent case study for how VS workflows are used in a practical sense, as well as how they evolve over time as additional information is collected about a target family. These Class C GPCRs have an extracellular orthosteric binding site that is conserved across the family [187], so targeting an allosteric region in the transmembrane domain represents a viable path for the identification of selective modulators [188].
2. Advances in LBVS were described by Noeske et al. in a pair of studies involving the use of topological pharmacophore descriptors to find $mGlu_1$ receptor NAMs. The first VS [189] resulted in a 26% hit-rate and identified a coumarin scaffold with sub-μM K_i, which was subsequently optimized to a compound with improved activity ($IC_{50} = 58$ nM) and sub-type selectivity. The second VS [190] found a quinoline structure as part of a scaffold hopping exercise, which was then optimized to a 24 nM compound with >1000-fold selectivity against $mGlu_5$ receptor.
3. Shape and electrostatics-based VS was used to find imidazo [1,2-α]pyridine positive allosteric modulators (PAMs) with nM activity on $mGlu_2$ receptor [191]. VS has also been combined with HTS output to train an ANN to identify $mGlu_4$ receptor PAMs [192]. Using an HTS training set of 434 PAMs

found within 155,000 compounds screened, the ANN was able to find 67 hits from 1100 compounds tested, which was a 22-fold enrichment relative to the HTS campaign.

4. The increased availability of structural information for Class A GPCRs also represents an opportunity to attempt SBVS on $mGlu_1$ receptor. A recent report outlined the identification of novel NAMs for this target using a sequential screening approach [193]. Initial rounds of screening were performed using pharmacophore and naive Bayesian models, and the results were docked against an MD-refined homology model built from a D_3 receptor template. This protocol was tested via retrospective VS and five hits with μM IC_{50}'s were found out of 35 compounds tested.
5. The identification of allosteric modulators by VS is not limited to the metabotropic glutamate receptor family. Integrated VS protocols have been used to find allosteric binders for glucagon receptor [194], D_3 receptor [195], and M_2 receptor [196].
6. A glucagon receptor VS campaign started with an extensive homology model construction and validation effort was performed on CRF_1 receptor, due to data availability. The actual VS against the glucagon receptor [194] started with property and shape-based similarity filters, followed by docking to a homology model built from the CRF_1 receptor template. Docking scores and IFPs, followed by visual inspection, were used to generate a hit list for experimental testing. Of 23 compounds tested, four were allosteric antagonists and two had different functional activity.
7. A VS against D_3 receptor [195] used two structural models: an apo and a dopamine bound model. Following docking, the apo model resulted in a 53% hit-rate with antagonists apparently spanning the orthosteric and extended pockets. The dopamine-bound model resulted in a 32% hit-rate, but in this case the compounds were predicted to bind an allosteric site in the extracellular extension of the pocket and contained no anchoring amino group; these compounds also exhibited a variety of functional activities.
8. In a recent study [196], the extracellular vestibule region of M_2 receptor was targeted for VS. Starting from the X-ray structure, a receptor ensemble was constructed from MD simulations and used for docking. A combination of docking scores and IFPs were used to select compounds for testing, which resulted in a 50% hit-rate. Additional compound profiling showed that four novel NAMs were identified along with one PAM.

4.6 Challenges

1. There has been a great deal of progress in GPCR VS over the past ~20 years, but there are still challenges to be met. While some of these are general issues for VS against any target class, some are more specific or have an outsized impact in application to GPCRs. One fundamental issue is the consistency of SAR information collected during data mining. Even though information in a company database may have been collected using a consistent assay format, the inclusion of public data sources, with contributions from many different labs, will be more difficult to reconcile.
2. In LBVS, the weighting of features in pharmacophore and shape-based methods is also a challenge, and often these settings are chosen not for data-driven reasons, but based on user experience and anecdotal evidence.
3. There are a number of challenges related to SBVS, including the fact that even with the impressive progress in X-ray crystallography there is still limited structural coverage of the GPCR phylogenetic tree, which leads to difficulties in homology modeling. There are also many potential binding regions that ligands can occupy in addition to the orthosteric site and information on these sites is still growing. The intracellular binding of vircirnon to CCR9 is a recent example [101]. Lack of structural information is also a problem in modeling different functional states, especially activated states, of many GPCRs. Even though there is a growing understanding of the relationship between structure and function of GPCRs, there are still very few structural examples with which to build high quality models that can be used to find functionally relevant binders.
4. Another set of issues involves the inherent dynamics of GPCRs and flexibility within the orthosteric site resulting in uncertainty in side chain positions as well as difficulties in loop modeling. This can also make it difficult to accurately model allosteric sites (including PAMs and NAMs). The addition of water molecules adds an additional layer of complexity to VS, but the inclusion of key water molecules, either from X-ray data or MD simulations, can be critical to the success of a VS.
5. Improvements in protein-ligand affinity prediction and incorporation of desolvation effects into scoring are ongoing issues not just in the VS of GPCRs, but for structure-based design generally. Even marginal improvements in scoring can have a large effect on VS success.

4.7 Best Practices

VS has proven to be a valuable tool for GPCR drug discovery and an effective method for finding novel chemical starting points for medicinal chemistry optimization. It can be used on its own or in conjunction with other experimental screening methods (HTS, fragment screening) to deliver a larger pool of hit compounds. LBVS has long been used on GPCR targets, and with the increasing number of GPCR X-ray structures, SBVS, whether using these structures or homology models, is a viable VS path. Best practices and recommendations include:

1. Collect all available SAR from public and private databases, including for other nearby targets in the GPCR family. There may be opportunities to use privileged structures or promiscuous ligands to initiate medicinal chemistry.
2. Curation of the ligand screening library will have an effect on VS success. Remove undesirable compounds during library construction, but strict physicochemical property filtering may be applied just in advance of the screening step or during hit selection.
3. Circular FPs are recommended to maximize enrichment. Pilot retrospective studies may be necessary to adequately tune pharmacophore models.
4. Apply multiple ligand-based methods when sufficient data or models are available. Running multiple queries using the same technique can lead to complementary hit lists.
5. Use all available protein structural data and include as much SAR and mutagenesis information as possible to improve docking models. For close orthologs, a single template may suffice, but for distant targets multi-template homology modeling may be required.
6. The treatment of water molecules in SBVS can have a significant effect on enrichment. Consider using models with waters from X-ray structures or multiple MD snapshots.
7. Predicting functional activity of ligands is still difficult. Apply IFPs, preferably with atom-atom resolution rather than ligand-residue, to docking results to increase the likelihood of achieving the desired result.
8. Run as many models and techniques as is computationally feasible and use data fusion to select a ligand hit list for experimental testing. Sequential workflows can improve throughput, but at the possible cost of lost diversity. Parallel selection of hits from different VS methods has been shown to improve enrichment.
9. Use visual inspection of the final hit list to verify compounds are chemically acceptable and to create ownership within the medicinal chemistry team.

There are still challenges in applying VS to GPCRs, such as complications due to protein dynamics, lack of structural coverage, and uncertainties in the physical description of binding. However, ongoing improvements in computer hardware and the development of new algorithms will almost certainly make it possible to expand the scope of VS against this important class of pharmaceutical targets in the future.

References

1. Filmore D (2004) It's a GPCR world. Mod Drug Discov 7:24–28
2. Garland SL (2013) Are GPCRs still a source of new targets? J Biomol Screen 18:947–966. https://doi.org/10.1177/1087057113498418
3. Muegge I (2008) Synergies of virtual screening approaches. Mini Rev Med Chem 8:927–933. https://doi.org/10.2174/138955708785132792
4. Ripphausen P, Nisius B, Peltason L, Bajorath J (2010) Quo vadis, virtual screening? A comprehensive survey of prospective applications. J Med Chem 53:8461–8467. https://doi.org/10.1021/jm101020z
5. Tanrikulu Y, Krüger B, Proschak E (2013) The holistic integration of virtual screening in drug discovery. Drug Discov Today 18:358–364. https://doi.org/10.1016/j.drudis.2013.01.007
6. Irwin JJ (2008) Community benchmarks for virtual screening. J Comput Aided Mol Des 22:193–199. https://doi.org/10.1007/s10822-008-9189-4
7. Rohrer SG, Baumann K (2009) Maximum unbiased validation (MUV) data sets for virtual screening based on pubchem bioactivity data. J Chem Inf Model 49:169–184. https://doi.org/10.1021/ci8002649
8. Xia J, Jin H, Liu Z, Zhang L, Wang XS (2014) An unbiased method to build benchmarking sets for ligand-based virtual screening and its application to GPCRs. J Chem Inf Model 54:1433–1450. https://doi.org/10.1021/ci500062f
9. Weiss DR, Bortolato A, Tehan B, Mason JS (2016) GPCR-bench: a benchmarking set and practitioners' guide for G protein-coupled receptor docking. J Chem Inf Model 56:642–651. https://doi.org/10.1021/acs.jcim.5b00660
10. Tiikkainen P, Markt P, Wolber G, Kirchmair J, Distinto S, Poso A, Kallioniemi O (2009) Critical comparison of virtual screening methods against the MUV data set. J Chem Inf Model 49:2168–2178. https://doi.org/10.1021/ci900249b
11. Venkatraman V, Pérez-Nueno VI, Mavridis L, Ritchie DW (2010) Comprehensive comparison of ligand-based virtual screening tools against the DUD data set reveals limitations of current 3D methods. J Chem Inf Model 50:2079–2093. https://doi.org/10.1021/ci100263p
12. Warren GL, Andrews CW, Capelli A-M, Clarke B, LaLonde J, Lambert MH, Lindvall M, Nevins N, Semus SF, Senger S, Tedesco G, Wall ID, Woolven JM, Peishoff CE, Head MS (2006) A critical assessment of docking programs and scoring functions. J Med Chem 49:5912–5931. https://doi.org/10.1021/jm050362n
13. Onodera K, Satou K, Hirota H (2007) Evaluations of molecular docking programs for virtual screening. J Chem Inf Model 47:1609–1618. https://doi.org/10.1021/ci7000378
14. Cross JB, Thompson DC, Rai BK, Baber JC, Fan KY, Hu Y, Humblet C (2009) Comparison of several molecular docking programs: pose prediction and virtual screening accuracy. J Chem Inf Model 49:1455–1474. https://doi.org/10.1021/ci900056c
15. McGaughey GB, Sheridan RP, Bayly CI, Culberson JC, Kreatsoulas C, Lindsley S, Maiorov V, Truchon J-F, Cornell WD (2007) Comparison of topological, shape, and docking methods in virtual screening. J Chem Inf Model 47:1504–1519. https://doi.org/10.1021/ci700052x
16. Hu G, Kuang G, Xiao W, Li W, Liu G, Tang Y (2012) Performance evaluation of 2D fingerprint and 3D shape similarity methods in virtual screening. J Chem Inf Model 52:1103–1113. https://doi.org/10.1021/ci300030u
17. Jain AN, Nicholls A (2008) Recommendations for evaluation of computational methods. J Comput Aided Mol Des 22:133–139. https://doi.org/10.1007/s10822-008-9196-5
18. Truchon J-F, Bayly CI (2007) Evaluating virtual screening methods: good and bad metrics for the "early recognition" problem. J Chem Inf Model 47:488–508. https://doi.org/10.1021/ci600426e
19. Triballeau N, Acher F, Brabet I, Pin J-P, Bertrand H-O (2005) Virtual screening workflow development guided by the "receiver

operating characteristic" curve approach. Application to high-throughput docking on metabotropic glutamate receptor subtype 4. J Med Chem 48:2534–2547. https://doi.org/10.1021/jm049092j

20. Sheridan RP, Singh SB, Fluder EM, Kearsley SK (2001) Protocols for bridging the peptide to nonpeptide gap in topological similarity searches. J Chem Inf Comput Sci 41:1395–1406. https://doi.org/10.1021/ci0100144
21. Gaulton A, Bellis LJ, Bento AP, Chambers J, Davies M, Hersey A, Light Y, McGlinchey S, Michalovich D, Al-Lazikani B, Overington JP (2011) ChEMBL: a large-scale bioactivity database for drug discovery. Nucleic Acids Res 40(Database issue):D1100–D1107. https://doi.org/10.1093/nar/gkr777
22. Li Q, Cheng T, Wang Y, Bryant SH (2010) PubChem as a public resource for drug discovery. Drug Discov Today 15:1052–1057. https://doi.org/10.1016/j.drudis.2010.10.003
23. Liu T, Lin Y, Wen X, Jorissen RN, Gilson MK (2007) BindingDB: a web-accessible database of experimentally determined protein–ligand binding affinities. Nucleic Acids Res 35: D198–D201. https://doi.org/10.1093/nar/gkl999
24. Molecular Operating Environment (MOE). Chemical Computing Group Inc., Montreal, Canada
25. Schrödinger Release 2016–4: LigPrep. Schrödinger, LLC, New York, NY
26. Hawkins PCD, Skillman AG, Warren GL, Ellingson BA, Stahl MT (2010) Conformer generation with OMEGA: algorithm and validation using high quality structures from the protein databank and cambridge structural database. J Chem Inf Model 50:572–584. https://doi.org/10.1021/ci100031x
27. Biovia Pipeline Pilot. Dassault Systemes BIOVIA, San Diego, CA
28. Berthold MR, Cebron N, Dill F, Gabriel TR, Kötter T, Meinl T, Ohl P, Sieb C, Thiel K, Wiswedel B (2008) KNIME: the Konstanz information miner. In: Data analysis, machine learning and applications. Springer, Berlin, Heidelberg, pp 319–326
29. (2016) Schrödinger Release 2016–4: Canvas. Schrödinger, LLC, New York, NY
30. RDKit: Open-source cheminformatics
31. Discovery Studio Modeling Environment: CATALYST. Dassault Systemes BIOVIA, San Diego, CA
32. Dixon SL, Smondyrev AM, Knoll EH, Rao SN, Shaw DE, Friesner RA (2006) PHASE: a new engine for pharmacophore perception, 3D QSAR model development, and 3D database screening: 1. Methodology and preliminary results. J Comput Aided Mol Des 20:647–671. https://doi.org/10.1007/s10822-006-9087-6
33. Dixon SL, Smondyrev AM, Rao SN (2006) PHASE: a novel approach to pharmacophore modeling and 3D database searching. Chem Biol Drug Des 67:370–372. https://doi.org/10.1111/j.1747-0285.2006.00384.x
34. Wolber G, Langer T (2005) LigandScout: 3-D pharmacophores derived from protein-bound ligands and their use as virtual screening filters. J Chem Inf Model 45:160–169. https://doi.org/10.1021/ci049885e
35. Grant JA, Gallardo MA, Pickup BT (1996) A fast method of molecular shape comparison: a simple application of a Gaussian description of molecular shape. J Comput Chem 17:1653–1666. https://doi.org/10.1002/(SICI)1096-987X(19961115)17:14<1653::AID-JCC7>3.0.CO;2-K
36. Sastry GM, Dixon SL, Sherman W (2011) Rapid shape-based ligand alignment and virtual screening method based on atom/feature-pair similarities and volume overlap scoring. J Chem Inf Model 51:2455–2466. https://doi.org/10.1021/ci2002704
37. Mavridis L, Hudson BD, Ritchie DW (2007) Toward high throughput 3D virtual screening using spherical harmonic surface representations. J Chem Inf Model 47:1787–1796. https://doi.org/10.1021/ci7001507
38. Cheeseright T, Mackey M, Rose S, Vinter A (2007) Molecular field technology applied to virtual screening and finding the bioactive conformation. Expert Op Drug Discov 2:131–144. https://doi.org/10.1517/17460441.2.1.131
39. Jain AN (2004) Ligand-based structural hypotheses for virtual screening. J Med Chem 47:947–961. https://doi.org/10.1021/jm030520f
40. R Core Team (2017) R: a language and environment for statistical computing. R Foundation for Statistical Computing, Vienna, Austria
41. Shoichet BK, Kuntz ID, Bodian DL (1992) Molecular docking using shape descriptors. J Comput Chem 13:380–397. https://doi.org/10.1002/jcc.540130311
42. Morris GM, Huey R, Lindstrom W, Sanner MF, Belew RK, Goodsell DS, Olson AJ (2009) AutoDock4 and AutoDockTools4: automated docking with selective receptor flexibility. J Comput Chem 30:2785–2791. https://doi.org/10.1002/jcc.21256

43. Jones G, Willett P, Glen RC, Leach AR, Taylor R (1997) Development and validation of a genetic algorithm for flexible docking. J Mol Biol 267:727–748. https://doi.org/10.1006/jmbi.1996.0897
44. Kramer B, Rarey M, Lengauer T (1999) Evaluation of the FLEXX incremental construction algorithm for protein–ligand docking. Proteins 37:228–241. https://doi.org/10.1002/(SICI)1097-0134(19991101)37:2<228::AID-PROT8>3.0.CO;2-8
45. Jain AN (2003) Surflex: fully automatic flexible molecular docking using a molecular similarity-based search engine. J Med Chem 46:499–511. https://doi.org/10.1021/jm020406h
46. Friesner RA, Banks JL, Murphy RB, Halgren TA, Klicic JJ, Mainz DT, Repasky MP, Knoll EH, Shelley M, Perry JK, Shaw DE, Francis P, Shenkin PS (2004) Glide: a new approach for rapid, accurate docking and scoring. 1. Method and assessment of docking accuracy. J Med Chem 47:1739–1749. https://doi.org/10.1021/jm0306430
47. Halgren TA, Murphy RB, Friesner RA, Beard HS, Frye LL, Pollard WT, Banks JL (2004) Glide: a new approach for rapid, accurate docking and scoring. 2. Enrichment factors in database screening. J Med Chem 47:1750–1759. https://doi.org/10.1021/jm030644s
48. McGann MR, Almond HR, Nicholls A, Grant JA, Brown FK (2003) Gaussian docking functions. Biopolymers 68:76–90. https://doi.org/10.1002/bip.10207
49. Kiefer F, Arnold K, Kunzli M, Bordoli L, Schwede T (2009) The SWISS-MODEL repository and associated resources. Nucleic Acids Res 37:D387–D392. https://doi.org/10.1093/nar/gkn750
50. Zhang Y (2008) I-TASSER server for protein 3D structure prediction. BMC Bioinformatics 9:40. https://doi.org/10.1186/1471-2105-9-40
51. Yang J, Yan R, Roy A, Xu D, Poisson J, Zhang Y (2014) The I-TASSER suite: protein structure and function prediction. Nat Methods 12:7–8. https://doi.org/10.1038/nmeth.3213
52. Eswar N, Webb B, Marti-Renom MA, Madhusudhan MS, Eramian D, Shen M, Pieper U, Sali A (2007) Comparative protein structure modeling using MODELLER, Curr. Protoc. Protein Sci. John Wiley & Sons, Inc., Hoboken, NJ, pp 2.9.1–2.9.31
53. Schrödinger Release 2016–4: Prime. Schrödinger, LLC, New York, NY
54. Salomon-Ferrer R, Case DA, Walker RC (2013) An overview of the Amber biomolecular simulation package: Amber biomolecular simulation package. WIRE Comp Mol Sci 3:198–210. https://doi.org/10.1002/wcms.1121
55. Phillips JC, Braun R, Wang W, Gumbart J, Tajkhorshid E, Villa E, Chipot C, Skeel RD, Kalé L, Schulten K (2005) Scalable molecular dynamics with NAMD. J Comput Chem 26:1781–1802. https://doi.org/10.1002/jcc.20289
56. Bowers KJ, Sacerdoti FD, Salmon JK, Shan Y, Shaw DE, Chow E, Xu H, Dror RO, Eastwood MP, Gregersen BA, Klepeis JL, Kolossvary I, Moraes MA (2006) Molecular dynamics–scalable algorithms for molecular dynamics simulations on commodity clusters. ACM Press, New York, p 84
57. Goodford PJ (1985) A computational procedure for determining energetically favorable binding sites on biologically important macromolecules. J Med Chem 28:849–857. https://doi.org/10.1021/jm00145a002
58. Sciabola S, Stanton RV, Mills JE, Flocco MM, Baroni M, Cruciani G, Perruccio F, Mason JS (2010) High-throughput virtual screening of proteins using GRID molecular interaction fields. J Chem Inf Model 50:155–169. https://doi.org/10.1021/ci9003317
59. SZMAP 1.2.1.4. OpenEye Scientific Software, Santa Fe, NM
60. Young T, Abel R, Kim B, Berne BJ, Friesner RA (2007) Motifs for molecular recognition exploiting hydrophobic enclosure in protein–ligand binding. Proc Natl Acad Sci 104:808–813. https://doi.org/10.1073/pnas.0610202104
61. Abel R, Young T, Farid R, Berne BJ, Friesner RA (2008) Role of the active-site solvent in the thermodynamics of factor Xa Ligand binding. J Am Chem Soc 130:2817–2831. https://doi.org/10.1021/ja0771033
62. Beuming T, Che Y, Abel R, Kim B, Shanmugasundaram V, Sherman W (2012) Thermodynamic analysis of water molecules at the surface of proteins and applications to binding site prediction and characterization. Proteins 80:871–883. https://doi.org/10.1002/prot.23244
63. Misin M, Fedorov MV, Palmer DS (2015) Communication: accurate hydration free energies at a wide range of temperatures from 3D-RISM. J Chem Phys 142:91105. https://doi.org/10.1063/1.4914315
64. Evans BE, Rittle KE, Bock MG, DiPardo RM, Freidinger RM, Whitter WL, Lundell GF,

Veber DF, Anderson PS (1988) Methods for drug discovery: development of potent, selective, orally effective cholecystokinin antagonists. J Med Chem 31:2235–2246. https://doi.org/10.1021/jm00120a002

65. Lewell XQ, Judd DB, Watson SP, Hann MM (1998) RECAP-retrosynthetic combinatorial analysis procedure: a powerful new technique for identifying privileged molecular fragments with useful applications in combinatorial chemistry. J Chem Inf Comput Sci 38:511–522. https://doi.org/10.1021/ci970429i
66. Irwin JJ, Sterling T, Mysinger MM, Bolstad ES, Coleman RG (2012) ZINC: a free tool to discover chemistry for biology. J Chem Inf Model 52:1757–1768. https://doi.org/10.1021/ci3001277
67. Walters WP, Namchuk M (2003) Designing screens: how to make your hits a hit. Nat Rev Drug Discov 2:259–266. https://doi.org/10.1038/nrd1063
68. Walters WP, Stahl MT, Murcko MA (1998) Virtual screening—an overview. Drug Discov Today 3:160–178. https://doi.org/10.1016/S1359-6446(97)01163-X
69. Baell JB, Holloway GA (2010) New substructure filters for removal of pan assay interference compounds (PAINS) from screening libraries and for their exclusion in bioassays. J Med Chem 53:2719–2740. https://doi.org/10.1021/jm901137j
70. McGovern SL, Caselli E, Grigorieff N, Shoichet BK (2002) A common mechanism underlying promiscuous inhibitors from virtual and high-throughput screening. J Med Chem 45:1712–1722. https://doi.org/10.1021/jm010533y
71. Ryan AJ, Gray NM, Lowe PN, Chung C (2003) Effect of detergent on "promiscuous" inhibitors. J Med Chem 46:3448–3451. https://doi.org/10.1021/jm0340896
72. Walters WP, Murcko AA, Murcko MA (1999) Recognizing molecules with drug-like properties. Curr Opin Chem Biol 3:384–387. https://doi.org/10.1016/S1367-5931(99)80058-1
73. Teague SJ, Davis AM, Leeson PD, Oprea T (1999) The design of leadlike combinatorial libraries. Angew Chem Int Ed 38:3743–3748. https://doi.org/10.1002/(SICI)1521-3773(19991216)38:24<3743::AID-ANIE3743>3.0.CO;2-U
74. Lipinski CA, Lombardo F, Dominy BW, Feeney PJ (1997) Experimental and computational approaches to estimate solubility and permeability in drug discovery and development settings. Adv Drug Deliv Rev 23:3–25. https://doi.org/10.1016/S0169-409X(96)00423-1
75. Oprea TI (2000) Current trends in lead discovery: are we looking for the appropriate properties? Mol Divers 5:199–208. https://doi.org/10.1023/A:1021368007777
76. Oprea TI, Davis AM, Teague SJ, Leeson PD (2001) Is there a difference between leads and drugs? A historical perspective. J Chem Inf Comput Sci 41:1308–1315. https://doi.org/10.1021/ci010366a
77. Kelder J, Grootenhuis PDJ, Bayada DM, Delbressine LPC, Ploemen J-P (1999) Polar molecular surface as a dominating determinant for oral absorption and brain penetration of drugs. Pharm Res 16:1514–1519. https://doi.org/10.1023/A:1015040217741
78. Norinder U, Haeberlein M (2002) Computational approaches to the prediction of the blood–brain distribution. Adv Drug Deliv Rev 54:291–313. https://doi.org/10.1016/S0169-409X(02)00005-4
79. Wager TT, Hou X, Verhoest PR, Villalobos A (2010) Moving beyond rules: the development of a central nervous system multiparameter optimization (CNS MPO) approach to enable alignment of druglike properties. ACS Chem Neurosci 1:435–449. https://doi.org/10.1021/cn100008c
80. Ripphausen P, Nisius B, Bajorath J (2011) State-of-the-art in ligand-based virtual screening. Drug Discov Today 16:372–376. https://doi.org/10.1016/j.drudis.2011.02.011
81. Willett P (2006) Similarity-based virtual screening using 2D fingerprints. Drug Discov Today 11:1046–1053. https://doi.org/10.1016/j.drudis.2006.10.005
82. Cereto-Massagué A, Ojeda MJ, Valls C, Mulero M, Garcia-Vallvé S, Pujadas G (2015) Molecular fingerprint similarity search in virtual screening. Methods 71:58–63. https://doi.org/10.1016/j.ymeth.2014.08.005
83. O'Boyle NM, Sayle RA (2016) Comparing structural fingerprints using a literature-based similarity benchmark. J Cheminformatics 8:36. https://doi.org/10.1186/s13321-016-0148-0
84. Vogt I, Ahmed HEA, Auer J, Bajorath J (2008) Exploring structure–selectivity relationships of biogenic amine GPCR antagonists using similarity searching and dynamic compound mapping. Mol Divers 12:25–40. https://doi.org/10.1007/s11030-008-9071-2

85. Hu Y, Stumpfe D, Bajorath J (2016) Recent advances in scaffold hopping. J Med Chem 60 (4):1238–1246. https://doi.org/10.1021/acs.jmedchem.6b01437
86. Horvath D (2011) Pharmacophore-based virtual screening. In: Bajorath J (ed) Chemoinformatics and computational chemical biology. Humana Press, New York, pp 261–298
87. Sanders MPA, Verhoeven S, de Graaf C, Roumen L, Vroling B, Nabuurs SB, de Vlieg J, Klomp JPG (2011) Snooker: a structure-based pharmacophore generation tool applied to class A GPCRs. J Chem Inf Model 51:2277–2292. https://doi.org/10.1021/ci200088d
88. Hawkins PCD, Skillman AG, Nicholls A (2007) Comparison of shape-matching and docking as virtual screening tools. J Med Chem 50:74–82. https://doi.org/10.1021/jm0603365
89. Tawa GJ, Baber JC, Humblet C (2009) Computation of 3D queries for ROCS based virtual screens. J Comput Aided Mol Des 23:853. https://doi.org/10.1007/s10822-009-9302-3
90. Kirchmair J, Distinto S, Markt P, Schuster D, Spitzer GM, Liedl KR, Wolber G (2009) How to optimize shape-based virtual screening: choosing the right query and including chemical information. J Chem Inf Model 49:678–692. https://doi.org/10.1021/ci8004226
91. Melville JL, Burke EK, Hirst JD (2009) Machine learning in virtual screening. Comb Chem High Throughput Screen 12:332–343. https://doi.org/10.2174/138620709788167980
92. Geppert H, Vogt M, Bajorath J (2010) Current trends in ligand-based virtual screening: molecular representations, data mining methods, new application areas, and performance evaluation. J Chem Inf Model 50:205–216. https://doi.org/10.1021/ci900419k
93. Plewczynski D, Spieser SAH, Koch U (2009) Performance of machine learning methods for ligand-based virtual screening. Comb Chem High Throughput Screen 12:358–368. https://doi.org/10.2174/138620709788167962
94. Palczewski K, Kumasaka T, Hori T, Behnke CA, Motoshima H, Fox BA, Trong IL, Teller DC, Okada T, Stenkamp RE, Yamamoto M, Miyano M (2000) Crystal structure of rhodopsin: a G protein-coupled receptor. Science 289:739–745. https://doi.org/10.1126/science.289.5480.739
95. Cherezov V, Rosenbaum DM, Hanson MA, Rasmussen SGF, Thian FS, Kobilka TS, Choi H-J, Kuhn P, Weis WI, Kobilka BK, Stevens RC (2007) High-resolution crystal structure of an engineered human β2-Adrenergic G protein–coupled receptor. Science 318:1258–1265. https://doi.org/10.1126/science.1150577
96. Rasmussen SGF, Choi H-J, Fung JJ, Pardon E, Casarosa P, Chae PS, DeVree BT, Rosenbaum DM, Thian FS, Kobilka TS, Schnapp A, Konetzki I, Sunahara RK, Gellman SH, Pautsch A, Steyaert J, Weis WI, Kobilka BK (2011) Structure of a nanobody-stabilized active state of the β2 adrenoceptor. Nature 469:175–180. https://doi.org/10.1038/nature09648
97. Berman HM, Westbrook J, Feng Z, Gilliland G, Bhat TN, Weissig H, Shindyalov IN, Bourne PE (2000) The protein data bank. Nucleic Acids Res 28:235–242. https://doi.org/10.1093/nar/28.1.235
98. Wu B, Chien EYT, Mol CD, Fenalti G, Liu W, Katritch V, Abagyan R, Brooun A, Wells P, Bi FC, Hamel DJ, Kuhn P, Handel TM, Cherezov V, Stevens RC (2010) Structures of the CXCR4 chemokine GPCR with small-molecule and cyclic peptide antagonists. Science 330:1066–1071. https://doi.org/10.1126/science.1194396
99. Tan Q, Zhu Y, Li J, Chen Z, Han GW, Kufareva I, Li T, Ma L, Fenalti G, Li J, Zhang W, Xie X, Yang H, Jiang H, Cherezov V, Liu H, Stevens RC, Zhao Q, Wu B (2013) Structure of the CCR5 chemokine receptor–HIV entry inhibitor maraviroc complex. Science 341:1387–1390. https://doi.org/10.1126/science.1241475
100. Zheng Y, Qin L, Zacarías NVO, de Vries H, Han GW, Gustavsson M, Dabros M, Zhao C, Cherney RJ, Carter P, Stamos D, Abagyan R, Cherezov V, Stevens RC, IJzerman AP, Heitman LH, Tebben A, Kufareva I, Handel TM (2016) Structure of CC chemokine receptor 2 with orthosteric and allosteric antagonists. Nature 540:458–461. https://doi.org/10.1038/nature20605
101. Oswald C, Rappas M, Kean J, Doré AS, Errey JC, Bennett K, Deflorian F, Christopher JA, Jazayeri A, Mason JS, Congreve M, Cooke RM, Marshall FH (2016) Intracellular allosteric antagonism of the CCR9 receptor. Nature 540:462–465. https://doi.org/10.1038/nature20606
102. Chien EYT, Liu W, Zhao Q, Katritch V, Han GW, Hanson MA, Shi L, Newman AH, Javitch JA, Cherezov V, Stevens RC (2010) Structure of the human dopamine D3

receptor in complex with a D2/D3 selective antagonist. Science 330:1091–1095. https://doi.org/10.1126/science.1197410

103. Shimamura T, Shiroishi M, Weyand S, Tsujimoto H, Winter G, Katritch V, Abagyan R, Cherezov V, Liu W, Han GW, Kobayashi T, Stevens RC, Iwata S (2011) Structure of the human histamine H1 receptor complex with doxepin. Nature 475:65–70. https://doi.org/10.1038/nature10236

104. Haga K, Kruse AC, Asada H, Yurugi-Kobayashi T, Shiroishi M, Zhang C, Weis WI, Okada T, Kobilka BK, Haga T, Kobayashi T (2012) Structure of the human M2 muscarinic acetylcholine receptor bound to an antagonist. Nature 482:547–551. https://doi.org/10.1038/nature10753

105. Wacker D, Wang C, Katritch V, Han GW, Huang X-P, Vardy E, McCorvy JD, Jiang Y, Chu M, Siu FY, Liu W, Xu HE, Cherezov V, Roth BL, Stevens RC (2013) Structural features for functional selectivity at serotonin receptors. Science 340:615–619. https://doi.org/10.1126/science.1232808

106. Manglik A, Kruse AC, Kobilka TS, Thian FS, Mathiesen JM, Sunahara RK, Pardo L, Weis WI, Kobilka BK, Granier S (2012) Crystal structure of the μ-opioid receptor bound to a morphinan antagonist. Nature 485:321–326. https://doi.org/10.1038/nature10954

107. Wu H, Wacker D, Mileni M, Katritch V, Han GW, Vardy E, Liu W, Thompson AA, Huang X-P, Carroll FI, Mascarella SW, Westkaemper RB, Mosier PD, Roth BL, Cherezov V, Stevens RC (2012) Structure of the human κ-opioid receptor in complex with JDTic. Nature 485:327–332. https://doi.org/10.1038/nature10939

108. Thompson AA, Liu W, Chun E, Katritch V, Wu H, Vardy E, Huang X-P, Trapella C, Guerrini R, Calo G, Roth BL, Cherezov V, Stevens RC (2012) Structure of the nociceptin/orphanin FQ receptor in complex with a peptide mimetic. Nature 485:395–399. https://doi.org/10.1038/nature11085

109. Granier S, Manglik A, Kruse AC, Kobilka TS, Thian FS, Weis WI, Kobilka BK (2012) Structure of the δ-opioid receptor bound to naltrindole. Nature 485:400–404. https://doi.org/10.1038/nature11111

110. Hollenstein K, Kean J, Bortolato A, Cheng RKY, Doré AS, Jazayeri A, Cooke RM, Weir M, Marshall FH (2013) Structure of class B GPCR corticotropin-releasing factor receptor 1. Nature 499:438–443. https://doi.org/10.1038/nature12357

111. Wu H, Wang C, Gregory KJ, Han GW, Cho HP, Xia Y, Niswender CM, Katritch V, Meiler J, Cherezov V, Conn PJ, Stevens RC (2014) Structure of a class C GPCR metabotropic glutamate receptor 1 bound to an allosteric modulator. Science 344:58–64. https://doi.org/10.1126/science.1249489

112. Wang C, Wu H, Katritch V, Han GW, Huang X-P, Liu W, Siu FY, Roth BL, Cherezov V, Stevens RC (2013) Structure of the human smoothened receptor bound to an antitumour agent. Nature 497:338–343. https://doi.org/10.1038/nature12167

113. Yuriev E, Holien J, Ramsland PA (2015) Improvements, trends, and new ideas in molecular docking: 2012–2013 in review. J Mol Recognit 28:581–604. https://doi.org/10.1002/jmr.2471

114. Andrews SP, Brown GA, Christopher JA (2014) Structure-based and fragment-based GPCR drug discovery. ChemMedChem 9:256–275. https://doi.org/10.1002/cmdc.201300382

115. Beuming T, Lenselink B, Pala D, McRobb F, Repasky M, Sherman W (2015) Docking and virtual screening strategies for GPCR drug discovery. In: Filizola M (ed) G protein-coupled receptors drug discovery. Springer, New York, pp 251–276

116. Beuming T, Sherman W (2012) Current assessment of docking into GPCR crystal structures and homology models: successes, challenges, and guidelines. J Chem Inf Model 52:3263–3277. https://doi.org/10.1021/ci300411b

117. Cheng T, Li Q, Zhou Z, Wang Y, Bryant SH (2012) Structure-based virtual screening for drug discovery: a problem-centric review. AAPS J 14:133–141. https://doi.org/10.1208/s12248-012-9322-0

118. Michino M, Abola E, Participants GD, Brooks CL, Dixon JS, Moult J, Stevens RC (2009) Community-wide assessment of GPCR structure modelling and ligand docking: GPCR dock 2008. Nat Rev Drug Discov 8:455–463. https://doi.org/10.1038/nrd2877

119. Kufareva I, Rueda M, Katritch V, Stevens RC, Abagyan R (2011) Status of GPCR modeling and docking as reflected by community-wide GPCR dock 2010 assessment. Structure 19:1108–1126. https://doi.org/10.1016/j.str.2011.05.012

120. Kufareva I, Katritch V, Stevens RC, Abagyan R (2014) Advances in GPCR modeling evaluated by the GPCR dock 2013 assessment: meeting new challenges. Structure

22:1120–1139. https://doi.org/10.1016/j.str.2014.06.012

121. Kiss R, Jójárt B, Schmidt É, Kiss B, Keserű GM (2014) Identification of novel histamine H4 ligands by virtual screening on molecular dynamics ensembles. Mol Inform 33:264–268. https://doi.org/10.1002/minf.201300072

122. Virtanen SI, Niinivehmas SP, Pentikäinen OT (2015) Case-specific performance of MM-PBSA, MM-GBSA, and SIE in virtual screening. J Mol Graph Model 62:303–318. https://doi.org/10.1016/j.jmgm.2015.10.012

123. Vass M, Kooistra AJ, Ritschel T, Leurs R, de Esch IJ, de Graaf C (2016) Molecular interaction fingerprint approaches for GPCR drug discovery. Curr Opin Pharmacol 30:59–68. https://doi.org/10.1016/j.coph.2016.07.007

124. Lenselink EB, Jespers W, van Vlijmen HWT, IJzerman AP, van Westen GJP (2016) Interacting with GPCRs: using interaction fingerprints for virtual screening. J Chem Inf Model 56:2053–2060. https://doi.org/10.1021/acs.jcim.6b00314

125. de Graaf C, Rognan D (2009) Customizing G protein-coupled receptor models for structure-based virtual screening. Curr Pharm Des 15:4026–4048. https://doi.org/10.2174/138161209789824786

126. Mobarec JC, Sanchez R, Filizola M (2009) Modern homology modeling of G-protein coupled receptors: which structural template to use? J Med Chem 52:5207–5216. https://doi.org/10.1021/jm9005252

127. Mysinger MM, Weiss DR, Ziarek JJ, Gravel S, Doak AK, Karpiak J, Heveker N, Shoichet BK, Volkman BF (2012) Structure-based ligand discovery for the protein–protein interface of chemokine receptor CXCR4. Proc Natl Acad Sci 109:5517–5522. https://doi.org/10.1073/pnas.1120431109

128. Carlsson J, Coleman RG, Setola V, Irwin JJ, Fan H, Schlessinger A, Sali A, Roth BL, Shoichet BK (2011) Ligand discovery from a dopamine D3 receptor homology model and crystal structure. Nat Chem Biol 7:769–778. https://doi.org/10.1038/nchembio.662

129. Vass M, Schmidt É, Horti F, Keserű GM (2014) Virtual fragment screening on GPCRs: a case study on dopamine D3 and histamine H4 receptors. Eur J Med Chem 77:38–46. https://doi.org/10.1016/j.ejmech.2014.02.034

130. Tarcsay Á, Paragi G, Vass M, Jójárt B, Bogár F, Keserű GM (2013) The impact of molecular dynamics sampling on the performance of virtual screening against GPCRs. J Chem Inf Model 53:2990–2999. https://doi.org/10.1021/ci400087b

131. Manglik A, Kobilka B (2014) The role of protein dynamics in GPCR function: insights from the β2AR and rhodopsin. Curr Opin Cell Biol 27:136–143. https://doi.org/10.1016/j.ceb.2014.01.008

132. Simpson LM, Wall ID, Blaney FE, Reynolds CA (2011) Modeling GPCR active state conformations: the β2-adrenergic receptor. Proteins 79:1441–1457. https://doi.org/10.1002/prot.22974

133. Arany A, Bolgar B, Balogh B, Antal P, Matyus P (2013) Multi-aspect candidates for repositioning: data fusion methods using heterogeneous information sources. Curr Med Chem 20:95–107

134. Liu X, Xu Y, Li S, Wang Y, Peng J, Luo C, Luo X, Zheng M, Chen K, Jiang H (2014) In Silico target fishing: addressing a "big data" problem by ligand-based similarity rankings with data fusion. J Cheminform 6:33. https://doi.org/10.1186/1758-2946-6-33

135. Drwal MN, Griffith R (2013) Combination of ligand- and structure-based methods in virtual screening. Drug Discov Today Technol 10:e395–e401. https://doi.org/10.1016/j.ddtec.2013.02.002

136. Riniker S, Fechner N, Landrum GA (2013) Heterogeneous classifier fusion for ligand-based virtual screening: or, how decision making by committee can be a good thing. J Chem Inf Model 53:2829–2836. https://doi.org/10.1021/ci400466r

137. Willett P (2013) Combination of similarity rankings using data fusion. J Chem Inf Model 53:1–10. https://doi.org/10.1021/ci300547g

138. Willett P (2013) Fusing similarity rankings in ligand-based virtual screening. Comput Struct Biotechnol J 5:1–6. https://doi.org/10.5936/csbj.201302002

139. Zhang Q, Muegge I (2006) Scaffold hopping through virtual screening using 2D and 3D similarity descriptors: ranking, voting, and consensus scoring. J Med Chem 49:1536–1548. https://doi.org/10.1021/jm050468i

140. Charifson PS, Corkery JJ, Murcko MA, Walters WP (1999) Consensus scoring: a method for obtaining improved hit rates from docking databases of three-dimensional structures into proteins. J Med Chem 42:5100–5109. https://doi.org/10.1021/jm990352k

141. Kelemen ÁA, Kiss R, Ferenczy GG, Kovács L, Flachner B, Lőrincz Z, Keserű GM (2016) Structure-based consensus scoring scheme for selecting class A aminergic GPCR fragments. J Chem Inf Model 56:412–422. https://doi.org/10.1021/acs.jcim.5b00598
142. Houston DR, Walkinshaw MD (2013) Consensus docking: improving the reliability of docking in a virtual screening context. J Chem Inf Model 53:384–390. https://doi.org/10.1021/ci300399w
143. Sastry GM, Inakollu VSS, Sherman W (2013) Boosting virtual screening enrichments with data fusion: coalescing hits from two-dimensional fingerprints, shape, and docking. J Chem Inf Model 53:1531–1542. https://doi.org/10.1021/ci300463g
144. Bologa CG, Revankar CM, Young SM, Edwards BS, Arterburn JB, Kiselyov AS, Parker MA, Tkachenko SE, Savchuck NP, Sklar LA, Oprea TI, Prossnitz ER (2006) Virtual and biomolecular screening converge on a selective agonist for GPR30. Nat Chem Biol 2:207–212. https://doi.org/10.1038/nchembio775
145. Pérez-Nueno VI, Pettersson S, Ritchie DW, Borrell JI, Teixidó J (2009) Discovery of novel HIV entry inhibitors for the CXCR4 receptor by prospective virtual screening. J Chem Inf Model 49:810–823. https://doi.org/10.1021/ci800468q
146. Svensson F, Karlén A, Sköld C (2012) Virtual screening data fusion using both structure- and ligand-based methods. J Chem Inf Model 52:225–232. https://doi.org/10.1021/ci2004835
147. Tan L, Geppert H, Sisay MT, Gütschow M, Bajorath J (2008) Integrating structure- and ligand-based virtual screening: comparison of individual, parallel, and fused molecular docking and similarity search calculations on multiple targets. ChemMedChem 3:1566–1571. https://doi.org/10.1002/cmdc.200800129
148. Patchett AA, Nargund RP (2000) Privileged structures — an update. In: Doherty AM (ed) Annual reports in medicinal chemistry. Academic Press, London, pp 289–298
149. Ungashe S, Wei Z, Basak A, Charvat T, Jin J, Moore J, Zang Y, Punna S, Dairaghi D, Hansen D, Pennell A, Wright J (2006) Heteroaryl sulfonamides and CCR2. US Patent Application 20060173019 A1
150. Zheng C, Cao G, Xia M, Feng H, Glenn J, Anand R, Zhang K, Huang T, Wang A, Kong L, Li M, Galya L, Hughes RO, Devraj R, Morton PA, Rogier DJ, Covington M, Baribaud F, Shin N, Scherle P, Diamond S, Yeleswaram S, Vaddi K, Newton R, Hollis G, Friedman S, Metcalf B, Xue C-B (2011) Discovery of INCB10820/PF-4178903, a potent, selective, and orally bioavailable dual CCR2 and CCR5 antagonist. Bioorg Med Chem Lett 21:1442–1446. https://doi.org/10.1016/j.bmcl.2011.01.015
151. Baba M, Takashima K, Miyake H, Kanzaki N, Teshima K, Wang X, Shiraishi M, Iizawa Y (2005) TAK-652 inhibits CCR5-mediated human immunodeficiency virus type 1 infection in vitro and has favorable pharmacokinetics in humans. Antimicrob Agents Chemother 49:4584–4591. https://doi.org/10.1128/AAC.49.11.4584-4591.2005
152. van der Horst E, van der Pijl R, Mulder-Krieger T, Bender A, IJzerman AP (2011) Substructure-based virtual screening for adenosine A2A receptor ligands. ChemMedChem 6:2302–2311. https://doi.org/10.1002/cmdc.201100369
153. Taylor CM, Rockweiler NB, Liu C, Rikimaru L, Tunemalm A-K, Kisselev OG, Marshall GR (2010) Using ligand-based virtual screening to allosterically stabilize the activated state of a GPCR. Chem Biol Drug Des 75:325–332. https://doi.org/10.1111/j.1747-0285.2009.00944.x
154. Low CMR, Buck IM, Cooke T, Cushnir JR, Kalindjian SB, Kotecha A, Pether MJ, Shankley NP, Vinter JG, Wright L (2005) Scaffold hopping with molecular field points: identification of a cholecystokinin-2 (CCK2) receptor pharmacophore and its use in the design of a prototypical series of pyrrole- and imidazole-based CCK2 antagonists. J Med Chem 48:6790–6802. https://doi.org/10.1021/jm049069y
155. Saeh JC, Lyne PD, Takasaki BK, Cosgrove DA (2005) Lead hopping using SVM and 3D pharmacophore fingerprints. J Chem Inf Model 45:1122–1133. https://doi.org/10.1021/ci049732r
156. Bock JR, Gough DA (2005) Virtual screen for ligands of orphan G protein-coupled receptors. J Chem Inf Model 45:1402–1414. https://doi.org/10.1021/ci050006d
157. Jacob L, Hoffmann B, Stoven V, Vert J-P (2008) Virtual screening of GPCRs: an in silico chemogenomics approach. BMC Bioinformatics 9:363. https://doi.org/10.1186/1471-2105-9-363
158. Ananthan S, Zhang W, Hobrath JV (2009) Recent advances in structure-based virtual screening of G-protein coupled receptors. AAPS J 11:178–185. https://doi.org/10.1208/s12248-009-9094-3

159. Rodríguez D, Ranganathan A, Carlsson J (2015) Discovery of GPCR ligands by molecular docking screening: novel opportunities provided by crystal structures. Curr Top Med Chem 15:2484–2503
160. Rodríguez D, Gao Z-G, Moss SM, Jacobson KA, Carlsson J (2015) Molecular docking screening using agonist-bound GPCR structures: probing the A2A adenosine receptor. J Chem Inf Model 55:550–563. https://doi.org/10.1021/ci500639g
161. Weiss DR, Ahn S, Sassano MF, Kleist A, Zhu X, Strachan R, Roth BL, Lefkowitz RJ, Shoichet BK (2013) Conformation guides molecular efficacy in docking screens of activated β-2 adrenergic G protein coupled receptor. ACS Chem Biol 8:1018–1026. https://doi.org/10.1021/cb400103f
162. Taylor CM, Barda Y, Kisselev OG, Marshall GR (2008) Modulating G-protein coupled receptor/G-protein signal transduction by small molecules suggested by virtual screening. J Med Chem 51:5297–5303. https://doi.org/10.1021/jm800326q
163. Sato M, Hirokawa T (2014) Extended template-based modeling and evaluation method using consensus of binding mode of GPCRs for virtual screening. J Chem Inf Model 54:3153–3161. https://doi.org/10.1021/ci500499j
164. de Graaf C, Kooistra AJ, Vischer HF, Katritch V, Kuijer M, Shiroishi M, Iwata S, Shimamura T, Stevens RC, de Esch IJP, Leurs R (2011) Crystal structure-based virtual screening for fragment-like ligands of the human histamine H1 receptor. J Med Chem 54:8195–8206. https://doi.org/10.1021/jm2011589
165. Istyastono EP, Kooistra AJ, Vischer HF, Kuijer M, Roumen L, Nijmeijer S, Smits RA, de Esch IJP, Leurs R, de Graaf C (2015) Structure-based virtual screening for fragment-like ligands of the G protein -coupled histamine H 4 receptor. MedChemComm 6:1003–1017. https://doi.org/10.1039/C5MD00022J
166. Kooistra AJ, Vischer HF, McNaught-Flores-D, Leurs R, de EIJP, de GC (2016) Function-specific virtual screening for GPCR ligands using a combined scoring method. Sci Rep 6:28288. https://doi.org/10.1038/srep28288
167. de Graaf C, Rognan D (2008) Selective structure-based virtual screening for full and partial agonists of the β2 adrenergic receptor. J Med Chem 51:4978–4985. https://doi.org/10.1021/jm800710x
168. Kooistra AJ, Leurs R, de Esch IJP, de Graaf C (2015) Structure-based prediction of G-protein-coupled receptor ligand function: a β-adrenoceptor case study. J Chem Inf Model 55:1045–1061. https://doi.org/10.1021/acs.jcim.5b00066
169. Heifetz A, Barker O, Verquin G, Wimmer N, Meutermans W, Pal S, Law RJ, Whittaker M (2013) Fighting obesity with a sugar-based library: discovery of novel MCH-1R antagonists by a new computational–VAST approach for exploration of GPCR binding sites. J Chem Inf Model 53:1084–1099. https://doi.org/10.1021/ci4000882
170. Radestock S, Weil T, Renner S (2008) Homology model-based virtual screening for GPCR ligands using docking and target-biased scoring. J Chem Inf Model 48:1104–1117. https://doi.org/10.1021/ci8000265
171. Chen J-Z, Wang J, Xie X-Q (2007) GPCR structure-based virtual screening approach for CB2 antagonist search. J Chem Inf Model 47:1626–1637. https://doi.org/10.1021/ci7000814
172. Evers A, Klabunde T (2005) Structure-based drug discovery using gpcr homology modeling: successful virtual screening for antagonists of the Alpha1A adrenergic receptor. J Med Chem 48:1088–1097. https://doi.org/10.1021/jm0491804
173. Cavasotto CN, Orry AJW, Murgolo NJ, Czarniecki MF, Kocsi SA, Hawes BE, O'Neill KA, Hine H, Burton MS, Voigt JH, Abagyan RA, Bayne ML, Monsma FJ (2008) Discovery of novel chemotypes to a G-protein-coupled receptor through ligand-steered homology modeling and structure-based virtual screening. J Med Chem 51:581–588. https://doi.org/10.1021/jm070759m
174. Levoin N, Labeeuw O, Billot X, Calmels T, Danvy D, Krief S, Berrebi-Bertrand I, Lecomte J-M, Schwartz J-C, Capet M (2017) Discovery of nanomolar ligands with novel scaffolds for the histamine H4 receptor by virtual screening. Eur J Med Chem 125:565–572. https://doi.org/10.1016/j.ejmech.2016.09.074
175. Vilar S, Ferino G, Phatak SS, Berk B, Cavasotto CN, Costanzi S (2011) Docking-based virtual screening for ligands of G protein-coupled receptors: not only crystal structures but also in silico models. J Mol Graph Model 29:614–623. https://doi.org/10.1016/j.jmgm.2010.11.005
176. Kołaczkowski M, Bucki A, Feder M, Pawłowski M (2013) Ligand-optimized

homology models of D1 and D2 dopamine receptors: application for virtual screening. J Chem Inf Model 53:638–648. https://doi.org/10.1021/ci300413h

177. Thomas T, McLean KC, McRobb FM, Manallack DT, Chalmers DK, Yuriev E (2014) Homology modeling of human muscarinic acetylcholine receptors. J Chem Inf Model 54:243–253. https://doi.org/10.1021/ci400502u
178. McRobb FM, Capuano B, Crosby IT, Chalmers DK, Yuriev E (2010) Homology modeling and docking evaluation of aminergic G protein-coupled receptors. J Chem Inf Model 50:626–637. https://doi.org/10.1021/ci900444q
179. Langmead CJ, Andrews SP, Congreve M, Errey JC, Hurrell E, Marshall FH, Mason JS, Richardson CM, Robertson N, Zhukov A, Weir M (2012) Identification of novel adenosine A2A receptor antagonists by virtual screening. J Med Chem 55:1904–1909. https://doi.org/10.1021/jm201455y
180. Lam VM, Rodríguez D, Zhang T, Koh EJ, Carlsson J, Salahpour A (2015) Discovery of trace amine-associated receptor 1 ligands by molecular docking screening against a homology model. MedChemComm 6:2216–2223. https://doi.org/10.1039/C5MD00400D
181. Higgs C, Beuming T, Sherman W (2010) Hydration site thermodynamics explain SARs for triazolylpurines analogues binding to the A2A receptor. ACS Med Chem Lett 1:160–164. https://doi.org/10.1021/ml100008s
182. Mason JS, Bortolato A, Congreve M, Marshall FH (2012) New insights from structural biology into the druggability of G protein-coupled receptors. Trends Pharmacol Sci 33:249–260. https://doi.org/10.1016/j.tips.2012.02.005
183. Lenselink EB, Beuming T, Sherman W, van Vlijmen HWT, IJzerman AP (2014) Selecting an optimal number of binding site waters to improve virtual screening enrichments against the adenosine A2A receptor. J Chem Inf Model 54:1737–1746. https://doi.org/10.1021/ci5000455
184. Tömöri T, Hajdú I, Barna L, Lőrincz Z, Cseh S, Dormán G (2012) Combining 2D and 3D in silico methods for rapid selection of potential PDE5 inhibitors from multimillion compounds' repositories: biological evaluation. Mol Divers 16:59–72. https://doi.org/10.1007/s11030-011-9335-0
185. Hert J, Willett P, Wilton DJ, Acklin P, Azzaoui K, Jacoby E, Schuffenhauer A (2006) New methods for ligand-based virtual screening: use of data fusion and machine learning to enhance the effectiveness of similarity searching. J Chem Inf Model 46:462–470. https://doi.org/10.1021/ci050348j
186. Baber JC, Shirley WA, Gao Y, Feher M (2006) The use of consensus scoring in ligand-based virtual screening. J Chem Inf Model 46:277–288. https://doi.org/10.1021/ci050296y
187. Gregory KJ, Dong EN, Meiler J, Conn PJ (2011) Allosteric modulation of metabotropic glutamate receptors: structural insights and therapeutic potential. Neuropharmacology 60:66–81. https://doi.org/10.1016/j.neuropharm.2010.07.007
188. Bennett KA, Doré AS, Christopher JA, Weiss DR, Marshall FH (2015) Structures of mGluRs shed light on the challenges of drug development of allosteric modulators. Curr Opin Pharmacol 20:1–7. https://doi.org/10.1016/j.coph.2014.09.022
189. Noeske T, Jirgensons A, Starchenkovs I, Renner S, Jaunzeme I, Trifanova D, Hechenberger M, Bauer T, Kauss V, Parsons CG, Schneider G, Weil T (2007) Virtual screening for selective allosteric mGluR1 antagonists and structure–activity relationship investigations for coumarine derivatives. ChemMedChem 2:1763–1773. https://doi.org/10.1002/cmdc.200700151
190. Noeske T, Trifanova D, Kauss V, Renner S, Parsons CG, Schneider G, Weil T (2009) Synergism of virtual screening and medicinal chemistry: identification and optimization of allosteric antagonists of metabotropic glutamate receptor 1. Bioorg Med Chem 17:5708–5715. https://doi.org/10.1016/j.bmc.2009.05.072
191. Tresadern G, Cid JM, Macdonald GJ, Vega JA, de Lucas AI, García A, Matesanz E, Linares ML, Oehlrich D, Lavreysen H, Biesmans I, Trabanco AA (2010) Scaffold hopping from pyridones to imidazo[1,2-α]pyridines. New positive allosteric modulators of metabotropic glutamate 2 receptor. Bioorg Med Chem Lett 20:175–179. https://doi.org/10.1016/j.bmcl.2009.11.008
192. Mueller R, Dawson ES, Niswender CM, Butkiewicz M, Hopkins CR, Weaver CD, Lindsley CW, Conn PJ, Meiler J (2012) Iterative experimental and virtual high-

throughput screening identifies metabotropic glutamate receptor subtype 4 positive allosteric modulators. J Mol Model 18:4437–4446. https://doi.org/10.1007/s00894-012-1441-0

193. Jang JW, Cho N-C, Min S-J, Cho YS, Park KD, Seo SH, No KT, Pae AN (2016) Novel scaffold identification of mGlu1 receptor negative allosteric modulators using a hierarchical virtual screening approach. Chem Biol Drug Des 87:239–256. https://doi.org/10.1111/cbdd.12654

194. de Graaf C, Rein C, Piwnica D, Giordanetto F, Rognan D (2011) Structure-based discovery of allosteric modulators of two related class B G-protein-coupled receptors. ChemMedChem 6:2159–2169. https://doi.org/10.1002/cmdc.201100317

195. Lane JR, Chubukov P, Liu W, Canals M, Cherezov V, Abagyan R, Stevens RC, Katritch V (2013) Structure-based ligand discovery targeting orthosteric and allosteric pockets of dopamine receptors. Mol Pharmacol 84:794–807. https://doi.org/10.1124/mol.113.088054

196. Miao Y, Goldfeld DA, Moo EV, Sexton PM, Christopoulos A, McCammon JA, Valant C (2016) Accelerated structure-based design of chemically diverse allosteric modulators of a muscarinic G protein-coupled receptor. Proc Natl Acad Sci 113:E5675–E5684. https://doi.org/10.1073/pnas.1612353113

Chapter 12

Approaches for Differentiation and Interconverting GPCR Agonists and Antagonists

Przemysław Miszta, Jakub Jakowiecki, Ewelina Rutkowska, Maria Turant, Dorota Latek, and Sławomir Filipek

Abstract

Predicting the functional preferences of the ligands was always a highly demanding task, much harder that predicting whether a ligand can bind to the receptor. This is because of significant similarities of agonists, antagonists and inverse agonists which are binding usually in the same binding site of the receptor and only small structural changes can push receptor toward a particular activation state. For G protein-coupled receptors, due to a large progress in crystallization techniques and also in receptor thermal stabilization, it was possible to obtain a large number of high-quality structures of complexes of these receptors with agonists and non-agonists. Additionally, the long-time-scale molecular dynamics simulations revealed how the activation processes of GPCRs can take place. Using both theoretical and experimental knowledge it was possible to employ many clever and sophisticated methods which can help to differentiate agonists and non-agonists, so one can interconvert them in search of the optimal drug.

Key words GPCRs, Agonists, Activation, Ligand docking, Fingerprints, Molecular dynamics

1 Introduction

The approaches used for GPCR agonist/antagonist differentiation, which is a prerequisite for their interconversion, can be divided into several categories based on specific methods: (1) the molecular fingerprints; (2) ligand docking to active and inactive receptor structures; (3) long time-scale molecular dynamics simulations and network correlation analysis; (4) metadynamics simulations and free-energy profiles; (5) methods using specific parameters of the ligand binding site or the vibrational modes; and (6) and methods based on ligands alone properties. In this chapter the above approaches will be described and illustrated by usage of several examples.

G protein-coupled receptors (GPCRs) represent the largest family of surface receptors, with approximately 800 members in humans [1]. The participation of GPCRs in numerous

Alexander Heifetz (ed.), *Computational Methods for GPCR Drug Discovery*, Methods in Molecular Biology, vol. 1705, https://doi.org/10.1007/978-1-4939-7465-8_12,

physiological and pathological processes entails a potential role for their modulation by ligands of various functions: agonists (increase receptor activity), antagonists (block the receptor not changing the activity), and inverse agonists (decrease activity). GPCRs are extremely important as molecular targets for drugs in medicine since their ligands are used in the treatment of many diseases, including cardiovascular and mental disorders [2], cancer [3], and viral infections [4]. Currently, approximately 30–50% of drugs in clinical use exert their effects by acting on GPCR-mediated signaling pathways [5]. From the practical point of view, for drug design purposes, the major need is a differentiation of GPCR ligands into two groups: agonists and antagonists/inverse agonists, the latter referred to as non-agonists in the text.

According to the GRAFS classification [6], the human GPCRs are grouped into five main classes: Glutamate (previous class C), Rhodopsin (previous class A—the most populated), Adhesion (previous class G), Frizzled/taste2 (previous class F), and Secretin (previous class B). From a structural viewpoint, all the members of the GPCR family share a common architecture represented by seven membrane-spanning helices connected by three intracellular and three extracellular loops (ICLs and ECLs, respectively) with the N-terminal domain exposed toward the EC side (Fig. 1). The EC area includes N-terminal domain, ECLs, and upper transmembrane

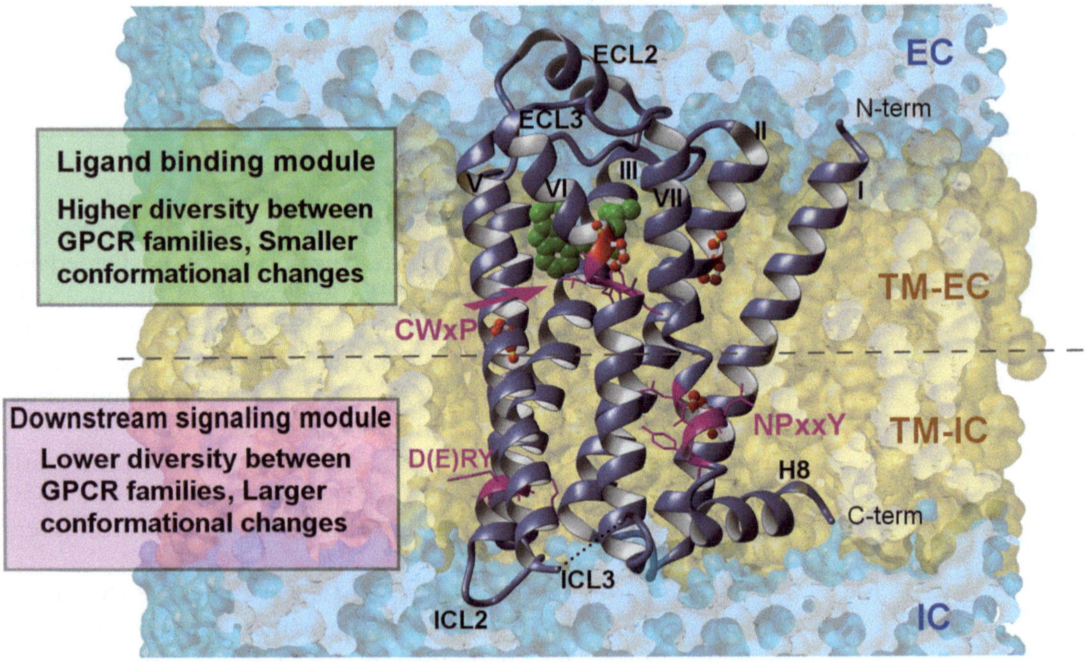

Fig. 1 General scheme showing modularity of GPCRs. Purple ribbon patches highlight highly conserved, functionally relevant motifs in the TM helices of class A GPCRs. Prolines, which induce kinks in helices, are shown in ball-and-stick representation

(TM) region, where the ligand recognition and binding occur. The IC area includes lower TM region, ICLs, and C-terminal domain—this area experiences the largest conformational changes during receptor activation [7].

Our current understanding of the function of GPCRs was changed from simple On-Off switches to microprocessor-like action [8]. Especially, the phenomenon of functional selectivity, whereby certain ligands initiate only portions of the signaling mechanisms mediated by a given receptor, opened new horizons for drug discovery. Each receptor undergoes a series of conformational rearrangements controlled by molecular switches leading to partial or full activation and the dynamic character of GPCRs is thought to be essential for their diverse physiological functions. Transition between these intermediate states involves the disruption of intramolecular interactions that stabilize the basal state of a receptor. Such profound changes are evoked by the action of molecular switches (Fig. 1) [9]. The major switches proposed so far for different GPCRs include the "Trp rotamer toggle switch" also called a "transmission switch" involving the CWxP sequence on transmembrane helix TM6, the "Tyr rotamer toggle switch" based on the NPxxY sequence on TM7, the "ionic lock" linking transmembrane helices TM3 and TM6 and employing the (D/E) RY motif on TM3, and the "3–7 lock" interaction connecting TM3 and TM7 (involving, e.g., Schiff base-counterion interaction in rhodopsin).

As a result of their broad influence on human physiology and behavior, GPCRs are promising candidates for the development of new and more effective small-molecule therapeutics. However, the development of selective GPCR drugs is challenging for several reasons: there is a high degree of homology among many closely related receptor subtypes that can regulate diverse physiological functions; additionally, a single GPCR may couple to more than one G protein, or signal through G protein-independent pathways. Although this functional versatility is important for normal physiological signaling, it makes identifying effective therapeutics very challenging. New data support a multi-state activation of GPCRs, where the receptor can adopt multiple conformations, including active, inactive, and other intermediate ones. In such multi-state model the ligands have the propensity to stabilize a unique conformation leading to a specific signaling response.

2 The Molecular Fingerprints

2.1 Definition of Molecular Fingerprints

Protein–ligand interaction fingerprints (IFPs) are binary 1D representations of the 3D structure of protein–ligand complexes encoding the presence or absence of specific interactions between the binding pocket amino acids and the ligand [10]. For example, if a ligand forms a hydrogen bond (H-bond) with a specific amino acid of the binding pocket, the respective bit in the fingerprint will be one, if the interaction is missing, it will be zero. Fingerprints derived from ligands, proteins, or protein–ligand complexes are computer-digestible representations of (bio)chemical structures and are particularly well suited for working with large amounts of data allowing for rapid processing and comparisons [10–17]. Machine learning methods are increasingly used to derive complex relationships between bioactivity data and fingerprint descriptors of chemical and structural information of protein-ligand interactions [18]. The fingerprints obtained for docking poses of certain compounds can be compared with the fingerprints obtained for other group of ligands with different properties, or a reference ligand from the crystal structure. Some methods and coefficients used for the comparison of binary fingerprints are presented in Table 1. Very often the Tanimoto coefficient (*Tc*) is used for that purpose which is the number of common bits in the two fingerprints divided by the number of bits present in at least one of the fingerprints. *Tc* ranges from zero for dissimilar binding interactions to one for identical interactions [11].

The first structural interaction fingerprint (SIFt) algorithm was developed by Deng et al. in 2004 for the clustering of kinase-inhibitor complexes [12]. This fingerprint contains seven bits for each interacting amino acid for predefined interaction types (any, backbone, sidechain, polar, hydrophobic, H-bond donor/acceptor). A more recent implementation of SIFt, was described by Mordalski et al. extending the interaction fingerprint implementation by two bits to encode aromatic and charged interactions and

Table 1
Some similarity coefficients and distances used with fingerprints

Measure	Expression	Range
Tanimoto/Jaccard coefficient	$\frac{c}{a+b-c}$	0–1
Euclidean distance	$\sqrt{a+b-2c}$	0–N
City-block/Manhattan/hamming distance	$a+b-2c$	0–N
Dice coefficient	$\frac{2c}{a+b}$	0–1
Cosine similarity	$\frac{c}{\sqrt{ab}}$	0–1
Russell–RAO coefficient	$\frac{c}{m}$	0–1
Forbes coefficient	$\frac{cm}{ab}$	0–1
Soergel distance	$\frac{a+b-2c}{a+b-c}$	0–1

Where, given the fingerprints of two compounds, A and B, m equals the total amount of bits present in the fingerprints, a equals the amount of bit set to 1 in A, b equals the amount of bits set to 1 in B, and c equals the amount of bits set to 1 in both A and B [11].

	D3.32	S5.42	S5.43	K5.58	N6.55	Y7.35	N7.39	Tc score
Reference	1 0 0 0 1 0 1	0 0 0 1 0 0 0	0 0 0 0 0 0 0	1 0 0 1 0 0 0	1 0 0 0 0 0 0	1 0 0 0 0 0 0	1 0 0 1 1 0 0	-
Ligand 1	1 0 0 0 1 0 1	0 0 0 1 0 0 0	0 0 0 0 0 0 0	1 0 0 0 0 0 0	1 0 0 0 0 0 0	1 0 0 0 0 0 0	1 0 0 0 1 0 0	0.818
Ligand 2	1 0 0 0 1 0 1	0 0 0 0 0 0 0	0 0 0 0 0 0 0	1 0 0 0 0 0 0	0 0 0 0 0 0 0	1 1 0 0 0 0 0	1 0 0 1 0 0 0	0.583
Ligand 3	1 0 0 0 1 0 1	0 0 0 0 0 0 0	0 0 0 0 0 0 0	1 0 0 0 0 1 0	0 0 0 0 0 0 0	1 0 1 0 0 0 0	1 0 0 0 1 0 0	0.538

1. Apolar (hydrophobic), 2. Aromatic face-to-face, 3. Aromatic edge-to-face, 4. H-bond donor (protein) - H-bond acceptor (ligand), 5. H-bond donor (ligand) - H-bond acceptor (protein), 6. Ionic interaction ligand (negative) - protein (positive, 7. Ionic interaction protein (negative) - ligand (positive)

Fig. 2 Exemplary interaction fingerprints of β_2AR hits from each of the scoring approaches compared to the X-ray structure. Adapted from [22]

implementing technical improvements [19]. SIFt allowed pointing out crucial amino acids involved in interactions with antagonists docked into serotonin 5-HT7 receptor homology models. LIFt, a similar method to SIFt, was described by Cao and Wang with 10 bits per amino acid, who used it to predict kinase targets for ligands [20].

A commonly used variant of a binary interaction fingerprint has been developed by Marcou and Rognan: a 7-bit fingerprint encoding (1) hydrophobic, (2) aromatic face-to-face, (3) aromatic edge-to-face, (4) H-bond donor, (5) H-bond acceptor, (6) cationic, and (7) anionic interaction types [13]. That variant of molecular fingerprint (Fig. 2) has been used for ligands functional activity prediction by Kooistra et al. [21, 22] and the results were compared to experimental data proving the high accuracy of the obtained predictions [10].

2.2 Usage of Molecular Fingerprints from Crystal Structures and Docking Poses

Kooistra et al. [21, 22] employed unique protein-ligand interaction fingerprints (IFPs) derived from all the ligand-bound β-adrenergic crystal structure monomers to post-process the docking poses of known β_1AR/β_2AR ligands, and physicochemically similar decoys in each of the β_1AR/β_2AR structures. The analysis of 1920 unique IFP-structure combinations using IFP scoring was employed to virtual screening (VS) for selecting ligands with a specific agonist/non-agonist functional effect [21]. IFP rescoring was shown to be essential to obtain high enrichment factors and at the same time a high selectivity. The analysis showed that the IFPs of non-agonists were more similar to each other than to agonist IFPs (75% versus 19% similar pairs, respectively). Analogously, the pairwise similarity between agonist IFPs was higher (62%) than similarity to non-agonists (21%) (Fig. 3). By using the correctly chosen agonist reference IFP (a complex with epinephrine), it was possible to selectively retrieve agonists compared to non-agonists with a high efficiency ($EF_{1\%} = 43.6$ and 9.4, respectively).

Protein-ligand interaction fingerprints have been successfully used for post-docking processing of ligand poses, and that method proved to be superior to the conventional energy-based docking scores or RMSD calculations. To improve the predictive value of docking even further, interaction fingerprints are often used

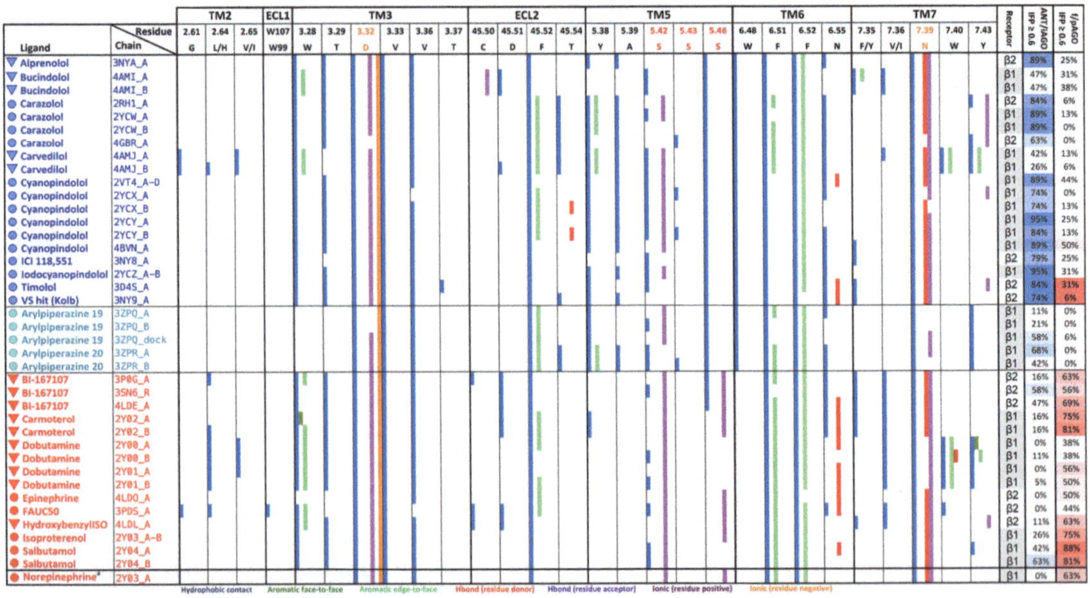

Fig. 3 Overview of interaction fingerprints of all cocrystallized ligands in β_1AR and β_2AR. The colors indicate the presence of a particular interaction according to the 7-bit fingerprints (colors described at the bottom of the figure). The last two columns describe the amount of times (as a percentage of the total comparisons) an IFP comparisons having score $\geq$0.6 when compared with non-agonist IFPs (ANT/iAGO, antagonist/inverse agonist, names in *blue*, a *blue* background indicates a high percentage), and the f/pAGO IFPs (full/partial agonists, names in *red*, a *red* background indicates a high percentage) [21]

together with conventional scoring methods. To test the usefulness of IFPs for docking results scoring, the three data sets of ligand poses were generated using popular docking tools: FlexX, Glide, Gold, and Surflex. In all the cases, scoring by the similarity of interaction fingerprints to a given reference was statistically superior to conventional scoring functions. RMSD values are based on the ligand coordinates only and thus lose information about the kind of intermolecular interactions which have been reproduced or not. In many cases, a low RMSD correlates with a high *Tc*-IFP and vice versa; however, there are a significant number of cases for which the RMSD value is misleading. The *Tc*-IFP metric is clearly better than the RMSD criterion for handling false positives and false negatives in virtual screening [13].

2.3 Usage of Molecular Fingerprints from MD Simulations

To address a question on specific requirements for agonist and antagonist with nearly identical structures, the all-atom molecular dynamics (MD) simulations were employed to investigate how two diastereomers (epimers) of dihydrofuroaporphine bind to the serotonin 5-HT1A receptor and exert opposite effects [23]. The POPC lipids and TIP3P water molecules were used, and 3 × 1.2 μs MD simulations were performed per each ligand, 7.2 μs in total. It was discovered that the agonist could mobilize located nearby amino

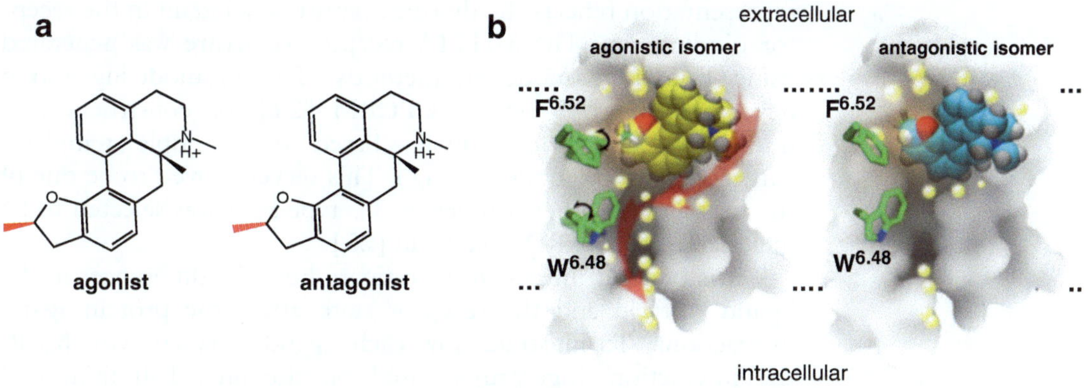

Fig. 4 Stereoselective pair of 5-HT1A receptor ligands. (**a**) agonist and antagonist epimers. (**b**) Action of molecular switches and water influx in agonist-bound receptor [23]

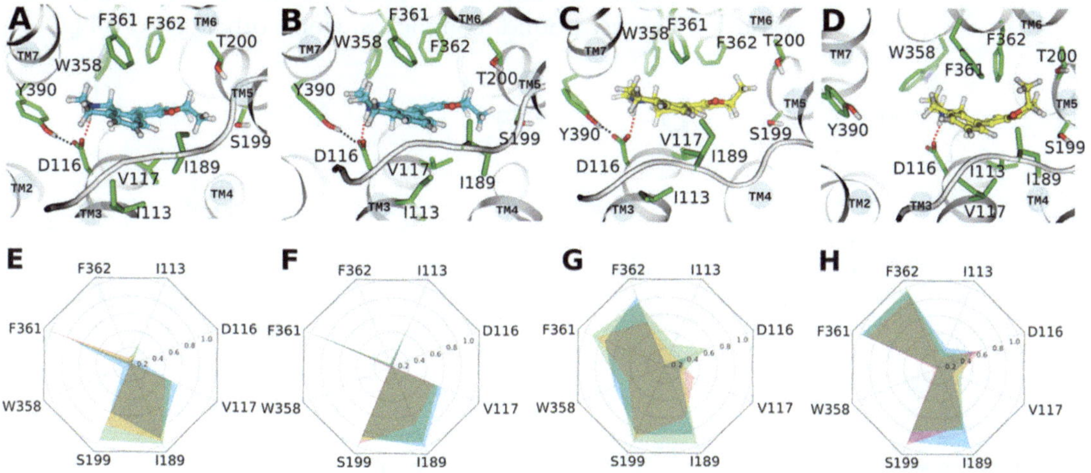

Fig. 5 The binding sites and the interaction fingerprints of 5-HT1A receptor ligands. (ABEF) antagonist-bound receptor. (CDGH) Agonist-bound receptor. Start (ACEG) and end (BDFH) of simulations. *Blue*, *green*, and *red* areas on spider plots represent the three different MD simulations presented in this work [23]

acid residues to act as molecular switches for the formation of a continuous water channel. In contrast, the antagonist epimer remained firmly stabilized in the binding pocket (Fig. 4). The molecular fingerprints of ligands in the binding site revealed a difference between antagonist- and agonist-bound receptor and also the changes in the agonist-bound receptor due to activation events (Fig. 5). The method of molecular fingerprints based on MD simulations was performed using the IChem toolkit [13]. IChem converts protein-ligand interactions into a fingerprint (TIFP) of 210 integers/bits registering the corresponding molecular interaction pattern. Because of averaging over 100 snapshots from the final 50 ns MD simulation the obtained interaction fingerprints are not 0/1 values but the real numbers from the [0–1] range. Such a representation reflects the dynamic nature of a ligand in the receptor binding site. The 5-HT1A receptor structure was generated using homology modeling methods. The Biomodeling group developed the web service GPCRM [24] for construction of homology models of GPCRs based on multiple templates (http://gpcrm.biomodellab.eu/). This server proved to be one of the best among other services of this type so it was selected to be included into GPCRDB platform [25].

It is of great interest to investigate how the differences in the ligand scaffold and the receptor state affect the protein-ligand interactions. To illustrate how each ligand interacts with β_2AR, the interaction fingerprints based on the final 100 frames of 100 ns MD simulations of 12 crystal structures of β_2AR with agonists, antagonists, and inverse agonists were generated [26]. The analysis of interactions in the binding site and also between transmembrane helices provided clear clues for differentiation of agonist-bound complexes from other types (Fig. 6).

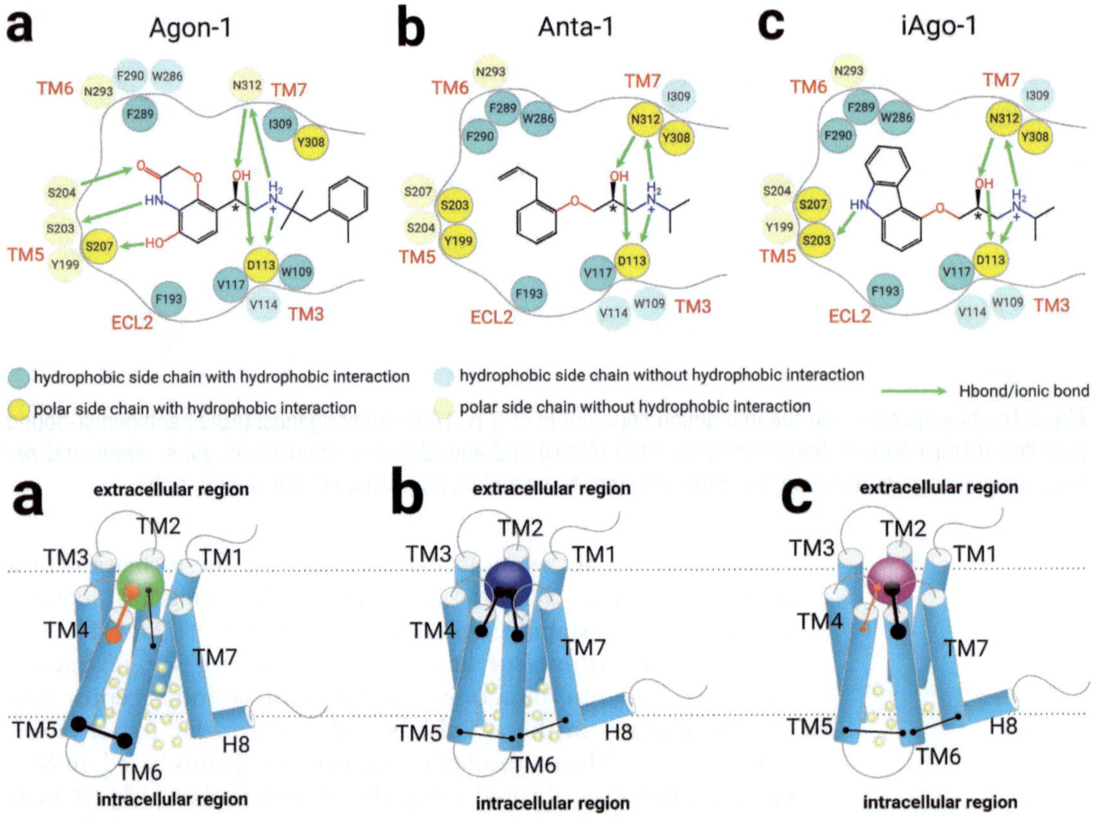

Fig. 6 (*Upper panels*) The 16 important residues that form polar and/or hydrophobic interactions with exemplary ligands in binding site of β_2AR. (*Lower panels*) The relationship of protein-ligand interactions and the state of β_2AR. *Thick lines*–dominant interactions; *thin lines*–rare interactions; *orange*–hydrogen bonds; *black*–non-polar interactions. *Yellow spheres*: water molecules; (**a**) agonist; (**b**) antagonist; (**c**) inverse agonist [26]

Protein-ligand interaction fingerprints generated from MD trajectories helped identify the important residues and the type of interactions required for designing ligands with desired properties. In this study, using the interaction fingerprints, a dynamic behavior of 16 residues important in the binding pockets, among which $D113^{3.32}$ and $N312^{7.39}$ are essential for ligand binding, was analyzed (the superscript numbers are based on the Ballesteros-Weinstein numbering Scheme [27] developed for GPCRs). These 16 residues were selected based on frequency of interactions with any of the ligands tested (interaction present at least 30 times among 100 frames). It was demonstrated that the molecular fingerprints can be a powerful tool for capturing the specific profile of protein-ligand interactions, and can be employed together with MD simulations in predicting the nature of a ligand. The polar interactions of ligands with residues in TM5, particularly $S203^{5.42}$ and $S207^{5.46}$, were assigned to the agonistic properties, whereas hydrophobic interactions with residues in TM5 and TM6 helped stabilize the receptor. Agonists of β_2AR which predominantly form H-bonds with TM5 disrupt the interactions between helices in the extracellular region and then in the rest of TM area leading to increasing a flexibility of the receptor. As a result, TM5 as well as TM6 form frequent nonpolar interactions in the intracellular region and move away from TM7, causing the expansion of intracellular pocket and a water influx (Fig. 6a). This also explains why the residues of TM5 and TM6 in this region share the same interaction domain. In contrast, antagonists form prominently nonpolar interactions with both TM5 and TM6 (Fig. 6b), whereas inverse agonists mainly form nonpolar interactions with TM6 only (Fig. 6c).

In the above analysis the interaction fingerprints between the receptor and the ligand were also done with IChem [13]. The default parameters of IChem were kept and two types of interactions, polar interaction and hydrophobic contacts, were focused on. The former comprises ionic bonds and H-bonds, while the latter incorporates hydrophobic contacts, the face-to-face and edge-to-face between aromatic rings. Using fingerprints it was possible to characterize interactions specific for differentiation of not only agonists from non-agonists but also inverse agonists from antagonists (Fig. 7). In general, the fingerprints have been successfully used in three scenarios: (1) interaction-biased alignment of protein-ligand complexes, (2) postprocessing of docking poses according to known interaction patterns for a particular target, and (3) virtual screening for bioisosteric scaffolds sharing similar interaction patterns. The frame-invariant interaction pattern TIFP fingerprints have been calculated for ca. 10,000 druggable protein-ligand complexes enabling a wide comparison of relationships between interaction pattern similarity and ligand binding site similarity [14].

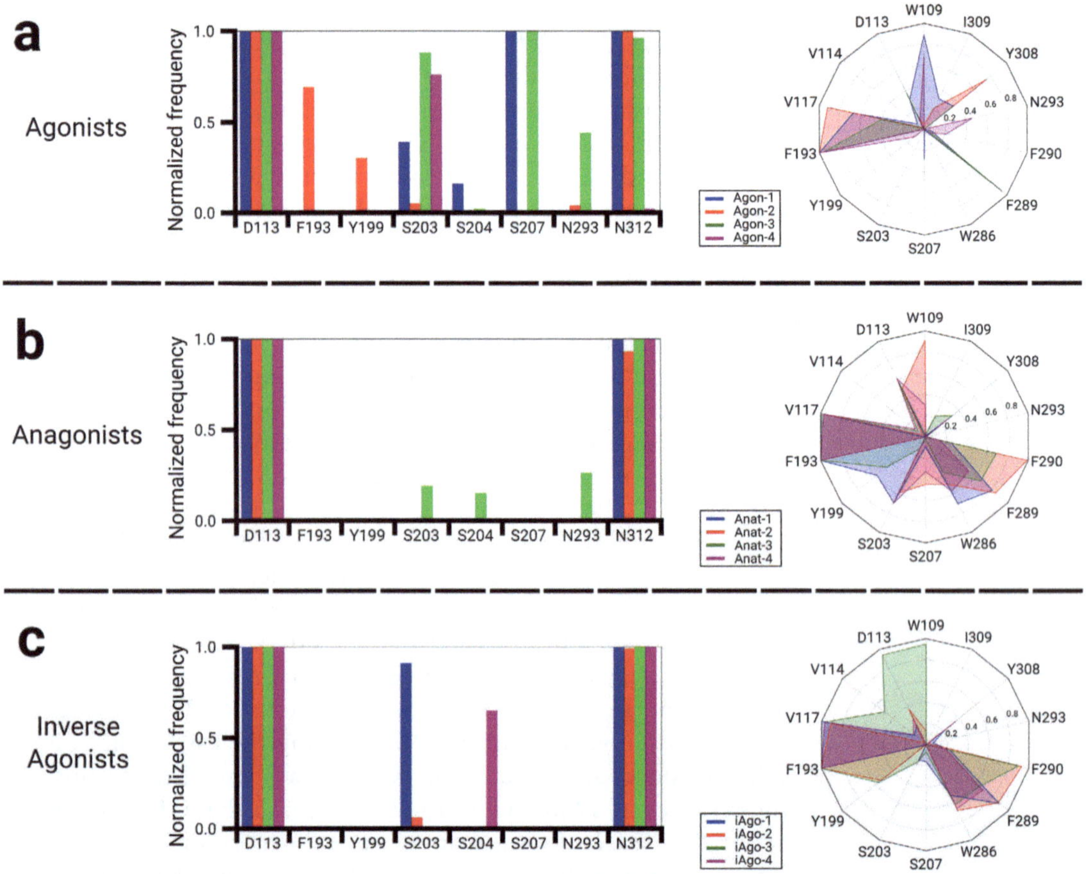

Fig. 7 The interaction fingerprints of the twelve ligands with β_2AR. (**a**) The four agonists. (**b**) The four antagonists. (**c**) The four inverse agonists. (*Left panels*) The normalized frequency of H-bonds/ionic bonds in the final 100 frames. (*Right panels*) The normalized frequency of hydrophobic interactions in the final 100 frames [26]

3 Ligand Docking to Active and Inactive Receptor Structures

3.1 Structure-Based Virtual Screening

The recent abundance of crystal structures of GPCRs has stimulated the structure-based drug discovery studies against these targets. The increased amount of high-resolution structural information on GPCRs has opened up new opportunities for the identification of novel GPCR ligands by structure-based virtual screening (SBVS). SBVS can be employed for the efficient identification of chemically novel ligands with high hit rates and also for the structure-based prediction of GPCR ligand function. Increased computational power enabled docking screenings of very large libraries of small compounds to identify ligands that complement GPCR binding sites, which may lead to identifying drug-like molecules with tailored pharmacological properties [28]. The recent review on screening of adenosine receptor ligands [29] also describes SBVS, among other methods, for such purposes. The determined structures of GPCRs in activated states revealed surprisingly subtle changes in the orthosteric binding sites supporting the idea that agonist binding and receptor activation require only small conformational changes in this region. Taking into account the small differences between the active and inactive binding site conformations, the obtained high functional fidelity of docking hits to the state of the receptor was surprising. The first such results were obtained for four different GPCR families: adrenergic β_2AR, adenosine A_{2A}R, dopamine D_3R, and chemokine CXCR4 receptors. It was possible to obtain high hit-rates in each case despite using several different ligand-docking programs. It could suggest that SBVS is a fruitful strategy in drug development for other GPCRs. It was proposed that such success originated from (1) the druggability of the orthosteric sites and (2) a bias in chemical libraries toward GPCR ligands. The adrenergic, adenosine, and dopamine receptors have evolved to recognize small endogenous ligands and their relatively well-covered orthosteric sites, which involve only a few polar interactions for ligand binding, which were proved to be particularly well suited for the molecular docking algorithms. In contrast, the more open binding site of chemokine receptor may explain the lower hit-rate and compound affinities obtained in this particular case. The high hit-rates were also suggested to stem from a biogenic bias in compound libraries together with the interest in GPCRs as drug targets, which may have led to an accumulation of "GPCR-like" ligand chemotypes in commercially available chemical space. Kolb et al. [30] estimated that GPCR ligands were from 3- to 12-fold more common in chemical libraries compared to other targets such as kinases, proteases, and ligand-gated ion channels.

3.2 Limitations of SBVS and a Combination of SBVS with Molecular Fingerprints

The development of structure-based functions specific to virtual screening methods is hampered by the fact that for most crystallized GPCRs only one or a few agonist-bound structures are available which leads to limited diversity of binding modes and the resulting functional effects. The functional properties of a ligand bound to the receptor in most cases reflect a bias in the receptor structure or homology model used in VS [31]. Selective SBVS for ligands with a specific function (e.g., inverse agonist/antagonist, partial/full agonist) probably requires a customized modeling protocol. Recently, Kooistra et al. [22] applied a novel docking scoring approach for the identification of novel fragment-like GPCR ligands and the prediction of their functional effect using GPCR crystal structures. This docking scoring approach combined a conventional docking scoring function (ChemPLP) using PLANTS [32] docking with the molecular interaction fingerprint (IFP) rescoring approach [13, 33]. It was assumed that the combination of two basically different scoring functions can result in increased performance by combining the strengths of each scoring function [34]. ChemPLP is an empirical energy-based scoring function whereas IFP compares the interaction pattern between a docking pose of a ligand in the protein to the reference binding mode, most often the crystal pose of a known ligand. The aim of this idea was to use the strength of PLANTS to identify compounds with energetically favorable docking poses with the strength of IFP to select the most probable binding modes by selecting those with an interaction profile resembling the reference IFP in the crystal structures. The consensus scoring method was evaluated by: (1) the discovery of chemically novel, fragment-like, high affinity histamine H_1 receptor (H_1R) antagonists/inverse agonists, and (2) the selective structure-based identification of β_2-adrenoceptor (β_2AR) agonists, including the experimental validation and comparison of the combined and individual scoring approaches. In the previous studies, the researchers using SBVS method reported successful prospective study for H_1R [33] and retrospective for β_2AR [21]. In the recent work the VS performances of the different scoring approaches and combinations for H_1R and β_2AR were compared in a prospective manner [22]. The SBVS protocol for the first target, H_1R structure (determined in a complex with doxepin–an inverse agonist [35]), has been developed and experimentally validated [33]. For both the PLANTS and IFP approaches the top 500 scoring compounds were selected and subsequently the compounds matching the combined approach were removed. The remaining compounds were processed in the same fashion as the compounds from the combined approach. The hits similar to known H_1R ligands (ECFP-4 Tanimoto score [36] >0.4) were excluded. The compounds were visually clustered based on the scaffold similarity, and compounds with polar groups that were placed in the hydrophobic parts of the H_1R binding site were discarded. In SBVS of β_2AR the same approach and fragment library was used as for H_1R. Based on the cutoffs used for the H_1R study a set of only 318 compounds was selected for the combined scoring approach. Due to limited chemical diversity of known β_2AR ligands the selected compounds were filtered with - ECFP-4 (circular Extended Connectivity FingerPrint) Tanimoto score of ≤ 0.5 compared to any known β_2AR ligand. The combined IFP and PLANTS scoring approach proved to be the most successful approach, followed by rankings of IFP-score and PLANTS-score. For the H_1R VS hit rates of 73%, 61%, and 45% were obtained for the combined, IFP, and PLANTS approach, respectively. For the β_2AR VS study the hit rates were 53%, 44%, and 39% for the combined, IFP, and PLANTS approach, respectively (Fig. 8).

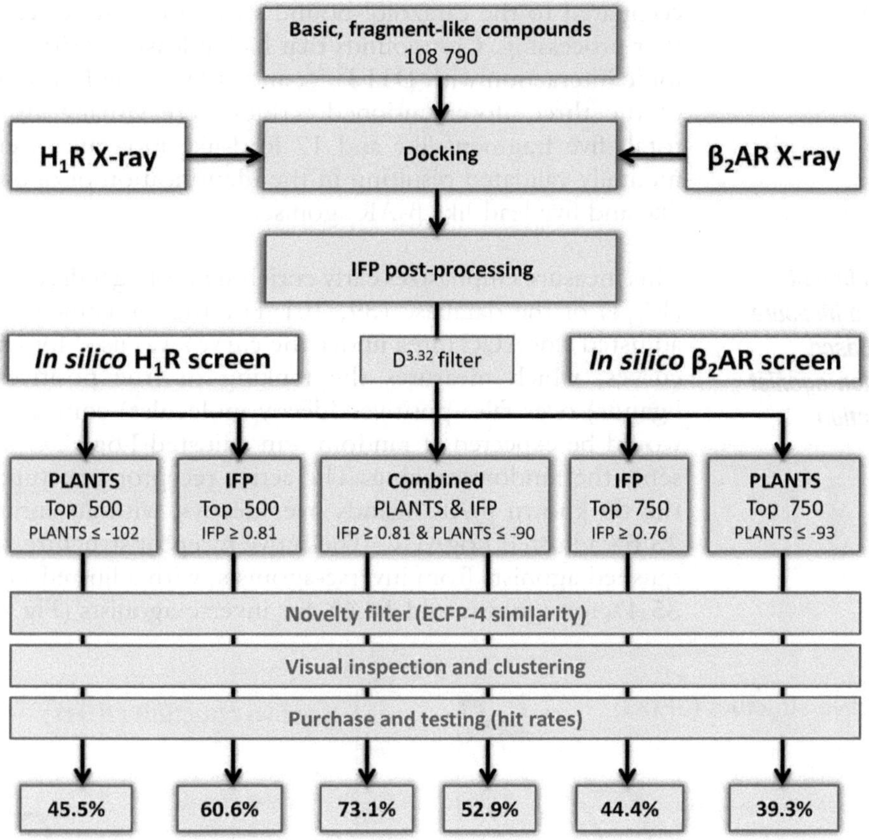

Fig. 8 Workflow of the virtual-screening approaches performed on the H_1R and the β_2AR. Only compounds within the top 500 (H_1R) or top 750 (β_2AR) compounds were selected for further processing. Figure adapted from [22]

3.3 Enhanced Docking in SBVS

A prospective SBVS using a large library of 3.4 million molecules aiming at β_2AR, as well as dopamine D_2 receptor (D_2R) agonists, was done by Weiss et al. [37]. In this study, a library of lead-like and fragment-like molecules from the ZINC database was screened against the active state (PDB id: 3P0G) and inactive state, carazolol-bound crystal structure (PDB id:2RH1). Using a set of 30 agonists and 30 inverse-agonists of β_2AR, they tested the ability of receptor's active structure to recognize known β_2AR ligands against a background of property matched decoys and to preferentially score agonists over inverse-agonists. To ensure that the reasonable poses were obtained in docking, the dipole moment of $S203^{5.42}$, $S204^{5.43}$, or $S207^{5.46}$ was increased to enhance docking scores for poses in polar contact with these residues. Serine residues were proposed to be important for interactions with agonists and for the activation [38], and the largest change between the active and inactive β_2AR structures was associated with those residues. Compounds ranking within the top 0.2% of the active-state structure and ranking at least 5000 positions higher for the active-state compared to the carazolol-bound structure were selected for further processing. Compounds that had at least a positive charge, an ionic interaction with D113$^{3.32}$, and at least one H-bond with any of the three aforementioned serines were visually inspected. In total, five fragment-like and 17 lead-like molecules were experimentally validated resulting in the identification of one fragment-like and five lead-like β_2AR agonists.

3.4 Enrichment Factor as a Measure of SBVS-Based Agonist/Non-agonist Differentiation

This measure emphasizes early enrichment of ligands, at the first 1% ($EF_{1\%}$) of the database [39, 40]. For the enrichment metric the adjusted LogAUC (area under the curve) was used for enrichment curves, which measures the ranking of true positives (known ligands) over false positives (decoy molecules) compared to what would be expected at random—an adjusted LogAUC of 0 represents the random ranking. The active receptor structure enriched the 60 known β_2AR ligands over decoys, with an enrichment of 23.6% adjusted LogAUC. The active receptor structure also distinguished agonists from inverse-agonists, with adjusted LogAUC of 35.4% for agonists and 10.6% for inverse-agonists (Fig. 9). In the

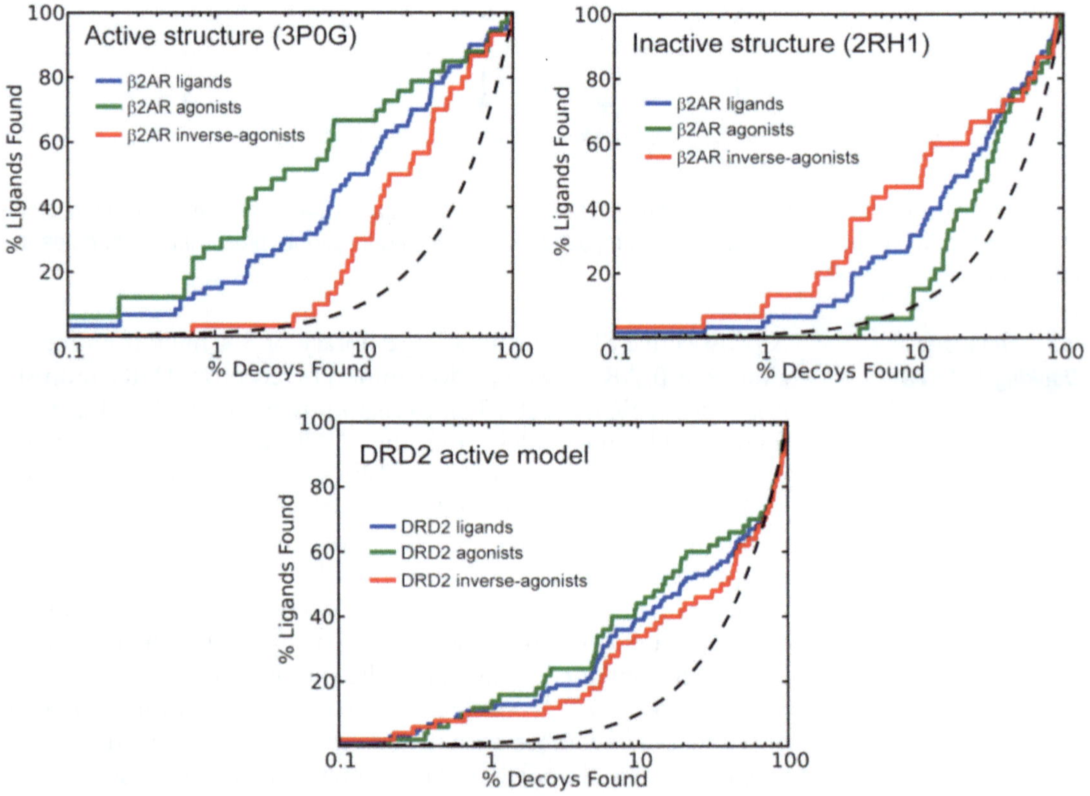

Fig. 9 Semilogarithmic enrichment curves for retrospective enrichment of known GPCR ligands from a set of computational decoys employing β_2AR active structure (*top*), inactive β_2AR with the same set of known ligands (*middle*), and the active dopamine D_2R model with 100 known ligands (50 agonists and 50 antagonists). Figure adapted from [37]

top 1% of the database, 20% of the agonists were found (6/30 docked agonists), while at 10% of the database, 75% (22/30 agonists) were found. Using the same set of agonists and inverse-agonists with the same decoys, the inactive crystal structure found no agonists in the top 1% of the database and 13% (4/30 agonists) in the top 10% of the database. SBVS of active dopamine D_2R homology model based on the active β_2AR structure as a template predicted only two weak agonists, as well as an inverse-agonist. These results indicated that despite 42% sequence identity, structural information from the active β_2AR was not transferrable.

3.5 Employing Induced Fit Docking for Changing GPCR Preferences for Ligand Binding

An important issue to resolve is to what extent the structures solved with a ligand of a certain class of functionality can be used in docking studies focused on another class of ligands. Constanzi and Vilar [41] evaluated crystal structures of β_2AR for their ability to discriminate between agonist and antagonist compounds. The results clearly showed that inactive crystal structures favored the retrieval of antagonists, while the active-state crystal structure prioritized agonists over antagonists. With the aim of evaluating the effect of a ligand-induced optimization on the receptor's ability to recognize agonist or antagonist compounds, they built three β_2AR models by refining the binding site of an inverse agonist-bound structure (solved in complex with carazolol) with three full agonists, i.e., epinephrine, isoproterenol, and fenoterol. To optimize the recognition of agonist compounds in VS the induced fit docking (IFD) protocol has been applied to GPCR crystal structures. Interestingly, all three agonist-induced β_2AR models reverted their initial preferences, being as effective in prioritizing agonists over antagonists as the active-state crystal structure of the β_2AR. Moreover, Constanzi and Vilar demonstrated that the induced-fit docking of agonists is a viable way of modifying an inactive crystal structure and bias it toward the in silico recognition of agonists rather than blockers.

3.6 The Effect of Water Molecules on SBVS Enrichment

Due to the pharmacological importance [42] and availability of crystal structures solved in both the active and inactive states, the adenosine A_{2A} receptor has been a widely studied target using structure-based computational approaches. Once a A_{2A} receptor crystal structure with a resolution of 1.8 Å was released in 2012 (PDB id:4EIY) [43], it revealed several interesting and novel features, including a large number of water molecules located deep in the binding site. These water molecules have been shown to play a pivotal role in binding of ligands to the A_{2A} receptor [44]. In particular, the "unhappy" waters trapped between the ligand and the protein lead to the short residence time of a ligand and can be used for ligand binding kinetics prediction and also to generate working hypotheses how to improve binding in the lead optimization program. The effect of presence of water molecules on virtual screening enrichment was tested in docking studies by Lenselink et al. [45]. They showed that including crystal waters greatly improves VS enrichment but the optimization of water hydrogen positions is needed in order to achieve the best results. The waters derived from MD simulations, without any knowledge of crystallographic waters, can improve enrichments to a similar degree as the crystallographic waters. They also employed decision trees algorithm to select an ensemble of structures with different water molecule positions and orientations that outperformed any single structure with water molecules. In addition to the retrospective study, the validated protocol was also employed to a prospective application for virtual screening enrichments against the adenosine A_{2A} receptor. For the above studies the structural modeling was performed using tools in the Schrödinger small-molecule discovery suite. Over 2.5 million drug-like and lead-like compounds from the commercially available eMolecules database were docked in Glide [46]. To retrieve novel scaffolds an explicit similarity filter was included. Tanimoto similarities between all computational hits and all tested compounds were calculated based on Molprint2D fingerprints in Canvas (Schrödinger suite). In this study, Lenselink et al. [45] selected compounds that bore no resemblance to any compound previously tested against the A_{2A} receptor. Hence, they explored an unchartered chemical space for this GPCR, despite a very low hit rate (1.4%). Predicting of function of a ligand could also benefit from inclusion of water molecules into the binding site.

3.7 Collective Measure for Agonist/Non-agonist Differentiation

To investigate whether a collective measure of ligand fitting in the receptor binding site using Autodock could be useful for ligand-functional selectivity we employed the β_1- and β_2-adrenergic receptors. The molecular docking with Autodock VINA was performed in two modes, with a rigid receptor and with a partially flexible receptor with several residues inside the ligand binding site set as movable. It turned out that for the ligand function recognition the flexible receptor docking performed much better. All the obtained results were assessed with the Welch's t-test with the statistical significance α equal to 0.05, however, only the β_2AR docking results passed the t-test. In case of β_1AR the difference between the estimated free energy of binding represented by the Autodock VINA scoring function for agonists and antagonists was not statistically significant. The main reason for this effect is the similarity of the active and inactive structures of β_1AR compared to β_2AR, so the docking to similar β_1AR structures gave similar results in terms of the estimated free energy of binding, regardless the ligand type (data not shown). In that study two sets of ligands were used. The first set included 21 ligands from all available crystal structures of adrenergic receptors. The second set included selected agonists

Table 2
The percentage of the β_2AR agonist/antagonist pairs for which the Eq. 1 was fulfilled

Inactive β_2AR structure	Ligands from crystal structures 3P0G Active β_2AR	Ligands from databases 3P0G Active β_2AR
2RH1	100%	76.3%
3D4S	100%	74.0%
4GBR	100%	75.3%
3NY8	100%	74.5%
3NYA	92.9%	73.7%

and antagonists deposited in the GLIDA database (http://pharminfo.pharm.kyoto-u.ac.jp/services/glida/) and DrugBank (https://www.drugbank.ca/)—24 antagonists and 14 agonists of β_1AR, 22 antagonists and 18 agonists of β_2AR. For both sets of ligands Autodock VINA provided the estimated free energy of ligand binding. In principle, agonists should be better fitted to the active conformation of the receptor while antagonists should fit to the inactive receptor conformation. The above statement was transferred to a mathematical formula which compared free energies of binding for all the tested ligand/receptor pairs estimated via scoring function s:

$$s_{\text{active}}^{\text{agonist}} - s_{\text{active}}^{\text{antagonist}} < s_{\text{inactive}}^{\text{agonist}} - s_{\text{inactive}}^{\text{antagonist}} \tag{1}$$

Here, s was selected as the lowest estimated free energy of binding out of all energies computed for 20 ligand poses obtained in one docking round for each ligand/receptor pair. Given the above formula, we computed the percentage of cases for which Eq. 1 was fulfilled. Here, one case included four docking runs: docking of agonist and antagonist to an active receptor conformation, and docking of agonist and antagonist to an inactive receptor conformation. It was proved that the above formula was indeed fulfilled in most cases for the β_2AR receptor when employing 3P0G active structure of β_2AR (Table 2). Thus, the type of the ligand was correctly predicted for nearly all β_2AR ligands from crystal structures and for majority of ligands from databases.

4 Usage of Long Time-Scale MD Simulations and Network Correlation Analysis

4.1 Correlation Analysis of Network Interactions

Using molecular dynamics (MD) simulations it was possible to reveal distinct conformational transitions of the adenosine A_{2A} receptor [47]. It was found that the conserved W246$^{6.48}$ residue in transmembrane helix TM6 performs a key rotamer toggle switch

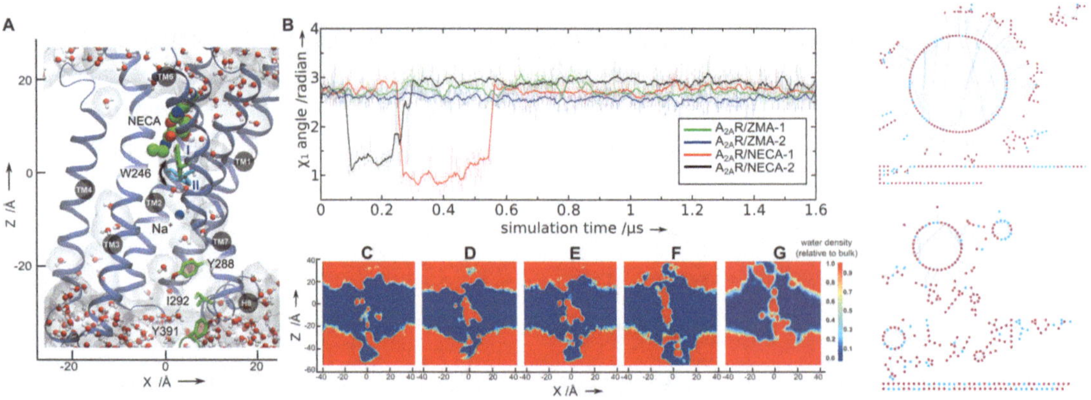

Fig. 10 Conformational fluctuations of W246$^{6.48}$ and distribution of water molecules in the complex A_{2A}R/NECA. (**a**) 3D structure of A_{2A}R with agonist NECA. (**b**) W246$^{6.48}$ fluctuates between two distinct conformations but only in complex with agonist. (**c**) Average water density during MD simulation showing formation of intrinsic water pathway. *Right panels*: correlation analysis of network interactions in antagonist- (*upper panel*) and agonist-bound (*lower panel*) A_{2A} receptor. Residues in the helix and in the loop are *red* and *cyan* dots, respectively. Line connections indicate residues' contacts [47]

(Fig. 10, left panels). Agonist binding induces the sidechain of W6.48 to fluctuate between two distinct conformations enabling the diffusion of water molecules from the bulk into the center of the receptor. After passing the W6.48 gate, the internal water molecules induce another conserved residue, Y7.53, to switch to a distinct rotamer conformation establishing a continuous transmembrane water pathway across the receptor. The correlation analysis of network interactions in antagonist- and agonist-bound receptor revealed its suitability for ligand function recognition (Fig. 10, right panels).

Another version of correlation analysis, called *the community residue interaction network* or *the domain interaction network*, was employed for complexes of β_2AR (Fig. 11) [26]. Each node represents a cluster of residues in close interaction, while the thickness of the line connecting the nodes is weighed by the correlation values between the two clusters. It was found that agonist-bound systems form less domains than the antagonist- or inverse agonist-bound systems. Noticeably, the position of TM6 in the intracellular region varies with the state of the receptor, leading to a different interaction network. Based on MD simulation the cross-correlation analysis was done to identify correlated and anticorrelated pairs of residues. To characterize correlated atomic fluctuations, a second method, a correlation network analysis, is usually performed as implemented in the Bio3D package [48, 49]. Community analysis and node centrality with Bio3D as well as a suboptimal path calculation with the WISP software [50] were performed on each network to characterize network properties and to identify residues involved in the dynamic coupling of distant sites.

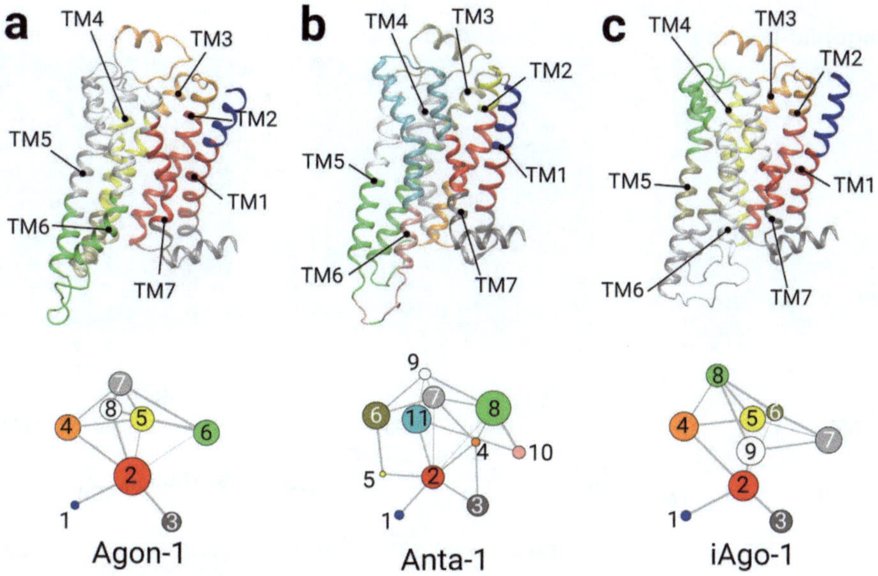

Fig. 11 Simultaneous view of the community residue interaction network and 3D structure of β_2AR. (**a**) Agonist-bound β_2AR; (**b**) antagonist-bound β_2AR; (**c**) inverse agonist-bound β_2AR [26]

4.2 The Network of Binary Switches

Another kind of signaling network was proposed by Lee et al. [51]. Based on 5 μs all-atom MD simulations of A_{2A} adenosine receptor in its apo, antagonist-bound, and agonist-bound complexes, they examined the corresponding dynamics and correlation between the 10 key structural motifs that serve as the allosteric hotspots in the intramolecular signaling network (Fig. 12). For this purpose they identified 10 molecular interactions that switch between two distinct states and they called them "binary switches". Such switches are able to yield in total 2^{10} microstates and the communication over the network of binary switches regulates the activation of A_{2A} adenosine receptor. Their cross-correlation analysis showed that W6.48, located deep inside the binding cleft can serve as both an agonist sensor and actuator of intramolecular signaling during the receptor activation. A signal of rotameric change of W6.48, triggered by a direct hydrophobic interaction with an agonist was transmitted to six other binary switches (S1, S2, S3, S5, S8, S9). Statistical analyses on the dynamics of and correlation among the 10 binary switches reveal that the three receptor states retain distinct dynamic properties. The antagonist- and agonist-bound form of the receptors explore vastly different conformational space, and the apo form lies between them, yet located closer to the antagonist-bound form.

4.3 Allosteric Effects of Sodium Ions in GPCRs

To investigate the activation process of opioid receptors, the MD studies on μ-opioid receptor (μOR) crystal structure [52] were performed [53] which revealed distinct consecutive mobility

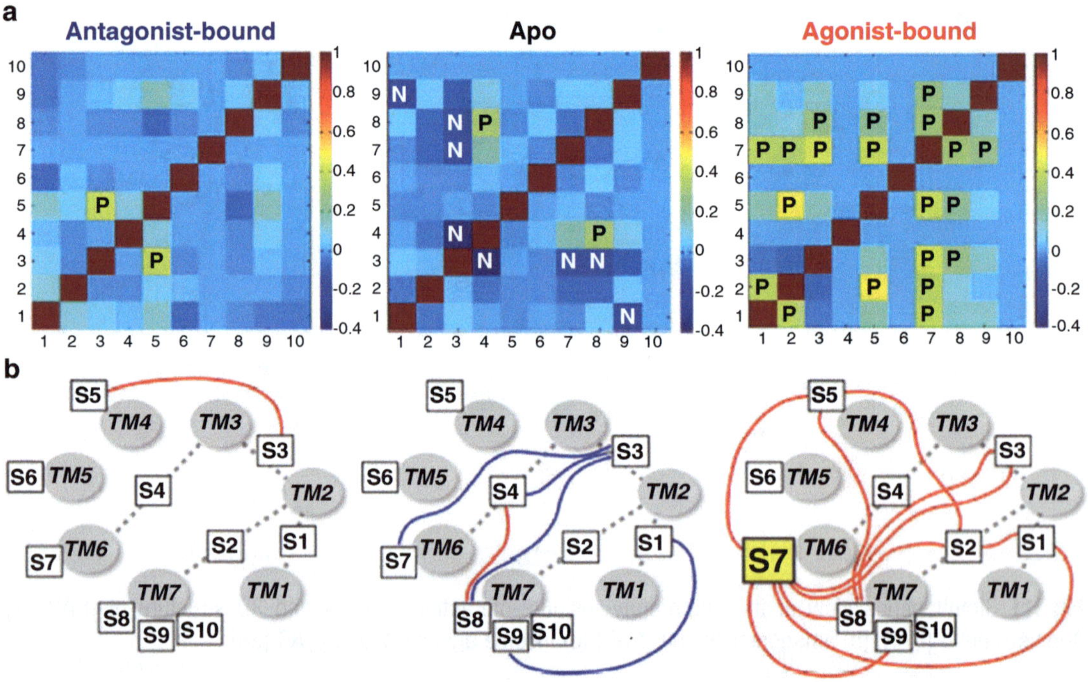

Fig. 12 Cross-correlations among binary switches. (**a**) Cross-correlation matrices between the changes in 10 binary switches: symbols P means a positive correlation ($C_{ij} > 0.25$); N is for the negative correlation ($C_{ij} < -0.25$). (**b**) Diagram of the cross-correlation between the switches. TM1–TM7 helices are displayed in *gray* circles, and the ten switches are specified with the boxes. The *positive* and *negative* correlations are depicted using *red* and *blue* lines, respectively. S7 switch (W6.48) is as a central switch located at the *bottom* of the orthosteric binding cleft [51]

patterns in the trajectory pathway of the sodium ions entering the receptor from the extracellular side toward the internal conserved residue D114$^{2.50}$. The distinct stages in the pathway of the ions are correlated with distinct local structural changes within the receptor as well as the distribution of internal waters. The MD simulations resolved the experimentally found dual role of sodium ions (1) to decrease the binding of ligands, and (2) to facilitate G protein activation (Fig. 13). Sodium ion may be necessary for modulation of interactions for ligand docking and performing MD simulations for most of class A GPCRs. It is assumed that sodium ion is present in its allosteric site in non-agonist-bound receptors and absent in agonist-bound receptors; therefore, unbinding of sodium ion could indicate the activation process. However, removal of Na^+ by binding of agonist was not encountered in MD simulations so far.

4.4 Elements of MD Methodology

The long-time-scale molecular dynamics simulations are usually performed employing all-atom approach using mostly programs NAMD [54], GROMACS [55], AMBER [56], ACEMD [57], and DESMOND [58]. Standard all-atom forcefield CHARMM [59] is usually employed, which is available in all the above

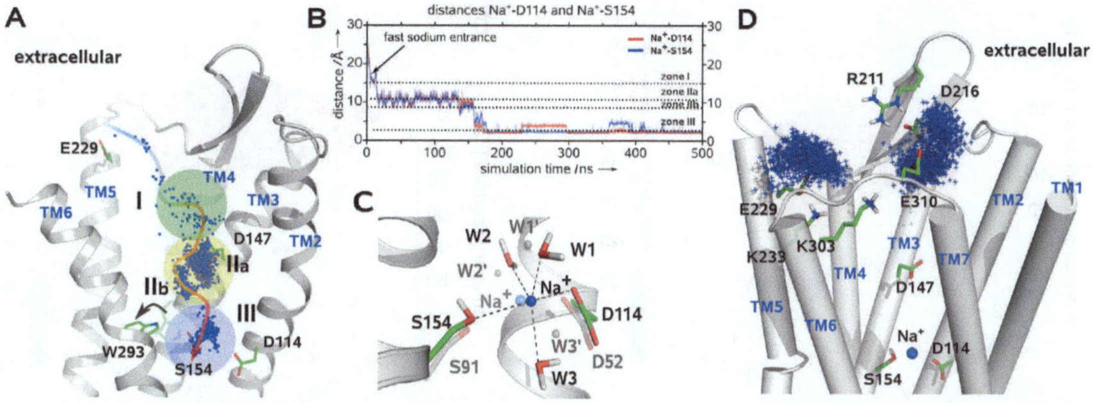

Fig. 13 Sodium ion entrance pathway. (**a**) Cross-section of μOR showing a pathway of a particular sodium ion during the initial 200 ns of Apo μOR simulation. *Blue* dots are consecutive positions of the sodium ion along its pathway. An arrow indicates a rotamer switch of W293$^{6.48}$. (**b**) Distances Na$^+$-D114$^{2.50}$ (*blue*) and Na$^+$-S154$^{3.39}$ (*red*) during two separate 500 ns simulations. (**c**) Superimposed crystal structure of A_{2A}R (*light colors*) and final Apo μOR MD structure (*bold colors*). (**d**) Positions of second sodium ion (*blue crosses*) during simulation of unliganded receptor [53]

programs. For the receptor-membrane system building one can use a convenient g_membed tool with receptor crystal structure prealigned in the OPM (Orientations of Proteins in Membranes) database. For GPCR monomers a standard periodic box of 7 nm × 7 nm × 10 nm is usually employed. Number of atoms in typical MD simulation of single GPCR in the membrane and water system is about 50,000–70,000. Simulations of empty receptors as well as complexes with ligands are usually conducted in typical phospholipid bilayer composed of POPC (1-palmitoyl-2-oleoyl-phosphatidylcholine). Simulations with antagonists are conducted to stabilize ligands in the binding site and to show that there is no action of molecular switches. A *simulated annealing* method (continual heating and cooling) is used for the precise determination of ligand binding modes after the procedure of automatic docking. All the simulations should be repeated several times (depending on the system) because activation events are statistical and rather rare so they do not have to happen in a particular simulation. A whole process of GPCR activation is completed on a millisecond time scale; however, using 1–10 μs MD simulations one can see action of particular switches and movements of helices associated with initial activation steps of GPCRs.

5 Usage of Metadynamics Simulations and Free-Energy Profiles

5.1 Probing Active, Inactive, and Meta-State Conformations

For adenosine A_{2A} receptor it was possible to find that a hydrophobic layer of amino acid residues next to the characteristic NPxxY motif forms a gate that opens to form a continuous water channel

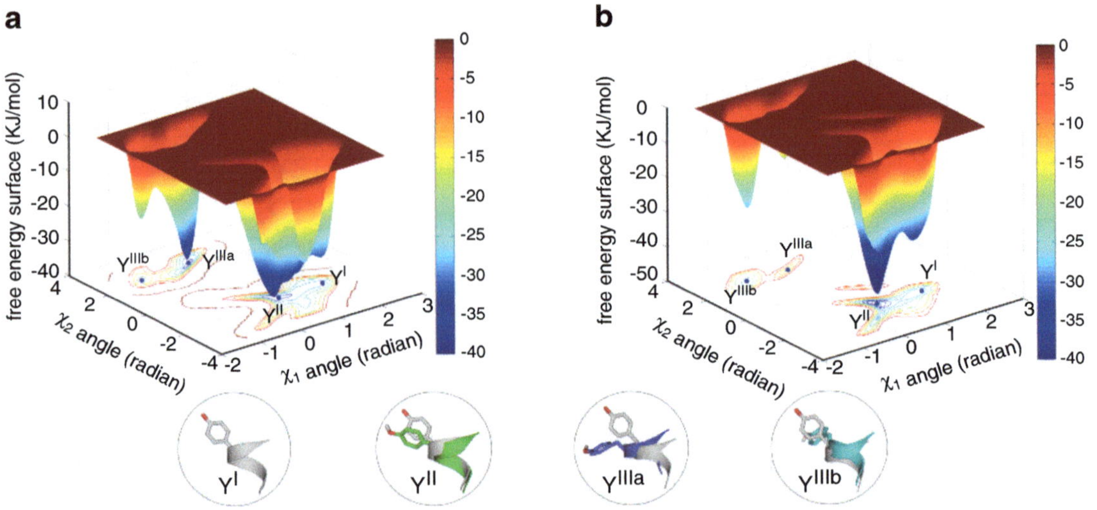

Fig. 14 Three distinct rotamer states of Y288$^{7.53}$ at the NPxxY motif of adenosine A_{2A} receptor. (**a**) Free-energy surface of the agonist NECA bound to the $A_{2A}R$ in absence of Gα. (**b**) Free-energy surface of agonist NECA bound to the $A_{2A}R$ in presence of Gα at the cytoplasmic site [60]

only upon receptor activation. The highly conserved tyrosine residue Y288$^{7.53}$ undergoes transitions between three distinct conformations (Fig. 14) representing inactive (Y^{I}), G protein activated (Y^{II}), and GPCR metastates (Y^{IIIa} and Y^{IIIb}), respectively [60]. Additional analyses of the available GPCR crystal structures, including rhodopsin and β_2AR, revealed general principles governing the functional roles of internal waters in GPCRs. On the basis of a total of 32 μs all-atom MD simulations for three different receptors of family A of GPCRs, it was found that water from the bulk can enter into the receptor during activation from two different directions depending on the receptor type: from the intracellular side (in case of rhodopsin) and from the extracellular side (A_{2A}R and β_2AR).

Free-energy surfaces (free-energy profiles) for GPCR activation by agonists can be generated employing well-tempered metadynamics in GROMACS with Plumed patches [61]. Metadynamics adds a history-dependent potential V(s,t) to accelerate sampling of the specific collective variables (CV's) [62]. V(s,t) is usually constructed as the sum of multiple Gaussians centered along the trajectory of the CV's. Periodically during the simulation, another Gaussian potential, whose location is dictated by the current values of the CVs, is added to V(s,t) [63]. For instance, in the simulations of adenosine A_{2A} receptor the dihedral angles of residue Y288$^{7.53}$, χ_1 and χ_2, were assigned as the CV's s_1 and s_2 [60]. It was also applied to 5-HT3 ion channel to study opening of the hydrophobic gate upon binding of agonist serotonin [64].

5.2 Modulation of the Free-Energy Landscape of the Receptor by the Ligand

Taking advantage of the recently published inactive and active crystal structures of GPCRs, Provasi et al. [65] developed a computational strategy that enabled the identification of the specific conformations taken by β_2AR when interacting with ligands that elicit different physiological responses. This methodology can be also used for virtual screening, and possibly lead to the structure-based rational discovery of novel "biased" ligands that are capable of selectively activating one cellular signaling pathway over another. They showed that ligands with different efficacies (either inverse agonists, neutral antagonists, or agonists) modulate the free-energy landscape of the receptor by shifting the conformational equilibrium toward active or inactive conformations. Using metadynamics simulations they estimated the free-energy surface of the complexes as a function of three important descriptors of receptor activation, namely the distance between R3.50 and E6.30 (the "ionic lock"), the rotamer of residue W6.48 (the so-called toggle switch), and the outward displacement of the intracellular segment of TM6. Specifically, the receptor was studied in its unliganded form as well as in complex with the full agonist epinephrine, the weak partial agonist dopamine, the very weak partial agonist catechol, the inverse agonist ICI-118-551, the inverse agonist carazolol, and the neutral antagonist alprenolol. The ligands with varied efficacies are believed to modulate the free-energy landscape of a GPCR shifting the conformational equilibrium toward active or inactive conformations of the receptor, depending on their pharmacological action.

In case of agonist, epinephrine, it was found that full agonist was capable of stabilizing a state of β_2AR presenting structural features that have been found in the nanobody-stabilized agonist-bound crystal structure of β_2AR. They also obtained a relatively stable agonist-bound inactive state that was structurally similar to the inverse agonist-bound crystal structure of β_2AR. This is in line with the absence of outward location of TM6 noted in both the β_2AR crystal structure with a covalently-bound agonist [66], and the agonist-bound β_1AR crystal structures. The obtained relatively small difference in free energy between the fully active and the inactive agonist-bound conformations was probably due to the lack of the G-protein in the simulations because a ligand alone was not sufficient to stabilize a fully active state of the receptor.

6 Other Methods Based on Receptor Structure

6.1 Usage of Parameters of the Ligand Binding Site (SASA)

To investigate why certain molecules act as activators whereas others, with similar structures, block GPCR activation, the levallorphan and other 69 agonists and antagonists of opioid receptors κOR and μOR were employed [67]. Using all-atom MD simulations it was found that levallorphan behaved as an agonist for κOR by inducing unstable binding with D3.32 and subsequent water

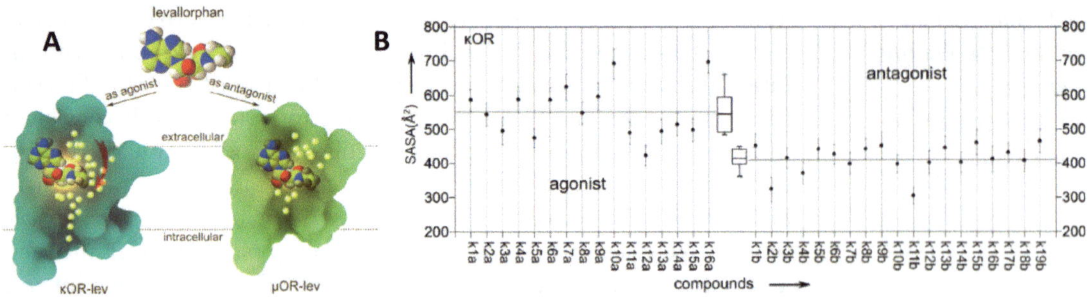

Fig. 15 Binding of ligands to opioid receptors. (**a**) Levallorphan in different receptors. (**b**) Solvent accessible surface areas (SASA) values for agonist-bound κOR (*left panel*) and antagonist-bound κOR (*right panel*). Error bars represent standard deviations obtained from statistical evaluation of 200 snapshots extracted from the final parts of the MD simulations [67]

penetration owing to the large SASA (solvent accessible surface area) of the κOR-levallorphan complex. By contrast, levallorphan was stabilized in the binding site of μOR by residues D3.32 and Y7.43 (Fig. 15a). It was also found that water molecules penetrating into the receptor interior mediate the activating versus blocking effects in a particular ligand-receptor complex. Both the size and the flexibility of the bound ligand regulated water influx into the receptor. The SASA values of the binding site were found to be a parameter that can help predict the function (agonist/antagonist) of the bound ligand (Fig. 15b). Similar dependences were found for both κOR and μOR.

6.2 Analysis of a Volume of G Protein Binding Site

Another method requiring docking and MD simulations to distinguish agonists and antagonists could be the analysis of a volume of G protein binding site [26]. The analysis of intracellular pockets of twelve studied ligand-receptor complexes coming from β_2AR crystal structures after conducting 100 ns MD simulations revealed that there is a clear difference in agonist-bound receptor structures compared to the rest (Fig. 16)–3PDS structure that includes the whole trimeric G protein was excluded from these calculations. Since rather long MD simulations are required to obtain the visible movements of TM helices, this method is of little usefulness; however, the obtained volume-ligand function correlations indicate feasibility of using fingerprints or docking score or combined for the ligand function prediction since the ligand-receptor interactions are allosterically transmitted to the G protein binding site and change its volume.

6.3 Principal Component Analysis of the Ligand-Receptor Vibrational Modes

This method was used for visualization of helix movements for agonists, antagonists and inverse agonists bound to β_2AR [26]. Fig. 17 shows the lowest frequency mode calculated for the ligands using the alpha carbons. The helix movements in the intracellular region are more consistent than in the extracellular region.

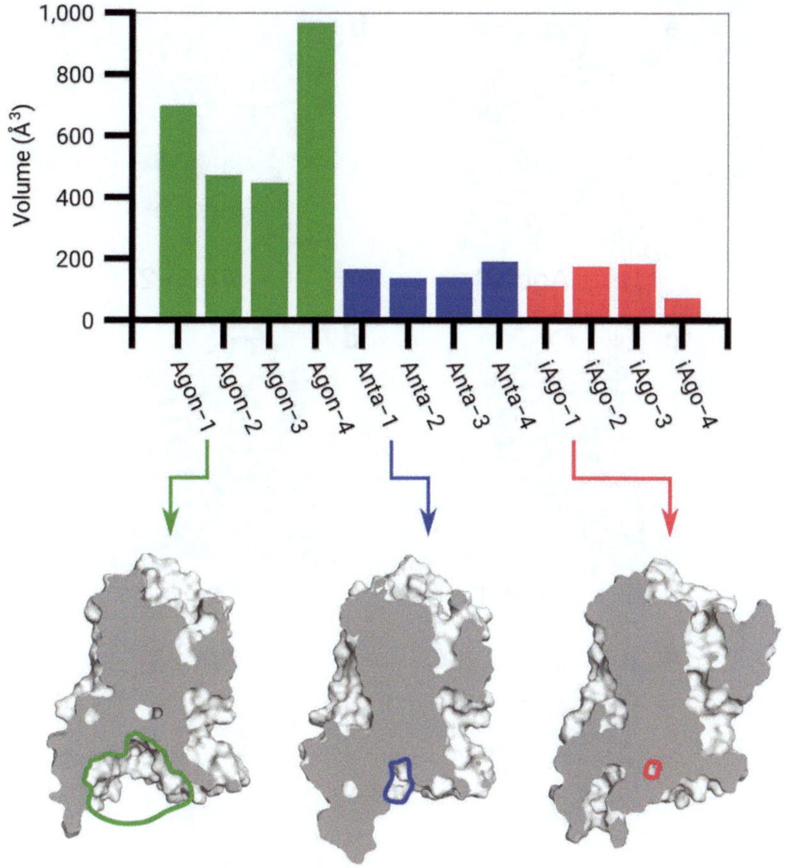

Fig. 16 The volumes of the intracellular pockets in the twelve β_2AR studied systems and the cross sections of selected complexes [26]

In agonist bound systems, TM6 and TM7 move away from each other and thus a large void space is created. In antagonist and inverse agonist bound systems, the movement of TM6 is more diverse and consequently the helix keeps the intracellular pocket closed. The principal component analysis for the final frames of MD simulations was calculated in VMD [68].

7 Methods Based on Ligand Properties

7.1 Simple QSAR Methods

For the particular receptor types it was possible to discriminate agonists and non-agonists based on ligand properties using even simple parameters: the molecular weight (MW), calculated logarithm of octanol/water partition coefficient (clogP), molar refraction, dipole moment, E_{LUMO} (the energy of the lowest unoccupied molecular orbital, a measure of the electron affinity of a molecule and its reactivity as an electrophile), E_{HOMO} (the energy of the highest occupied molecular orbital, related to the ionization

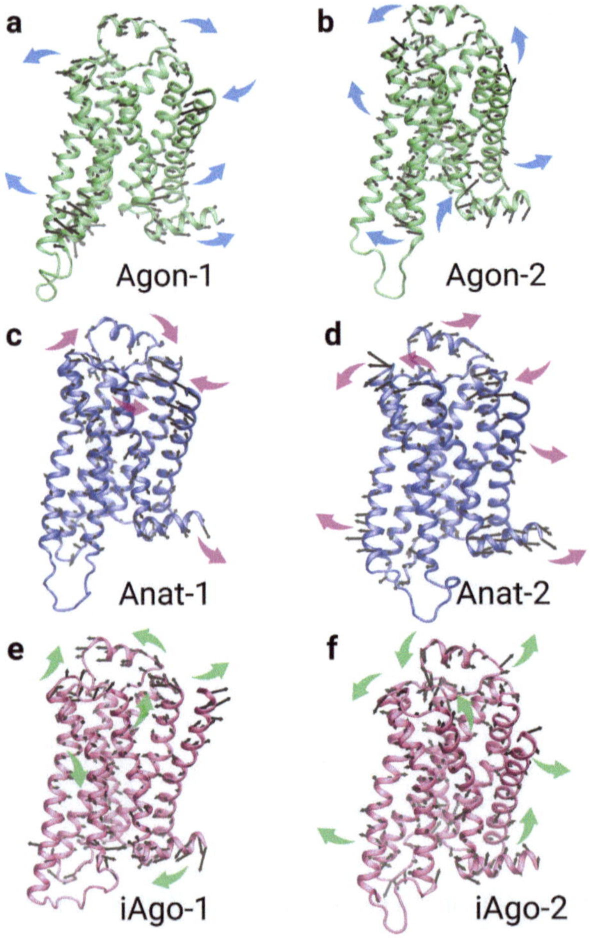

Fig. 17 Principal component analysis of the vibrational modes in the presence of agonists, antagonists, and inverse agonists of β_2AR [26]

potential of a molecule, and its reactivity as a nucleophile), and the total number of hydrogen bonds (donors and acceptors) [69]. Such molecular descriptors (among others) were chosen for Quantitative Structure-Activity Relationships (QSAR), sometimes combined with machine learning methods; however, such analyses were done for ligands of single receptors and no VS of ligand databases were done in search for novel ligands with particular function.

7.2 Classification of Ligands Using QSAR and Machine Learning

The problem with the classification methods based on ligands alone is that the methods are not associated with the receptor, because they do not include influence of the environment. Some compounds can play different roles depending on the type of receptor they bind to, and in such cases the ligand-based classification methods that are not taking into account the environment (the receptor binding site) will fail. For example, flibanserin is a full agonist of serotonin 5-HT1A receptor and, with lower affinity, is an antagonist of 5-HT2A receptor as well as an antagonist or very weak partial agonist of dopamine D4 receptor. Therefore, lack of provisions for the ligand environment can, in some cases, lead to misleading classification of the ligand as an agonist or antagonist. It should be noted, however, that if the crystal structure of the receptor is not known, the ligand-based differentiation methods may contribute to the initial classification of the ligand.

More reliable methods employ the impact of the environment on the ligand. In this case, due to the rapid growth of number of parameters that describe the system one can use QSAR and 3D-QSAR methods, which from the sea of parameters allow drawing constructive conclusions. These methods are often combined with neural networks and other machine learning algorithms. To construct a computational model for the classification of agonists and antagonists of serotonin 5-HT1A receptor, Zhu et al. [70] used the support vector machine (SVM), a machine learning method, to build a prediction model, while the genetic algorithm (GA) was used to select the most relevant descriptors (among 292 molecular descriptors including topological, graph-theoretical, quantum-chemical, and electro-topological) and to optimize two important parameters, C and r of the SVM model. The overall dataset used in this study comprised 284 ligands of the 5-HT1A receptor with diverse structures reported in the literature. The SVM model successfully classified ligands being agonists and antagonists of 5-HT1A receptor with the predictive accuracy for training (207 ligands) and test (52 ligands) sets 0.942 and 0.865, respectively.

7.3 Usage of Vibrational Frequency Calculations for Ligands

The set of 47 ligands of the histamine receptors H_1-H_4 was analyzed [71] by structural similarity and molecular vibrational frequency patterns using by the quantum calculations method, namely density functional theory (DFT) for geometry optimization, and vibrational frequency calculations in the GAMESS program. Then, the radial tree was produced by clustering analysis of molecular vibrational frequency patterns. The "corralled" intensity of molecular vibrational frequency (CIMVF) allowed creating a hierarchical clustering of all ligands where eight agonists were located close together, except impromidine, and all antagonists were clustered close to each other in the radial tree. The same method was used later for ligands of adenosine receptors A_1R, $A_{2A}R$, $A_{2B}R$, and A_3R [72]. The molecular vibrational frequency may play a role in the classification of agonists/antagonists for GPCR ligands as a possible molecular descriptor. Employing a larger set of adenosine receptor ligands they performed molecular vibration calculations followed by clustering and they employed a novel classification method based on information gain (IG) function and machine learning [73]. The IG measures the amount of information (relative entropy) about the class prediction in bits, if the only

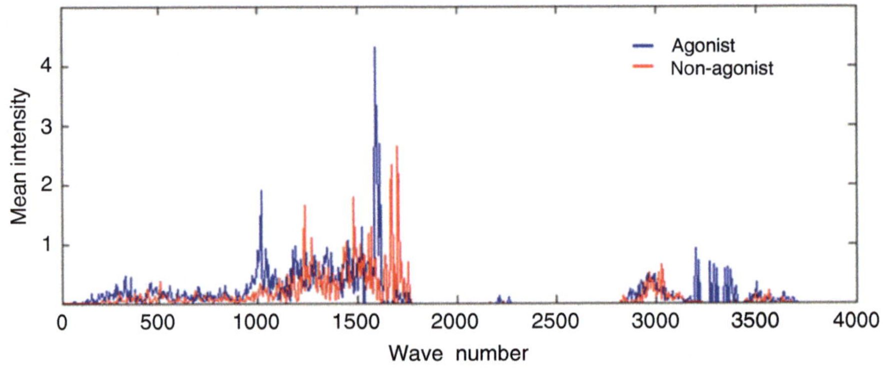

Fig. 18 The mean intensities of the two corralled intensities of molecular vibrational frequency (CIMVFs) of adenosine receptor agonists (*blue*) and non-agonists (*red*) according to the wave number of the molecular vibration [73]

information available is the presence of a feature and the corresponding class distribution (Eq. 2).

$$IG(S_x, x_i) = H(S_x) - \sum_{v=\text{values}(x_i)} \frac{|S_{x_i=v}|}{|S_x|} \cdot H(S_{x_i=v}) \quad (2)$$

where H is the entropy function, Sx is the set of training examples, x_i is the vector of the ith variable in the set, and $|Sx_i = v|/|Sx|$ is the fraction of examples of the ith variable having value v. They applied IG-based feature selection to identify the corrals of molecular vibrational frequency that were the most informative among the 800 elements for binary classification of adenosine receptor ligands as agonists or non-agonists. They trained and tested the procedure by applying leave-one-out cross-validation to each ligand. The calculation of IG was performed using the Weka machine learning package [74]. The calculated vibrational spectra in this study did not simulate actual IR or Raman spectra, but their patterns were consistent for the molecular properties of each ligand (Fig. 18). Selecting subsets of highly informative features from different spectra would be beneficial and useful for developing and presenting new approaches for ligand design and contribute to ligand and drug discovery.

Acknowledgments

Figure 3 is reproduced from *J. Chem. Inf. Model.* 2015 (http://pubs.acs.org/doi/full/10.1021/acs.jcim.5b00066) with permission from American Chemical Society. Figure 9 is reproduced from *ACS Chem. Biol.* 2013 (http://pubs.acs.org/doi/full/10.1021/cb400103f) with permission from American Chemical Society. Figure 18 is reproduced from *FEBS Lett.* 2015 with permission from John Wiley and Sons.

References

1. Pierce KL, Premont RT, Lefkowitz RJ (2002) Seven-transmembrane receptors. Nat Rev Mol Cell Biol 3:639–650
2. Moreno JL, Holloway T, Gonzalez-Maeso J (2013) G protein-coupled receptor heterocomplexes in neuropsychiatric disorders. Prog Mol Biol Transl Sci 117:187–205
3. O'Hayre M, Degese MS, Gutkind JS (2014) Novel insights into G protein and G protein-coupled receptor signaling in cancer. Curr Opin Cell Biol 27:126–135
4. Sodhi A, Montaner S, Gutkind JS (2004) Viral hijacking of G-protein-coupled-receptor signalling networks. Nat Rev Mol Cell Biol 5:998–1012
5. Lundstrom K (2006) Latest development in drug discovery on G protein-coupled receptors. Curr Protein Pept Sci 7:465–470
6. Schioth HB, Fredriksson R (2005) The GRAFS classification system of G-protein coupled receptors in comparative perspective. Gen Comp Endocrinol 142:94–101
7. Katritch V, Cherezov V, Stevens RC (2011) Diversity and modularity of G protein-coupled receptor structures. Trends Pharmacol Sci 33 (1):17–27
8. Kenakin T, Miller LJ (2010) Seven transmembrane receptors as shapeshifting proteins: the impact of allosteric modulation and functional selectivity on new drug discovery. Pharmacol Rev 62:265–304
9. Trzaskowski B, Latek D, Yuan S, Ghoshdastider U, Debinski A, Filipek S (2012) Action of molecular switches in GPCRs–theoretical and experimental studies. Curr Med Chem 19:1090–1109
10. Vass M, Kooistra AJ, Ritschel T, Leurs R, de Esch IJ, de Graaf C (2016) Molecular interaction fingerprint approaches for GPCR drug discovery. Curr Opin Pharmacol 30:59–68
11. Cereto-Massague A, Ojeda MJ, Valls C, Mulero M, Garcia-Vallve S, Pujadas G (2015) Molecular fingerprint similarity search in virtual screening. Methods 71:58–63
12. Deng Z, Chuaqui C, Singh J (2004) Structural interaction fingerprint (SIFt): a novel method for analyzing three-dimensional protein-ligand binding interactions. J Med Chem 47:337–344
13. Marcou G, Rognan D (2007) Optimizing fragment and scaffold docking by use of molecular interaction fingerprints. J Chem Inf Model 47:195–207
14. Desaphy J, Raimbaud E, Ducrot P, Rognan D (2013) Encoding protein-ligand interaction patterns in fingerprints and graphs. J Chem Inf Model 53:623–637
15. Da C, Kireev D (2014) Structural protein-ligand interaction fingerprints (SPLIF) for structure-based virtual screening: method and benchmark study. J Chem Inf Model 54:2555–2561
16. Baroni M, Cruciani G, Sciabola S, Perruccio F, Mason JS (2007) A common reference framework for analyzing/comparing proteins and ligands. Fingerprints for ligands and proteins (FLAP): theory and application. J Chem Inf Model 47:279–294
17. Wood DJ, de Vlieg J, Wagener M, Ritschel T (2012) Pharmacophore fingerprint-based approach to binding site subpocket similarity and its application to bioisostere replacement. J Chem Inf Model 52:2031–2043
18. Lavecchia A (2015) Machine-learning approaches in drug discovery: methods and applications. Drug Discov Today 20:318–331
19. Mordalski S, Kosciolek T, Kristiansen K, Sylte I, Bojarski AJ (2011) Protein binding site analysis by means of structural interaction fingerprint patterns. Bioorg Med Chem Lett 21:6816–6819
20. Cao R, Wang Y (2016) Predicting molecular targets for small-molecule drugs with a ligand-based interaction fingerprint approach. ChemMedChem 11:1352–1361
21. Kooistra AJ, Leurs R, de Esch IJ, de Graaf C (2015) Structure-based prediction of G-protein-coupled receptor ligand function: a beta-adrenoceptor case study. J Chem Inf Model 55:1045–1061
22. Kooistra AJ, Vischer HF, McNaught-Flores D, Leurs R, de Esch IJ, de Graaf C (2016) Function-specific virtual screening for GPCR ligands using a combined scoring method. Sci Rep 6:28288
23. Yuan S, Peng Q, Palczewski K, Vogel H, Filipek S (2016) Mechanistic studies on the stereoselectivity of the serotonin 5-HT1A receptor. Angew Chem Int Ed 55:8661–8665
24. Latek D, Pasznik P, Carlomagno T, Filipek S (2013) Towards improved quality of GPCR models by usage of multiple templates and profile-profile comparison. PLoS One 8: e56742
25. Vroling B, Sanders M, Baakman C, Borrmann A, Verhoeven S, Klomp J, Oliveira L, de Vlieg J, Vriend G (2011) GPCRDB: information system for G protein-coupled receptors. Nucleic Acids Res 39: D309–D319

26. Chan HC, Filipek S, Yuan S (2016) The principles of ligand specificity on beta-2-adrenergic receptor. Sci Rep 6:34736
27. Ballesteros JA, Weinstein H (1995) Integrated methods for the construction of three-dimensional models and computational probing of structure-function relations in G protein-coupled receptors. Methods Neurosci 25:366–428
28. Rodriguez D, Ranganathan A, Carlsson J (2015) Discovery of GPCR ligands by molecular docking screening: novel opportunities provided by crystal structures. Curr Top Med Chem 15:2484–2503
29. Gutierrez-de-Teran H, Sallander J, Sotelo E (2017) Structure-based rational design of adenosine receptor ligands. Curr Top Med Chem 17:40–58
30. Kolb P, Rosenbaum DM, Irwin JJ, Fung JJ, Kobilka BK, Shoichet BK (2009) Structure-based discovery of beta2-adrenergic receptor ligands. Proc Natl Acad Sci U S A 106:6843–6848
31. Bissantz C, Bernard P, Hibert M, Rognan D (2003) Protein-based virtual screening of chemical databases. II. Are homology models of G-protein coupled receptors suitable targets? Proteins 50:5–25
32. Korb O, Stutzle T, Exner TE (2009) Empirical scoring functions for advanced protein-ligand docking with PLANTS. J Chem Inf Model 49:84–96
33. de Graaf C, Kooistra AJ, Vischer HF, Katritch V, Kuijer M, Shiroishi M, Iwata S, Shimamura T, Stevens RC, de Esch IJ, Leurs R (2011) Crystal structure-based virtual screening for fragment-like ligands of the human histamine H(1) receptor. J Med Chem 54:8195–8206
34. O'Boyle NM, Liebeschuetz JW, Cole JC (2009) Testing assumptions and hypotheses for rescoring success in protein-ligand docking. J Chem Inf Model 49:1871–1878
35. Shimamura T, Shiroishi M, Weyand S, Tsujimoto H, Winter G, Katritch V, Abagyan R, Cherezov V, Liu W, Han GW, Kobayashi T, Stevens RC, Iwata S (2011) Structure of the human histamine H1 receptor complex with doxepin. Nature 475:65–70
36. Rogers D, Hahn M (2010) Extended-connectivity fingerprints. J Chem Inf Model 50:742–754
37. Weiss DR, Ahn S, Sassano MF, Kleist A, Zhu X, Strachan R, Roth BL, Lefkowitz RJ, Shoichet BK (2013) Conformation guides molecular efficacy in docking screens of activated beta-2 adrenergic G protein coupled receptor. ACS Chem Biol 8:1018–1026
38. Liapakis G, Ballesteros JA, Papachristou S, Chan WC, Chen X, Javitch JA (2000) The forgotten serine. A critical role for Ser-2035.42 in ligand binding to and activation of the beta 2-adrenergic receptor. J Biol Chem 275:37779–37788
39. Huang N, Shoichet BK, Irwin JJ (2006) Benchmarking sets for molecular docking. J Med Chem 49:6789–6801
40. Mysinger MM, Shoichet BK (2010) Rapid context-dependent ligand desolvation in molecular docking. J Chem Inf Model 50:1561–1573
41. Costanzi S, Vilar S (2012) In silico screening for agonists and blockers of the beta(2) adrenergic receptor: implications of inactive and activated state structures. J Comput Chem 33:561–572
42. Jazayeri A, Andrews SP, Marshall FH (2017) Structurally enabled discovery of adenosine A2A receptor antagonists. Chem Rev 117:21–37
43. Liu W, Chun E, Thompson AA, Chubukov P, Xu F, Katritch V, Han GW, Roth CB, Heitman LH, AP IJ, Cherezov V, Stevens RC (2012) Structural basis for allosteric regulation of GPCRs by sodium ions. Science 337:232–236
44. Bortolato A, Tehan BG, Bodnarchuk MS, Essex JW, Mason JS (2013) Water network perturbation in ligand binding: adenosine a (2A) antagonists as a case study. J Chem Inf Model 53:1700–1713
45. Lenselink EB, Beuming T, Sherman W, van Vlijmen HW, AP IJ (2014) Selecting an optimal number of binding site waters to improve virtual screening enrichments against the adenosine A2A receptor. J Chem Inf Model 54:1737–1746
46. Friesner RA, Banks JL, Murphy RB, Halgren TA, Klicic JJ, Mainz DT, Repasky MP, Knoll EH, Shelley M, Perry JK, Shaw DE, Francis P, Shenkin PS (2004) Glide: a new approach for rapid, accurate docking and scoring. 1. Method and assessment of docking accuracy. J Med Chem 47:1739–1749
47. Yuan S, Hu Z, Filipek S, Vogel H (2015) W246 (6.48) opens a gate for a continuous intrinsic water pathway during activation of the adenosine A2A receptor. Angew Chem Int Ed 54:556–559
48. Skjaerven L, Yao XQ, Scarabelli G, Grant BJ (2014) Integrating protein structural dynamics and evolutionary analysis with Bio3D. BMC Bioinformatics 15:399

49. Grant BJ, Rodrigues AP, ElSawy KM, McCammon JA, Caves LS (2006) Bio3d: an R package for the comparative analysis of protein structures. Bioinformatics 22:2695–2696
50. Van Wart AT, Durrant J, Votapka L, Amaro RE (2014) Weighted implementation of suboptimal paths (WISP): an optimized algorithm and tool for dynamical network analysis. J Chem Theory Comput 10:511–517
51. Lee Y, Choi S, Hyeon C (2015) Communication over the network of binary switches regulates the activation of A2A adenosine receptor. PLoS Comput Biol 11:e1004044
52. Manglik A, Kruse AC, Kobilka TS, Thian FS, Mathiesen JM, Sunahara RK, Pardo L, Weis WI, Kobilka BK, Granier S (2012) Crystal structure of the micro-opioid receptor bound to a morphinan antagonist. Nature 485:321–326
53. Yuan S, Vogel H, Filipek S (2013) The role of water and sodium ions in the activation of the mu-opioid receptor. Angew Chem Int Ed 52:10112–10115
54. Phillips JC, Braun R, Wang W, Gumbart J, Tajkhorshid E, Villa E, Chipot C, Skeel RD, Kale L, Schulten K (2005) Scalable molecular dynamics with NAMD. J Comput Chem 26:1781–1802
55. Pronk S, Pall S, Schulz R, Larsson P, Bjelkmar P, Apostolov R, Shirts MR, Smith JC, Kasson PM, van der Spoel D, Hess B, Lindahl E (2013) GROMACS 4.5: a high-throughput and highly parallel open source molecular simulation toolkit. Bioinformatics 29:845–854
56. Salomon-Ferrer R, Gotz AW, Poole D, Le Grand S, Walker RC (2013) Routine microsecond molecular dynamics simulations with AMBER on GPUs. 2. Explicit solvent particle mesh ewald. J Chem Theory Comput 9:3878–3888
57. Harvey MJ, Giupponi G, Fabritiis GD (2009) ACEMD: accelerating biomolecular dynamics in the microsecond time scale. J Chem Theory Comput 5:1632–1639
58. Bergdorf M, Baxter S, Rendleman CA, Shaw DE (2016) Desmond/GPU performance as of November 2016, D. E. Shaw Research Technical Report DESRES/TR
59. Klauda JB, Venable RM, Freites JA, O'Connor JW, Tobias DJ, Mondragon-Ramirez C, Vorobyov I, AD MK Jr, Pastor RW (2010) Update of the CHARMM all-atom additive force field for lipids: validation on six lipid types. J Phys Chem B 114:7830–7843
60. Yuan S, Filipek S, Palczewski K, Vogel H (2014) Activation of G-protein-coupled receptors correlates with the formation of a continuous internal water pathway. Nat Commun 5:4733
61. Bonomi M, Branduardi D, Bussi G, Camilloni C, Provasi D, Raiteri P, Donadio D, Marinelli F, Pietrucci F, Broglia RA, Parrinello M (2009) PLUMED: a portable plugin for free-energy calculations with molecular dynamics. Comput Phys Commun 180:1961–1972
62. Barducci A, Bussi G, Parrinello M (2008) Well-tempered metadynamics: a smoothly converging and tunable free-energy method. Phys Rev Lett 100:020603
63. Li JN, Jonsson AL, Beuming T, Shelley JC, Voth GA (2013) Ligand-dependent activation and deactivation of the human adenosine A (2A) receptor. J Am Chem Soc 135:8749–8759
64. Yuan S, Filipek S, Vogel H (2016) A gating mechanism of the serotonin 5-HT3 receptor. Structure 24:816–825
65. Provasi D, Artacho MC, Negri A, Mobarec JC, Filizola M (2011) Ligand-induced modulation of the free-energy landscape of G protein-coupled receptors explored by adaptive biasing techniques. PLoS Comput Biol 7:e1002193
66. Rosenbaum DM, Zhang C, Lyons JA, Holl R, Aragao D, Arlow DH, Rasmussen SGF, Choi H-J, Devree BT, Sunahara RK, Chae PS, Gellman SH, Dror RO, Shaw DE, Weis WI, Caffrey M, Gmeiner P, Kobilka BK (2011) Structure and function of an irreversible agonist-beta(2) adrenoceptor complex. Nature 469:236–240
67. Yuan S, Palczewski K, Peng Q, Kolinski M, Vogel H, Filipek S (2015) The mechanism of ligand-induced activation or inhibition of mu- and kappa-opioid receptors. Angew Chem Int Ed 54:7560–7563
68. Humphrey W, Dalke A, Schulten K (1996) VMD: visual molecular dynamics. J Mol Graph Model 14:33–38
69. Kuo CL, Wang RB, Shen LJ, Lien LL, Lien EJ (2004) G-protein coupled receptors: SAR analyses of neurotransmitters and antagonists. J Clin Pharm Ther 29:279–298
70. Zhu XL, Cai HY, Xu ZJ, Wang Y, Wang HY, Zhang A, Zhu WL (2011) Classification of 5-HT(1A) receptor agonists and antagonists using GA-SVM method. Acta Pharmacol Sin 32:1424–1430
71. Oh SJ (2012) Characteristics in molecular vibrational frequency patterns between agonists and antagonists of histamine receptors. Genomics Inform 10:128–132

72. Chee HK, Oh SJ (2013) Molecular vibration-activity relationship in the agonism of adenosine receptors. Genomics & informatics 11:282–288
73. Chee HK, Yang JS, Joung JG, Zhang BT, Oh SJ (2015) Characteristic molecular vibrations of adenosine receptor ligands. FEBS Lett 589:548–552
74. Frank E, Hall M, Trigg L, Holmes G, Witten IH (2004) Data mining in bioinformatics using Weka. Bioinformatics 20:2479–2481

Chapter 13

Opportunities and Challenges in the Discovery of Allosteric Modulators of GPCRs

Damian Bartuzi, Agnieszka A. Kaczor, and Dariusz Matosiuk

Abstract

From the pharmacological point of view, allosteric modulators may present numerous advantages over orthosteric ligands. Growing availability of novel tools and experimental data provides a tempting opportunity to apply computational methods to improve known modulators and design novel ones. However, recent progress in understanding of complexity of allostery increases awareness of problems involved in design of modulators with desired properties. Deeper insight into phenomena such as probe dependence, altering signaling bias with minor changes in ligand structure, as well as influence of subtle endogenous allosteric factors turns out to be fundamental. These effects make the design of a modulator with precise pharmacological outcome a very challenging task, and need to be taken into consideration throughout the design process. In this chapter, we focus on nuances of targeting GPCR allosteric sites in computational drug design efforts, in particular with application of docking, virtual screening, and molecular dynamics.

Key words Allostery, Allosteric modulation, GPCR, Molecular dynamics, Molecular docking, Virtual screening, Probe dependence, Signal transduction, Structure-based drug design, Protein dynamics

1 Introduction

In recent years, one can observe an enormous advance in understanding of protein structure and function. Elucidation of mechanisms governing these macromolecules results in novel strategies of the design of pharmacologically active ligands. There are new trends, including development of biased compounds, multi-target drugs, compounds affecting dimerization interfaces, or allosteric modulators. The latter seem to be a very promising option, with great practical potential. Although very useful as pharmacological probes, the most tempting application of allosteric modulators would be their medicinal use. Allosteric modulators offer several advantages over more traditional drugs, e.g., spatial and temporal selectivity or the ceiling effect [1]. However, similarly as in the case of other medicines, their application encounters a number of problems, and the more sophisticated the pharmacological agent is, the

Alexander Heifetz (ed.), *Computational Methods for GPCR Drug Discovery*, Methods in Molecular Biology, vol. 1705,
https://doi.org/10.1007/978-1-4939-7465-8_13,

more traps for its safe and efficient use in humans are set. On one hand, Cinacalcet [2], Maraviroc [3], and Plerixafor [4] serve as an example of successful application of such strategy into therapy. On the other hand, an example of LY2033298 underlines the complexity of probe dependence and subtype specificity issues [5], while an example of Oliceridine (TRV130), a biased ligand of the μ opioid receptors [6], clearly shows that some sophisticated mechanisms, like functional selectivity, are strongly dependent on very subtle influences of number of elusive factors, which manifested itself as pronounced species-dependent differences in pharmacology of that agonist [7], and due to similar nature of allostery and biased signaling, analogical problems can appear in design of allosteric modulators. Analysis of these cases could suggest that the only way to prevent such complications would be performing studies on the target species in vivo, which is obviously impossible. However, the present in silico techniques, together with growing amount of experimental data, allow for satisfactory reconstruction of native conditions to improve quality of computer-aided design and investigation of allosteric modulators [8–11]. In this chapter, we are going to highlight some key concepts crucial in planning in silico studies on allosteric modulation of G protein-coupled receptors (GPCRs). Although we focus on this particular important family of receptors, a number of issues apply to other classes of proteins as well.

2 On Interplay of Allosteric Factors

Allosteric modulation has found itself in the spotlight of the medicinal chemistry, which resulted in a large number of reports in the field. In the present literature, one can often find particular modulators being described as "positive modulator of an X receptor" or "negative modulator of an Y receptor". Although such statements are intuitively understood, it should be stressed that such nomenclature is just an abbreviation, and without appropriate description it can be misleading [12]. Modulators cannot be simply described as positive, negative or neutral allosteric factors [1]. This is because the resultant activity of the entire complex depends on all of its constituents—a receptor, its orthosteric ligand, an allosteric modulator and an intracellular coupling partner—and on a subtle interplay between all of them. Binding of a modulator can amplify a receptor's response to an agonist, which would mean that there is a positive cooperativity between them. However, the same modulator can affect another agonist in a completely different way, e.g., it can decrease a regular receptor's response, or simply not affect the other agonist's action at all. In such a situation, the modulator would exert negative or neutral allosteric cooperativity, respectively. The effect of the different influence of a modulator on various orthosteric ligands is known as probe dependence. It is clearly visible on the striking example of the LY2033298 compound on the M2 cholinergic receptor. The modulator does not affect affinity nor efficacy of the acetylcholine, which is endogenous, native ligand of the receptor. However, it positively modulates a metabolite of the acetylcholine—choline, to the considerable extent [5]. This clearly demonstrates that one has to be more specific when describing allosteric effects of a ligand. Moreover, some studies indicate that effects of some allosteric modulators can at least partially result not only from their influence on the orthosteric ligand, but indirectly, from their interplay with endogenous allosteric modulators. A compound named BMS986122, which is an allosteric modulator of μ opioid receptor, can serve as an example—its effect on morphine seems to result from its allosteric interplay with sodium ion bound at Asp 2.50 (Ballesteros-Weinstein notation [13]) [8, 14]. In such a case, a negative cooperativity between two allosteric modulators is partially responsible for a modulatory effect on an orthosteric ligand. Therefore, it can be concluded that properties of a modulator strongly depend on a context, e.g., features of a modulated orthosteric ligand or presence of other allosteric factors.

Underlying mechanisms of cooperativity can be understood intuitively (Fig. 1). An allosteric factor affects the structure of a receptor, biasing the conformational space explored by the protein to a new equilibrium. The structure affected by the modulator has altered propensity to bind an orthosteric ligand and to undergo further changes. On the other hand, the same applies to the orthosteric agent, in the inverse direction—if the agonist induces altered conformation of a receptor, this new conformation has altered affinity to the modulator. The influences are reciprocal—if binding of one ligand induces a conformation favorable for binding of the other one, it usually means that binding of the other ligand alone induces conformation favorable for binding of the first one.

It is well demonstrated by an example provided in a study of by Azzi et al., who investigated influence of excessive concentrations of G protein on affinity of GPCR agonists [15]. Also other authors investigated influence of G proteins or arrestins on interactions of GPCRs with their ligands [16, 17]. These studies have shown that the signal between the binding pocket and the receptor—intracellular receiver interface is bidirectional, and that just as agonists induce increased complexing with G proteins, the excess of the latter increases binding of agonists. It underlines the allosteric nature of signal transmission within GPCRs and, as such, it suggests that action of an allosteric modulator can be understood as addition of further allosteric factors to a complex intrinsically allosteric machinery.

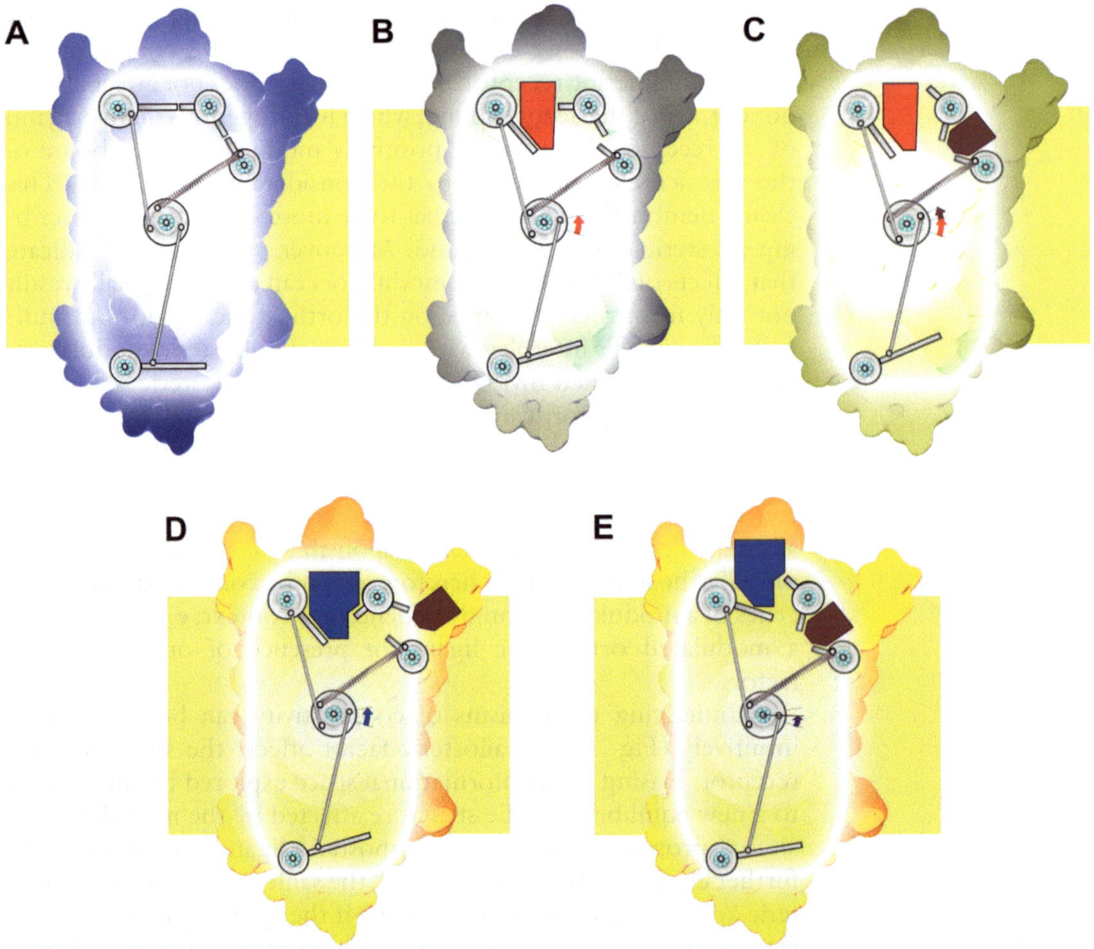

Fig. 1 The scheme depicting possible mechanisms of probe dependence. It presents interrelationships of signals transmitted through allosteric pathways. The illustrated mechanism helps to understand the need of native conditions in studies on allosteric ligands. *Central cog* represents a hypothetical molecular switch, *upper left cog* represent orthosteric binding pocket, *upper right cog*—allosteric site, *lower cog*—G-protein coupling region. (**a**) In the inactive receptor, all the molecular switches oscillate around their rest positions. (**b**) Binding of the *red* agonist activates the receptor to a certain extent, shifting the equilibrium toward the active conformation. (**c**) Additional binding of a *brown* allosteric modulator allows for shifting the equilibrium even farther, increasing the maximal activation level. Binding of one of them stabilizes the favorable structure, facilitating binding of the other, and therefore increasing its affinity. *Red* and *brown* compounds present positive cooperativity. (**d**) The *blue* agonist is capable of activating the receptor to similar extent as the *red* one. However, its specific structural features induce rearrangements that prevent the *brown* modulator binding. (**e**) The signal is reciprocal. In the presence of the *brown* modulator, *blue* agonist has decreased affinity. *Blue* and *brown* compounds present negative cooperativity

2.1 Allosteric Spider's Web

The above-mentioned example brings to mind another reflexion—there are more allosteric factors in a living cell, than just native transmitters or G proteins. Membrane receptors float in lipid bilayers that may vary in fluidness and contain numerous organic components [18]. A living cell is full of small organic compounds that can potentially be modulators. GPCRs can form homo- and heterodimers and oligomers. All such factors affect the receptor, a structure of which results from a balance between all influences. Therefore, regardless of the application of in silico or in vitro methodology, rational design of allosteric modulators requires considering a number of variables. Omission of one of the essential factors in the design of the artificial experimental environment can result in changes in the receptor behavior, so that the results won't reflect the actual in vivo properties of the investigated compound. This, in turn, can result in both false positives, capable of interacting with receptor in artificial experimental conditions but not in the native ones, and false negatives—compounds that could present favorable pharmacological properties in vivo, but were rejected due to their unsatisfactory performance in the artificial environment. While orthosteric ligands usually bind in a well-defined pocket with considerable affinity and their binding can be reproduced even in such suboptimal conditions, initial lead compounds for potential allosteric modulators are likely to present low affinity, and they often explore the more exposed binding sites, so they are more sensitive to such subtle changes in conformational ensemble.

There are a number of various endogenous allosteric modulators that can affect the conformation of GPCRs (Fig. 2). Interestingly, in contrast to most of artificial modulators, allosteric factors affecting GPCR structure can also be intracellular, like G proteins, arrestins, or phosphate groups attached during phosphorylation, or lipophilic, affecting the GPCR from the membrane side. It becomes even more fascinating since the recent reports on GPCR X-ray structures with drugs bound at the intracellular surface were published [19, 20], which prove that the intracellular surface can be druggable.

There are numerous important extracellular endogenous factors that can potentially affect the GPCR structure and, consequently, result of a study. The most broadly distributed would probably be the known influence of sodium ions, which are suggested to have their own allosteric pocket in the interior of many GPCRs, located in the neighborhood of the conserved Asp 2.50 residue [21, 22]. Other ions also can affect GPCR function allosterically. For instance, zinc and magnesium can modulate a number of receptors [23]. Other endogenous allosteric ligands affecting GPCRs are, for example, glutathione which modulates CaSR [24], prolinyl-leucyl-glycine tripeptide which positively modulates agonists action on D_2 dopamine receptor [25], and 5HT-moduline which decreases affinity of serotonin to its receptors [26]. Single aromatic amino acids are capable of modulating CaSR [27], while Phe, Leu, and Ile affect baclofen binding at $GABA_B$ [28]. Some amino acid metabolites can also modulate GPCRs, like homocystein modulating D_2 receptors [29] or agmatin, which is a factor influencing α_2 adrenergic receptors [30].

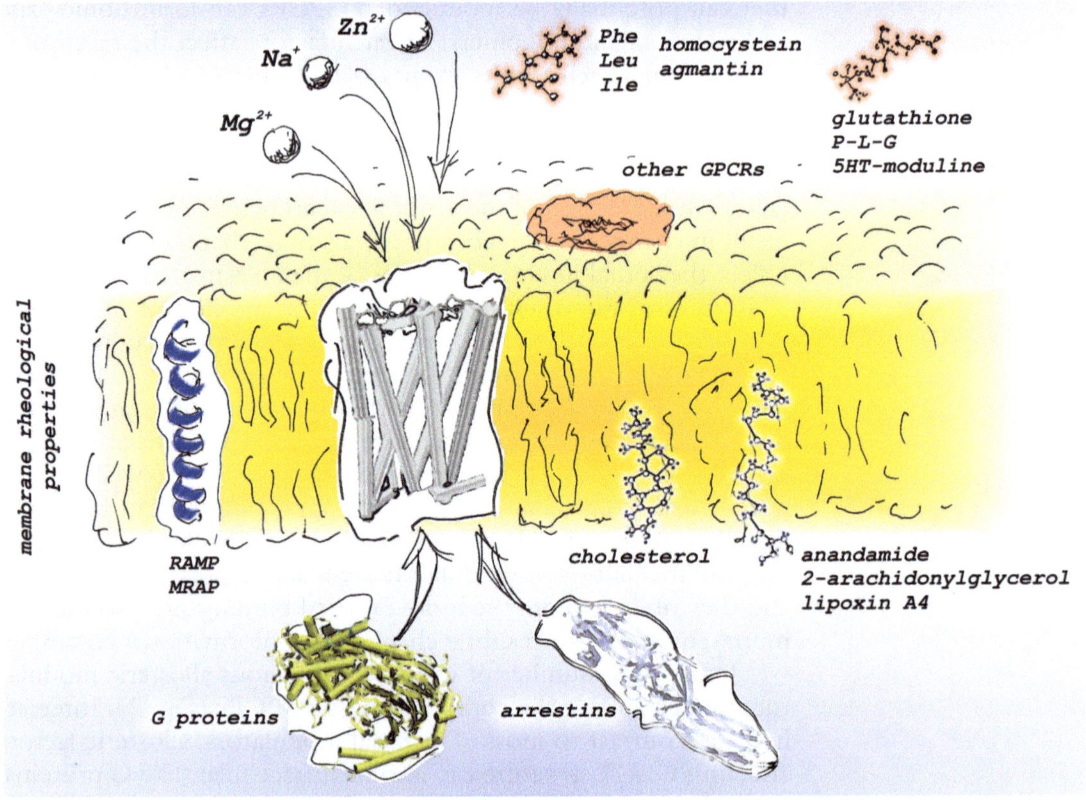

Fig. 2 Overview of known endogenous allosteric modulators of GPCRs (detailed description in the text)

There are known examples of endogenous GPCR modulators immersed in the cell membrane and modifying action of some of these receptors. They are, e.g., RAMP and MRAP proteins, which are single membrane-spanning helices associating with some GPCRs. For instance, the calcitonin receptor-like receptor (CLR) can form complexes with RAMP1, which makes it act as a receptor for the calcitonin gene-related peptide (CGRP), while complexing with RAMP2 results in affinity to adrenomedulin, and with RAMP3—to both adrenomedulin and CGRP [31]. RAMP proteins can also form complexes with calcitonin receptors, increasing their affinity to amylin [32], as well as with CaSR or CRH1R [23]. MRAP proteins also present specific modulatory activity toward melanocortin receptors [33].

Similarly to single-helix membrane-spanning proteins, seven-helix transmembrane-spanning receptors can affect function of other receptors as well. One has to be aware of possibility of dimerization and oligomerization, which can affect ligand binding [34–36] and/or signal processing [37]. As a consequence, a lead compound presenting favorable properties in vivo may exert its effect via binding to a functional dimer/oligomer, and its in silico refinement step should take the dimerization-affected, altered receptor structure into consideration. Moreover, sometimes the allosteric signal can be transmitted between two interacting monomers, so that ligand binding in one of them induces the signaling in the other [38, 39]. This clearly demonstrates the allosteric nature of GPCR dimers, and explains the necessity of appropriate in silico reconstruction of dimeric systems in order to obtain valid results.

Except of proteins, membrane-side modulators can belong to lipids. Among examples of modulation by lipophilic substances there would be influence of anandamide on serotonin and muscarinic receptors [40], 2-arachidonylglycerol on the adenosine A3 receptor [41] and lipoxin A4 on cannabinoid CB1 receptor [42]. Moreover, cholesterol content of the membrane is known to affect a number of GPCR properties. Some data suggest that there are some cholesterol binding or high-occupancy sites on the receptor-membrane interface [43, 44]. Such specific sites are important for two reasons—they are potential allosteric sites that could be targeted, and binding of a native modulator present in the membranes and participating in physiological receptor's function should be considered during in silico studies.

Except of binding or high occupancy sites of specific lipids, more general membrane properties can play an important role in protein behavior [18, 45–47]. The membrane environment can be characterized by varying thickness and rheological properties depending on the exact lipid content. It is known that lipid rafts are present in biological membranes, and it is suggested that they can be spontaneously formed and destroyed [48, 49]. Lipid rafts are characterized by more ordered structure, greater density and thickness, as well as their resistance to some detergents. There were reports on receptors migrating to lipid rafts upon activation [50], which could suggest that their active-state conformations exhibit a more favorable fit to such an environment, which in turn could be described as positive cooperativity. Since, as mentioned, allosteric interactions are reciprocal, one can expect that putting such a receptor into a lipid raft will result in greater affinity to agonists. Existence of such phenomena should be taken into consideration in studies on allosteric modulators, just as another allosteric factor affecting receptor structure.

Taken together, when designing a computational study on allosteric modulators of GPCRs, except of the orthosteric ligand and the investigated allosteric modulator one should consider ion content of the solvent, presence of the allosteric sodium, lipid content of the membrane, presence of appropriate intracellular coupling protein, as well as all other known endogenous allosteric modulators of the investigated system. Oversimplifications at this step can result in inappropriate receptor structure and, e.g., skipping some druggable allosteric sites that could be revealed in physiological conditions, which can further result in false negative results in e.g., VS approach.

3 In Search for a Place to Bind

GPCRs are generally considered a good target for virtual screening (VS) approaches. Well-defined orthosteric binding pocket of rhodopsin-like receptors facilitates design of their classical ligands. Indeed, there are a number of successful VS studies that resulted in a number of active compounds. There are even some reports on successful VS for allosteric modulators of GPCRs [51]. This could result in impression that in silico design would be a relatively simple, well-defined process with a significant chance for success. Yet, this would be a dramatic oversimplification. There are a number of traps set on the path of computer-aided allosteric modulator design, with some of them being universal and applying to allosteric modulators in general, while some others being specific for GPCRs.

While members of glutamate, adhesion, frizzled, and secretin receptor families are characterized by a relatively large extracellular domain, the largest family of GPCRs, rhodopsin-like receptors have significantly shorter N-termini and loops, which makes their extracellular solvent accessible surface smaller. This, in turn, should intuitively result in easier identification of putative ligand binding sites. However, some reports indicate that ligands can enter GPCRs from their intracellular side [19, 20] or even from the membrane side [52–55]. While entrance of orthosteric ligands from the lipid bilayer would apply only to receptors whose ligands are lipophilic, things change dramatically when we consider possibility of allosteric modulation. Keeping in mind the possibility of membrane-side ligand entrance, and considering the fact that there is known influence of membrane cholesterol on the function of GPCRs with a number of suggested preferred occupancy sites, possibility of allosteric modulation by compounds approaching from the lipid bilayer should also be taken into consideration.

3.1 Hidden Sites

It should be stressed that all the endogenous allosteric factors described in the previous section can affect a GPCR behavior. Therefore, their action contributes to mechanisms implied in the creation of another class of binding sites—the hidden ones. The receptor structure can be significantly altered under allosteric influence, so that potential allosteric binding pockets can be revealed. It is known that proteins are not static entities, but they could be more appropriately represented by ensembles of structures balancing around an equilibrium. Therefore, some potential binding sites can be buried in static snapshots caught in X-ray crystals, but revealed, e.g., during molecular dynamics simulations [56–60]. Presence of appropriate endogenous modulators would affect resultant conformation derived from such simulation, and could greatly contribute to revealing novel druggable surfaces.

3.2 Allosteric Mechanisms within GPCR Dimers and Oligomers

The picture of allosteric modulation of GPCRs is even more complex thanks to the possibility of allosteric modulation within dimers and oligomers and the allosteric effect of membrane components. It can be thus concluded that allosteric mechanisms enable an integrative activity to emerge either intramolecularly in G protein-coupled receptor monomers or intermolecularly via receptor–receptor interactions in GPCR homodimers, heterodimers, and receptor mosaics [61].

Allostery can be generally termed a mode of long distance communication between distal sites in proteins [61]. According to the classical point of view one binding site influences the activity of another site via a conformational change. The Monod–Wyman–-Changeux model describes allosteric proteins as symmetric oligomers with identical protomers existing in "at least" two conformational states (tense, T; relaxed, R) with different affinities for ligands [61]. According to this model the protein interconverts between two conformations, R and T, in a concerted manner. In the Koshland–Nemethy–Filmer sequential model subunits change conformation, one at a time. The new concept allostery assumes that proteins may have an ensemble of conformations and dynamic states. Allostery is thus a thermodynamic phenomenon governed by enthalpy and/or entropy. In the receptor heteromers the allosteric communication between the two receptors occurs via the receptor interface, which has a key role in mediating the receptor–-receptor interaction. This takes place thanks to modulation of the orthosteric and allosteric binding sites of the neighboring protomer, its G protein activation and selectivity, its signaling cascades, and through occurrence of new allosteric pockets which may affect, e.g., G protein coupling and selectivity [61].

The experimental proofs for allosterism or cooperativity within homomeric proteins are mainly based on radioligand binding studies [62]. They can be also derived from receptor heteromers, in that one of the interacting partners has been changed in such a way that it can be distinguished from its dimeric partner. Negative cooperativity has been shown by radioligand binding for numerous GPCR homomers. Changes in ligand binding affinity or dissociation kinetics have been reported in cells and tissues co-expressing pairs of GPCRs that can form heteromers and supply further evidence of allosteric communication across GPCRs [62]. The studies on CXCR2-δ-opioid receptor heteromers demonstrate another pharmacological effect typical for allosterism at heteromers: increased signaling of an orthosteric ligand as a result of the presence of another receptor with or without a bound ligand [62]. However, more common pharmacological outcome of heteromerization is the cross-inhibition of signaling in an effect of allosteric communication.

Discovery of compounds that allosterically modulate GPCR dimers is problematic as nowadays design of molecules binding to

GPCR dimers is almost limited to bivalent ligands and to the lesser extent, dimer-specific monovalent ligands. However, the development of GPCR allosteric modulators has supplied an experimental and conceptual framework for better understanding of allosteric effects within GPCR complexes. It can be expected that during next years ligand- and structure-based design methods will be utilized to design allosteric modulators of GPCR dimers.

The above considerations on possible location of allosteric sites demonstrate that no GPCR surface should be rejected a priori—modulators can possibly bind to the extracellular, intracellular, and membrane-side surfaces, including di/oligomerization interfaces. Moreover, lack of a visible pocket in a region hypothetically implied in allosteric signal transmission does not exclude a possibility of existence of a hidden pocket, which could be revealed during dynamic protein motion. All these possibilities make the investigation of allosteric modulators much more complex. However, they also provide an opportunity of very precise targeting of particular GPCR functions.

4 Considerations on Applications of Computational Tools

Depending on the desired effect, different computational methods can be applied to investigate allostery in GPCRs. Dynamic nature of proteins enforces consideration of their structure as an ensemble of states. On the other hand, it increases complexity and computational cost of the in silico investigation. Therefore, combination of methods should be adjusted depending on the available input data and the aim of the particular study. Molecular dynamics (MD) is one of the tools useful for computer-aided drug design. One of its most important features is providing insights into dynamic protein behavior. As such, it can be used for discovery of novel modulators in many different ways. Harnessing MD into a workflow may significantly improve quality of results. For some applications, MD can serve as the main component of the protocol. However, unbiased, all-atom MD is a resource-demanding method, and in some cases it is better suited as a supplementary tool. For instance, it can deliver GPCR structures for further processing.

Success of docking and structure-based virtual screening approaches is strongly dependent on the target structure. Given all the factors affecting the GPCR conformation, structure refinement together with appropriate modulators can be beneficial. Molecular dynamics simulation of GPCR structure in adequate activation state, in complex with its specific endogenous modulators, e.g., sodium ions, immersed in a native-like membrane can adjust the structure of potential allosteric sites or reveal new ones, greatly affecting results of VS approach. Except for structure refinement, MD and its modifications can also be used as a source of GPCR conformational ensembles, subsequently targeted by docking [63].

When computational resources are available and the study focuses on a small number of ligands, MD and its modifications can be successfully used for the reconstruction of modulator binding or refinement of initial docking poses [8, 64, 65]. These applications, however, are more processor time-consuming and could not be efficiently used for screening with large library of compounds. Therefore, they are more useful in investigation of known modulators, e.g., in finding their binding sites, which may facilitate further refinement of the lead structure. In turn, in a hunt for novel allosteric lead structures, more processor-saving, high-throughput methods would be more appropriate. Molecular docking is one of such methods. Its philosophy allows for various degrees of simplifications, allowing for efficient screening of large compound libraries. In the following sections, we will focus on the application of molecular docking and molecular dynamics in the design and discovery of allosteric modulators.

4.1 Design of a Molecular Dynamics Study

As every study, an investigation of allosteric modulation of a GPCR should be preceded by collecting available literature data. Gathering all the possible information on the native receptor's environment, including possible complexing/binding partners and rheological properties of the membrane is especially important in studies aiming at finding novel, pharmacologically useful modulators—number of in vitro and in silico studies indicate that membrane environment and presence of, e.g., allosteric sodium ions or coupled G protein play an important role in GPCR signaling. Even availability of a solved X-ray structure of the receptor of interest does not allow for skipping this stage—crystals can be affected by the crystal packing forces, the protein structure may contain point mutations facilitating crystallization, while some crucial components affecting the receptor structure might be absent. For instance, while there are a few known GPCR structures solved together with arrestin, G-protein or a G-protein-mimicking nanobody, in most other cases these elements are not cocrystallized. Therefore, refinement of the homology models or X-ray structures according to the present state of knowledge, with consideration of all the components appropriate for the investigated activation state and signaling pathway, would be very beneficial.

It is worth mentioning that accurate in silico reproduction of the membrane environment depends not only on introducing appropriate lipid concentrations, but also on the choice of a well-performing force field. An excellent recent review by Lyubartsev and Rabinovich [66] summarizes most important differences between the most popular currently available lipid force fields. Notably, CHARMM is supported with convenient CHARMM-

GUI web-based interface [67]. It also provides parameters for a large number of various lipids. On the other hand, although Slipids supports less lipids, it represents them with remarkable quality, which is reflected in accurate reproduction of such parameters as, e.g., area per lipid [66, 68, 69]. Moreover, Slipids is compatible with a popular Amber force field, as well as with General Amber Force Field, which is useful in ligand parametrization [70–72]. Fortunately, CHARMM-GUI can produce structure files containing membranes described mostly by the same atom names as Slipids, which provides a convenient way for membrane generation for those researchers who want to use Amber as a protein force field, or whose membrane would be better represented in Slipids. Minor differences in atom names between those tools are usually easy to fix with simple scripts, or even with free text tools capable of batch string replacement.

Obviously, there are also other automated membrane builders worth consideration, e.g., MemBuilder II [73] which supports CHARMM, AMBER/Slipids, and GROMOS force fields. Unfortunately, it does not support automated protein insertion, so employing this builder would involve subsequent manual receptor settlement, or utilizing, e.g., Inflategro script [74] or g_membed command of Gromacs [75]. Among other useful services, Lipid Builder 3.0 is a handy tool for the generation of parameters for non-standard lipids In CHARMM [76].

Except for correct membrane composition, its interactions with a receptor should also be appropriate. Proper insertion of the receptor into membrane is greatly facilitated by the Orientations of Proteins in Membranes database [77]. The database contains crystallographic structures of membrane proteins. The structure files contain additional dummy atoms representing the membrane plane. Such atoms can be recognized e.g., by CHARMM-GUI Membrane Builder. Therefore, preprocessed OPM structures can serve as an input for further processing, which greatly facilitates preparation of simulation boxes. Unfortunately, as mentioned before, X-ray structures often contain point mutations or loop replacements, which need corrections. Moreover, not all GPCR structures were elucidated by crystallography, and even if structure of the receptor of one's interest was solved, its activation state might not be appropriate. However, since GPCRs share the same structural scaffold, and both inactive and active-state receptors are present in the database, the OPM-derived boxes can serve as a template and help in the preparation of boxes containing homology models or refined crystallographic structures.

Particular care should be maintained during the preparation of ligand parameters. Perfect representation of a novel small-molecule ligand would be considerably time-consuming and, as such, not suitable for drug design purposes. Therefore, the best solution would be careful application of reliable simplified protocols.

There is a tempting possibility of using freely available on-line servers or standalone tools capable of generating ligand topologies on the basis of an input structure file, like PRODRG [78], ATB [79], ACPYPE [80], or TopolGen. However, one should not trust unconditionally in their output, which is often stated even by some of their authors. Manual corrections are usually needed, and in most cases charge values could be improved by employing additional software. In particular, calculating RESP charges might improve ligand representation in simulations [81].

Taken together, MD is a versatile in silico tool useful in allosteric modulator design. Depending on the particular research protocol, it can provide refined input protein structures for other techniques, help in identifying allosteric sites [8, 9, 65] or finding hidden binding sites [58–60], as well as for investigation of underlying mechanisms of allostery [8] or biased signaling [82]. The latter applications provide significant amount of data, and therefore require thoughtful analysis [83]. The principal component analysis (PCA) [8, 84, 85] and information theory-based [86, 87] methods seem to be most powerful of analysis tools. They are able to sift relevant information from the plethora of measurable variables possible to gather from MD trajectories. In particular, PCA is capable of returning the most apparent relationships as the highest order results, called principal components. In both cases proper definition of the examined region is an important issue. In PCA calculations performed on the entire GPCR structure, together with termini and loop regions, their chaotic movement is likely to dominate the entire analysis, and the results may not be conclusive. In general, decreasing noise is a favorable PCA strategy, so splitting the receptor and separate analyses of its different regions may be productive [8]. In all the cases, careful fitting of frames is essential, and the issue is especially important in analysis of various simulations of the same receptor—lack of appropriate fitting can produce artifacts, which can easily be over-interpreted. Analogically to PCA, N-body information theory (NbIT), an information theory-based method, requires defining functional residue clusters, which also should be carefully thought out [86].

4.2 Molecular Docking

Molecular docking to allosteric binding sites of GPCRs is in principle technically identical as docking to their orthosteric sites, conditioned that an allosteric site is known and well defined. It is usually not the case of family A GPCRs which can be allosterically modulated by ligands bound not only to sites in the upper part of the transmembrane bundle and the extracellular domain but also to the sites bordering with the cell membrane (as in the case of GPR40 X-ray structure [52]) or in the intracellular site of G protein as it was reported for chemokine receptors [19, 20]. One receptor can contain a number of allosteric sites. It has been suggested that in case of muscarinic receptors there are at least two sites as certain indolocarbazole and androstane derivatives are predicted to bind to a non-orthosteric site different from that recognized by other allosteric muscarinic ligands such as gallamine or strychnine [88]. One allosteric binding site at muscarinic M_2 receptor has been identified using X-ray crystallography [89]. This structure demonstrates that allosteric modulator LY2119620 interacts in a largely pre-formed binding site in the extracellular vestibule of the iperoxo-bound receptor, and induces a slight contraction of this outer binding pocket [89]. In case of family C GPCRs, the known allosteric site is situated within the transmembrane bundle as found in structures of metabotropic glutamate receptors in complex with allosteric modulators [90–92]. There are also attempts to identify allosteric sites of family B GPCRs. Bhattacharya et al. modeled the corticotropin releasing factor receptor (CRF1-R) and identified the binding sites of representative small-molecule allosteric antagonists. The predicted binding sites of the investigated compounds are located within the transmembrane domain encompassing TM helices TM3, TM5, and TM6.

There are, however, special tools that are dedicated to improve molecular docking of allosteric ligands. One of them is Alloscore, a web server that predicts the binding affinities of allosteric ligand-protein interactions [93]. The results include the predicted energy terms, binding affinity, and interactive 3D representation of each allosteric ligand-protein complex. This method displays good performance in describing allosteric binding and could be useful in allosteric virtual screening and the structural optimization of allosteric agonists/antagonists [93]. There are also a number of tools for the prediction of allosteric binding sites, such as, e.g., Allosite [94]. An interesting approach was proposed by Ivetac and McCammon [59]. In order to facilitate structure-based drug design of GPCR allosteric modulators they elaborated a fragment-based molecular dynamics approach to map druggable allosteric space of GPCRs. They used the FTMAP algorithm to identify "hot spots" with affinity for a variety of organic probe molecules corresponding to drug fragments.

In an attempt to dock an unknown GPCR allosteric modulator, one can consider at least two strategies. The most straightforward is to perform molecular docking to known allosteric sites, however without experimental support that a given modulator binds to one of these sites, this strategy is risky. Thus, at least the whole extracellular region [8, 9] or even the whole receptor surface [95] should be considered a potential binding site. By considering the whole receptor Planeseas et al. [95] proposed two different allosteric binding sites at chemokine CXCR4 receptor, one located in the intracellular loops 1, 2, and 3 (ICL1, ICL2, and ICL3) which binds the pepducin allosteric agonist ATI-2341, and the other at a subsite of the main extracellular orthosteric binding pocket between the extracellular loops 1 and 2 and the N-terminus, which binds the allosteric antagonists AMD11070 and GSK812397. In case multiple allosteric binding sites are found from molecular docking, they may be ranked on the basis of short molecular dynamics simulations [8, 9] and/or literature data [96], e.g., mutations [97].

A case of molecular docking of atypical allosteric modulator, a snake toxin MT7 to dimeric hM1 muscarinic receptor, was described by Marquer et al. [98]. Their strategy was based on the combination of double mutant cycle experiments and molecular modeling calculations. Their docking calculations converged to the conformation of the MT7-hM1 dimer complex which was in accordance with experimental results: (a) the snake toxin binds at the extracellular side of the receptor, (b) the tips of MT7 loops II and III interact with one hM1 protomer, whereas the tip of loop I contacts with the other protomer, and (c) the hM1dimeric interface is formed by the transmembrane helices TM6 and TM7 [98].

Ragnarsson et al. [99] identified a novel allosteric site at adrenergic β_1 receptor which is a site of allosteric modulation by Conopeptide ρ-TIA. They found that peptide binding to the extracellular surface residues on TM6 and TM7 at the base of extracellular loop 3 (ECL3) is sufficient to allosterically inhibit agonist signaling at a GPCR.

Another noteworthy example was reported by Mukund et al. [100]. They identified a monoclonal antibody that inhibits the glucagon receptor (GCGR), a class B GPCR, through a unique allosteric mechanism. Moreover, they suggested that receptor inhibition is governed by antibody binding to two distinct sites which are situated outside of the glucagon binding cleft. One of these sites is formed by a patch of residues that are surface-exposed on the face of the extracellular domain (ECD) opposite to the ligand-binding pocket, while the second binding site consists of residues in the αA helix of the ECD.

At present structure-based drug design of allosteric modulators is limited to searching for small molecules which bind to extracellular pockets (family A) or transmembrane bundle (families B and C). Further studies are needed to determine the druggability of other allosteric sites identified for GPCRs.

4.3 Ligand- and Structure-Based Drug Design of Allosteric Modulators

Structure-based drug design of allosteric GPCR ligands has been hampered by the lack of structural data for allosteric binding sites, making a strong case for predictive computational methods [59]. In this context, ligand-based approaches have been often applied, in particular for designing family C GPCR allosteric modulators.

There are only a few reports of successful computational design of family A and family B allosteric modulators and a number of cases referring to family C ligands. Lane et al. [101] reported structure-based ligand discovery targeting orthosteric and allosteric pockets of dopamine receptors. In order to identify novel ligands they used D_3 receptor with an empty orthosteric pocket and this receptor in complex with dopamine. From the virtual screening using receptor in complex with dopamine they found compounds that are predicted to occupy an allosteric site at the extracellular extension of the pocket, which lack the anchoring amino group. They discovered that compounds targeting the allosteric site display a variety of functional activity profiles, and that behavior of at least two compounds is consistent with non-competitive allosteric modulation of dopamine signaling in the extracellular signal-regulated kinase 1 and 2 phosphorylation and β-arrestin recruitment assays [101]. Miao et al. [63] used accelerated structure-based drug design in order to identify chemically diverse allosteric modulators of muscarinic M_2 receptor. In their approach they used a combination of accelerated molecular dynamics and Glide induced fit docking which displayed better enrichment factors than Glide virtual screening workflow. As a result they identified 12 compounds with affinity of $\leq$30 μM. With final functional experiments on six selected compounds, they confirmed four of them as new negative allosteric modulators and one as a positive allosteric modulator of agonist-mediated response at the M_2 muscarinic receptor [63].

Regarding family B allosteric modulators, de Graaf et al. [102] used a combination of a ligand-based approach with a docking-based virtual screening approach to identify novel allosteric ligands which bind in the transmembrane bundle of the glucagon receptor (GLR) and the glucagon-like peptide 1 receptor (GLP-1R). They started from a knowledge-based three-dimensional model of the GLR, and docked a database of 1.9 million commercially available drug-like compounds to its transmembrane cavity. Next, they selected 23 compounds based on protein–ligand interaction fingerprints, which were then purchased and evaluated for in vitro binding to GLR and modulation of glucagon-induced cAMP release. They found that two of the 23 compounds inhibited the effect of glucagon in a dose-dependent manner, and that one inhibitor displayed the same potency as L-168049, a reference noncompetitive GLR antagonist, in a whole-cell-based functional assay [102].

As has been mentioned, there are a number of successful ligand- and structure-based virtual screening for family C GPCRs, in particular for different subtypes of metabotropic glutamate receptors. Ligand-based pharmacophore virtual screening was performed by Kubas et al. [103] in order to identify allosteric modulators of metabotropic glutamate receptor 5 (mGluR5). As a result they found 2-(m-tolylamino)-7,8-dihydroquinazolin-5(6H)-one as a moderately active negative allosteric modulator of this receptor and subjected it to optimization which results in a series of 2-aminoquinazoline derivatives, a structurally novel class of mGluR5 negative allosteric modulators. A similar approach, based on a combination of molecule encoding and clustering to predict compound selectivity which was realized by a topological pharmacophore descriptor and the SOM algorithm, was applied to identify mGluR1 allosteric modulators. Noeske et al. [104] applied alignment-free topological pharmacophore descriptors (CATS) to encode the screening compounds in order to search for new mGluR1 allosteric antagonists and identified one hit based on coumarine scaffold. Mueller et al. applied artificial neural networks which were trained based on a recently reported high-throughput screen in order to find novel allosteric modulators of mGluR4 [105] and mGluR5 receptor [106, 107].

Structure-based virtual screening for family C GPCRs has been performed by Omer and Prasad [108] and by Jang et al. [109] for mGluR1 receptor. In particular, Jang et al. used a hierarchical virtual screening approach and homology model based on dopamine D_3 receptor template. Jiang et al. applied ligand- and structure-based virtual screening and molecular dynamics to identify negative allosteric modulators of mGlur1 from Chinese herbs [110]. This study was based on the crystal structure of the seven-transmembrane domain of mGluR1.

In summary, ligand- and structure-based design of allosteric modulators of GPCRs is still a relatively unexplored field. Ligand-based approach is hampered by a number of available modulators and structure-based technique by limited structural data about allosteric binding sites. With the increasing availability of both, the field could become very active in near future.

5 Summary

Application of computer-aided drug design methods to invention of novel GPCR allosteric modulators brings many benefits, like possibility of finding novel binding pockets or gaining a deeper insight into mechanisms underlying allostery, just to mention a few. However, as in the case of any computational method, an algorithm can process any input except of dividing by zero, but processing itself does not guarantee that its results would be conclusive. Careful input preparation is particularly challenging in discovery of allosteric or biased compounds, because of delicate and elusive nature of mechanisms underlying these phenomena. Improvements in available force fields, availability of significant computational power of supercomputers or GPU stations, and growing amount of experimental data may support in silico modulator design efforts, as long as all these elements are appropriately applied. In turn, they may fail when one of essential factors is omitted and, e.g., signaling bias of a compound is to be calculated. Awareness of these issues is increasing, and, e.g., including moderate cholesterol concentrations into membrane bilayer constructs during simulation box preparation becomes a standard. Obviously, it is most likely that not all the relevant factors affecting the receptor of one's interest are already known. In turn, careful application of all the known ones to prepare more native-like environment can be very time-consuming. However, a significantly improved, reliable outcome might be a worthy reward. Moreover, besides the practical meaning, it gives intangible satisfaction of dealing with best-as-possible reconstruction of actual, tiny portion of the very basic, living matter.

References

1. Christopoulos A (2014) Advances in G protein-coupled receptor allostery: from function to structure. Mol Pharmacol 86:463–478. https://doi.org/10.1124/mol.114.094342
2. Sekercioglu N, Busse JW, Mustafa RA et al (2016) Cinacalcet versus standard treatment for chronic kidney disease: a protocol for a systematic review and meta-analysis. Syst Rev 5:2. https://doi.org/10.1186/s13643-015-0177-1
3. Woollard SM, Kanmogne GD (2015) Maraviroc: a review of its use in HIV infection and beyond. Drug Des Devel Ther 9:5447–5468. https://doi.org/10.2147/DDDT.S90580
4. Müller CE, Schiedel AC, Baqi Y (2012) Allosteric modulators of rhodopsin-like G protein-coupled receptors: opportunities in drug development. Pharmacol Ther 135:292–315. https://doi.org/10.1016/j.pharmthera.2012.06.002
5. Valant C, Felder CC, Sexton PM, Christopoulos A (2012) Probe dependence in the allosteric modulation of a G protein-coupled receptor: implications for detection and validation of allosteric ligand effects. Mol Pharmacol 81:41–52. https://doi.org/10.1124/mol.111.074872
6. Schneider S, Provasi D, Filizola M (2016) How oliceridine (TRV-130) binds and stabilizes a μ-opioid receptor conformational state that selectively triggers G protein-signaling pathways. Biochemistry (Mosc) 55 (46):6456–6466. https://doi.org/10.1021/acs.biochem.6b00948
7. Rankovic Z, Brust TF, Bohn LM (2016) Biased agonism: an emerging paradigm in GPCR drug discovery. Bioorg Med Chem Lett 26:241–250. https://doi.org/10.1016/j.bmcl.2015.12.024
8. Bartuzi D, Kaczor AA, Matosiuk D (2016) Interplay between two allosteric sites and their influence on agonist binding in human μ opioid receptor. J Chem Inf Model 56:563–570. https://doi.org/10.1021/acs.jcim.5b00705
9. Bartuzi D, Kaczor AA, Matosiuk D (2015) Activation and allosteric modulation of human μ opioid receptor in molecular dynamics. J Chem Inf Model 55:2421–2434. https://doi.org/10.1021/acs.jcim.5b00280
10. Sadiq SK, Guixa-Gonzalez R, Dainese E et al (2013) Molecular modeling and simulation of membrane lipid-mediated effects on GPCRs. Curr Med Chem 20:22–38
11. Guixà-González R, Ramírez-Anguita JM, Kaczor AA, Selent J (2013) Simulating G protein-coupled receptors in native-like membranes: from monomers to oligomers. Methods Cell Biol 117:63–90. https://doi.org/10.1016/B978-0-12-408143-7.00004-9
12. Fenton AW (2008) Allostery: an illustrated definition for the "second secret of life". Trends Biochem Sci 33:420–425. https://doi.org/10.1016/j.tibs.2008.05.009
13. Ballesteros JA, Weinstein H (1995) [19] integrated methods for the construction of three-dimensional models and computational probing of structure-function relations in G protein-coupled receptors. In: Sealfon SC (ed) Methods neuroscience. Academic Press, San Diego, CA, pp 366–428
14. Livingston KE, Traynor JR (2014) Disruption of the Na+ ion binding site as a mechanism for positive allosteric modulation of the mu-opioid receptor. Proc Natl Acad Sci U S A 111:18369–18374. https://doi.org/10.1073/pnas.1415013111
15. Azzi M, Piñeyro G, Pontier S et al (2001) Allosteric effects of G protein overexpression on the binding of beta-adrenergic ligands with distinct inverse efficacies. Mol Pharmacol 60:999–1007
16. Burstein ES, Spalding TA, Brann MR (1997) Pharmacology of muscarinic receptor subtypes constitutively activated by G proteins. Mol Pharmacol 51:312–319
17. Yan F, Mosier PD, Westkaemper RB, Roth BL (2008) Galpha-subunits differentially alter

the conformation and agonist affinity of kappa-opioid receptors. Biochemistry (Mosc) 47:1567–1578. https://doi.org/10.1021/bi701476b

18. Periole X (2017) Interplay of G protein-coupled receptors with the membrane: insights from supra-atomic coarse grain molecular dynamics simulations. Chem Rev 117:156–185. https://doi.org/10.1021/acs.chemrev.6b00344
19. Zheng Y, Qin L, Zacarías NVO et al (2016) Structure of CC chemokine receptor 2 with orthosteric and allosteric antagonists. Nature 540:458–461. https://doi.org/10.1038/nature20605
20. Oswald C, Rappas M, Kean J et al (2016) Intracellular allosteric antagonism of the CCR9 receptor. Nature 540:462–465. https://doi.org/10.1038/nature20606
21. Yuan S, Vogel H, Filipek S (2013) The role of water and sodium ions in the activation of the μ-opioid receptor. Angew Chem Int Ed Engl 52:10112–10115. https://doi.org/10.1002/anie.201302244
22. Selent J, Sanz F, Pastor M, De Fabritiis G (2010) Induced effects of sodium ions on dopaminergic G-protein coupled receptors. PLoS Comput Biol 6(8):e1000884. https://doi.org/10.1371/journal.pcbi.1000884
23. van der Westhuizen ET, Valant C, Sexton PM, Christopoulos A (2015) Endogenous allosteric modulators of G protein-coupled receptors. J Pharmacol Exp Ther 353:246–260. https://doi.org/10.1124/jpet.114.221606
24. Broadhead GK, Mun H, Avlani VA et al (2011) Allosteric modulation of the calcium-sensing receptor by gamma-glutamyl peptides: inhibition of PTH secretion, suppression of intracellular cAMP levels, and a common mechanism of action with L-amino acids. J Biol Chem 286:8786–8797. https://doi.org/10.1074/jbc.M110.149724
25. Ott MC, Mishra RK, Johnson RL (1996) Modulation of dopaminergic neurotransmission in the 6-hydroxydopamine lesioned rotational model by peptidomimetic analogues of L-prolyl-L-leucyl-glycinamide. Brain Res 737:287–291
26. Murdoch R, Morecroft I, MacLean MR (2003) 5-HT moduline: an endogenous inhibitor of 5-HT(1B/1D)-mediated contraction in pulmonary arteries. Br J Pharmacol 138:795–800. https://doi.org/10.1038/sj.bjp.0705123
27. Conigrave AD, Quinn SJ, Brown EM (2000) L-amino acid sensing by the extracellular Ca2+−sensing receptor. Proc Natl Acad Sci U S A 97:4814–4819
28. Kerr DIB, Ong J (2003) Potentiation of metabotropic GABAB receptors by L-amino acids and dipeptides in rat neocortex. Eur J Pharmacol 468:103–108
29. Agnati LF, Ferré S, Genedani S et al (2006) Allosteric modulation of dopamine D2 receptors by homocysteine. J Proteome Res 5:3077–3083. https://doi.org/10.1021/pr0601382
30. Molderings GJ, Menzel S, Kathmann M et al (2000) Dual interaction of agmatine with the rat alpha(2D)-adrenoceptor: competitive antagonism and allosteric activation. Br J Pharmacol 130:1706–1712. https://doi.org/10.1038/sj.bjp.0703495
31. McLatchie LM, Fraser NJ, Main MJ et al (1998) RAMPs regulate the transport and ligand specificity of the calcitonin-receptor-like receptor. Nature 393:333–339. https://doi.org/10.1038/30666
32. Muff R, Bühlmann N, Fischer JA, Born W (1999) An amylin receptor is revealed following co-transfection of a calcitonin receptor with receptor activity modifying proteins-1 or −3. Endocrinology 140:2924–2927. https://doi.org/10.1210/endo.140.6.6930
33. Novoselova TV, Jackson D, Campbell DC et al (2013) Melanocortin receptor accessory proteins in adrenal gland physiology and beyond. J Endocrinol 217:R1–11. https://doi.org/10.1530/JOE-12-0501
34. El Moustaine D, Granier S, Doumazane E et al (2012) Distinct roles of metabotropic glutamate receptor dimerization in agonist activation and G-protein coupling. Proc Natl Acad Sci U S A 109:16342–16347. https://doi.org/10.1073/pnas.1205838109
35. White JH, Wise A, Main MJ et al (1998) Heterodimerization is required for the formation of a functional GABAB receptor. Nature 396:679–682. https://doi.org/10.1038/25354
36. Schonenbach NS, Hussain S, O'Malley MA (2015) Structure and function of G protein-coupled receptor oligomers: implications for drug discovery. Wiley Interdiscip Rev Nanomed Nanobiotechnol 7:408–427. https://doi.org/10.1002/wnan.1319
37. Hasbi A, O'Dowd BF, George SR (2011) Dopamine D1-D2 receptor heteromer signaling pathway in the brain: emerging physiological relevance. Mol Brain 4:26. https://doi.org/10.1186/1756-6606-4-26
38. Brock C, Oueslati N, Soler S et al (2007) Activation of a dimeric metabotropic

glutamate receptor by intersubunit rearrangement. J Biol Chem 282:33000–33008. https://doi.org/10.1074/jbc.M702542200

39. Rivero-Müller A, Chou Y-Y, Ji I et al (2010) Rescue of defective G protein–coupled receptor function in vivo by intermolecular cooperation. Proc Natl Acad Sci 107:2319–2324. https://doi.org/10.1073/pnas.0906695106
40. Lanzafame AA, Guida E, Christopoulos A (2004) Effects of anandamide on the binding and signaling properties of M1 muscarinic acetylcholine receptors. Biochem Pharmacol 68:2207–2219. https://doi.org/10.1016/j.bcp.2004.08.005
41. Lane JR, Beukers MW, Mulder-Krieger T, Ijzerman AP (2010) The endocannabinoid 2-arachidonylglycerol is a negative allosteric modulator of the human A3 adenosine receptor. Biochem Pharmacol 79:48–56. https://doi.org/10.1016/j.bcp.2009.07.024
42. Pamplona FA, Ferreira J, Menezes de Lima O et al (2012) Anti-inflammatory lipoxin A4 is an endogenous allosteric enhancer of CB1 cannabinoid receptor. Proc Natl Acad Sci U S A 109:21134–21139. https://doi.org/10.1073/pnas.1202906109
43. Prasanna X, Sengupta D, Chattopadhyay A (2016) Cholesterol-dependent conformational plasticity in GPCR dimers. Sci Rep 6:31858. https://doi.org/10.1038/srep31858
44. Pluhackova K, Gahbauer S, Kranz F et al (2016) Dynamic cholesterol-conditioned dimerization of the G protein coupled chemokine receptor type 4. PLoS Comput Biol 12:e1005169. https://doi.org/10.1371/journal.pcbi.1005169
45. Marino KA, Prada-Gracia D, Provasi D, Filizola M (2016) Impact of lipid composition and receptor conformation on the spatio-temporal organization of μ-opioid receptors in a multi-component plasma membrane model. PLoS Comput Biol 12:e1005240. https://doi.org/10.1371/journal.pcbi.1005240
46. Koldsø H, Reddy T, Fowler PW et al (2016) Membrane compartmentalization reducing the mobility of lipids and proteins within a model plasma membrane. J Phys Chem B 120:8873–8881. https://doi.org/10.1021/acs.jpcb.6b05846
47. Dawaliby R, Trubbia C, Delporte C et al (2016) Allosteric regulation of GPCR activity by phospholipids. Nat Chem Biol 12:35–39. https://doi.org/10.1038/nchembio.1960
48. Pike LJ, Han X, Chung K-N, Gross RW (2002) Lipid rafts are enriched in arachidonic acid and plasmenylethanolamine and their composition is independent of caveolin-1 expression: a quantitative electrospray ionization/mass spectrometric analysis. Biochemistry (Mosc) 41:2075–2088
49. Langelier B, Linard A, Bordat C et al (2010) Long chain-polyunsaturated fatty acids modulate membrane phospholipid composition and protein localization in lipid rafts of neural stem cell cultures. J Cell Biochem 110:1356–1364. https://doi.org/10.1002/jcb.22652
50. Saulière-Nzeh Ndong A, Saulière-Nzeh AN, Millot C et al (2010) Agonist-selective dynamic compartmentalization of human mu opioid receptor as revealed by resolutive FRAP analysis. J Biol Chem 285:14514–14520. https://doi.org/10.1074/jbc.M109.076695
51. Kaczor AA, Silva AG, Loza MI et al (2016) Structure-based virtual screening for dopamine D2 receptor ligands as potential antipsychotics. ChemMedChem 11:718–729. https://doi.org/10.1002/cmdc.201500599
52. Srivastava A, Yano J, Hirozane Y et al (2014) High-resolution structure of the human GPR40 receptor bound to allosteric agonist TAK-875. Nature 513:124–127. https://doi.org/10.1038/nature13494
53. Stanley N, Pardo L, Fabritiis GD (2016) The pathway of ligand entry from the membrane bilayer to a lipid G protein-coupled receptor. Sci Rep 6:22639. https://doi.org/10.1038/srep22639
54. Hurst DP, Schmeisser M, Reggio PH (2013) Endogenous lipid activated G protein-coupled receptors: emerging structural features from crystallography and molecular dynamics simulations. Chem Phys Lipids 169:46–56. https://doi.org/10.1016/j.chemphyslip.2013.01.009
55. Hanson MA, Roth CB, Jo E et al (2012) Crystal structure of a lipid G protein-coupled receptor. Science 335:851–855. https://doi.org/10.1126/science.1215904
56. Hertig S, Latorraca NR, Dror RO (2016) Revealing atomic-level mechanisms of protein allostery with molecular dynamics simulations. PLoS Comput Biol 12:e1004746. https://doi.org/10.1371/journal.pcbi.1004746
57. Wassman CD, Baronio R, Demir Ö et al (2013) Computational identification of a transiently open L1/S3 pocket for

reactivation of mutant p53. Nat Commun 4:1407. https://doi.org/10.1038/ncomms2361

58. Bakan A, Nevins N, Lakdawala AS, Bahar I (2012) Druggability assessment of allosteric proteins by dynamics simulations in the presence of probe molecules. J Chem Theory Comput 8:2435–2447. https://doi.org/10.1021/ct300117j
59. Ivetac A, McCammon JA (2010) Mapping the druggable allosteric space of G-protein coupled receptors: a fragment-based molecular dynamics approach. Chem Biol Drug Des 76:201–217. https://doi.org/10.1111/j.1747-0285.2010.01012.x
60. Tan YS, Śledź P, Lang S et al (2012) Using ligand-mapping simulations to design a ligand selectively targeting a cryptic surface pocket of polo-like kinase 1. Angew Chem Int Ed Engl 51:10078–10081. https://doi.org/10.1002/anie.201205676
61. Fuxe K, Marcellino D, Borroto-Escuela DO et al (2010) The changing world of G protein-coupled receptors: from monomers to dimers and receptor mosaics with allosteric receptor-receptor interactions. J Recept Signal Transduct Res 30:272–283. https://doi.org/10.3109/10799893.2010.506191
62. Smith NJ, Milligan G (2010) Allostery at G protein-coupled receptor homo- and heteromers: uncharted pharmacological landscapes. Pharmacol Rev 62:701–725. https://doi.org/10.1124/pr.110.002667
63. Miao Y, Goldfeld DA, Moo EV et al (2016) Accelerated structure-based design of chemically diverse allosteric modulators of a muscarinic G protein-coupled receptor. Proc Natl Acad Sci U S A 113:E5675–E5684. https://doi.org/10.1073/pnas.1612353113
64. Dror RO, Green HF, Valant C et al (2013) Structural basis for modulation of a G-protein-coupled receptor by allosteric drugs. Nature 503:295–299. https://doi.org/10.1038/nature12595
65. Shang Y, Yeatman HR, Provasi D et al (2016) Proposed mode of binding and action of positive allosteric modulators at opioid receptors. ACS Chem Biol 11(5):1220–1229. https://doi.org/10.1021/acschembio.5b00712
66. Lyubartsev AP, Rabinovich AL (2016) Force field development for lipid membrane simulations. Biochim Biophys Acta 1858:2483–2497. https://doi.org/10.1016/j.bbamem.2015.12.033
67. Lee J, Cheng X, Swails JM et al (2016) CHARMM-GUI input generator for NAMD, GROMACS, AMBER, OpenMM, and CHARMM/OpenMM simulations using the CHARMM36 additive force field. J Chem Theory Comput 12:405–413. https://doi.org/10.1021/acs.jctc.5b00935
68. Jämbeck JPM, Lyubartsev AP (2013) Another piece of the membrane puzzle: extending slipids further. J Chem Theory Comput 9:774–784. https://doi.org/10.1021/ct300777p
69. Jämbeck JPM, Lyubartsev AP (2012) Derivation and systematic validation of a refined all-atom force field for phosphatidylcholine lipids. J Phys Chem B 116:3164–3179. https://doi.org/10.1021/jp212503e
70. Wang J, Wolf RM, Caldwell JW et al (2004) Development and testing of a general amber force field. J Comput Chem 25:1157–1174. https://doi.org/10.1002/jcc.20035
71. Jämbeck JPM, Lyubartsev AP (2012) An extension and further validation of an all-atomistic force field for biological membranes. J Chem Theory Comput 8:2938–2948. https://doi.org/10.1021/ct300342n
72. Paloncýová M, Fabre G, DeVane RH et al (2014) Benchmarking of force fields for molecule-membrane interactions. J Chem Theory Comput 10:4143–4151. https://doi.org/10.1021/ct500419b
73. Ghahremanpour MM, Arab SS, Aghazadeh SB et al (2014) MemBuilder: a web-based graphical interface to build heterogeneously mixed membrane bilayers for the GROMACS biomolecular simulation program. Bioinforma Oxf Engl 30:439–441. https://doi.org/10.1093/bioinformatics/btt680
74. Schmidt TH, Kandt C (2012) LAMBADA and inflateGRO2: efficient membrane alignment and insertion of membrane proteins for molecular dynamics simulations. J Chem Inf Model 52:2657–2669. https://doi.org/10.1021/ci3000453
75. Pronk S, Páll S, Schulz R et al (2013) GROMACS 4.5: a high-throughput and highly parallel open source molecular simulation toolkit. Bioinformatics 29:845–854. https://doi.org/10.1093/bioinformatics/btt055
76. Home | Lipid builder. http://lipidbuilder.epfl.ch/home. Accessed 20 Jan 2017
77. Lomize MA, Lomize AL, Pogozheva ID, Mosberg HI (2006) OPM: orientations of proteins in membranes database. Bioinformatics 22:623–625. https://doi.org/10.1093/bioinformatics/btk023
78. Schüttelkopf AW, van Aalten DMF (2004) PRODRG: a tool for high-throughput crystallography of protein-ligand complexes. Acta

Crystallogr D Biol Crystallogr 60:1355–1363. https://doi.org/10.1107/S0907444904011679

79. Koziara KB, Stroet M, Malde AK, Mark AE (2014) Testing and validation of the automated topology builder (ATB) version 2.0: prediction of hydration free enthalpies. J Comput Aided Mol Des 28:221–233. https://doi.org/10.1007/s10822-014-9713-7
80. Sousa da Silva AW, Vranken WF (2012) ACPYPE–antechamber python parser interfacE. BMC Res Notes 5:367. https://doi.org/10.1186/1756-0500-5-367
81. Dupradeau F-Y, Pigache A, Zaffran T et al (2010) The R.E.D. Tools: advances in RESP and ESP charge derivation and force field library building. Phys Chem Chem Phys PCCP 12:7821–7839. https://doi.org/10.1039/c0cp00111b
82. Perez-Aguilar JM, Shan J, LeVine MV et al (2014) A functional selectivity mechanism at the serotonin-2A GPCR involves ligand-dependent conformations of intracellular loop 2. J Am Chem Soc 136:16044–16054. https://doi.org/10.1021/ja508394x
83. Kaczor AA, Rutkowska E, Bartuzi D et al (2016) Computational methods for studying GPCRs. Methods Cell Biol 132:359–399. https://doi.org/10.1016/bs.mcb.2015.11.002
84. Ng HW, Laughton CA, Doughty SW (2014) Molecular dynamics simulations of the adenosine A2a receptor in POPC and POPE lipid bilayers: effects of membrane on protein behavior. J Chem Inf Model 54:573–581. https://doi.org/10.1021/ci400463z
85. Ng HW, Laughton CA, Doughty SW (2013) Molecular dynamics simulations of the adenosine A2a receptor: structural stability, sampling, and convergence. J Chem Inf Model 53:1168–1178. https://doi.org/10.1021/ci300610w
86. LeVine MV, Weinstein H (2014) NbIT–a new information theory-based analysis of allosteric mechanisms reveals residues that underlie function in the leucine transporter LeuT. PLoS Comput Biol 10:e1003603. https://doi.org/10.1371/journal.pcbi.1003603
87. McClendon CL, Friedland G, Mobley DL et al (2009) Quantifying correlations between allosteric sites in thermodynamic ensembles. J Chem Theory Comput 5:2486–2502. https://doi.org/10.1021/ct9001812
88. Wess J (2005) Allosteric binding sites on muscarinic acetylcholine receptors. Mol Pharmacol 68:1506–1509. https://doi.org/10.1124/mol.105.019141
89. Kruse AC, Ring AM, Manglik A et al (2013) Activation and allosteric modulation of a muscarinic acetylcholine receptor. Nature 504:101–106. https://doi.org/10.1038/nature12735
90. Christopher JA, Aves SJ, Bennett KA et al (2015) Fragment and structure-based drug discovery for a class C GPCR: discovery of the mGlu5 negative allosteric modulator HTL14242 (3-Chloro-5-[6-(5-fluoropyridin-2-yl)pyrimidin-4-yl]benzonitrile). J Med Chem 58:6653–6664. https://doi.org/10.1021/acs.jmedchem.5b00892
91. Doré AS, Okrasa K, Patel JC et al (2014) Structure of class C GPCR metabotropic glutamate receptor 5 transmembrane domain. Nature 511:557–562. https://doi.org/10.1038/nature13396
92. Wu H, Wang C, Gregory KJ et al (2014) Structure of a class C GPCR metabotropic glutamate receptor 1 bound to an allosteric modulator. Science 344:58–64. https://doi.org/10.1126/science.1249489
93. Li S, Shen Q, Su M et al (2016) Alloscore: a method for predicting allosteric ligand-protein interactions. Bioinformatics 32:1574–1576. https://doi.org/10.1093/bioinformatics/btw036
94. Huang W, Lu S, Huang Z et al (2013) Allosite: a method for predicting allosteric sites. Bioinformatics 29:2357–2359. https://doi.org/10.1093/bioinformatics/btt399
95. Planesas JM, Pérez-Nueno VI, Borrell JI, Teixidó J (2015) Studying the binding interactions of allosteric agonists and antagonists of the CXCR4 receptor. J Mol Graph Model 60:1–14. https://doi.org/10.1016/j.jmgm.2015.05.004
96. Hui W-Q, Cheng Q, Liu T-Y, Ouyang Q (2016) Homology modeling, docking, and molecular dynamics simulation of the receptor GALR2 and its interactions with galanin and a positive allosteric modulator. J Mol Model 22:90. https://doi.org/10.1007/s00894-016-2944-x
97. Garcia-Perez J, Rueda P, Alcami J et al (2011) Allosteric model of maraviroc binding to CC chemokine receptor 5 (CCR5). J Biol Chem 286:33409–33421. https://doi.org/10.1074/jbc.M111.279596
98. Marquer C, Fruchart-Gaillard C, Letellier G et al (2011) Structural model of ligand-G protein-coupled receptor (GPCR) complex based on experimental double mutant cycle data: MT7 snake toxin bound to dimeric

hM1 muscarinic receptor. J Biol Chem 286:31661–31675. https://doi.org/10.1074/jbc.M111.261404

99. Ragnarsson L, Wang C-IA, Andersson Å et al (2013) Conopeptide ρ-TIA defines a new allosteric site on the extracellular surface of the α1B-adrenoceptor. J Biol Chem 288:1814–1827. https://doi.org/10.1074/jbc.M112.430785
100. Mukund S, Shang Y, Clarke HJ et al (2013) Inhibitory mechanism of an allosteric antibody targeting the glucagon receptor. J Biol Chem 288:36168–36178. https://doi.org/10.1074/jbc.M113.496984
101. Lane JR, Chubukov P, Liu W et al (2013) Structure-based ligand discovery targeting orthosteric and allosteric pockets of dopamine receptors. Mol Pharmacol 84:794–807. https://doi.org/10.1124/mol.113.088054
102. de Graaf C, Rein C, Piwnica D et al (2011) Structure-based discovery of allosteric modulators of two related class B G-protein-coupled receptors. ChemMedChem 6:2159–2169. https://doi.org/10.1002/cmdc.201100317
103. Kubas H, Meyer U, Krueger B et al (2013) Discovery, synthesis, and structure-activity relationships of 2-aminoquinazoline derivatives as a novel class of metabotropic glutamate receptor 5 negative allosteric modulators. Bioorg Med Chem Lett 23:4493–4500. https://doi.org/10.1016/j.bmcl.2013.06.049
104. Noeske T, Jirgensons A, Starchenkovs I et al (2007) Virtual screening for selective allosteric mGluR1 antagonists and structure-activity relationship investigations for coumarine derivatives. ChemMedChem 2:1763–1773. https://doi.org/10.1002/cmdc.200700151
105. Mueller R, Dawson ES, Niswender CM et al (2012) Iterative experimental and virtual high-throughput screening identifies metabotropic glutamate receptor subtype 4 positive allosteric modulators. J Mol Model 18:4437–4446. https://doi.org/10.1007/s00894-012-1441-0
106. Mueller R, Rodriguez AL, Dawson ES et al (2010) Identification of metabotropic glutamate receptor subtype 5 potentiators using virtual high-throughput screening. ACS Chem Neurosci 1:288–305. https://doi.org/10.1021/cn9000389
107. Mueller R, Dawson ES, Meiler J et al (2012) Discovery of 2-(2-benzoxazoyl amino)-4-aryl-5-cyanopyrimidine as negative allosteric modulators (NAMs) of metabotropic glutamate receptor 5 ($mGl\mu_5$): from an artificial neural network virtual screen to an in vivo tool compound. ChemMedChem 7:406–414. https://doi.org/10.1002/cmdc.201100510
108. Omer A, Prasad CS (2012) Designing allosteric modulators for active conformational state of m-glutamate G-protein coupled receptors. Bioinformation 8:170–174. https://doi.org/10.6026/97320630008170
109. Jang JW, Cho N-C, Min S-J et al (2016) Novel scaffold identification of mGlu1 receptor negative allosteric modulators using a hierarchical virtual screening approach. Chem Biol Drug Des 87:239–256. https://doi.org/10.1111/cbdd.12654
110. Jiang L, Zhang X, Chen X et al (2015) Virtual screening and molecular dynamics study of potential negative allosteric modulators of mGluR1 from Chinese herbs. Molecules 20:12769–12786. https://doi.org/10.3390/molecules200712769

Chapter 14

Challenges and Opportunities in Drug Discovery of Biased Ligands

Ismael Rodríguez-Espigares, Agnieszka A. Kaczor, Tomasz Maciej Stepniewski, and Jana Selent

Abstract

The observation of biased agonism in G protein-coupled receptors (GPCRs) has provided new approaches for the development of more efficacious and safer drugs. However, in order to rationally design biased drugs, one must understand the molecular basis of this phenomenon. Computational approaches can help in exploring the conformational universe of GPCRs and detecting conformational states with relevance for distinct functional outcomes. This information is extremely valuable for the development of new therapeutic agents that promote desired conformational receptor states and responses while avoiding the ones leading to undesired side-effects.

This book chapter intends to introduce the reader to powerful computational approaches for sampling the conformational space of these receptors, focusing first on molecular dynamics and the analysis of the produced data through methods such as dimensionality reduction, Markov State Models and adaptive sampling. Then, we show how to seek for compounds that target distinct conformational states via docking and virtual screening. In addition, we describe how to detect receptor-ligand interactions that drive signaling bias and comment current challenges and opportunities of presented methods.

Key words G protein-coupled receptor, Receptor plasticity, Conformational space, Signaling bias, Drug discovery

1 Introduction

Biased agonism (or functional selectivity) of G protein-coupled receptors (GPCRs) is related to their ability to preferentially elicit a subset of responses of all possible receptor responses. This observation has opened new avenues for producing more efficacious and safer drugs [1]. However, in order to rationally design biased drugs, one must understand the molecular basis of this phenomenon. GPCRs are of extremely flexible nature which explains the existence of countless conformational receptor states. Among them, distinct conformational populations can be assigned to inactive, intermediate, or active receptor states of distinct signaling

Alexander Heifetz (ed.), *Computational Methods for GPCR Drug Discovery*, Methods in Molecular Biology, vol. 1705, https://doi.org/10.1007/978-1-4939-7465-8_14, © Springer Science+Business Media LLC 2018

pathways (e.g., G protein or arrestin-mediated pathways). Exploring the entire conformational space of GPCRs, as well as the transition between distinct conformational populations, is a major challenge for the GPCR research community. Targeting only receptor populations that are linked to beneficial therapeutic signaling pathways, and at the same time, avoiding conformational populations related to undesired side effects is a promising strategy for obtaining a new class of drugs with an improved therapeutic profile. Just to mention one example that emphasizes the advantage of drugging distinct receptor states: balanced agonists of the angiotensin type 1 receptor reduce blood pressure via G protein antagonism, but also lower undesirably cardiomyocyte contractility by their β-arrestin antagonism [2]. In this context, new biased agonists have been described to selectively couple to β-arrestin (increased cardiomyocyte contractility) while maintaining G protein antagonism (reduced blood-pressure). Such biased molecules are a favorable starting point for more efficacious drug candidates for the treatment of cardiovascular diseases.

Currently, high-resolution insights into receptor architecture are mainly provided by X-ray crystallography capturing mostly the inactive receptor state. An important milestone represents the active structure of the β_2-adrenergic receptor (β2AR) in complex with the stimulatory G protein (Gs) for adenylyl cyclase (PDB ID: 3SN6) [3] and rhodopsin coupling to arrestin (PDB ID: 4ZWJ) [4]. Despite these advances in elucidating different receptor states, many conformational states remain elusive.

In this scenario, computational approaches can help in exploring the black holes of the conformational universe of GPCRs. Our book chapter intends to introduce the reader to powerful computational approaches to sample the conformational space of GPCRs. Then, we focus on how to seek for compounds that target distinct conformational states via docking and virtual screening, as well as how to detect receptor-ligand interactions that drive signaling bias.

2 Sampling the Conformational Universe of GPCRs

To understand molecular mechanisms behind GPCR activation and pharmacological action, computational methods as molecular dynamics (MD) provide models that explain how different conformations correlate in time while generating information about receptor activation kinetics. However, two of the biggest challenges of molecular dynamics are: (1) the treatment of the huge quantities of dynamics data generated by MD and, (2) the statistical relevance or sampling quality of the obtained MD data. Here, we will focus on dimensionality reduction methods and Markov State Models (MSM) which can be combined to tackle these problems [5].

Dimensionality reduction methods are mathematical tools that reduce the number of dimensions of a specific dataset taking

2.1 Dimensionality Reduction

advantage of the interdependence among the different variables or reaction coordinates. Most popular methods are linear methods. These are based on linear transformations that create a new space with new bases (known as independent components, ICs) where the bases with less relevance are discarded considering different criteria. In this chapter, we are going to focus on two linear reduction methods: Principal Component Analysis (PCA) and Time-dependent Independent Component Analysis (tICA).

PCA is based on eigenvalue decomposition of the covariance or correlation matrix of the input data that yields several eigenvectors (or ICs), in this case, known as principal components (PC). These represent processes in our system that best explain variance of the data. PCA has been continuously used for the study of the conformational space of proteins [6–9]. In the case of GPCRs, PCA is useful to analyze conformational changes on the whole protein or on specific protein regions [10, 11]. However, PCA is not able to take advantage of the kinetic information obtained from MD simulations as it is variance-based.

tICA is another linear method but kinetic-based instead of variance-based. Using as input the chosen reaction coordinates, tICA obtain ICs by solving the following eigenvalues problem (Eq. 1):

$$C_\tau r_i = C_0 \lambda_i(\tau) r_i \tag{1}$$

$$C_0 = (X_t - \langle X \rangle_t)^T (X_t - \langle X \rangle_t) \frac{1}{N-1} \tag{2}$$

$$C_\tau = (X_t - \langle X \rangle_t)^T (X_{t+\tau} - \langle X \rangle_t) \frac{1}{N-1-\tau} \tag{3}$$

where X_t is a sequence of multivariate data, C_0 is the covariance matrix (Eq. 2), C_τ is the time-lagged covariance matrix (Eq. 3), r_i are the ICs, $\lambda_i(\tau)$ are the associated eigenvalues, and τ is the time-lag (in discrete time), an arbitrary parameter that must be adjusted in every case [12, 13]. In tICA, ICs are ordered from slowest to faster processes being a useful reduction method for data pre-processing on Markov State Model (MSM) construction (*see* below) [14]. Furthermore, it can be also used on MD data in the same fashion as a PCA. However, kinetic data is necessary for generating the time-lagged covariant matrix, thus tICA cannot be easily used for comparing MD data with non-kinetic data (e.g., X-ray crystal structures).

2.2 Markov State Models and Adaptive Sampling

Although dimensionality reduction methods are useful in MD analysis, they have problems when obtaining long-living relevant metastable states from the protein conformational space as they cannot confirm their energy landscape convergence. Grossfield et al. have studied the convergence of PCA on membrane proteins, especially GPCRs, and they concluded that 26 simulations of 100 ns are not enough sample for the convergence of the simulation principal nodes or quantifying fluctuations in several protein regions of rhodopsin [15]. Hence, in order to study activation/inactivation pathway rates or activation/inactivation models, one must ensure enough sampling. Unfortunately, the above-mentioned methods do not tell us anything about energetic convergence or sampling quality. For this reason, we propose Markov State Models (MSMs) to accomplish this purpose. An MSM is a mathematical model that assumes "Markovianity," which means that the kinetics are modeled by a memoryless jump process between states. A summary of the MSM construction process can be seen at Fig. 1.

The process of building an MSM usually begins with a dimensionality reduction applied to a set of reaction coordinates (Fig. 1a–c). This first step is recommended because (1) it reduces the computational time and memory needed for MSM estimation, (2) it reduces the noise created by not so important processes, and (3) it improves further clustering of data. Then, continuous data is discretized using one of the available data clustering algorithms, such as regular spatial clustering [16, 17], k-means [18], or Mini Batch K-Means [19] (Fig. 1d, e). By slicing the time dimension in intervals, which size is called time-lag, time is also discretized. Finally, the transition probability matrix is estimated considering every jump between clusters (states) as transition produced at each discontinuous time unit (Fig. 1f) [20]. Once our MSM is computed, we can obtain coarse-grained states (or macro-states) by methods such as Robust Perron Cluster Cluster Analysis (PCCA+) [21], which are candidates to become metastable states (Fig. 1g). Furthermore, possible pathways between states or metastable states, and their fluxes and reaction rates can also be analyzed through transition path theory (TPT) algorithms (Fig. 1h) [22]. The reliability of our model, energetic convergence and sampling quality can be tested by several methods such as Swope-Pitera eigenvalue test [23], information theory approaches [24], Chapman-Kolmogorov tests (Fig. 1i) [22], and Bayesian Model selection approaches [25]. Finally, with adaptive sampling, we can improve our model iteratively. The MSM transition matrix is able to resample our MD data for underexplored conformational states/ macro-states from which it generates new starting points in order to increase the sampling of poorly visited conformational space regions (Fig. 1j). Thus, several shorter MD trajectories can be simulated (~10–100 ns) instead of longer ones improving computational power parallelization (Fig. 1b) [26].

Disadvantages of MSM are that systems with slow dynamics are hard to sample, and may need huge quantities of MD data and longer trajectories. In addition, a proper initial state clustering is needed as it determines in a great measure the quality of the future MSM [20]. Some python implementations of these methods are freely available, such as PyEMMA 2 [13], and the MD analysis framework HTMD which is freely available for non-commercial use [27].

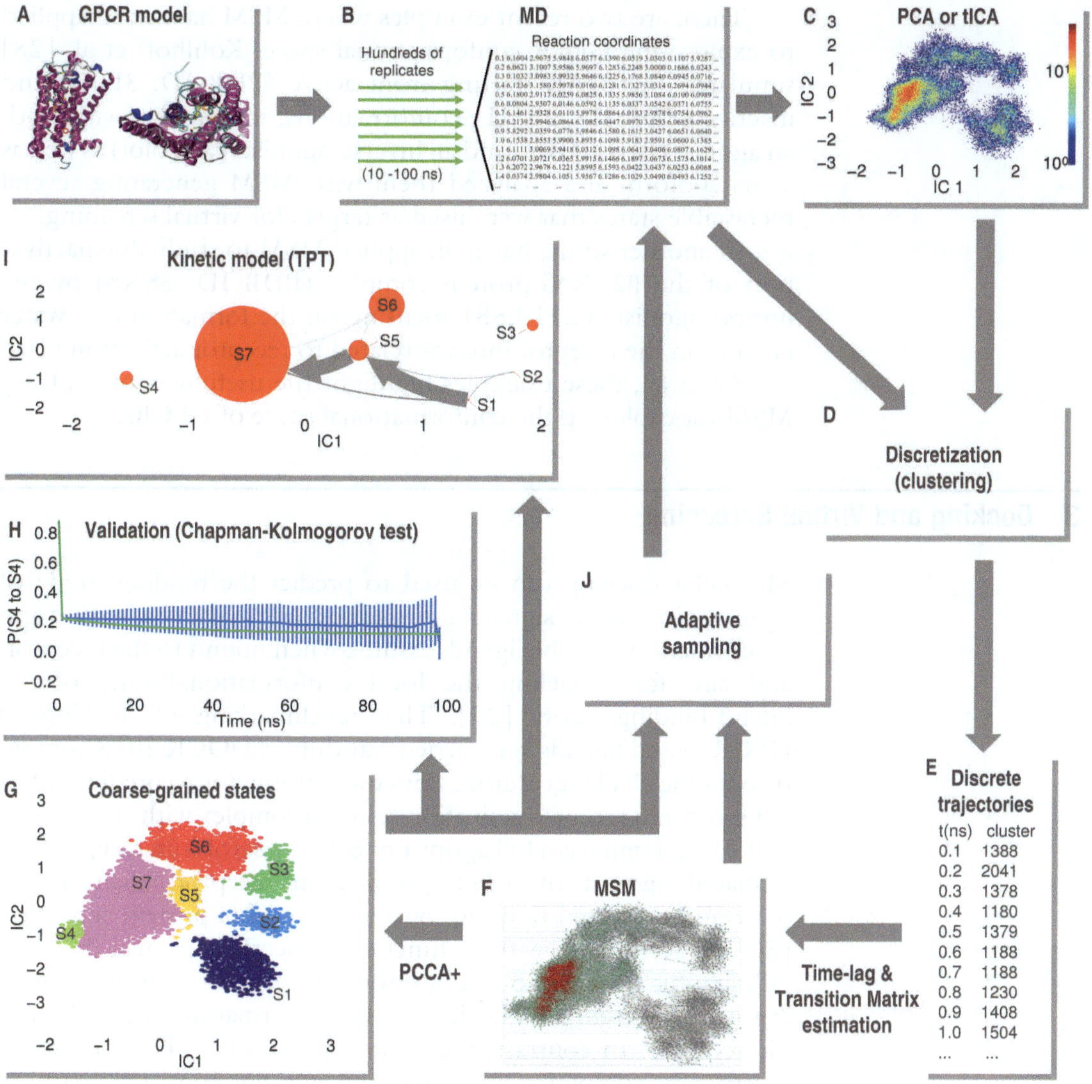

Fig. 1 Workflow of GPCR conformational analysis via MSM and adaptive sampling methods. (**a**) 3D representation of GPCR structural model backbone and residues selected for their study through MSM. (**b**) Relevant reaction coordinates are obtained from MD simulation of the GPCR model. (**c**) Bidimensional histogram of two first IC obtained from dimensionality reduction of reaction coordinates. (**d**) Discretization of reaction coordinates or relevant ICs obtained in dimensionality reduction. (**e**) Example of discretized trajectories. (**f**) Connectivity between the different clusters/states (nodes, **d**–**e**) of the build MSM. Equilibrium probabilities shown in a color-scale from *blue* to *red*. (**g**) States with high probability membership at coarse-grained states (macro-states) obtained from PCCA+. (**h**) Chapman-Kolmogorov test for transition probability from macro-state S4 to itself. Transition matrix propagation (*green*) against estimated from MD data with standard error bars (*blue*). (**i**) Kinetic model obtained by TPT analysis. Equilibrium probabilities of macro-states as node width. *Arrow* width is proportional to net fluxes between macro-states. (**j**) Adaptive sampling is initiated if more input data from MD is needed for MSM convergence. Reproduced from Rodríguez-Espigares, I., Kaczor, A. A., and Selent, J. (2016). In silico Exploration of the Conformational Universe of GPCRs. Molecular Informatics, 35 (6–7), 227–237

There are two recent examples where MSM have been applied to explore the GPCR conformational space. Kohlhoff et al. [28] simulated the β2AR starting from active (PDB ID: 3P0G) and inactive (PDB ID: 2RH1) conformations, in complex with both an agonist (BI-167107) and an inverse-agonist (carazolol) as well as in its apoform and analyzed them with MSM generating several metastable states that were used as targets for virtual screening.

In another study, Bai et al. applied MSM to study the inactivation of the β2AR-G-protein complex (PDB ID: 3SN6) by the inverse-agonist ICI118,551 focusing on the formation of a water channel in the receptor interior related to receptor activation [10].

All in all, these examples highlight the usefulness of applying MSM for exploring the conformational space of GPCRs.

3 Docking and Virtual Screening

Molecular docking can be used to predict the binding mode of ligands with various signaling properties such as the orientation and conformation that the ligands assume when bound to the receptor, and also for modeling the local conformational state of the ligand-binding cavity [29]. The benchmarking of docking of GPCR biased ligands was carried out during DOCK 2013 competition. One challenge during this competition was to predict different serotonin receptor activation states in complex with ergotamine [30]. Ergotamine is a full agonist of 5-HT_{1B} serotonin receptor but a biased agonist of 5-HT_{2B} serotonin receptor, eliciting full β-arrestin-mediated response but only partial G protein activation [30]. In accordance with its functional selectivity profile, the crystallographic model of 5-HT_{1B} serotonin receptor in complex with ergotamine resembles classical active conformation of β_2-adrenergic receptor. In contrast, the conformation of 5-HT_{2B} serotonin receptor in complex with this ligand is more consistent with an inactive state in the transmembrane helix (TM) and 6 region but an active state in the TM7 [30]. Accordingly, many of the submitted models successfully predicted the activation state of 5-HT_{1B}, but not the biased state of 5-HT_{2B} which demonstrates that application of molecular docking for predicting functional selectivity of GPCR ligands can be challenging.

In recent years, a limited number of successful and in some cases experimentally verified molecular docking experiments to construct complexes of GPCRs with biased ligands have been published. These efforts focus either on indicating receptor residues involved in biased signaling [31, 32] or deciphering which ligand moieties govern signaling to particular intracellular partners [33].

Yiu-Ho Woo et al. [31] determined that tyrosine 308 is necessary for ligand-directed Gs protein-biased signaling of β_2-

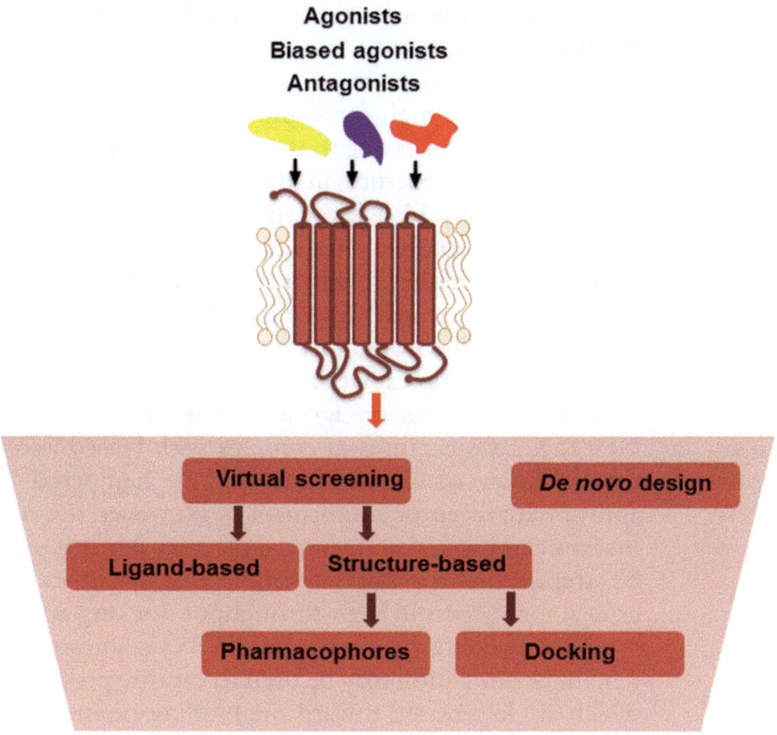

Fig. 2 Identification of biased ligands using virtual screening and de novo design

adrenergic receptor. Zhang et al. [32] used molecular docking and site-directed mutagenesis to identify specific interactions between angiotensin AT_1 receptor and different ligands with inverse agonist, neutral antagonist, or agonist activities. They found that mutation $N111^{3.35}A$ in the putative sodium binding site affects binding of the endogenous peptide agonist angiotensin II but not the β-arrestin-biased peptide TRV120027. These two encouraging examples demonstrate that molecular docking, in particular, in combination with respective experimental techniques can be useful in the identification of receptor molecular determinants of biased signaling. Similarly, molecular docking supported by molecular dynamics simulations was successfully used by Gmeiner's group to determine ligand structural determinants of biased signaling for dopamine D_2 receptor [33].

Biased signaling and functional selectivity can be exploited to design novel GPCRs ligands with desired signaling properties which may lead to elaboration of more selective drugs with fewer side effects. As an example, compound PZM21, which is an agonist of μ-opioid receptor, activates selectively Gi protein pathway and is almost inactive toward β-arrestin pathway [34]. Compound PZM21, in contrast to morphine, does not cause respiratory depression and reinforcing activity. Discovery of biased ligands can be done mainly using virtual screening or de novo design (Fig. 2) [5, 35]. When the receptor 3D structure is known from

X-ray studies or homology modeling, high-throughput docking is a method of choice.

In general, most successful structure-based virtual screening campaigns using high-throughput docking resulted in the identification of antagonists due to the availability of an increasing number of GPCR X-ray structures in inactive conformation in complex with antagonists [36]. First X-ray structures of GPCRs in active conformation and in complex with agonists enabled searching for agonists using this technique. In general, it should be stressed that receptor conformation strongly determines the properties of ligands identified using high-throughput docking [37]. Using receptor X-ray structures in complex with antagonists or agonists favors identification of antagonists or agonists, respectively. Thus, due to few GPCR X-ray structures in complex with biased agonists available, the discovery of biased ligand may be hampered by the lack of appropriate receptor conformation. However, the receptor conformation can be adjusted using molecular dynamics [37]. It should be stressed that application of molecular dynamics seems nowadays crucial for successful high-throughput docking application.

In this context, Tarcsay et al. [38] suggested that molecular dynamics simulation may be used to capture protein conformations which are less biased toward the binding points of the chemotype that has been crystalized to obtain the X-ray structure.

Kohlhoff et al. [28] applied Markov State Models on molecular dynamics data obtained by cloud-computing techniques to gain insights into the conformational space and activation mechanisms of GPCRs (*see* also Subheading 2). Virtual screening (high-throughput docking) on obtained receptor conformations with known agonists, antagonists and decoys resulted in higher accuracy on predicting possible targets than following the same procedure only with crystal structures.

Bhattacharya and Vaidehi [39] applied coarse-grain molecular dynamics simulations to decipher the changes in the potential energy landscape of the β_2-adrenergic receptor by two full agonists, two partial agonists, and an inverse agonist, starting from the receptor X-ray structure in complex with an inverse agonist, carazolol. Virtual screening with a salbutamol-stabilized conformation demonstrated enrichment of non-catechol agonists over a norepinephrine-stabilized conformation which was produced by a different activation pathway.

Kakarala and Jamil [40] used in silico methods, including molecular docking and molecular dynamics to identify protease activated receptor 2 (PAR2)-biased ligands-specific conformations and agonists and antagonists from the GPCR ligand library (GLL), which may induce biased signaling in PAR2 using the concept of existence of multiple ligand-stabilized receptor conformations.

There are also examples of application of high-throughput docking only. Gandhimathi and Sowdhamini [41] performed virtual screening and docking studies with both active and inactive state models of serotonin 5-HT_{1A} receptors identifying agonist-like and antagonist-like compounds. Molecular docking can be also supported by using protein-ligand interaction fingerprints (IFPs) to post-process the docking poses as it was done by Kooistra et al. [42] for adrenergic β_1 and β_2 receptors. Finally, de novo design techniques can be applied to develop novel ligands as it was done for novel selective nanomolar and ligand-efficient serotonin 5-HT_{2B} receptor ligands [43].

In summary, molecular docking and subsequent successful structure-based virtual screening to identify biased ligands of GPCRs require the availability of appropiate receptor conformations responsible for functional selectivity. When an adequate receptor X-ray structure is not available, molecular dynamics supported by experimental data can be used to adjust receptor conformation. It can be expected that more and more successful applications of molecular docking, in particular high-throughput docking to identify functionally selective ligands, will be reported in next years along with new X-ray structures, new site-directed mutagenesis data and development of molecular dynamics techniques.

4 Detecting Receptor-Ligand Interaction that Drives Signaling Bias Using MD

As mentioned in the previous paragraph, the current understanding of the signaling processes is based on the fact that a GPCR can exist in a multitude of conformations. Biased ligands have the propensity to establish specific interaction with the receptor which stabilizes a distinct receptor conformation linked to signaling outcome. A recent study successfully used molecular dynamics to identify receptor-ligand interactions at the serotonin 5-HT_{2A} receptor that are relevant to signaling bias [44]. The 5-HT_{2A} receptor is targeted by hallucinogenic drugs of abuse [45] as well as by second generation antipsychotic drugs [46], which function as antagonists at these receptors [47]. The studied receptor can initiate many diverse signaling pathways, among which is release of arachidonic acid (AA) and accumulation of inositol phosphate (IP) [48]. The search for biased signaling molecules was driven by the finding that a full receptor inactivation has been suggested to be counterproductive for the treatment of schizophrenia [49].

In order to detect ligand-receptor contacts that are relevant for the type of signaling outcome, Martí-Solano et al. compared the natural and balanced agonist serotonin (AA and IP pathway) to a known biased molecule 2C-N (biased toward the AA pathway) [44]. Complexes of serotonin and 2C-N with the active 5-HT_{2A} receptor were subjected to 100 replicates of 100 ns each. A dynamic interaction fingerprint was extracted from the accumulated simulation data and revealed preferred interaction sites. Interestingly, the balanced ligand serotonin established polar contacts with TM5 and TM6 whereas the AA-biased ligand 2C-N forms them only with TM6. Based on this observation, the authors envisaged that they could tune the natural agonist serotonin into an IP-biased signaling molecule by avoiding the AA-biased TM6 interaction. Intriguingly, using this strategy, the authors were able to rationally discover new molecules with an unprecedented level of IP bias [44].

Another challenging goal of describing signaling bias using MD simulation is to understand how protein-ligand interactions propagate through an allosteric network of interactions in the protein to yield a distinct signaling response. Recent attempts have displayed the promise of these techniques in obtaining relevant data [50]. With the Carma package it is possible to calculate a covariance matrix between defined groups of atoms in a simulation [51, 52]. If two residues fluctuate in a correlated way, especially if they are distantly placed in the protein, it is highly probable that there are allosteric interactions between them. Analyzing this matrix can help to detect propagating signals from the extracellular side to the intracellular one. The covariance data can be conveniently visualized and analyzed using the Network Viewer module, available in VMD [53]. This approach was recently utilized to study biased signaling in the μ-opioid receptor [54]. The authors found that in the structure bound to the unbiased agonist—morphine, there is intense allosteric communication between the orthosteric binding site and the intracellular ends of helixes TM3 and TM6. In contrast, when replacing morphine with the G-protein-biased agonist—oliceridine, they no longer observed such communication with the end of the TM6 helix while retaining it between the binding site and end of helix TM3. Such results may help rationalize, what phenomena are responsible for such a divergent signaling response between the two molecules.

In summary, the field of MD simulations is rapidly progressing. Because of utilizing massive parallel graphical processor unit (GPU) computation techniques, the current computational speeds progression slightly exceeds Moore's law, doubling every 1,3 years [55]. Also various algorithms are developed, to enhance the sampling of the energetic landscape [56, 57], and are successfully applied to GPCR studies [5, 58]. Thus, it appears that MD studies will progressively reflect natural phenomena such as signaling bias with better quality. Utilizing those techniques can provide valuable data for structure-based drug design, ranging from obtaining molecules with a tailored signaling profile, to predicting their affinity and binding kinetics.

5 Concluding Remarks About Challenges and Opportunities

The existence of multiple receptor conformational states linked to distinct signaling outcomes has changed our perception of drug design and is the basis for producing more efficacious and safer drugs. This book chapter describes the advances on how computational chemists and biologists try to capture these different receptor conformations and transitions among them starting from available structural information.

For instance, molecular dynamics techniques combined with dimensional reduction methods and MSM have been proven to be useful to explore the conformational space of GPCRs. Once stable/metastable conformational states have been detected by experiment/computation, several success stories show that distinct states can be exploited for the search of bias ligands via docking and virtual screening. It should be mentioned that the selection of relevant conformational receptor states for virtual screening can be complicated, as we often do not know the implication of specific receptor states for the signaling outcome. Obviously, more structural information on complexes between GPCRs and diverse signal transducers is necessary for easing the selection of physiologically relevant receptor states. In addition to high-throughput screening, also structural approaches that take advantage of dynamic fingerprints of ligand-receptor interaction obtained by MD simulations have been proven to be efficient in rationally designing novel biased agonists.

Despite those success stories, the discovery of specific molecules that stabilize relevant receptor states is not trivial at all. This can be explained by the fact that only subtle structural differences exist between complexes that elicit a balanced or biased signal. To address this challenge and to be more successful in detecting new biased ligands, more structural information on different receptor states related to signaling bias is needed. In this respect, we envisage that in near future receptor structures with definite coupling and signaling capacities will be available thanks to the current developments in solving distinct receptor states.

Considering available structural insights on GPCRs, the on-going process in obtaining more structural information on different receptor states combined with the potential of computational-assisted drug discovery turns our present into an exciting era for developing novel drugs with higher efficacy and safer therapeutic profile.

Acknowledgments

I.R.-E. acknowledges financial support from Secretaria d'Universitats i Recerca del Departament d'Economia i Coneixement de la Generalitat de Catalunya (2015 FI_B00145). The paper was developed using the equipment purchased within the project "The equipment of innovative laboratories doing research on new medicines used in the therapy of civilization and neoplastic diseases" within the Operational Program Development of Eastern Poland 2007-2013, Priority Axis I Modern Economy, operations I.3 Innovation promotion.

T.M.S. acknowledges financial support from Hospital del Mar Medical Research Institute.

Finally, J.S. acknowledges financial support from Instituto de Salud Carlos III FEDER (PI15/00460).

References

1. Martí-Solano M, Guixà-González R, Sanz F et al (2013) Novel insights into biased agonism at G protein-coupled receptors and their potential for drug design. Curr Pharm Des 19:5156–5166
2. Violin JD, Dewire SM, Yamashita D et al (2010) Selectively engaging B-arrestins at the angiotensin II type 1 receptor reduces blood pressure and increases cardiac performance. Pharmacol Ther 335:572–579. https://doi.org/10.1124/jpet.110.173005
3. Rosenbaum DM, Cherezov V, Hanson MA et al (2007) GPCR engineering yields high-resolution structural insights into 2-adrenergic receptor function. Science 318:1266–1273. https://doi.org/10.1126/science.1150609
4. Kang Y, Zhou XE, Gao X et al (2015) Crystal structure of rhodopsin bound to arrestin by femtosecond X-ray laser. Nature 523:561–567. https://doi.org/10.1038/nature14656
5. Rodríguez-Espigares I, Kaczor AA, Selent J (2016) *In silico* exploration of the conformational universe of GPCRs. Mol Inform 35:227–237. https://doi.org/10.1002/minf.201600012
6. Altis A, Nguyen PH, Hegger R, Stock G (2007) Dihedral angle principal component analysis of molecular dynamics simulations. J Chem Phys 126:244111. https://doi.org/10.1063/1.2746330
7. Brown WM, Martin S, Pollock SN et al (2008) Algorithmic dimensionality reduction for molecular structure analysis. J Chem Phys 129:64118
8. Lange OF, Grubmüller H (2006) Generalized correlation for biomolecular dynamics. Proteins 62:1053–1061. https://doi.org/10.1002/prot.20784
9. Teodoro ML, Phillips GN, Kavraki LE (2003) Understanding protein flexibility through dimensionality reduction. J Comput Biol 10:617–634. https://doi.org/10.1089/10665270360688228
10. Bai Q, Pérez-Sánchez H, Zhang Y et al (2014) Ligand induced change of β2 adrenergic receptor from active to inactive conformation and its implication for the closed/open state of the water channel: insight from molecular dynamics simulation, free energy calculation and Markov state model analysis. Phys Chem Chem Phys 16:15874–15885. https://doi.org/10.1039/c4cp01185f
11. Ng HW, Laughton CA, Doughty SW (2013) Molecular dynamics simulations of the adenosine A2a receptor: structural stability, sampling, and convergence. J Chem Inf Model 53:1168–1178. https://doi.org/10.1021/ci300610w
12. Pérez-Hernández G, Paul F, Giorgino T et al (2013) Identification of slow molecular order parameters for Markov model construction. J Chem Phys 139:15102. https://doi.org/10.1063/1.4811489
13. Scherer MK, Trendelkamp-Schroer B, Paul F et al (2015) PyEMMA 2: a software package for estimation, validation, and analysis of Markov models. J Chem Theory Comput 11:5525–5542. https://doi.org/10.1021/acs.jctc.5b00743

14. Razavi AM, Wuest WM, Voelz VA (2014) Computational screening and selection of cyclic peptide hairpin mimetics by molecular simulation and kinetic network models. J Chem Inf Model 54:1425–1432. https://doi.org/10.1021/ci500102y
15. Grossfield A, Feller SE, Pitman MC (2007) Convergence of molecular dynamics simulations of membrane proteins. Proteins 67:31–40. https://doi.org/10.1002/prot.21308
16. Hartigan AJ (1975) Clustering algorithms. John Wiley & Sons, Inc, Hoboken, NJ
17. Prinz J-H, Wu H, Sarich M et al (2011) Markov models of molecular kinetics: generation and validation. J Chem Phys 134:174105. https://doi.org/10.1063/1.3565032
18. Arthur D, Vassilvitskii S (2007) K-means++: the advantages of careful seeding. In: Proceedings of the Eighteenth Annual ACM-SIAM Symposium on Discrete Algorithms. Society for Industrial and Applied Mathematics, Philadelphia, PA, pp 1027–1035
19. Sculley D (2010) Web-scale k-means clustering. In: Proceedings of the 19th international conference on World wide web–WWW '10. ACM Press, New York, NY, p 1177
20. Pande VS, Beauchamp K, Bowman GR (2010) Everything you wanted to know about Markov state models but were afraid to ask. Methods 52:99–105. https://doi.org/10.1016/j.ymeth.2010.06.002
21. Röblitz S, Weber M (2013) Fuzzy spectral clustering by PCCA+: application to Markov state models and data classification. Adv Data Anal Classif 7:147–179. https://doi.org/10.1007/s11634-013-0134-6
22. Noé F, Schütte C, Vanden-Eijnden E et al (2009) Constructing the equilibrium ensemble of folding pathways from short off-equilibrium simulations. Proc Natl Acad Sci U S A 106:19011–19016. https://doi.org/10.1073/pnas.0905466106
23. Swope WC, Pitera JW, Suits F (2004) Describing protein folding kinetics by molecular dynamics simulations. 1. Theory. J Phys Chem B 108:6571–6581. https://doi.org/10.1021/jp037421y
24. Park S, Pande VS (2006) Validation of Markov state models using Shannon's entropy. J Chem Phys 124:54118. https://doi.org/10.1063/1.2166393
25. Bacallado S, Chodera JD, Pande V (2009) Bayesian comparison of Markov models of molecular dynamics with detailed balance constraint. J Chem Phys 131:45106. https://doi.org/10.1063/1.3192309
26. Bowman GR, Ensign DL, Pande VS (2010) Enhanced modeling via network theory: adaptive sampling of Markov state models. J Chem Theory Comput 6:787–794. https://doi.org/10.1021/ct900620b
27. Doerr S, Harvey MJ, Noé F, De Fabritiis G (2016) HTMD: high-throughput molecular dynamics for molecular discovery. J Chem Theory Comput 12:1845–1852. https://doi.org/10.1021/acs.jctc.6b00049
28. Kohlhoff KJ, Shukla D, Lawrenz M et al (2013) Cloud-based simulations on Google Exacycle reveal ligand modulation of GPCR activation pathways. Nat Chem 6:15–21. https://doi.org/10.1038/nchem.1821
29. Bruno A, Costantino G (2012) Molecular dynamics simulations of G protein-coupled receptors. Mol Inform 31:222–230. https://doi.org/10.1002/minf.201100138
30. Kufareva I, Katritch V, Participants of GPCR Dock 2013 et al (2014) Advances in GPCR modeling evaluated by the GPCR Dock 2013 assessment: meeting new challenges. Structure 22:1120–1139. https://doi.org/10.1016/j.str.2014.06.012
31. Woo AY-H, Jozwiak K, Toll L et al (2014) Tyrosine 308 is necessary for ligand-directed Gs protein-biased signaling of β2-adrenoceptor. J Biol Chem 289:19351–19363. https://doi.org/10.1074/jbc.M114.558882
32. Zhang H, Unal H, Desnoyer R et al (2015) Structural basis for ligand recognition and functional selectivity at angiotensin receptor. J Biol Chem 290:29127–29139. https://doi.org/10.1074/jbc.M115.689000
33. Weichert D, Banerjee A, Hiller C et al (2015) Molecular determinants of biased agonism at the dopamine D_2 receptor. J Med Chem 58:2703–2717. https://doi.org/10.1021/jm501889t
34. Manglik A, Lin H, Aryal DK et al (2016) Structure-based discovery of opioid analgesics with reduced side effects. Nature 537:185–190. https://doi.org/10.1038/nature19112
35. Kaczor AA, Rutkowska E, Bartuzi D et al (2016) Chapter 17 – computational methods for studying G protein-coupled receptors (GPCRs). Methods Cell Biol 132:359–399. https://doi.org/10.1016/bs.mcb.2015.11.002
36. Topiol S, Sabio M (2015) The role of experimental and computational structural approaches in 7TM drug discovery. Expert Opin Drug Discovery 10:1071–1084. https://doi.org/10.1517/17460441.2015.1072166

37. Costanzi S (2014) Modeling G protein-coupled receptors in complex with biased agonists. Trends Pharmacol Sci 35:277–283. https://doi.org/10.1016/j.tips.2014.04.004
38. Tarcsay A, Paragi G, Vass M et al (2013) The impact of molecular dynamics sampling on the performance of virtual screening against GPCRs. J Chem Inf Model 53:2990–2999. https://doi.org/10.1021/ci400087b
39. Bhattacharya S, Vaidehi N (2010) Computational mapping of the conformational transitions in agonist selective pathways of a G-protein coupled receptor. J Am Chem Soc 132:5205–5214. https://doi.org/10.1021/ja910700y
40. Kakarala KK, Jamil K (2016) Biased signaling: potential agonist and antagonist of PAR2. J Biomol Struct Dyn 34:1363–1376. https://doi.org/10.1080/07391102.2015.1079556
41. Gandhimathi A, Sowdhamini R (2015) Molecular modelling of human 5-hydroxytryptamine receptor (5-HT 2A) and virtual screening studies towards the identification of agonist and antagonist molecules. J Biomol Struct Dyn 34(5):952–970. https://doi.org/10.1080/07391102.2015.1062802
42. Kooistra AJ, Roumen L, Leurs R et al (2013) From heptahelical bundle to hits from the haystack: structure-based virtual screening for GPCR ligands. In: Conn PM (ed) G protein coupled receptors modeling, activation, interactions and virtual screening. Academic Press, New York, pp 279–336
43. Rodrigues T, Hauser N, Reker D et al (2015) Multidimensional de novo design reveals 5-HT2B receptor-selective ligands. Angew Chem Int Ed Engl 54(5):1551. https://doi.org/10.1002/anie.201410201
44. Marti-Solano M, Iglesias A, de Fabritiis G et al (2015) Detection of new biased agonists for the serotonin 5-HT2A receptor: modeling and experimental validation. Mol Pharmacol 87:740–746. https://doi.org/10.1124/mol.114.097022
45. Nichols DE (2004) Hallucinogens. Pharmacol Ther 101:131–181. https://doi.org/10.1016/j.pharmthera.2003.11.002
46. Meltzer H (1999) The role of serotonin in antipsychotic drug action. Neuropsychopharmacology 21:106S–115S. https://doi.org/10.1016/S0893-133X(99)00046-9
47. González-Maeso J, Sealfon SC (2009) Psychedelics and schizophrenia. Trends Neurosci 32:225–232. https://doi.org/10.1016/j.tins.2008.12.005
48. Berg KA, Stout BD, Cropper JD et al (1999) Novel actions of inverse agonists on 5-HT2C receptor systems. Mol Pharmacol 55(5):863–872
49. Kurita M, Holloway T, García-Bea A et al (2012) HDAC2 regulates atypical antipsychotic responses through the modulation of mGlu2 promoter activity. Nat Neurosci 15:1245–1254. https://doi.org/10.1038/nn.3181
50. Hertig S, Latorraca NR, Dror RO (2016) Revealing atomic-level mechanisms of protein allostery with molecular dynamics simulations. PLoS Comput Biol 12:e1004746. https://doi.org/10.1371/journal.pcbi.1004746
51. Glykos NM (2006) Software news and updates carma: a molecular dynamics analysis program. J Comput Chem 27:1765–1768. https://doi.org/10.1002/jcc.20482
52. Koukos PI, Glykos NM (2013) Grcarma: a fully automated task-oriented interface for the analysis of molecular dynamics trajectories. J Comput Chem 34:2310–2312. https://doi.org/10.1002/jcc.23381
53. Humphrey W, Dalke A, Schulten K (1996) VMD: visual molecular dynamics. J Mol Graph 14:33–38. https://doi.org/10.1016/0263-7855(96)00018-5
54. Schneider S, Provasi D, Filizola M (2016) How oliceridine (TRV-130) binds and stabilizes a μ-opioid receptor conformational state that selectively triggers G protein signaling pathways. Biochemistry 55:6456–6466. https://doi.org/10.1021/acs.biochem.6b00948
55. Perez A, Morrone JA, Simmerling C, Dill KA (2016) Advances in free-energy-based simulations of protein folding and ligand binding. Curr Opin Struct Biol 36:25–31. https://doi.org/10.1016/j.sbi.2015.12.002
56. Barducci A, Bonomi M, Parrinello M (2011) Metadynamics. WIRE Comput Mol Sci 1:826–843. https://doi.org/10.1002/wcms.31
57. Hamelberg D, Mongan J, McCammon JA (2004) Accelerated molecular dynamics: a promising and efficient simulation method for biomolecules. J Chem Phys 120:11919–11929. https://doi.org/10.1063/1.1755656
58. Miao Y, McCammon JA (2016) G-protein coupled receptors: advances in simulation and drug discovery. Curr Opin Struct Biol 41:83–89. https://doi.org/10.1016/j.sbi.2016.06.008

Chapter 15

Synergistic Use of GPCR Modeling and SDM Experiments to Understand Ligand Binding

Andrew Potterton, Alexander Heifetz, and Andrea Townsend-Nicholson

Abstract

There is a substantial amount of historical ligand binding data available from site-directed mutagenesis (SDM) studies of many different GPCR subtypes. This information was generated prior to the wave of GPCR crystal structure, in an effort to understand ligand binding with a view to drug discovery. Concerted efforts to determine the atomic structure of GPCRs have proven extremely successful and there are now more than 80 GPCR crystal structure in the PDB database, many of which have been obtained in the presence of receptor ligands and associated G proteins. These structural data enable the generation of computational model structures for all GPCRs, including those for which crystal structures do not yet exist. The power of these models in designing novel ligands, especially those with improved residence times, and for better understanding receptor function can be enhanced tremendously by combining them synergistically with historic SDM ligand binding data. Here, we describe a protocol by which historic SDM binding data and receptor models may be used together to identify novel key residues for mutagenesis studies.

Key words GPCRs, Adenosine receptors, Homology modeling, Ligand binding, Binding kinetics, Receptor, Site-directed mutagenesis

1 Introduction

1.1 Site-Directed Mutagenesis (SDM) Binding Studies

Despite their shared seven transmembrane helix structure, GPCRs recognize a wide array of ligands in many different signaling pathways [1]. Ligand specificity stems from sequence variance between receptors, at key residues. The mechanism of specificity is important to understand so that structure-based drug design can achieve high efficacy and avoid off-target side effects. To determine these key amino acid residues, site-directed mutagenesis (SDM) studies are performed. By comparing binding values for mutant compared to wild-type receptors, the influence of a given residue on ligand binding affinity or kinetics can be determined. Many mutagenesis studies have been conducted in a shotgun approach, but careful targeting of informative mutations for these experiments will

Alexander Heifetz (ed.), *Computational Methods for GPCR Drug Discovery*, Methods in Molecular Biology, vol. 1705, https://doi.org/10.1007/978-1-4939-7465-8_15,

enable more precise information about the role of a given residue to be determined.

Extensive mutagenesis studies were performed on the four adenosine receptor subtypes with the aim of determining residues that confer binding specificity for each of the receptor subtypes [2]. In general, these were based on multiple sequence alignments that had been used to identify residues, which differed between the four receptors, with the view that they might contribute to subtype-specific ligand binding. These studies led to the design of a number of different subtype-specific ligands [3]. SDM binding studies are still carried out on the adenosine receptors, mostly performing kinetic binding experiments to rationalize residence time.

1.2 Using GPCR Models to Elucidate Binding

Advances in techniques that stabilize GPCRs, which have many different conformations, have allowed a greater number of these receptors to be crystallized [4]. Further, these stabilization methods have enable receptors to be cocrystallized, generating structures with agonist bound to the receptor [5]. These methods involve specific thermostabilizing mutations and often include the engineering of a fusion domain between transmembrane helix 5 and 6. If these modifications are restored to the wild-type residues, homology modeling can be used to further increase the number of receptors for which accurate models can be obtained. In homology modeling, the model of the receptor is largely treated as static during drug docking, ignoring ligand flexibility and any conformational changes that could take place upon ligand binding. The loop regions, particularly extracellular loop 2, have been found to be involved in ligand binding [6], causing a problem for traditional GPCR homology modeling as the loop regions tend to be inaccurately predicted. Hierarchical GPCR modeling protocol (HGMP), described in Chapter 19, is a more advanced modeling workflow that addresses these problems.

The use of residue engineering and the introduction of non-GPCR sequences to stabilize receptors for improved crystallization means that computational models based on crystal structure are the best means of exploring structure-function relationships for GPCRs. Using accurate models allows the mutagenesis data to be put in the 3D context of the binding site, allowing for indirect interactions to be more easily noted. For the A_{2A} adenosine receptor, viewing mutagenesis studies in the context of a model has enabled the identification of a hydrophobic pocket, which holds the adenine ring of agonists [7]. Historic mutagenesis binding data, therefore, when explored in the context of a computational model can be of great help in understanding ligand binding at the atomic level.

2 Methods

2.1 Models Can Inform Which Residues to Mutate

The overall process of this workflow is outlined in Fig. 1. The first stage of a SDM binding experiment is to plan which residues to mutate, something that requires careful attention in order to maximize the amount of information that can be obtained. Identification of a suitable model, either crystal or homology, should be used as a starting point in planning:

1. Search for a crystal structure for the GPCR to be studied, using the PDB [8].
   (a) If there are multiple entries for the receptor, make a quick table comparing the resolution, the ligand bound, if any, and any crystallization techniques that may have altered

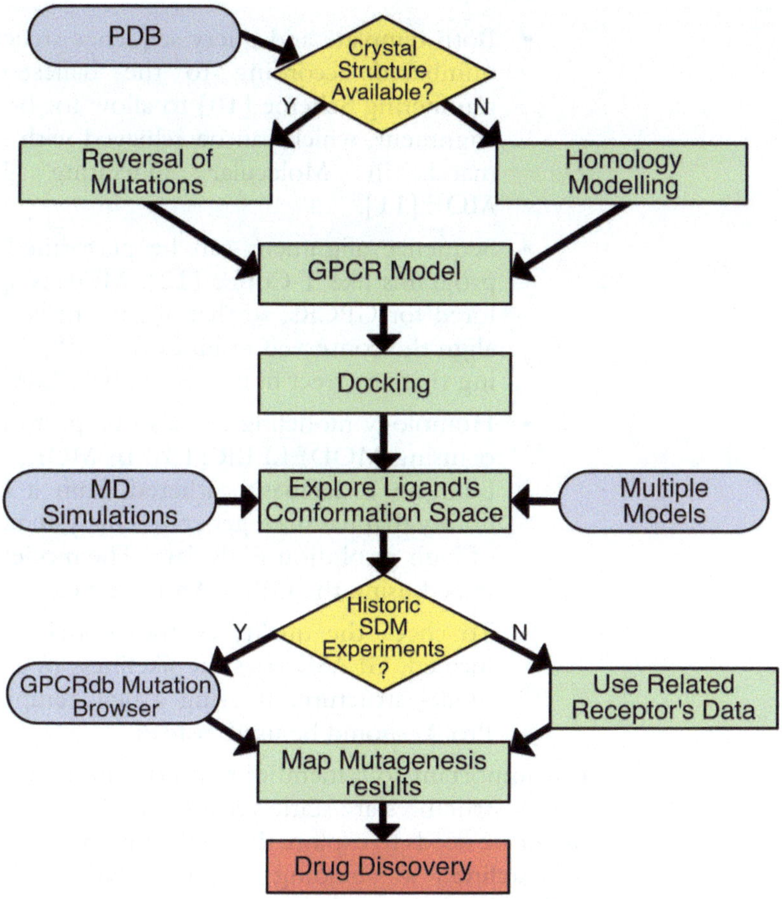

Fig. 1 Flow diagram of a simplified version of the methodology detailed in this chapter. *Diamond-shaped boxes* represent decisions that need to be made and the *rounded rectangle boxes* represent inputs that help carry out a task

the structure of the receptor, including the use of fusion domains and stabilizing mutations.

(b) If there is a single entry for the receptor of interest, it is important to note these structural features, as they will allow one to assess the reliability of using the structure for SDM binding experiment design. In cases where the receptor has been engineered to look at how the receptor functions, for example by fusion of the receptor with a G_s [9], it may be more suitable to make a homology model for experimental design.

(c) If there are no crystal structures, conduct a literature search to determine whether an appropriate template structure has previously been used for homology modeling; otherwise, a template can be generated by homology modeling using the crystal structure of a closely related GPCR:

- Both template and query sequence structure must be numbered according to the Ballesteros-Weinstein numbering Scheme [10] to allow for better sequence alignment, which can be achieved with a single command in Molecular Operating Environment, MOE [11].
- Sequence alignment can be performed in MOE or programs like T-Coffee [12]. MOE is specifically tailored for GPCRs, so that alignment is constrained to align the conserved residues and GPCR motifs, ensuring that a correct fit is likely to be achieved.
- Homology modeling can also be performed in MOE or using MODELLER [13]. In MOE, sidechains are built on a database gathered from a large rotamer library that has been generated by systematic clustering of high-resolution PDB data. The model is then minimized using the OPLS-AA force-field.
- To check the quality of the model, scoring is performed. GLAS© tests the likeliness of a GPCR to the model structure. If using a new template, however, ProS© should be used instead.

2. It is important to remember that both homology models and crystal structures are static versions of a dynamic system. One therefore needs to explore the conformation space of the ligand to define a binding pocket. Molecular dynamics (MD) simulations can be used to achieve this. An alternative to using MD simulations is to look at multiple models and docking results, as the variability in structures will give a picture of a ligand's conformational space.
3. Search for interactions that are present in the model or MD simulations. Hydrogen bonds can easily be found by predictor tools available in most molecular visualization software

packages, such as PyMOL [14] or UCSF's Chimera [15]. These software packages can also be used to view the hydrophobicity surface of the receptor, which may indicate hydrophobic interactions or buried hydrophilic interactions. Make a list of all these interactions, detailing their nature and in which model/simulation the interaction was seen (*see* **Note 1**).

4. Look at the conservation of an interaction, not just in the different models or simulations, but in other related receptors. To look at the equivalent residue in other receptors, the Ballesteros-Weinstein numbering scheme is very useful. This is a numbering system devised for class A GPCRs, allowing a given amino acid to be compared across different GPCRs. In this system, the number before the decimal point indicates what transmembrane helix the amino acid is in, the numbers after it specify the location of this residue compared to the helixes conserved residue which, by convention, is defined as 0.50. For example, the position that corresponds to 5.58 is on the fifth transmembrane helix and is eight positions away from the conserved position [10].
5. Identify the level of structural conservation at a given position, to ascertain whether the interaction is likely to be conserved. It should be noted that the lack of conservation of an interaction may also be valuable when looking at receptor specificity, particularly for models based on nearest neighbors that bind a different endogenous ligand.
6. Once a residue has been selected for mutagenesis, decide which amino acid the residue will be mutated to. The usual choice is alanine, as it prevents that residue from making interactions, is a residue that favors helix formation and is unlikely to cause steric clashes, due to its size. Alanine mutations can, on occasion, prevent stable expression of the receptor in which case other amino acids will need to be considered. In those cases, one can mutate to an amino acid that loses the functional group but retains the relative bulk of the R group. Mutating a position to two or more different amino acids can detail the nature of the interaction. To maximize the information gained from making two or more mutants, one of the mutants should retain the moiety that can make the interaction and the other should lose the interaction but retain the relative bulk of the R group. Mutation to cysteine can be useful for cross-linking experiments.

2.2 Selecting Ligands for Binding Studies Using Models and SDM Binding Data

Tailoring the selection of ligands for each mutagenesis study will maximize the information gained from the binding experiments. Four ligands are usually selected for a mutagenesis study that is intended to determine whether a residue interacts with a specific category of ligand. For radioligand binding studies, at least one of the ligands must be available as a radiolabeled ligand to allow for competition binding studies. The following steps will help make this selection of ligand:

1. Look at different interactions made between the receptor and various ligands; to achieve this crystal structures and homology models with ligands docked into their structure must exist (*see* **Note 2**).
2. From these structures, identify pairs of similar ligands that differ by a single interaction. This is the residue that will be mutated. These ligands may have similar binding affinity values because of their similar structure (*see* **Note 3**).

2.3 Models and Historic SDM Binding Data Add Value to the Analyses of New SDM Binding Experiments

After mutagenesis binding studies have been performed, the results should be analyzed in conjunction with models of the receptor to gain a more informed understanding of the nature of the binding and the changes that mutagenesis elicits. One should similarly evaluate any existing historic mutagenesis data with a view to integrating all sources of information needed to put the recently obtained results in context. This will allow for a comprehensive 3D analysis of binding.

1. Gather all mutagenesis binding study data for the receptor of interest. To find previous mutagenesis data for a given receptor, one can use GPCRdb's mutation browser: http://gpcrdb.org/mutations/ [16]. This will show details of all the mutants that have been made in a specified receptor which can be made into a database. The entries of the database must be checked to indicate whether the mutation was made as part of a binding study or if it was mutated for some other purpose, these latter entries should be removed. The database will allow one to check for historic data for the residue of interest. After the removal of nonbinding study data, the ligands used in each study and the fold differences in ligand binding values to the mutant compared to the wildtype can be added by searching the reference associated with each entry. If no mutants have been made for binding studies with the receptor of interest, another closely related receptor can be used. In these cases, the models of the two receptors should be superimposed (*see* **Note 4**), to check for equivalent residues.
2. Map the historic mutagenesis study results to the models of the receptors that have the ligands used in the binding study docked. This can be done by editing the color of residues that have significantly different binding values for a given ligand using a molecular visualization software package, such as PyMOL or Chimera. The current experimental results should also be mapped to the model.
3. Check the position of the residue that has been mutated in comparison with other residues that are colored because of a

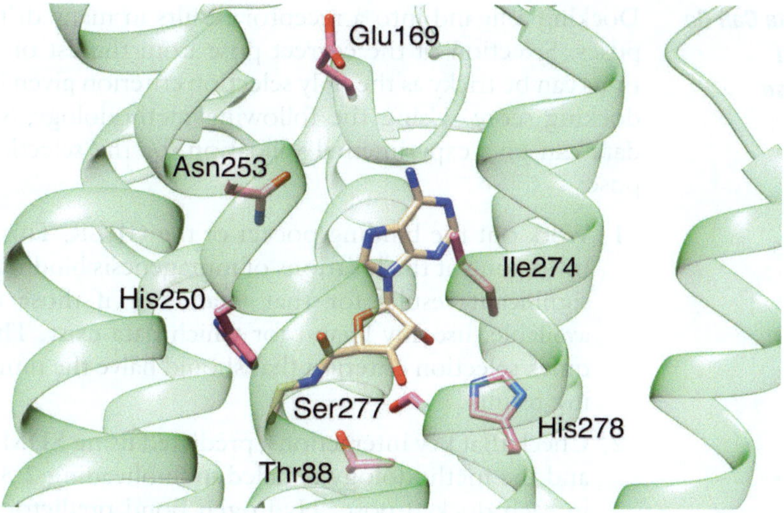

Fig. 2 A model of the A_{2A} adenosine receptor with SDM binding data for NECA mapped, using the methodology described in this protocol. Residues, which when mutated have significantly different binding affinity value compared to the wildtype, are colored in *pink* and are in stick representation. NECA, the ligand, is also shown in *stick* representation and is tan colored

significant binding difference. This comparison with other residues is not obvious when looking solely at the sequence or the residue numbers because structurally adjacent residues may be on different transmembrane helices, so the model to which the mutagenesis studies were mapped must be used. Significant results located some distance from the predicted binding pocket, as seen from docking or a crystal structure, may indicate a secondary binding site. It is important to look at interactions between the residue of interest and other residues to check for possible indirect interactions with the ligand. The mapped model will indicate whether the residues of indirect interaction play a significant role in binding. The properties of the residues surrounding the mutant residue, in the model, are also important to consider as these may reveal a hydrophobic pocket or a buried hydrophilic interaction.

Figure 2 shows an output of this workflow for the example case of understanding NECA binding to the A_{2A} adenosine receptor. SDM NECA binding data for the A_{2A} receptor was used to make the mapped model [17, 18]. This mapped model can be used to help explain the SDM binding data. Looking at Ser277 and Thr88, in the context of the mapped model, rationalized why these mutants caused loss of binding only to agonists because of their proximity to the ribose ring not present in adenosine receptor antagonists.

2.4 SDM Data Can Be Used to Select a Docking Pose

Docking a ligand into a receptor results in many distinct docking poses. Selection of the correct pose from the list of docked positions can be tricky as the only selection criterion given is an arbitrary docking score. Using the following methodology, SDM binding data can give experimental validation for the selection of a given pose:

1. Work out the binding pocket of the GPCR. This can be done by looking at the locations of mutagenesis binding studies with significant results for that ligand, or if those data are not available, use any ligand for which data exist. This will give a quick selection criterion that should halve the number of docking results.
2. Check that key interactions, predicted using SDM binding data and the methodology detailed in Subheading 2.3, are possible in each docked pose. Hydrogen bond predictor tools can be useful to help determine this but it is important to remember that the receptor is flexible, which is something that is not accounted for in docking.
3. Make certain that the orientation of the ligand corresponds to most the significant mutagenesis binding studies data; these significant residues can be colored so that they can be easily identified.

3 Notes

1. Distance measurements can be used to increase the conformational space of the ligand, overcoming the issue of having a static model. Residues within a specified distance from the ligand can be selected.
2. If one is docking an agonist, the model must be an active structure. This means that the original crystal structure must have an agonist bound to the receptor.
3. For the adenosine receptors, there are position-substituted ligands that are particularly useful for blocking a potential interaction; these can be compared to the unsubstituted ligand (e.g., CPA and *R*-PIA are both N^6-substituted agonists).
4. UCSF's Chimera has a "MatchMaker" feature [15], found under "Structure Comparison" in the "Tools" dropdown, which is good at quickly overlaying structures. This finds the best overall fit between the two receptors; however, selection can be made to just include specific residues, like those the binding site.

Acknowledgments

This work was supported by the Biotechnology and Biological Sciences Research Council (grant numbers BB/M009513/1 and BB/P004245/1) and by the EU H2020 CompBioMed project (http://www.compbiomed.eu/, 675451).

References

1. Marinissen MJ, Gutkind JS (2001) G-protein-coupled receptors and signaling networks: emerging paradigms. Trends Pharmacol Sci 22:368–376. https://doi.org/10.1016/S0165-6147(00)01678-3
2. Fredholm BB, IJzerman AP, Jacobson KA et al (2001) International Union of Pharmacology. XXV. Nomenclature and classification of adenosine receptors. Pharmacol Rev 53:527–552. https://doi.org/10.1124/pr.110.003285.1
3. Franchetti P, Cappellacci L, Marchetti S et al (1998) 2′-C-methyl analogues of selective adenosine receptor agonists: synthesis and binding studies. J Med Chem 41:1708–1715
4. Heydenreich F, Vuckovic Z, Matkovic M, Veprintsev D (2015) Stabilization of G protein-coupled receptors by point mutations. Front Pharmacol 6:1–15. https://doi.org/10.3389/fphar.2015.00082
5. Lebon G, Bennett K, Jazayeri A, Tate CG (2011) Thermostabilisation of an agonist-bound conformation of the human adenosine A2A receptor. J Mol Biol 409:298–310. https://doi.org/10.1016/j.jmb.2011.03.075
6. Nguyen ATN, Baltos J-A, Thomas T et al (2016) Extracellular loop 2 of the adenosine A1 receptor has a key role in orthosteric ligand affinity and agonist efficacy. Mol Pharmacol 90:703–714. https://doi.org/10.1124/mol.116.105007
7. Olah ME, Stiles GL (2000) The role of receptor structure in determining adenosine receptor activity. Pharmacol Ther 85:55–75. https://doi.org/10.1016/S0163-7258(99)00051-0
8. Berman HM, Westbrook J, Feng Z et al (2000) The protein data bank. Nucleic Acids Res 28:235–242. https://doi.org/10.1093/nar/28.1.235
9. Carpenter B, Nehmé R, Warne T et al (2016) Structure of the adenosine A2A receptor bound to an engineered G protein. Nature 536:104–107. https://doi.org/10.1038/nature18966
10. Ballesteros J, Weinstein H (1995) Integrated methods for the construction of three-dimensional models and computational probing of structure-function relations in G protein-coupled receptors. Methods Neurosci 25:366–428. https://doi.org/10.1016/S1043-9471(05)80049-7
11. Chemical Computing Group Inc. (2017) Molecular operating Environement (MOE)., Version 2015.10
12. Notredame C, Higgins DG, Heringa J (2000) T-coffee: a novel method for fast and accurate multiple sequence alignment. J Mol Biol 302:205–217. https://doi.org/10.1006/jmbi.2000.4042
13. Fiser A, Sali A (2003) MODELLER: generation and refinement of homology-based protein structure models. Methods Enzymol 374:461–491. https://doi.org/10.1016/S0076-6879(03)74020-8
14. Schrodinger LLC (2015) The PyMOL molecular graphics system. Version 1.8
15. Pettersen EF, Goddard TD, Huang CC et al (2004) UCSF chimera - a visualization system for exploratory research and analysis. J Comput Chem 25:1605–1612. https://doi.org/10.1002/jcc.20084
16. Munk C, Isberg V, Mordalski S et al (2016) GPCRdb: the G protein-coupled receptor database - an introduction. Br J Pharmacol 16:2195–2207. https://doi.org/10.1111/bph.13509
17. Kim J, Wess J, van Rhee AM et al (1995) Site-directed mutagenesis identifies residues involved in ligand recognition in the human A2a adenosine receptor. J Biol Chem 270:13987–13997
18. Jiang Q, Rhee AM v, Kim J et al (1996) Hydrophilic side chains in the third and seventh transmembrane helical domains of human A2A adenosine receptors are required for ligand recognition. Mol Pharmacol 50:512–521

Chapter 16

Computational Support of Medicinal Chemistry in Industrial Settings

Daniel F. Ortwine

Abstract

The practice of computational chemistry in an industrial setting poses unique opportunities and challenges. Industrial computational chemists must manage large amounts of data, master modeling software, write scripts to perform custom calculations, and stay abreast of scientific advances in the field. Just as importantly, because computational chemists are full partners in the drug discovery effort at companies, in order to influence and streamline the drug discovery process, they must communicate effectively with medicinal chemists and other scientists to deliver results of their calculations in a timely fashion. The skills necessary to play this role require education that emphasizes a combination of chemistry, programming, and communication skills. Professors are encouraged to incorporate such training in their curriculum.

Key words Computational chemistry, Industry, Integration, Education, Data, Environment

1 Introduction

Computational chemistry clearly has come of age, taking its place alongside organic, medicinal, analytical, and biophysical chemistry as a mainstream discipline in the support of drug discovery. Pharmaceutical and biotechnology companies have universally adopted computational approaches to accelerating drug discovery, and virtually all have individuals or groups dedicated to these tasks. There are several reasons why computational chemistry has remained in the forefront. Biological targets have become more challenging to prosecute as companies attempt to tackle traditionally difficult problems, such as inhibition of protein-protein interactions. There is an ever-increasing avalanche of biological, structural, and other data to manage, integrate, and interpret. Scientific advances in the field continue apace, resulting in an ever-increasing ability to accurately forecast physical and biological properties of small molecules ahead of synthesis. Computational horsepower, particularly the availability of fast graphics processing units (GPUs) and cloud computing, continues to rapidly increase, allowing large-scale

Alexander Heifetz (ed.), *Computational Methods for GPCR Drug Discovery*, Methods in Molecular Biology, vol. 1705, https://doi.org/10.1007/978-1-4939-7465-8_16, © Springer Science+Business Media LLC 2018

calculations such as molecular dynamics simulations to be performed in a fraction of the time they used to take. Finally, medicinal chemists have become increasingly computer savvy. Many have received some training in schools in computational chemistry techniques, and are in general interested in applying these techniques in their drug design decision making.

Industrial computational chemistry presents unique opportunities and challenges. Industrial computational chemists are expected to be fully integrated, collaborative project team members, driving analyses, generating hypotheses, and designing compounds and other experiments. They are expected to understand all aspects of their project, from all the data, to how assays are performed, structural biology, biophysical approaches, journal and patent literature, etc. This is an exciting opportunity for industrial computational chemists, as they get to help drive and not just support a drug discovery team, with a chance of seeing their project(s) progress to human clinical trials. This is rarely possible in academia. Finally, industrial computational chemists gain experience from multiple projects and are therefore well-positioned to identify or develop new computational approaches to improve how drugs are discovered. Specific opportunities and challenges can be divided into broad categories related to data, science and technology, environment, integration, collaboration, and education. Each will be covered in sections below.

1.1 Data

Unlike typical academic settings where data may be available for a limited number of compounds, the average industrial drug discovery project involves multiple results on hundreds to thousands of compounds. Data are generated from in vitro assay experiments on the biological target of interest or specific off-targets, cellular assays, and in vivo pharmacology experiments in animals. There is frequently panel screening done against a wide variety of enzymes and receptors to assess selectivity. Physiochemical measurements such as log P, solubility, and pKa are obtained. Many compounds undergo pharmacokinetic evaluation for hepatic stability, permeability, plasma protein binding, and other endpoints. Safety (i.e., toxicity) data may also be available. A number of small molecule and protein Xrays are often available to support structure-based drug design analyses. As a result, a substantial fraction of an industrial computational chemist's time is consumed by data management, from ensuring timely results are available to the project teams they support in an easy-to-consume format, to providing insightful analyses on that data to help drive decision making on what molecules to synthesize next.

1.2 Science and Technology

Computational chemistry as a field continues to evolve. Techniques such as free-energy perturbation calculations [1, 2] and deep learning [3] technology are two that have been at the forefront recently. In the competitive industrial pharmaceutical environment, it is important to stay abreast of developments in the field to ensure state-of-the-art methods are being applied. Industrial computational chemists are also bombarded with vendors trying to sell their software that run the latest and greatest calculations. One must pick and choose which techniques to invest time in evaluating, efficiently determining if the approach is worth pursuing in depth. This must be done while continuing to support therapeutic projects, so efficient time management for an industrial computational chemist is a must. Having direct access to data and technology coupled to the needs of project teams provides the industrial computational chemist with a unique perspective to assess gaps in science and technology. They can and do devote effort themselves, or work with vendors, to develop new scientific or technological approaches to address those gaps. Having access to significant internal datasets permits the industrial computational chemist to validate and compare third party scientific methods, for example to compare free energy perturbation methods for calculating free energy of binding against experimentally determined values.

1.3 Environment

The pharmaceutical industry continues to be a highly competitive business. The target or even chemical series being targeted by one company is probably also being pursued by others. Unlike an academic setting where one might analyze a certain aspect of a protein-ligand interaction in detail by applying a number of computational techniques, the need to rapidly prosecute chemical series into a clinical candidate molecule (or a reach a no-go decision) means an industrial computational chemist must be able identify the key issues at hand and rapidly bring to bear computational tools to address those issues. Tools must be employed to rapidly deliver results in a timeframe that can impact a rapidly moving project. One must also maintain an up-to-date awareness of developments in the field against the biological target of interest by tracking the literature (including patents) and attending conferences. Being first in class to the clinic means a lot to a pharmaceutical company.

1.4 Integration

With access to multiple sources of data and technology in industry, putting it all together in a form readily accessible to chemists and other scientists on the project team is essential. Three-dimensional X-ray and modeling information must be combined with potency and other data by making all available inside in a unified interface to permit all aspects of molecules' and chemical series' behavior to be considered when determining trends and what to synthesize next. Tracking virtually designed compounds, along with the reasons they were suggested for synthesis, is important [4]. Once tested, one learns if the hypothesis for the synthesis was correct or wrong by returning to the tracking tool to see what idea was being tested, and then moving on to the next design hypothesis. Most large pharmaceutical companies have developed internal statistical models to predict experimental properties such as log D, solubility, microsomal stability, and project-specific potency. Integration of these predictions into molecular modeling software is important and a focus in the industrial arena. Publications on these efforts have recently appeared [9, 10].

1.5 Collaboration

In academia, one might interact with a co-investigator or two, a handful of students, and perhaps an individual from an industrial collaborator. Academic drug discovery occurs in a fairly insulated environment. In industry, computational chemists are immersed in a highly collaborative environment. They interact with medicinal chemists, biochemists, biologists, and colleagues from structural biology, bioinformatics, formulations, pharmacokinetics, safety assessment, and legal representatives on a regular basis. A majority of companies employ external contract research organizations (CROs) to perform chemical syntheses and experiments. A few engage their CROs in true medicinal chemistry collaborations, expecting the CRO to provide insightful data analyses and compound designs. Drug design typically occurs in dynamic team settings, often at a distance via teleconferences. This puts a premium on teamwork, active listening, and clear communication skills. An insightful design or analysis, if poorly communicated, may be missed.

Training is an important part of effective collaboration. In industry, training in the use and interpretation of models, whether they are docking paradigms, quantum chemical calculations, dynamics simulations, or statistical models, is essential. A computational model that is poorly understood by a medicinal chemist will be rarely used. Training in the use of software is also paramount. Most companies deploy computational techniques to the desktops of their medicinal chemists and other scientists. Part of the computational chemist's job is to ensure that deployed techniques being deployed are properly used and results appropriately interpreted and presented.

1.6 Education

Most medicinal chemists perform their own docking and scoring, and are becoming proficient in understanding physical organic chemistry principles. The environment is waning where the computational chemist is asked to dock a chemist's idea and pass judgement on it, or explain existing structure-activity relationships in terms of protein-ligand interactions. Concurrently, new scientific methodology is constantly being developed that requires integration into a company's infrastructure. Industrial computational chemists are expected to be able to script, know modeling software, and perhaps generate statistical models. Just as importantly, they need to be able to communicate the results of their modeling clearly and effectively, working in a team setting toward the goal of delivering a clinical candidate molecule. As a result, pharmaceutical and biotech companies now actively seek candidates with chemistry, drug design, informatics, *and* programming backgrounds. In fact, many smaller companies have just a single individual that fill all these roles.

Unfortunately, in today's academic environment, programs that emphasize this combination of chemistry, programming, and communication skills are hard to find, leading to a shortage of qualified candidates. Companies such as Genentech and Novartis have active computational chemistry summer intern and postdoctoral programs that can fill some of this gap. Students are encouraged to seek out schools with such training, and universities are encouraged to offer such diversity in their coursework.

2 Conclusions and Perspective

Synthesizing and testing molecules is expensive and time consuming. It is estimated that the largest portion of the preclinical development cost of a drug comes from these activities, due to the large number of molecules that must be made to find a clinical candidate [5]. Computational chemistry plays a central role in the goal of making an informed decision about what molecules to make next, and to learn something from the results on each molecule synthesized. The field continues to evolve, as does the industrial computational chemistry environment along with it. Working in industry poses a unique set of advantages and challenges. Having access to a wealth of data requires knowledge of how to organize, store, report, and analyze it. One must keep abreast of new developments in the field while developing and testing new internal scientific capabilities. Add in the task of supporting therapeutic projects produces a job description that requires excellent time management and prioritization skills.

Industrial computational chemistry groups are well integrated in the drug discovery teams throughout the pharmaceutical industry. For more information, a perspective on the organization and function of a number of computational chemistry groups in industry can be found in a special issue of the Journal of Computer-Aided Molecular Design [6], including a description of how the Computational Drug Discovery group at Genentech functions [7]. Industrial computational chemistry groups require individuals with drug design, modeling, as well as programming (or at least scripting) skills. Students considering a career in industrial computational chemistry are advised to consult Chapter 16 in a recently published book [8], in which considerations when deciding on a career in this field are described. It is suggested that more schools and professors train students in chemistry, drug design, programming, and oral/ written communication skills, as there is a bright future and continued strong demand for such multitalented individuals.

References

1. Abel R, Mondal S, Masse C, Greenwood J, Harriman G, Ashwell MA, Bhat S, Wester R, Frye L, Kapeller R, Friesner RA (2017) Accelerating drug discovery through tight integration of expert molecular design and predictive scoring. Curr Opin Struct Biol 43:38–44. https://doi.org/10.1016/j.sbi.2016.10.007
2. Kuhn B, Tichý M, Wang L, Robinson S, Martin RE, Kuglstatter A, Benz J, Giroud M, Schirmeister T, Abel R, Diederich F, Hert J (2017) Prospective evaluation of free energy calculations for the prioritization of Cathepsin L inhibitors. J Med Chem 60:2485–2497. https://doi.org/10.1021/acs.jmedchem.6b01881
3. Gawehn E, Hiss JA, Schneider G (2016) Deep learning in drug discovery. Mol Inform 35:3–14. https://doi.org/10.1002/minf.201501008
4. Lee M-L, Aliagas I, Dotson J, a Feng J, Gobbi A, Heffron T (2012) DEGAS: sharing and tracking target compound ideas with external collaborators. J Chem Inf Model 52:278–284. https://doi.org/10.1021/ci2003297
5. Paul SM, Mytelka DS, Dunwiddie CT, Persinger CC, Munos BH, Lindborg SR, Schacht AL (2010) How to improve R&D productivity: the pharmaceutical industry's grand challenge. Nat Rev Drug Discov 9:203–214. https://doi.org/10.1038/nrd3078
6. Warr WA (2017) A CADD-alog of strategies in pharma. J Comput Aided Mol Des 31:245–247. https://doi.org/10.1007/s10822-017-0017-6
7. Tsui V, Ortwine DF, Blaney JM (2016) Enabling drug discovery project decisions with integrated computational chemistry and informatics. J Comput Aided Mol Des 31:1–5. https://doi.org/10.1007/s10822-016-9988-y
8. Miller SM, Moos WH, Munk BH, Munk SA (2016) Managing the drug discovery process: how to make it more efficient and cost-effective. Woodhead Publishing, Elsevier, United Kingdom
9. Feng JA, Aliagas I, Bergeron P, Blaney JM, Bradley EK, Koehler MFT, Lee M-L, Ortwine DF, Tsui V, Wu J, Gobbi A (2015) An integrated suite of modeling tools that empower scientists in structure- and property-based drug design.J Comput Aided Mol Des 29(6):511–523
10. Lee M-L, Aliagas I, Feng JA, Gabriel T, O'Donnell TJ, Sellers BD, Wiswedel B, Gobbi A (2017) chemalot and chemalot_knime: Command line programs as workflow tools for drug discovery. J Chem Inf 9(1)

Chapter 17

Investigating Small-Molecule Ligand Binding to G Protein-Coupled Receptors with Biased or Unbiased Molecular Dynamics Simulations

Kristen A. Marino and Marta Filizola

Abstract

An increasing number of G protein-coupled receptor (GPCR) crystal structures provide important—albeit static—pictures of how small molecules or peptides interact with their receptors. These high-resolution structures represent a tremendous opportunity to apply molecular dynamics (MD) simulations to capture atomic-level dynamical information that is not easy to obtain experimentally. Understanding ligand binding and unbinding processes, as well as the related responses of the receptor, is crucial to the design of better drugs targeting GPCRs. Here, we discuss possible ways to study the dynamics involved in the binding of small molecules to GPCRs, using long timescale MD simulations or metadynamics-based approaches.

Key words Molecular dynamics, Ligand binding, Small-molecule drugs, GPCRs, Enhanced-sampling methods, Interaction fingerprints, Allosteric communication

1 Introduction

One of the main challenges of developing novel small molecules which target G protein-coupled receptors (GPCRs) is that these proteins are highly dynamic and exist in an ensemble of conformations rather than single inactive and activated states. While the dynamics of GPCRs represent a challenge for the discovery of novel ligands using traditional methods of computer-aided drug discovery (CADD), they also represent an opportunity to develop improved therapeutics by virtue of exploiting them to achieve functional selectivity. Increases in computing power and advancements in parallelization of molecular dynamics (MD) simulations, as well as the application of enhanced sampling algorithms, have led to the increasing use of these approaches to capture, at an atomic level of detail, dynamical processes such as ligand binding and ligand-induced conformational changes in the receptor. When considering the ensemble of conformations in which GPCRs exist,

Alexander Heifetz (ed.), *Computational Methods for GPCR Drug Discovery*, Methods in Molecular Biology, vol. 1705, https://doi.org/10.1007/978-1-4939-7465-8_17, © Springer Science+Business Media LLC 2018

docking into an available crystal structure may not be able to capture the correct binding pose of a ligand, especially one that is chemically different from the cocrystallized ligand. This is even more true for allosteric ligands, which bind to a site other than the canonical orthosteric binding site, where endogenous ligands normally bind. As this site typically corresponds to variable and flexible loop regions, the bound conformation of an allosteric ligand may not be readily identified from a crystal structure with only an orthosteric ligand. For instance, while a positive allosteric modulator (PAM) was found to be bound at the extracellular side of the M2 muscarinic receptor crystal structure (PDB: 4MQT [1]), the negative allosteric modulators (NAMs) cocrystallized with the chemokine receptors CCR9 (PDB: 5LWE [2]) and CCR2 (PDB: 5T1A [3]) were found at the intracellular side. Notably, allosteric GPCR ligands are of great interest from a drug discovery perspective due to their expected greater subtype selectivity, their ability to maintain temporal and spatial characteristics of endogenous signals, and their potentially limited on-target overdosing risks [4].

Another hallmark of GPCRs that can be particularly useful for drug discovery is their ability to mediate several signaling pathways due to possible activation of various intracellular proteins, including G proteins and β-arrestins. Although functionally selective ligands may, in principle, bind to the orthosteric binding site, they are capable of inducing changes in the receptor conformation which lead to the activation of selected signaling pathways, thereby separating beneficial from adverse side effects. For example, the observation that analgesic properties of opioids are mostly due to the activation of the G protein signaling pathway while signaling through the β-arrestin pathway leads to adverse side effects, including constipation, tolerance, and respiratory depression, led to the development of the G protein-biased agonist TRV-130, which is currently in clinical trials for FDA approval [5].

MD simulations have been used by our group [6–8] and others (e.g., [9–11]) to answer two key questions with respect to functionally selective and allosteric ligands targeting GPCRs: (1) what are the energetically preferred binding pathways and modes of ligand binding, and (2) how does the ligand transfer information from the ligand binding site to the intracellular side of the receptor? The ability of long timescale, unbiased MD simulations to predict the crystallographic binding pose of a small molecule, as well as to characterize its binding pathway to a GPCR crystal structure, was first shown for the binding of several beta blockers to the β_2 adrenergic receptor (β_2AR) [10]. Long unbiased MD simulations have recently been used to predict poses of TRV-130 to the μ-opioid receptor (MOR) [7], allosteric ligands to the M2 muscarinic receptor (M2) [9], and ML056 to the sphingosine-1-phosphate receptor 1 ($S1P_1R$) [12]. We pioneered the use of metadynamics [13] to study ligand binding to GPCRs [14] and have recently used this enhanced MD algorithm to predict the binding pose of a PAM, BMS-986187, to the δ-opioid receptor (DOR) [8], as well as kurkinorin, a G protein-biased agonist, to MOR [6]. Notably, our experimentally validated allosteric binding site of BMS-986187 at DOR [8] was not similar to any of the top-ranked poses from docking with Glide XP version 6.2 [15]. While accelerated MD (aMD) [16] was able to characterize the binding of the partial agonist arecoline and the agonist acetylcholine to the M3 muscarinic receptor [17], neither aMD [17] nor unbiased MD simulations [11] were able to recapitulate the crystallographic binding pose of the much larger and flexible antagonist tiotropium to M3, which shows that there are still limitations to these methods.

While the determination of the binding pose and binding pathway is a crucial part of understanding how ligands and receptors interact, understanding how information is communicated between the binding site and intracellular parts of the receptor is also necessary. The dynamical information derived from MD simulations can lead to the identification of the specific GPCR residues and ligand-receptor interactions that are responsible for this allosteric transmission (see e.g. [7, 18, 19]). This information is likely to impact the design of allosteric ligands which modulate the action of endogenous ligands, as well as functionally selective ligands that only activate desired therapeutic pathways.

Here, we discuss the computational methods that we used to predict energetically preferred binding poses of two functionally selective ligands, TRV-130 [7] and kurkinorin [6], and an allosteric ligand, BMS-986187 [8], as well as the analysis that led us to identify interesting allosteric communication differences in MOR between the classical unbiased agonist morphine and the G protein-biased agonist TRV-130 [7].

2 Materials

While there are many alternatives, the software and web-servers that we have used to set up, run, and analyze the MD simulations discussed here are described below. Most of the software listed below is available free of charge to academic researchers.

1. Crystal structures of GPCRs frequently require modifications (e.g., adding hydrogens) before they can be used in simulations. If the crystal structure is missing loops and/or side-chains, they can be added using modeling software packages such as MODELLER [20] or ROSETTA [21].
2. Once the protein structure is complete, it needs to be embedded in a membrane. The CHARMM-GUI webserver (charmm-gui.org) [22] is very user friendly and can be used to build membranes of many different compositions. An

alternative is to use a tool known as InflateGRO2 [23], which can automatically and efficiently embed the receptor in a pre-equilibrated membrane.

3. In most cases, the parameters of a small-molecule ligand are not readily available in standard force fields and need to be generated by the user in a manner that is consistent with the force field used to describe the protein and lipids (*see* **Note 1**). We generally use the CHARMM force field and generate initial ligand parameters using the CHARMM General Force Field (CGenFF) webserver (cgenff.paramchem.org [24, 25]). The parameters must be validated according to the procedures described in [26]. Validation requires performing quantum mechanics calculations, which we usually carry out with Gaussian for which a license must be purchased.
4. There are several software packages available to perform MD simulations. We frequently use GROMACS [27] or NAMD [28], which are both freely available (*see* **Note 2**). Metadynamics simulations can be implemented in either of these packages using the PLUMED plug-in [29].
5. Two programs that can be used to visualize structures and trajectories from MD simulations are VMD [30] and PyMOL [31]. Postprocessing and simulation analysis can be performed with GROMACS tools and in-house scripts. We typically use the PyEMMA python library [32] to construct Markov State Models (MSMs).

3 Methods

3.1 Protein Setup

In our recently published work, we used the crystal structures of the inactive DOR (PDB: 4N6H [33]), inactive MOR (PDB: 4DKL [34]), and activated MOR (PDB: 5C1M [35]) (*see* **Note 3**). First, with the exception of the crystallographic waters, the non-receptor atoms, including the ligands, lipids, and some of the proteins required for crystallization (BRIL for DOR and T4L for inactive MOR), were removed. Many crystal structures of GPCRs are missing segments of intracellular or extracellular loops that are too flexible to be resolved crystallographically or were removed to insert fusion proteins necessary for crystallization. These segments such as most of the intracellular loop 3 (ICL3) missing from the inactive MOR crystal structure (PDB: 4DKL [34]), need to be built ab initio or modeled by homology using an available, close template structure, as a reference. In our recently published studies, we used MODELLER to perform homology modeling of MOR ICL3 based on the corresponding segment in the ultra-high-resolution DOR crystal structure (PDB: 4N6H [33]) (*see* **Notes 4** and **5**). To crystallize activated forms of GPCRs, G protein mimetic nanobodies have been used to maintain the conformational changes on the intracellular side of the receptor that accompany receptor activation. We retained the nanobody cocrystallized with the MOR activated crystal structure in simulations of TRV-130 binding to avoid deactivation of the receptor while the ligand was still in the bulk, but removed it in simulations where we studied allosteric communication in MOR induced by TRV-130 or morphine [7].

Once the protein construct to be simulated is complete, it needs to be embedded in a membrane and solvated. Since cholesterol (CHOL) is an integral part of the plasma membrane, we typically use a membrane with a 1-palmitoyl,2-oleoyl-sn-glycero-3-phosphocholine (POPC)/10%CHOL composition. A user-friendly way to construct the protein/membrane/solvent system is to use the MEMBRANE BUILDER module [36] of the CHARMM-GUI [22], which can output input files for several MD packages [37], including GROMACS, CHARMM, and NAMD.

The final step is to equilibrate the system. First, we typically perform an energy minimization to remove steric clashes. Then, we equilibrate the system in a constant-temperature, constant-pressure (NPT) ensemble at 300 K and 1 bar to equilibrate the box size. To ensure stability of the system it is generally best to start by placing position restraints on the heavy atoms of the lipids and protein, and reduce the position restraints in stages by changing the force constant (e.g., $k = 1000$, 500, 100, and 50 kJ/mol/nm^2). Finally, we typically perform an NPT run without position restraints.

3.2 Trajectory Generation with Unbiased MD

To generate binding trajectories and identify the bound pose(s) of ligands, we have used two approaches in recently published works: (1) long-timescale MD and (2) multiple-walker metadynamics. Here, we discuss how to set up and run these types of simulations.

Normally, unbiased MD is unable to capture the timescale on which GPCR ligand binding from the bulk occurs, but thanks to specially designed hardware, e.g., D. E. Shaw Research's Anton supercomputer [38], this problem is partially alleviated. To further enhance the probability of ligand binding, the concentration of the ligand is increased in the simulation box instead of only using one molecule. For example, we added ten TRV-130 molecules to our recently published simulations [7]. They can be manually placed in the simulation box using PyMOL [31] at a distance of at least 1 nm from the receptor. Multiple copies of the system need to be created to further enhance the probability of a binding event. In the case of TRV-130, eight starting conformations were generated by randomly assigning initial velocities. The individual MD trajectories can be run until a binding event occurs or when all ligands are bound to the membrane. Once a ligand binds to the membrane it is unlikely to be released back into the bulk during typical timescales of ligand binding (several microseconds on Anton). For the recently published TRV-130 simulations, the initial eight simulations were run for between 1 and 8.4 μs each (39.10 μs total) [7]. Despite initiating eight trajectories, binding of TRV-130 at the orthosteric site only occurred in one trajectory, which was not sufficient to derive conclusions about the sampled binding pathway and adopted ligand bound poses. To enhance sampling, new trajectories were started from conformations in which the ligand was not in the orthosteric site but was at a bound position along the binding pathway. This second set of 25 simulations were run for between 0.2 and 1.0 μs each (5.15 μs total) and resulted in eight additional binding events.

3.3 Trajectory Generation with Multiple Walker Metadynamics

A more computationally efficient method of identifying ligand binding pathways and binding poses is to use metadynamics, which applies a history-dependent bias term on collective variables (CVs) to enhance the sampling of rare transitions [13]. The CVs, which are defined by the user, should account for the slow degrees of freedom of the system and must be able to distinguish between states. Multiple walker metadynamics [39] increases the efficiency of metadynamics by running several copies of the system in parallel. For the recently published simulations of the binding of BMS-986187 to DOR [8], two CVs were biased: (1) the distance between the centers-of-mass of the transmembrane (TM) bundle of the protein and the heavy atoms of the ligand and (2) a measure of the polar and hydrophobic contacts formed between the ligand and receptor sidechains. CV2 was defined as:

$$\sum_{\substack{\text{Ligand}\\\text{Polar}}}\ \sum_{\substack{\text{Receptor}\\\text{Polar}}} \frac{1-\left(r_{ij}/r_0\right)^6}{1-\left(r_{ij}/r_0\right)^{12}} + \sum_{\substack{\text{Ligand}\\\text{Hydroph.}}}\ \sum_{\substack{\text{Receptor}\\\text{Hydroph.}}} \frac{1-\left(r_{ij}/r_0\right)^6}{1-\left(r_{ij}/r_0\right)^{12}}$$

where r_{ij} is the distance between the atoms of the ligand and the receptor and r_0 was set to 5 Å. The same two CVs were biased in our recently published simulations which predict the binding pose of kurkinorin [6]. While the starting configuration of the walkers should be independent of the final results, sampling is more efficient if each walker starts from a different initial structure, including structures in which the ligand is in the orthosteric binding site as well as in the bulk. An easy way to generate the starting structures is to perform a metadynamics simulation in which only CV1 is biased. To restrict sampling of the ligand in the bulk to the area of interest (i.e., close to the protein) and prevent the ligand from binding to the membrane, limits can be placed on the *xy*-position of the ligand.

Since all replicas in multiple-walker metadynamics experience a bias, the trajectories must be reweighed to recover the Boltzmann distribution using, for instance, the method developed by Tiwary et al. [40]. The reweighting procedure can also be used to reconstruct the free-energy surface as a function of other CVs to aid in

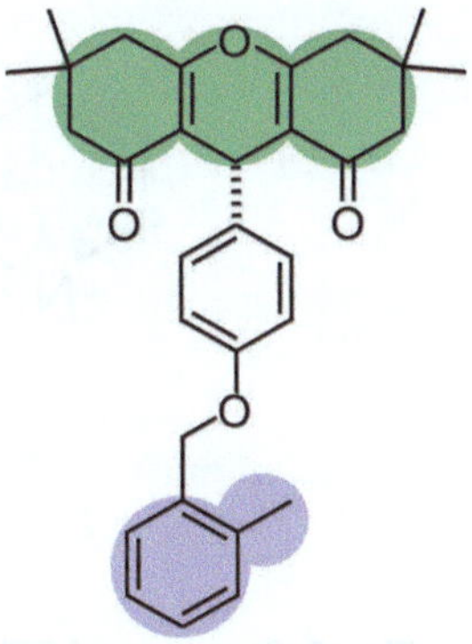

Fig. 1 The structure of BMS-986187, a DOR PAM, for which multiple-walker metadynamics was used to predict the binding pose at an allosteric site on DOR. The parts of the ligand which were used to calculate CV3 are the tricyclic moiety (*green*) and the ortho-substituted benzyl ring (*purple*). Adapted with permission from Shang Y, Yeatman HR, Provasi D, Alt A, Christopoulos A, Canals M, Filizola M (2016) Proposed mode of binding and action of positive allosteric modulators at opioid receptors. ACS Chemical Biology 11(5):1220–1229. Copyright 2016 American Chemical Society

discriminating between ligand binding poses, which have similar values of CV1 and CV2 but different orientations with respect to the receptor. For the simulations of the PAM binding to DOR, CV3 was defined as the z-component of a vector connecting centers of mass of the tricyclic moiety (green in Fig. 1) and the ortho-substituted benzyl ring (purple in Fig. 1) while CV4 was simply the *xy*-component of CV1. The 4-D free energy surface was then divided into microstates by dividing each CV into 75 bins. Only those microstates with energies less than 5 kJ/mol were included in further analysis.

3.4 Clustering to Identify the Binding Pose and Metastable States

To determine representative poses of the bound ligand and metastable states, the poses sampled during the simulations are clustered. In recently published work [6–8], we have used two types of interaction fingerprints, which describe the interactions between the ligand and the receptor. The first considers the number and type of ligand-receptor interaction with the interaction type classified as hydrophobic, polar, or aromatic. The polar interactions can be direct, between the ligand and the receptor in which the receptor is either the H-bond donor or acceptor, or water-mediated, in which one water molecule interacts with both the ligand and receptor at the same time. The aromatic interactions are divided into π-cation, edge-to-face, and edge-to-edge interactions. In the second type of interaction fingerprint, the ligand is divided into fragments and the interactions between the fragments and receptor residues are clustered regardless of the type of interaction. For example, TRV-130 was split into four fragments [7]: (1) the methoxy-thiophene moiety, (2) the pyridine, (3) the 6-oxaspiro

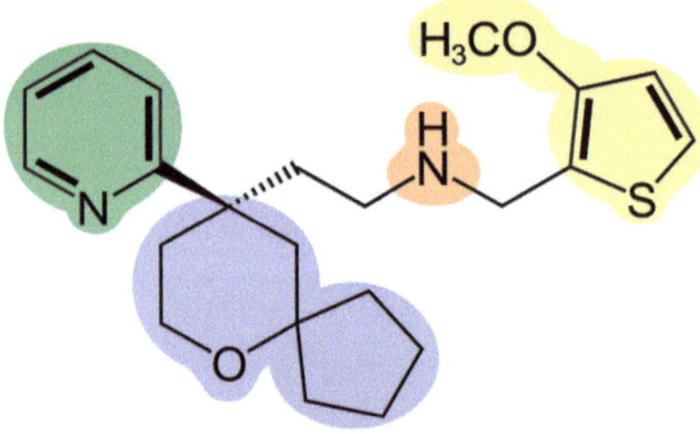

Fig. 2 Structure of TRV-130 which shows the four fragments into which the structure was broken to compute the interaction fingerprints to identify the ligand bound pose and metastable sites along the binding pathway: (1) the methoxy-thiophene moiety (*yellow*), (2) the pyridine moiety (*green*), (3) the 6-oxaspiro[4.5]decan-9-yl moiety (*purple*), and (4) the amine moiety (*orange*). Adapted with permission from Schneider S, Provasi D, Filizola M (2016) How oliceridine (TRV-130) binds and stabilizes a mu-opioid receptor conformational state that selectively triggers G protein signaling pathways. Biochemistry 55 (46):6456–6466. Copyright 2016 American Chemical Society

[4.5]decan-9-yl, and (4) the amine moiety (*see* Fig. 2). Sometimes, the definition of which interactions to cluster needs to be extended based on the problem of interest. For example, since we were only interested in defining the binding pose of kurkinorin [6], only the interactions between the ligand and receptor were considered. In the case of TRV-130, we were interested in the full binding pathway so interactions between the ligand and lipid headgroups were considered in addition to those between the ligand and the receptor since the ligand spent some time outside the extracellular vestibule. Finally, when determining the binding poses of the PAM BMS-986187 to DOR [8], the interactions between the PAM and the orthosteric ligand SNC-80 were also considered since they could come into contact. Using the Tanimoto dissimilarity coefficient as the distance metric, we then apply a density-based spatial clustering of applications with noise (DBSCAN) [41] algorithm to perform the clustering. This is our currently preferred method since it does not require the user to input the desired number of clusters as is necessary, for instance, in k-means clustering. Finally, the free energy of each cluster is calculated to determine which is the lowest energy ligand binding pose. The energy is directly proportional to the population of each cluster for unbiased simulations, but in the case of the metadynamics simulations, the free energy of a cluster α is calculated as

$$F_\alpha(t) = -k_B T log \int_\alpha ds \exp\left(-\frac{F(s,t)}{k_B T}\right) + k_B T \log Z$$

where s is a microstate within cluster α and Z is the partition function.

3.5 Characterizing Pathway Connectivity

Markov state models (MSM) can be used to derive kinetic information from MD simulations and are useful in characterizing transitions between states. While our simulations of TRV-130 binding [7] were not comprehensive enough to derive converged rate constants, we applied the PyEMMA python library [32] to determine likely transitions between identified metastable states (*see* **Note 6**). An alternative set of libraries for the construction of MSMs is MSMBUILDER [42].

A very extensive set of simulations totaling 831 μs was recently carried out by Stanley et al. [12] to examine the kinetics of binding of ML056 to $S1P_1R$. Specifically, a first set of 1000 trajectories totaling 579 μs was followed by two iterations of trajectory respawning to increase sampling of binding events. From the kinetic model, the simulations were able to show that the rate-limiting step of the binding of ML056, which occurs via the membrane, corresponds to entry into the vestibule region of the receptor and not to movement into the orthosteric site, an observation that is consistent with the work by Dror et al. [10].

3.6 Allosteric Communication Between the Orthosteric Ligand Binding Site and the Intracellular Side of the Receptor

While a comparison of the inactive and activated crystal structures of GPCRs can provide some clues as to how communication can travel from the orthosteric binding site to the intracellular side of the receptor, the use of computational analysis methods based on MD simulations of these crystal structures allows an assessment of these communication pathways based on dynamics. Extracting relevant allosteric pathways from simulations of proteins is a long-standing problem and a number of approaches have been developed (see e.g. [43–45] for reviews of these methods). Such methods have been applied to study various GPCRs, including the A_{2A}-adenosine receptor [46], β_2AR [18], dopamine receptors [47], luteinizing hormone receptor [48], MOR [7], rhodopsin [49, 50], and $5HT_{2A}$ serotonin receptor [51, 52].

We recently applied the N-body Information Theory (NbIT) analysis [53] of LeVine and Weinstein to study the allosteric communication between the MOR orthosteric binding site and the intracellular end of the receptor, in the presence of bound TRV-130 or bound morphine. NbIT provides a more detailed picture of allosteric communication because it is determined using an information theory-based analysis of N-body correlated motions derived from the configurational entropy of the system rather than simply pairwise atomic fluctuation correlations from MD.

To compare the allosteric communication between morphine bound to the activated MOR crystal structure and TRV-130 bound to activated MOR, we performed three 1 μs simulations for each ligand/MOR complex. While the nanobody used to crystallize activated MOR was retained in the simulations of TRV-130 binding, it was removed to study MOR communication in order to ensure we captured the communication in the receptor due to only the bound ligand. The first step was to define two sets of residues, the "transmitting" (T) region and the "receiving" (R) region. The T residues were selected as those within 5 Å of the ligand in the initial conformation of the ligand-protein complex. The selected R residues were those within 5 Å of the nanobody in the activated MOR crystal structure. Within the NbIT formalism [53], the mutual information (*MI*) between the T and R residues is defined as

$$MI(T, R) = H(R) + H(T) - H(R \cup T)$$

where *H* is the configurational entropy of the residues in the set X (either R or T).

$$H(X) = \frac{1}{2} \ln (2\pi e|\mathrm{CM}|)$$

Here, CM is the covariance matrix for all of the heavy atoms in the corresponding R or T region.

Furthermore, the co-information for the R and T regions, given the channel (*C*), is

$$CI(R, T, C) = MI(R, T) - MI(R, T|C)$$

in which the last term is the conditional mutual information,

$$MI(R, T|C) = H(R|C) + H(T|C) - H(R, T|C).$$

To determine the contribution of a specific residue in the *C* region, the *CI* is calculated with that residue removed from the *C* region and then normalized by the *MI* between the T and R regions, *MI(T,R)*. For further details on the NbIT method, please *see* Ref [53]. From the NbIT analysis, we were able to deduce key differences in the allosteric communication between morphine bound to MOR and TRV-130 bound to MOR [7].

4 Notes

1. It is important to ensure that ligands are parameterized with the same force field one is using for the protein and lipids to be simulated. A ligand parameterized with CGenFF cannot be used with the AMBER force field and vice-versa.
2. Some of the TRV-130 simulations discussed here were performed on the Anton supercomputer [38] at the Pittsburgh Supercomputing Center for which a very specific simulation setup is required.
3. Since an activated structure of DOR has not been solved yet, we elected to use the receptor's inactive crystal structure to simulate the binding of BMS-986187 to DOR. Also, at the time we started the kurkinorin simulations, the activated MOR

crystal structure was not available yet, requiring us to use the receptor's inactive structure. A comparison of the orthosteric binding sites of the inactive and activated MOR crystal structures shows that they are very similar, and we expect the same to be true for DOR.

4. In some cases, e.g., the muscarinic receptor, the missing loops from crystal structures are very long and we typically choose not to add them to the structure to be simulated. Instead the ends of the helices are capped.
5. When preparing the opioid receptor structures to simulate, we chose to remove the N-terminal fragment that is present in the activated MOR [35] and inactive DOR [33] crystal structures. The high flexibility of these fragments and the inability to sample their conformations thoroughly make the study of ligand binding more challenging in their presence than without them.
6. Building a MSM requires the selection of parameters (e.g., lag time, number of macrostates) which are dependent on the system of interest. Please see the documentation and tutorials on the http://emma-project.org webpage for more details.

Acknowledgments

This work was supported by National Institutes of Health grants MH107053, DA026434, and DA034049. Computations discussed here were run on resources available through (a) the Scientific Computing Facility at the Icahn School of Medicine at Mount Sinai, (b) the Extreme Science and Engineering Discovery Environment (XSEDE) under MCB080077, which is supported by National Science Foundation grant number ACI-1053575, and (c) the Pittsburgh Supercomputing Center which provided Anton computer time (under PSCA14006) through grant R01GM116961 from the National Institutes of Health. The Anton machine at PSC was generously made available by D.E. Shaw Research

References

1. Kruse AC, Ring AM, Manglik A, Hu J, Hu K, Eitel K, Hubner H, Pardon E, Valant C, Sexton PM, Christopoulos A, Felder CC, Gmeiner P, Steyaert J, Weis WI, Garcia KC, Wess J, Kobilka BK (2013) Activation and allosteric modulation of a muscarinic acetylcholine receptor. Nature 504(7478):101–106. https://doi.org/10.1038/nature12735
2. Oswald C, Rappas M, Kean J, Doré AS, Errey JC, Bennett K, Deflorian F, Christopher JA, Jazayeri A, Mason JS, Congreve M, Cooke RM, Marshall FH (2016) Intracellular allosteric antagonism of the CCR9 receptor. Nature 540(7633):462–465. https://doi.org/10.1038/nature20606

3. Zheng Y, Qin L, Zacarías NVO, de Vries H, Han GW, Gustavsson M, Dabros M, Zhao C, Cherney RJ, Carter P, Stamos D, Abagyan R, Cherezov V, Stevens RC, Ijzerman AP, Heitman LH, Tebben A, Kufareva I, Handel TM (2016) Structure of CC chemokine receptor 2 with orthosteric and allosteric antagonists. Nature 540(7633):458–461. https://doi.org/10.1038/nature20605
4. Wootten D, Christopoulos A, Sexton PM (2013) Emerging paradigms in GPCR allostery: implications for drug discovery. Nat Rev Drug Discov 12(8):630–644. https://doi.org/10.1038/nrd4052
5. DeWire SM, Yamashita DS, Rominger DH, Liu G, Cowan CL, Graczyk TM, Chen X-T, Pitis PM, Gotchev D, Yuan C, Koblish M, Lark MW, Violin JD (2013) A G protein-biased ligand at the μ-opioid receptor is potently analgesic with reduced gastrointestinal and respiratory dysfuncation compared with morphine. J Pharmacol Exp Ther 344:708–717. https://doi.org/10.1124/jpet.112.201616
6. Crowley RS, Riley AP, Sherwood AM, Groer CE, Shivaperumal N, Biscaia M, Paton K, Schneider S, Provasi D, Kivell BM, Filizola M, Prisinzano TE (2016) Synthetic studies of neoclerodane diterpenes from salvia divinorum: identification of a potent and centrally acting μ opioid analgesic with reduced abuse liability. J Med Chem 59(24):11027–11038. https://doi.org/10.1021/acs.jmedchem.6b01235
7. Schneider S, Provasi D, Filizola M (2016) How oliceridine (TRV-130) binds and stabilizes a mu-opioid receptor conformational state that selectively triggers G protein signaling pathways. Biochemistry 55(46):6456–6466. https://doi.org/10.1021/acs.biochem.6b00948
8. Shang Y, Yeatman HR, Provasi D, Alt A, Christopoulos A, Canals M, Filizola M (2016) Proposed mode of binding and action of positive allosteric modulators at opioid receptors. ACS Chem Biol 11(5):1220–1229. https://doi.org/10.1021/acschembio.5b00712
9. Dror RO, Green HF, Valant C, Borhani DW, Valcourt JR, Pan AC, Arlow DH, Canals M, Lane JR, Rahmani R, Baell JB, Sexton PM, Christopoulos A, Shaw DE (2013) Structural basis for modulation of a G-protein-coupled receptor by allosteric drugs. Nature 503 (7475):295–299. https://doi.org/10.1038/nature12595
10. Dror RO, Pan AC, Arlow DH, Borhani DW, Maragakis P, Shan Y, Xu H, Shaw DE (2011) Pathway and mechanism of drug binding to G-protein-coupled receptors. Proc Natl Acad Sci U S A 108(32):13118–13123
11. Kruse AC, Hu J, Pan AC, Arlow DH, Rosenbaum DM, Rosemond E, Green HF, Liu T, Chae PS, Dror RO, Shaw DE, Weis WI, Wess J, Kobilka BK (2012) Structure and dynamics of the M3 muscarinic acetylcholine receptor. Nature 482:552–556. https://doi.org/10.1038/nature10867
12. Stanley N, Pardo L, Fabritiis GD (2016) The pathway of ligand entry from the membrane bilayer to a lipid G protein-coupled receptor. Sci Rep 6:22639. https://doi.org/10.1038/srep22639
13. Laio A, Parrinello M (2002) Escaping free-energy minima. Proc Natl Acad Sci U S A 99 (20):12562–12566. https://doi.org/10.1073/pnas.202427399
14. Provasi D, Bortolato A, Filizola M (2009) Exploring molecular mechanisms of ligand recognition by opioid receptors with metadynamics. Biochemistry 48(42):10020–10029. https://doi.org/10.1021/bi901494n
15. Friesner RA, Murphy RB, Repasky MP, Frye LL, Greenwood JR, Halgren TA, Sanschagrin PC, Mainz DT (2006) Extra precision glide: docking and scoring incorporating a model of hydrophobic enclosure for protein–ligand complexes. J Med Chem 49(21):6177–6196. https://doi.org/10.1021/jm051256o
16. Hamelberg D, Mongan J, McCammon JA (2004) Accelerated molecular dynamics: a promising and efficient simulation method for biomolecules. J Chem Phys 120 (24):11919–11929
17. Kappel K, Miao Y, McCammon JA (2015) Accelerated molecular dynamics simulations of ligand binding to a muscarinic G-protein-coupled receptor. Q Rev Biophys 48 (4):479–487. https://doi.org/10.1017/S0033583515000153
18. Bhattacharya S, Vaidehi N (2014) Differences in allosteric communication pipelines in the inactive and active states of a GPCR. Biophys J 107(2):422–434. https://doi.org/10.1016/j.bpj.2014.06.015
19. Miao Y, Nichols SE, Gasper PM, Metzger VT, McCammon JA (2013) Activation and dynamic network of the M2 muscarinic receptor. Proc Natl Acad Sci U S A 110 (27):10982–10987
20. Fiser A, Do RKG, Sali A (2000) Modeling of loops in protein structures. Protein Sci 9:1753–1773
21. Rohl CA, Strauss CEM, Chivian D, Baker D (2004) Modeling structurally variable regions in homologous proteins with rosetta. Proteins 55(3):656–677. https://doi.org/10.1002/prot.10629

22. Jo S, Kim T, Iyer VG, Im W (2008) CHARMM-GUI: a web-based graphical user interface for CHARMM. J Comput Chem 29 (11):1859–1865. https://doi.org/10.1002/jcc.20945
23. Schmidt TH, Kandt C (2012) LAMBADA and InflateGRO2: efficient membrane alignment and insertion of membrane proteins for molecular dynamics simulations. J Chem Inf Model 52(10):2657–2669. https://doi.org/10.1021/ci3000453
24. Vanommeslaeghe K, MacKerell AD (2012) Automation of the CHARMM general force field (CGenFF) I: bond perception and atom typing. J Chem Inf Model 52(12):3144–3154. https://doi.org/10.1021/ci300363c
25. Vanommeslaeghe K, Raman EP, MacKerell AD (2012) Automation of the CHARMM general force field (CGenFF) II: assignment of bonded parameters and partial atomic charges. J Chem Inf Model 52(12):3155–3168. https://doi.org/10.1021/ci3003649
26. Vanommeslaeghe K, Hatcher E, Acharya C, Kundu S, Zhong S, Shim J, Darian E, Guvench O, Lopes P, Vorobyov I, Mackerell AD (2010) CHARMM general force field: a force field for drug-like molecules compatible with the CHARMM all-atom additive biological force fields. J Comput Chem 31 (4):671–690. https://doi.org/10.1002/jcc.21367
27. Abraham MJ, Murtola T, Schulz R, Páll S, Smith JC, Hess B, Lindahl E (2015) GROMACS: high performance molecular simulations through multi-level parallelism from laptops to supercomputers. SoftwareX 1–2:19–25. https://doi.org/10.1016/j.softx.2015.06.001
28. Phillips JC, Braun R, Wang W, Gumbart J, Tajkhorshid E, Villa E, Chipot C, Skeel RD, Kalé L, Schulten K (2005) Scalable molecular dynamics with NAMD. J Comput Chem 26 (16):1781–1802. https://doi.org/10.1002/jcc.20289
29. Tribello GA, Bonomi M, Branduardi D, Camilloni C, Bussi G (2014) PLUMED 2: new feathers for an old bird. Comput Phys Commun 185(2):604–613. https://doi.org/10.1016/j.cpc.2013.09.018
30. Humphrey W, Dalke A, Schulten K (1996) VMD: visual molecular dynamics. J Mol Graph 14(1):33–38. https://doi.org/10.1016/0263-7855(96)00018-5
31. Delano WL (2002) The PyMOL molecular graphics system. doi:citeulike-article-id:2816763
32. Scherer MK, Trendelkamp-Schroer B, Paul F, Pérez-Hernández G, Hoffmann M, Plattner N, Wehmeyer C, Prinz J-H, Noé F (2015) PyEMMA 2: a software package for estimation, validation, and analysis of Markov models. J Chem Theory Comput 11(11):5525–5542. https://doi.org/10.1021/acs.jctc.5b00743
33. Fenalti G, Giguere PM, Katritch V, Huang XP, Thompson AA, Cherezov V, Roth BL, Stevens RC (2014) Molecular control of delta-opioid receptor signalling. Nature 506 (7487):191–196. https://doi.org/10.1038/nature12944
34. Manglik A, Kruse AC, Kobilka TS, Thian FS, Mathiesen JM, Sunahara RK, Pardo L, Weis WI, Kobilka BK, Granier S (2012) Crystal structure of the micro-opioid receptor bound to a morphinan antagonist. Nature 485 (7398):321–326. https://doi.org/10.1038/nature10954
35. Huang W, Manglik A, Venkatakrishnan AJ, Laeremans T, Feinberg EN, Sanborn AL, Kato HE, Livingston KE, Thorsen TS, Kling RC, Granier S, Gmeiner P, Husbands SM, Traynor JR, Weis WI, Steyaert J, Dror RO, Kobilka BK (2015) Structural insights into micro-opioid receptor activation. Nature 524 (7565):315–321. https://doi.org/10.1038/nature14886
36. Wu EL, Cheng X, Jo S, Rui H, Song KC, Davila-Contreras EM, Qi Y, Lee J, Monje-Galvan V, Venable RM, Klauda JB, Im W (2014) CHARMM-GUI membrane builder toward realistic biological membrane simulations. J Comput Chem 35(27):1997–2004. https://doi.org/10.1002/jcc.23702
37. Lee J, Cheng X, Swails JM, Yeom MS, Eastman PK, Lemkul JA, Wei S, Buckner J, Jeong JC, Qi Y, Jo S, Pande VS, Case DA, Brooks CL 3rd, AD MK Jr, Klauda JB, Im W (2016) CHARMM-GUI input generator for NAMD, GROMACS, AMBER, OpenMM, and CHARMM/OpenMM simulations using the CHARMM36 additive force field. J Chem Theory Comput 12(1):405–413. https://doi.org/10.1021/acs.jctc.5b00935
38. Shaw DE, Deneroff MM, Dror RO, Kuskin JS, Larson RH, Salmon JK, Young C, Batson B, Bowers KJ, Chao JC, Eastwood MP, Gagliardo J, Grossman JP, Ho CR, Ierardi DJ, In K, Jl K, Layman T, Mcleavey C, Moraes MA, Mueller R, Priest EC, Shan Y, Spengler J, Theobald M, Towles B, Wang SC (2008) Anton, a special-purpose machine for molecular dynamics simulation. Commun ACM 51 (7):91–97. https://doi.org/10.1145/1364782.1364802

39. Raiteri P, Laio A, Gervasio FL, Micheletti C, Parrinello M (2006) Efficient reconstruction of complex free energy landscapes by multiple walkers metadynamics. J Phys Chem B 110 (8):3533–3539. https://doi.org/10.1021/jp054359r
40. Tiwary P, Parrinello M (2015) A time-independent free energy estimator for metadynamics. J Phys Chem B 119(3):736–742. https://doi.org/10.1021/jp504920s
41. Sander J, Ester M, Kriegel H-P, Xu X (1998) Density-based clustering in spatial databases: the algorithm GDBSCAN and its applications. Data Min Knowl Disc 2(2):169–194. https://doi.org/10.1023/A:1009745219419
42. Beauchamp KA, Bowman GR, Lane TJ, Maibaum L, Haque IS, Pande VS (2011) MSMBuilder2: modeling conformational dynamics on the picosecond to millisecond scale. J Chem Theory Comput 7 (10):3412–3419. https://doi.org/10.1021/ct200463m
43. Collier G, Ortiz V (2013) Emerging computational approaches for the study of protein allostery. Arch Biochem Biophys 538(1):6–15. https://doi.org/10.1016/j.abb.2013.07.025
44. Feher VA, Durrant JD, Van Wart AT, Amaro RE (2014) Computational approaches to mapping allosteric pathways. Curr Opin Struct Biol 25:98–103. https://doi.org/10.1016/j.sbi.2014.02.004
45. Stolzenberg S, Michino M, LeVine MV, Weinstein H, Shi L (2016) Computational approaches to detect allosteric pathways in transmembrane molecular machines. Biochim Biophys Acta 1858(7, Part B):1652–1662. https://doi.org/10.1016/j.bbamem.2016.01.010
46. Fanelli F, Felline A (2011) Dimerization and ligand binding affect the structure network of A2A adenosine receptor. Biochim Biophys Acta Biomembr 1808(5):1256–1266. https://doi.org/10.1016/j.bbamem.2010.08.006
47. Michino M, Free RB, Doyle TB, Sibley DR, Shi L (2015) Structural basis for Na+−sensitivity in dopamine D2 and D3 receptors. Chem Commun 51(41):8618–8621. https://doi.org/10.1039/C5CC02204E
48. Angelova K, Felline A, Lee M, Patel M, Puett D, Fanelli F (2011) Conserved amino acids participate in the structure networks deputed to intramolecular communication in the lutropin receptor. Cell Mol Life Sci 68 (7):1227–1239. https://doi.org/10.1007/s00018-010-0519-z
49. Kong Y, Karplus M (2007) The signaling pathway of rhodopsin. Structure 15(5):611–623. https://doi.org/10.1016/j.str.2007.04.002
50. Isin B, Schulten K, Tajkhorshid E, Bahar I (2008) Mechanism of signal propagation upon retinal isomerization: insights from molecular dynamics simulations of rhodopsin restrained by normal modes. Biophys J 95 (2):789–803. https://doi.org/10.1529/biophysj.107.120691
51. LeVine MV, Perez-Aguilar JM 2014, Weinstein H N-body information theory (NbIT) analysis of rigid-body dynamics in intracellular loop 2 of the 5-HT2A receptor. In: Ortuño F, Rojas I (eds) International Work-Conference on Bioinformatics and Biomedical Engineering, Granada
52. Perez-Aguilar JM, Shan J, LeVine MV, Khelashvili G, Weinstein H (2014) A functional selectivity mechanism at the serotonin-2A GPCR involves ligand-dependent conformations of intracellular loop 2. J Am Chem Soc 136(45):16044–16054. https://doi.org/10.1021/ja508394x
53. LeVine MV, Weinstein H (2014) NbIT - a new information theory-based analysis of allosteric mechanisms reveals residues that underlie function in the leucine transporter LeuT. PLoS Comput Biol 10(5):e1003603. https://doi.org/10.1371/journal.pcbi.1003603

Chapter 18

Ligand-Based Methods in GPCR Computer-Aided Drug Design

Paul C.D. Hawkins and Gunther Stahl

Abstract

This chapter describes two powerful 3D ligand-based shape similarity and scoring methods called ROCS and EON, their basic operation and selected validation data. The steps required to prepare a database of molecules for successful use with ROCS and EON are described and selected examples of their application in prospective lead discovery experiments are summarized.

Key words Lead discovery, Shape similarity, OMEGA, ROCS, EON, LBLD

1 Introduction

In recent years, virtual screening or computational lead discovery has become an important part of the armamentarium of modern drug discovery [1]. Much of the drive to use virtual screening (VS) or computational lead discovery (CLD) has arisen from increased pressure to put more compounds into the development pipeline and to reduce the costs of getting suitable compounds to this point. Experimental high-throughput screening (HTS) can be expensive and time-consuming [2], so computational methods for lead identification have become a routine part of drug discovery. Given that the time and costs associated with HTS can be substantially reduced by correctly applied CLD, a great many methods have been developed. They can be broadly divided into ligand-based lead discovery (LBLD) and structure-based lead discovery (SBLD).

Ligand-based methods offer a number of advantages over SBLD; they are usually much less computationally intensive, they require less user input to correctly set up and, perhaps most importantly, they do not require an atomic resolution structure of the target protein of interest in order to be used successfully. In cases where structural information is available LBLD and SBLD methods can often be productively used together [3]. However, even with

Alexander Heifetz (ed.), *Computational Methods for GPCR Drug Discovery*, Methods in Molecular Biology, vol. 1705,
https://doi.org/10.1007/978-1-4939-7465-8_18, © Springer Science+Business Media LLC 2018

the exponential increase in the number of protein structures available in internal collections and publicly in the PDB [4] there are many targets without structural characterization (many GPCRs and nearly all ion channels still do not have atomic resolution structures). In these cases, LBLD can be used immediately, while SBLD methods would first require the time-consuming generation and validation of homology models. Similarly, in the case of novel targets LBLD can be used as soon as an active ligand has been identified, which can provide substantial time-saving for a project over waiting for a crystal structure to be solved.

There are a great many LBLD methods available, some that require only the molecular graph to function, so-called 2D methods (fingerprints, substructure searches, etc.) and others that require a 3D representation of molecules (pharmacophore searches, shape similarity, 3D–QSAR etc.). In this chapter, we will focus on two particular methods for 3D LBLD, ROCS, and EON, which are both based on comparison of molecular shape.

2 Materials and Methods

In principle, the requirements for LBLD with ROCS are simple; a 3D conformation of the query molecule (usually a molecule active against the target protein) and a database of molecules to be screened as can be seen in Fig. 1. There are, however, a few steps that should be considered carefully before embarking on a LBLD calculation with ROCS.

2.1 Database Preparation

Before any lead discovery campaign is undertaken the database of molecules to be screened must be appropriately prepared. To generate a database suitable for use with ROCS tautomer and protonation state(s) must be assigned and 3D conformers generated. For ligand-based lead discovery methods the assignment of tautomer and protonation state is often simpler than for structure-based methods, as ligand-based methods do not depend on having the correct tautomer or protonation state, but only one consistent state across all molecules to be screened. It has been shown that generation of a single, stable tautomer for each molecule in the database produces equivalent results to enumerating sets of tautomers [5], and is obviously much faster for downstream processing. As such in general, it is recommended that for LBLD with ROCS a single tautomer state and a single protonation state be generated for each database molecule. After tautomer and protonation state assignment conformations must be generated.

2.2 Conformer Generation

Since ROCS performs its overlays rigidly, reasonable 3D conformations for both the query and database molecules are required. In our internal experiments, and in most published examples, conformer databases to be searched with ROCS are generated

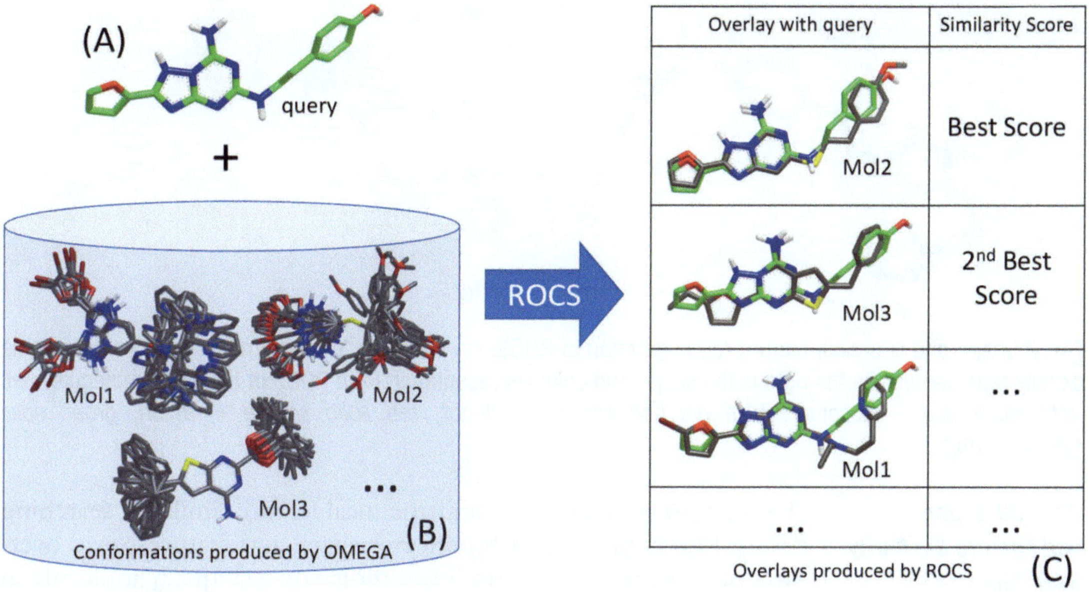

Fig. 1 Basic schema of 3D Ligand-based shape similarity searching: a query molecule (**A**) in one or more 3D conformations is compared to every conformer of every molecule in a pre-generated database of conformations (**B**). The optimized overlay between the single best conformer of each database molecule and each conformation of the query is reported together with a similarity score (**C**)

with OMEGA, as the two have been extensively tested together. We describe the OMEGA algorithm briefly here, more details can be found elsewhere [6, 7]. OMEGA uses two built-in knowledge bases to generate conformer ensembles; a library of allowed torsion angles and a library of fragment conformations in 3D. As such, conformer sampling with OMEGA requires only a connected molecular graph, i.e., a 3D structure of the molecule to be sampled is not required. The molecular graph is fragmented according to a set of rules and the fragments generated are compared against the internal library of 3D fragment conformations. If a fragment is not found in the built-in fragment library, then its conformations are generated on-the-fly. Once geometries for all the fragments of the molecule are available, they are assembled to produce one, or a few, starting 3D geometries for the molecule. From this starting 3D structure torsion driving is performed based on the settings in the torsion library to provide a possibly very large ensemble of conformations. This ensemble is subsequently sampled based on energetic and geometric criteria to provide a smaller ensemble suitable for use in downstream calculations. OMEGA has been extensively and carefully validated by the reproduction of solid state structures [6, 7] and has been successfully used to provide input conformations for retrospective validations in both pose prediction and lead discovery [8, 9].

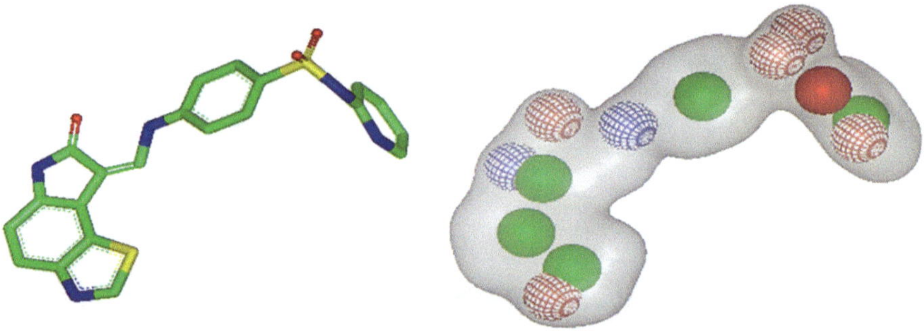

Fig. 2 Shape and chemical feature representation in ROCS. On the left a 3D conformation for a molecule as usually represented. On the right is the shape and color representation from ROCS of the same molecule. *Red hatched sphere* = acceptor, *blue hatched sphere* = donor, *red solid sphere* = anion, *green solid sphere* = ring

2.3 3D Shape and Feature Similarity Searching

Here, we introduce shape and chemical feature similarity searching using ROCS [10]. ROCS performs shape and feature-based overlays of conformers of a candidate molecule to a query molecule in one or more conformations. The overlays can be performed very quickly based on a description of molecular shape as the sum of atom-centered Gaussian functions [11]; on modern hardware speeds of 50–100 molecules/CPU/second can be attained. ROCS maximizes the rigid overlap of these Gaussian functions and thereby maximizes the shared volume and shared features between a single conformation of the query and a single conformation of a database molecule. The chemical features used in ROCS are based on the work of Mills and Dean [12] and are termed "color" features. An example of abstraction from the usual representation of a molecule to the shape and color feature used in ROCS is shown in Fig. 2.

Similarity in shape and color between the query and the database molecule is calculated from the best match of any conformation of the query to any conformation of the database molecule. Similarity is measured by a set of Tanimoto coefficients; by shape alone as ShapeTanimoto, by color alone (ColorTanimoto), and by the sum of these two measures (TanimotoCombo). The differences between these metrics are discussed under Subheading 3.

ROCS has successfully been used in a great many prospective lead discovery experiments. In cases where protein-ligand structures exist the cocrystal ligand is often used directly from the crystal structure as the query [13] but in some experiments the ligand has been substantially manually altered to improve relevance [14]. An interesting recent trend is to use results from molecular simulation on a protein-ligand cocrystal structure to generate structurally novel queries that may not be related to the structure of the original cocrystal ligand [15, 16]. Since in retrospective experiments SBLD methods, such as docking, and ROCS have been shown to identify different molecules [4, 17] ROCS has also been used

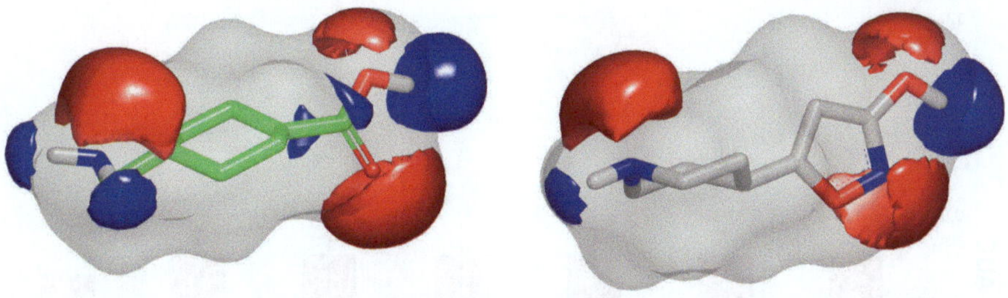

Fig. 3 Shapes and electrostatic potential for query (*left*) and database hit (*right*) as described in [24]. *Red color* denotes electronegative areas, whereas *blue* shows electropositive areas

prospectively in combination with docking [18, 19]. In cases where protein-ligand structures do not exist, ROCS has also proven useful [20, 21]. In these cases, the conformation for the query ligand must be computed and in spite of that ROCS still performs well (see Subheading 3 for further data on the use of computed query conformations in ROCS).

2.4 3D Shape and Electrostatic Similarity Searching

An alternate way to compare small molecules is to use their shapes in combination with electrostatic similarity using EON [22]. EON combines the shape similarity score from ROCS (ShapeTanimoto) with a field-based measure of similarity to compare the electrostatic potential of two small molecules. This electrostatic potential is calculated internally using Zap [23], OpenEye's Poisson-Boltzman (PB) electrostatics toolkit. Two ElectrostaticTanimoto (ET) measures are calculated using different outer dielectrics in the PB calculation (outer dielectric of 80.0 and 2.0). The rationale for using a PB electrostatic field is that the external potential is dampened by orientation of aqueous solvent.

A visualization of electrostatic similarity calculations with EON can be seen in Fig. 3. More examples of prospective EON use can be found in the literature [25–27].

3 Notes

In this section, we present some validation and performance data for ROCS as a guide to users of ROCS in ligand-based lead discovery. As a metric of performance we use the AUC of a ROC plot [28], which is bounded by 1 (perfect recovery) and 0, with 0.5 indicating random performance. The single largest effect that the user can have on ROCS is selection of the conformer sampling regime to be used in OMEGA, both for the query and the database, so here we present data on the influence of query conformation and database sampling to guide the user in their experiments. For all the data discussed below OMEGA's parameters were set at the defaults, except as noted.

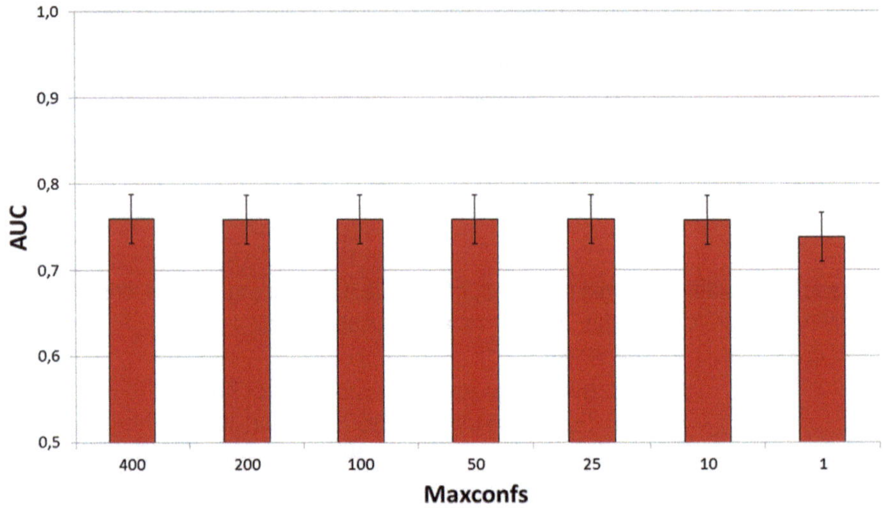

Fig. 4 Median AUC on DUD [29] with different levels of conformer sampling in the database. Maxconfs = maximum number of conformations per database molecule from OMEGA

3.1 Effect of Conformer Sampling

First, we examine the amount of conformer sampling in the database that is required. In Fig. 4, we show the effect of changing the number of conformations generated by OMEGA for each database molecule (by changing the *maxconfs* flag in OMEGA). The median performance for ROCS is constant as the maximum number of conformations allowed declines from 400 to 25, with a small decline at 10 and a further decline at only 1 conformation per database molecule. Since performance is not degraded by using small numbers of conformers, this allows the use of smaller databases, thereby increasing search speed (ROCS' speed is usually limited by disk reading speed, not computation capacity) and decreasing the burden of storage and data transfer across networks. A good balance of speed and performance is provided by setting *maxconfs* to 50; this is the setting used in all subsequent experiments.

Further investigations on ROCS' performance were carried out on a newer and larger dataset for virtual screening evaluation, the Database of Useful Decoys Enhanced or DUD-E [30]. DUD-E is a large dataset of over 100 diverse protein targets, and for many targets there are hundreds of active ligands and thousands of decoys. As such, DUD-E provides high statistical power (the ability to detect small, but genuine differences in performance between methods [31], which is rarely considered in CADD [32]) and low error rates (accurate prediction of performance on datasets other than DUD-E). While not designed for evaluating ligand-based lead discovery tools, DUD-E is appropriate for discriminating among different *settings* for the same tool, while it is perhaps not appropriate for discriminating among different *tools*. We used the recovery of active compounds from their background of presumed decoys in DUD-E to assess the influence of a variety of parameters in ROCS.

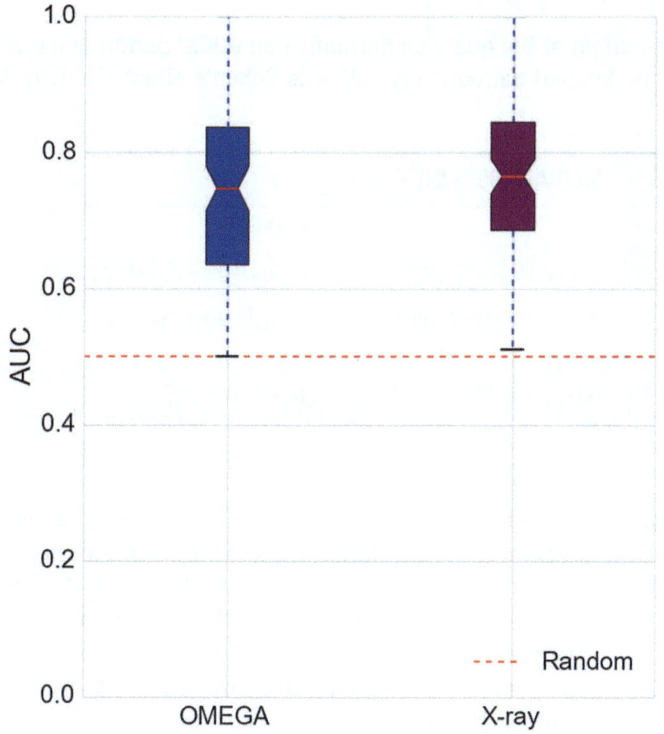

Fig. 5 The effect of using experimental ligand conformation (X-ray) or a low-energy computed conformation (OMEGA) on virtual screening on the DUD-E database (TanimotoCombo used as similarity measure)

3.2 Effect of Query Conformation

An obvious problem when performing lead discovery on classes of proteins that have few, if any, atomic resolution crystal structures is how to select the conformation of the protein (for SBLD) or the ligand (for 3D LBLD). In Fig. 5 we show the effect of using an experimentally derived conformation (from the DUD-E X-ray structure) or the lowest energy conformation found by OMEGA as the query when ranking the database molecules by their TanimotoCombo (TC) score to the query. The results for all three similarity measures are given in Table 1. (ShapeTanimoto (ST) alone, ColorTanimoto (CT) alone, or TanimotoCombo (TC)). The X-ray conformation is statistically and substantively significantly better than the OMEGA conformation when using TanimotoCombo, in accord with intuition. However, the performance of the OMEGA conformation is still good (median AUC is far above 0.5), indicating that the use of a computed conformation of the query molecule in ROCS in the absence of an X-ray conformation will likely still provide good results. In contrast, when ranking by ST and CT there is no substantive difference in performance whether the X-ray or the OMEGA conformation is used. The origin of the difference in sensitivity to the query conformation for the three metrics is unclear.

Table 1
Effect of changing the origin of the query conformation on ROCS' performance on the DUD-E dataset. The *p*-value is from the Student paired *t*-test [33], d is Cohen's effect size [31]. NS = not significant at $p = 0.05$

	MEDIAN [95% CI]			
	X-ray	OMEGA	*p*-value	d
TanimotoCombo	0.704 [0.667,0.720]	0.681 [0.652,0.712]	<0.001	0.48
ShapeTanimoto	0.627 [0.605,0.659]	0.61 [0.581,0.636]	<0.03	<0.2
ColorTanimoto	0.702 [0.671,0.725]	0.679 [0.649,0.718]	NS	<0.2

3.3 Effect of Scoring Function

The data in Table 1 show the effect of using different scoring measures for similarity on ROCS' performance. TC is both statistically and substantively significantly better than ST, and while TC is numerically superior to CT, the difference is not statistically or substantively significant (data not shown). As such, it is recommended that TC be used as the similarity measure in ROCS unless prior experimentation indicates otherwise.

The results above show that 3D similarity searching with ROCS is remarkably insensitive to the details of the conformer sampling used to generate either the query or the database conformations; an X-ray conformation of the query provides only a small, though significant, benefit over using the lowest energy conformer from OMEGA and even light sampling of the database molecules (a maximum of 25–50 conformers per molecule) provides identical performance to much heavier, and therefore much more time-consuming, sampling. This makes ROCS a fast and powerful LBLD tool, applicable in both high information projects, where one or more atomic resolution crystal structures are available, and in low information projects, where perhaps only one active ligand is known.

Recently, a new database for evaluating the performance of lead discovery tools specifically on GPCR targets, GPCR-Bench, was released [34]. The DUD-E dataset contains only 3 GPCR datasets, so the comparison of ROCS' performance on GPCR-Bench to that on DUD-E provides a useful estimate of how well predictions from general datasets like DUD-E transfer to other more target-specific sets. The performance of ROCS on GPCR-Bench and DUD-E is shown in Table 2. The results from GPCR-Bench are numerically slightly worse than from DUD-E; however, there is no statistically or substantively significant difference between the two sets. As such, DUD-E can be used to estimate ROCS' performance on other sets of targets, few of which might be represented in DUD-E.

Table 2
Comparison of ROCS' performance on the DUD-E and GPCR-Bench datasets

	Median AUC [95% CI]
DUD-E	0.704 [0.667,0.720]
GPCR-bench	0.682 [0.626,0.739]

4 Conclusion

Ligand-based computational methods are part of the armamentarium of Computational Chemistry for more than 20 years. The methods are applied in projects for Virtual Screening, Lead Hopping, Molecular Alignment as well as Pose Generation and Prediction. Ligand-based methods can be used when no structural information about the target is available as they don't require knowledge of the active site or the bioactive conformation of a query molecule. Even when structural information is available it was shown that ligand-based searches are efficient, fast and they perform consistently good over a large variety of target classes—including GPCRs.

References

1. Tanrikulu Y, Kruger BJ, Proschak E (2013) The holistic integration of virtual screening in drug discovery. Drug Discov Today 18:358–364
2. Kraemer O, Hazemann I, Podjarny AD, Klebe G (2004) Virtual screening for inhibitors of aldose reductase. Proteins 55:814–823
3. Kruger DM, Evers A (2010) Comparison of structure- and ligand-based virtual screening protocols considering hit list complementarity and enrichment factors. ChemMedChem 5:148–158
4. Berman HM, Westbrook J, Feng Z, Gilliland G, Bhat TN, Weissig H, Shindyalov IN, Bourne PE (2000) The protein data bank. Nucleic Acids Res 28:235–242. http://www.rcsb.org
5. Milletti F, Vulpetti A (2010) Tautomer preferences in PDB complexes and its impact on structure-based drug discovery. J Chem Inf Model 50:1062–1107
6. Hawkins PCD, Skillman AG, Warren GL, Ellingson BA, Stahl MT (2010) Conformer generation with OMEGA: algorithm and validation using high quality structures from the Protein Databank and Cambridge Structural Database. J Chem Inf Model 50:572–584
7. Hawkins PCD, Nicholls A (2012) Conformer generation with OMEGA: learning from the data set and the analysis of failures. J Chem Inf Model 52:2919–2936
8. McGann M (2012) FRED and HYBRID docking performance on standardized datasets. J Comput Aided Mol Des 26:897–906
9. Svensson F, Karlén A, Sköld C (2012) Virtual screening data fusion using both structure- and ligand-based methods. J Chem Inf Model 52:225–232
10. Hawkins PCD, Skillman AG, Nicholls A (2007) Comparison of shape-matching and docking as virtual screening tools. J Med Chem 50:74–82
11. Grant AJ, Gallardo MA, Pickup BT (1996) A fast method of molecular shape comparison: a simple application of a Gaussian description of molecular shape. J Comput Chem 17:1653–1666

12. Mills JEJ, Dean PM (1996) Three-dimensional hydrogen-bond geometry and probability information from a crystal survey. J Comput Aided Mol Des 10:607. https://doi.org/10.1007/BF00134183
13. Geldenhuys WJ, Funk MO, Van dr Schyf CJ, Carroll RT (2012) A scaffold hopping approach to identify novel monoamine oxidase B inhibitors. Bioorg Med Chem Lett 22:1380–1383
14. Waldner BJ, Fuchs JE, Schauperl M, Kramer C, Liedl KR (2016) Protease inhibitors in view of peptide substrate databases. J Chem Inf Model 56:1228–1235
15. Hall DR, Enyedy IJ (2016) The use of fake ligands from computational solvent mapping in ligand and structure-based virtual screening. Future Med Chem 8:1815–1822
16. Metz A, Schanda J, Grez M, Wichmann C, Gohlke H (2013) From determinants of RUNX1/ETO tetramerization to small-molecule protein-protein interaction inhibitors targeting acute myeloid leukemia. J Chem Inf Model 53:2196–2202
17. Swann SL, Brown SP, Muchmore SW, Patel H, Merta P, Locklear J, Hajduk PJ (2013) A unified, probabilistic framework for structure- and ligand-based virtual screening. J Med Chem 54:1223–1232
18. Vasudevan SR, Singh N, Churchill GC (2014) Scaffold hopping with virtual screening from IP3 to a drug-like partial agonist of the inositol trisphosphate receptor. Chembiochem 15:2774–2782
19. Santa Cruz EC, Carecho AR, Saidel ME, Montanari CA, Leitao A (2017) In silico selection and cell-based characterization of selective and bioactive compounds for androgen-dependent prostate cancer cell. Bioorg Med Chem Lett 27:546–550
20. Santos-Sierra S, Kirchmair J, Perna AM, Reiss D, Kemter K, Roschinger W, Glossmann H, Gersting SW, Muntau AC, Wolber G, Lagler FB (2012) Novel pharmacological chaperones that correct phenylketonuria in mice. Hum Mol Genet 21:1877–1887
21. Vasudevan SR, Moore JB, Schymura Y, Churchill GC (2012) Shape-based reprofiling of FDA-approved drugs for the H1 histamine receptor. J Med Chem 55:7054–7059
22. EON 2.2.0.5: OpenEye Scientific Software, Santa Fe, NM. http://www.eyesopen.com
23. Zap Toolkit 2017.Feb.1, (2016) OpenEye Scientific Software, Santa Fe, NM. http://www.eyesopen.com
24. Boström J, Grant JA, Fjellström O, Thelin A, Gustafsson D (2013) Potent fibrinolysis inhibitor discovered by shape and electrostatic complementarity to the drug tranexamic acid. J Med Chem 56:3273–3280
25. Muchmore SW, Souers AJ, Akritopoulou-Zanze I (2006) The use of three-dimensional shape and electrostatic similarity searching in the identification of a melanin-concentrating hormone receptor 1 antagonist. Chem Biol Drug Des 67:174–176
26. Markt P, Petersen RK, Flindt EN, Kristiansen K, Kirchmair J, Spitzer G, Distino S, Schuster D, Wolber G, Laggner C, Langer T (2008) Discovery of novel PPAR ligands by a virtual screening approach based on pharmacophore modeling, 3D shape and electrostatic similarity screening. J Med Chem 51:6303–6317
27. Naylor E, Arredouani A, Vasudevan SR, Lewis AM, Parkesh R, Mizote A, Rosen D, Thomas JM, Izumi M, Ganesan A, Galione A, Churchill GC (2009) Identification of a chemical probe for NAADP by virtual screening. Nat Chem Biol 5:220–226
28. Hanley JA, McNeil BJ (1982) The meaning and use of the area under a receiver operating characteristic (ROC) curve. Radiology 143:29
29. Huang N, Shoichet BK, Irwin JJ (2006) Benchmarking sets for molecular docking. J Med Chem 49:6789–6801
30. Mysinger MM, Carchia M, Irwin JJ, Shoichet BK (2012) Directory of useful decoys, enhanced (DUD-E): better ligands and decoys for better benchmarking. J Med Chem 55:6582–6594
31. Cohen J (1988) Statistical power analysis for the behavioral sciences, 2nd. Edition. Lawrence Erlbaum Associates: Mahwah NJ.
32. Hawkins PCD, Kelley BP, Warren GL (2014) The application of statistical methods to cognate docking: a path forward? J Chem Inf Model 54:1339–1355
33. Student (1908) The probable error of a mean. Biometrika 6:1–25
34. Weiss DR, Bortolato A, Tehan B, Mason JS (2016) GPCR-Bench: a benchmarking set and practitioner's guide for G Protein-Coupled Receptor docking. J Chem Inf Model 56:642–651

Chapter 19

Computational Methods Used in Hit-to-Lead and Lead Optimization Stages of Structure-Based Drug Discovery

Alexander Heifetz, Michelle Southey, Inaki Morao, Andrea Townsend-Nicholson, and Mike J. Bodkin

Abstract

GPCR modeling approaches are widely used in the hit-to-lead (H2L) and lead optimization (LO) stages of drug discovery. The aims of these modeling approaches are to predict the 3D structures of the receptor-ligand complexes, to explore the key interactions between the receptor and the ligand and to utilize these insights in the design of new molecules with improved binding, selectivity or other pharmacological properties. In this book chapter, we present a brief survey of key computational approaches integrated with hierarchical GPCR modeling protocol (HGMP) used in hit-to-lead (H2L) and in lead optimization (LO) stages of structure-based drug discovery (SBDD). We outline the differences in modeling strategies used in H2L and LO of SBDD and illustrate how these tools have been applied in three drug discovery projects.

Key words Structure-based drug design, Molecular dynamics, Simulation, Hit-to-lead, Lead optimization, G protein-coupled receptor, Docking

1 Introduction

1. GPCRs are cell surface receptors that contain seven transmembrane helices and constitute the largest superfamily of membrane proteins, regulating almost every aspect of cellular activity [1]. GPCRs have enormous physiological and biomedical importance, being the primary site of action of 40% of all prescribed drugs today [2]. There are over 800 human GPCRs known today [3, 4], involved in a diversity of diseases, including cancer, pain, inflammation, depression, anxiety [5]. Despite this, drugs have been developed just for 50 of these GPCRs. This renders GPCRs as one of the most important classes of current pharmacological targets [3, 5].
2. Recent advances in X-ray crystallography of GPCR experience its "renaissance" [2, 6–10], however, crystal structures are still not currently feasible for every receptor or receptor-ligand

Alexander Heifetz (ed.), *Computational Methods for GPCR Drug Discovery*, Methods in Molecular Biology, vol. 1705,
https://doi.org/10.1007/978-1-4939-7465-8_19,

complex [11]. This significantly limits the ability of the crystallography to guide SBDD for GPCR targets in "real-time" [11]. Furthermore, the experimentally determined structures represent just a few snapshots of what we know are very dynamic receptors and as a consequence offer only limited insights into the overall conformational space and related function of GPCR [11].

3. In the absence of crystallographic data, GPCR modeling is often the only practical alternative to guide SBDD [1, 12–14]. Modern computational approaches can address such key issues as GPCR flexibility [15] and ligand-induced dynamics, ligand kinetics (k_{on}/k_{off} rates) [16–19], prediction of water positions [20] and their role in ligand binding and prediction of the effects of mutations on ligand binding. However, the ultimate goals of any GPCR modeling protocol are: (1) to predict the structures of the complexes between the ligands and the target receptor, (2) to explore the key interactions between the ligand, surrounding residues, and water molecules, and (3) to utilize these insights in the design of the next generation of the lead compounds with improved binding, selectivity, or other pharmacological properties. The success of any GPCR modeling protocol applied in SBDD is always measured by decreased time and cost of the synthetic effort [14, 21].

4. *Hit to lead (H2L)* [22] is defined as early stage of drug discovery also known as lead generation (Fig. 1a). In H2L small molecule hits from a high-throughput screen (HTS) or from virtual screening (VS) are evaluated and undergo limited optimization to identify promising lead compounds, as illustrated in Fig. 1a. Through the limited H2L optimization steps, the affinities of these primary hits are often improved by several orders of magnitude to the nanomolar (10^{-9} M) range [22]. To achieve improvement in affinity it is usually sufficient to modify the hit in such a way that it will generate additional interaction/s with the target receptor compared to the primary hit.

5. *Lead optimization (LO)* [21] phase of drug discovery (Fig. 1b) is usually defined as the process of bringing a chemical series to clinical trials through iterative steps of design and testing. Compared to H2L, the initial lead compound(s) in LO often already demonstrated significant potency against the target. However, the affinity, selectivity or other pharmacological properties might need further optimization. The key challenge in LO is to improve of what are often already potent compounds. This requires detailed information on the interactions between the ligand and its corresponding target and off-target

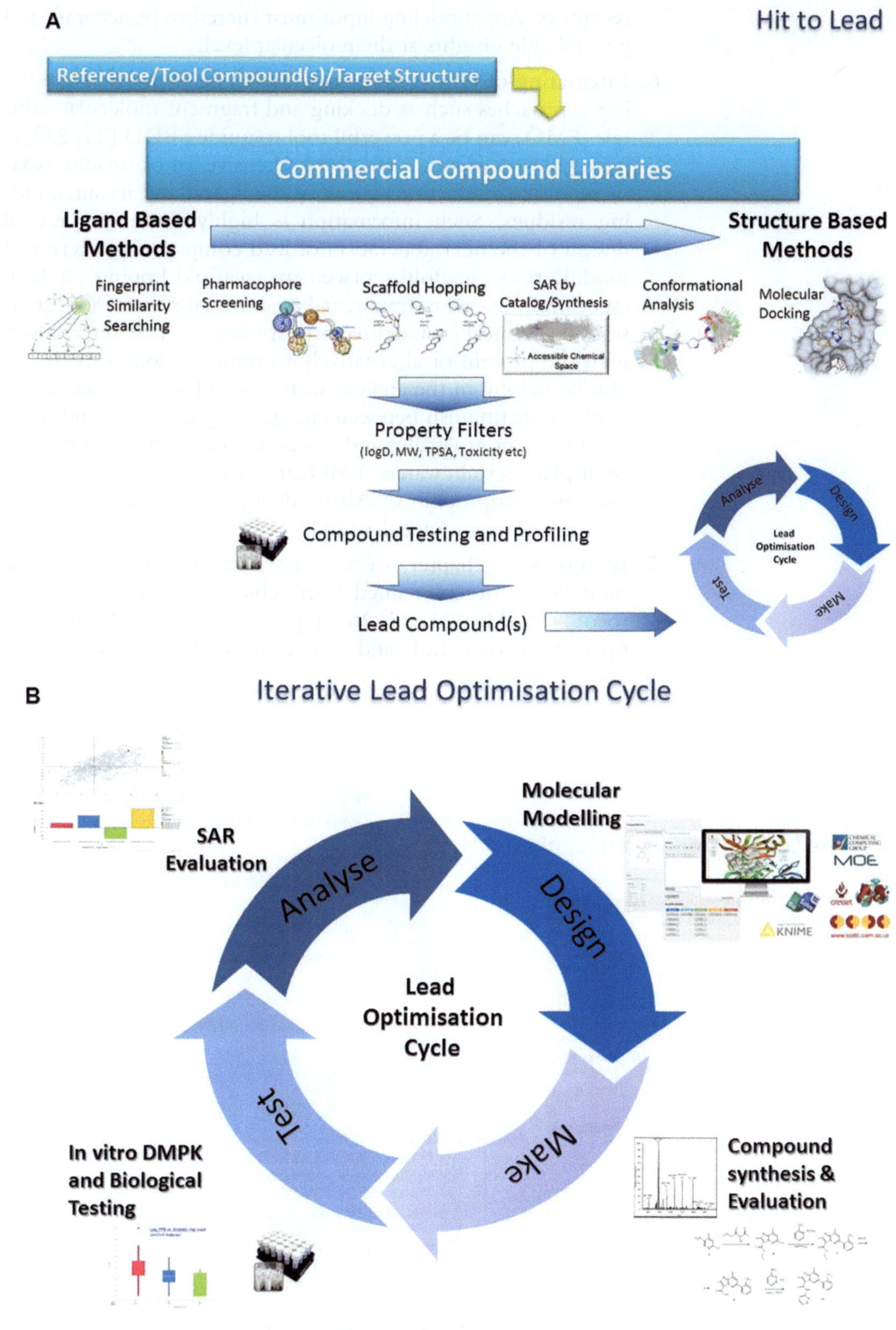

Fig. 1 (**a**) Optimization cycle for H2L. (**b**) Optimization cycle for LO

receptors. Any modeling input must therefore be accurate and give reliable insights at the molecular level.

6. Integrating of GPCR homology modeling with other modeling approaches such as docking and fragment molecular orbitals (FMO) can be a powerful tool to guide SBDD [21, 23], as it provides an accurate and comprehensive list of strong, weak, or repulsive interactions between the ligand and its surrounding residues. Such information is highly useful in rational design of the next generation of lead compounds in terms of modifications, scaffold replacement (scaffold hoping), linking (specifically in case of fragment-based drug discovery) or extension of chemical moieties to form stronger or new interactions with the protein or alternatively to remove repulsions. It can also be helpful in the analysis of the ligand-water-protein network, to distinguish between energetically favorable and unfavorable water molecules and to design ligands that can interact or displace certain waters. FMO energy terms can be efficiently used as descriptors in QSAR modeling to predict the binding affinities of new molecules [24].
7. In this book chapter, we will describe one of many GPCR modelling protocols named "hierarchical GPCR modeling protocol" (HGMP [21, 25, 26], Fig. 2). HGMP has been developed by Evotec Ltd. and University of Oxford to support

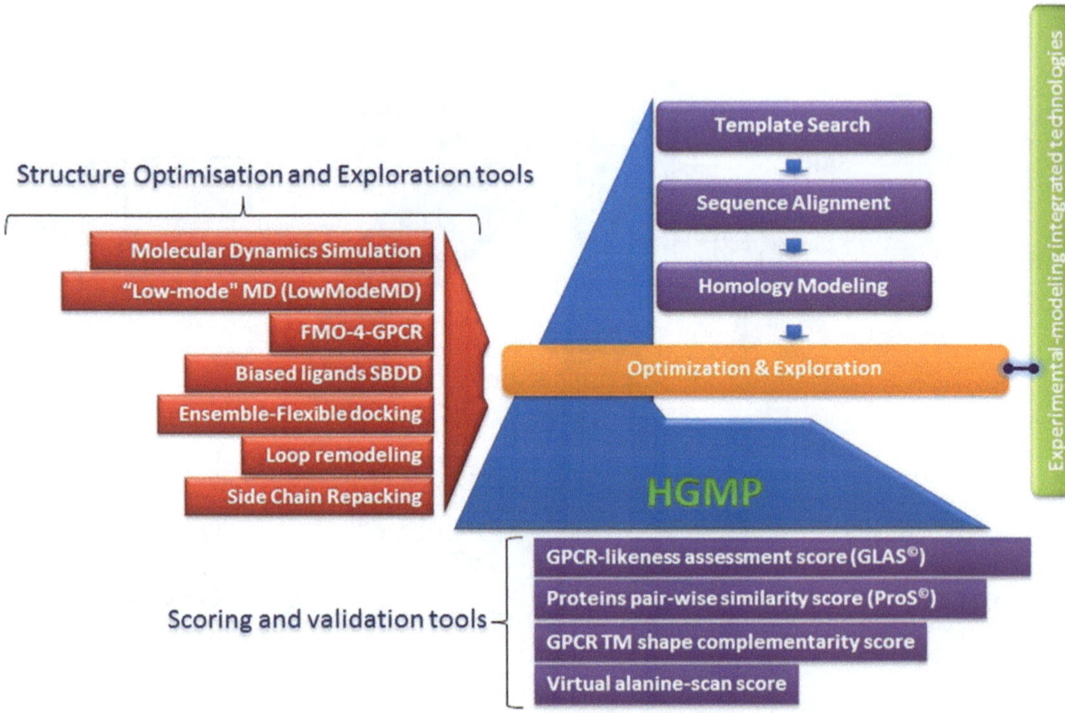

Fig. 2 A summary schematic of the Hierarchical GPCR Modeling Protocol (HGMP)

SBDD programs. HGMP generates a 3D model of GPCR structures and its complexes with small molecules by applying computational methods. In Subheading 2, we will describe how HGMP is integrated with other SBDD tools like: docking, molecular dynamic simulations, FMO, water molecules predictors, and KNIME. In Subheading 3, we will illustrate how these tools were used in 3 H2L and LO projects.

2 Methods

2.1 Constructing GPCR Models

1. *Traditional GPCR homology modeling approaches* [13] often involve the following steps: (1) sequence alignment between the modeled receptor and an appropriate template, (2) homology modeling and model refinement and (3) docking of ligands into the binding site. The key cons of such "static" approaches is that the modeled receptor is practically a "copy" of the original template and therefore some of the critical structural features are often lost. This significantly reduces the relevance of such models and their ability to guide SBDD. This is particularly problematic in the LO when information on the fine details of the system is highly important.
2. *Modern (dynamic) GPCR modeling protocols* [13, 27] have moved beyond the use of static homology modeling approaches by performing the type of extensive refinement and exploration of both structure and flexibility that is required to drive SBDD. To address the various challenges of GPCR drug discovery programs, these contemporary approaches are encapsulated as toolkits that can be flexibly assembled into workflows tailored to the specific needs of each project. The ability to incorporate experimental data during the modeling is another important factor that can enhance the effectiveness of these workflows. An example of such a workflow is the hierarchical GPCR modeling protocol (HGMP—Fig. 2).
3. *Hierarchical GPCR modeling protocol (HGMP)* [25] (Fig. 2)—generates a GPCR model and its potential complexes with small molecules by applying a series of computational methods incorporated mainly in molecular operating system (MOE, Chemical Computing Group, version 2016.08). The protocol makes use of homology modeling followed by MD simulations and docking (flexible docking if required) to predict binding poses and functions of ligands. The HGMP is practically a toolbox for GPCR modeling that can be "tailored" for project needs where experimental data can be easily fed in. It is equipped with GPCR-specific "plugins", including a GPCR-likeness assessment score (GLAS) to evaluate model quality and a pairwise protein comparison method (ProS) used to cluster

structural data and distinguish between different activation substrates. The HGMP has been applied in a number of industrial drug design projects, which have also led to further refinements of the protocol (*see* **Notes 1–3**). Even in cases where the sequence identity to the target is very low, careful model building in conjunction with site-directed mutagenesis and binding assays can be very useful in aiding the future direction of a drug discovery program.

2.2 Generating of the GPCR-Ligand Complex

1. Having the model of the receptor in hand, the next step is often predicting of the receptor-ligand complex, this process is called molecular docking. Predicting this complex is highly important if we want to study the interactions between the ligand and the receptor and to guide the SBDD. As numerous docking approaches have been reviewed in the literature [28] quite recently, we here survey briefly the unique challenges and docking protocols relevant to GPCRs.
2. *Docking protocols* [28, 29] are the molecular modelling processes aimed to explore the interaction between the ligand and protein. The ultimate goal of any docking protocol is to predict the bioactive conformation of the ligand and its place and orientation inside of the receptor binding site named as "docking pose" or "binding mode". The docking procedure consists of two sequential tasks: first, flexible placement of the ligand in a predefined binding site of the receptor and then scoring the poses of the docked ligands. Both posing and scoring phases are equally important and can be carried out by very different methodologies depending on how exhaustive the conformational sampling of both the ligand and protein is considered.
3. Some commercial available **docking suites of programs** are AutoDock [30], AutoDock Vina [31], MOE [21], FlexX [32], GOLD [33], and Glide [34]. Different search algorithms are designed to predict the bioactive conformation of the studied compounds through the evaluation of the interactions between ligands and targets [29]. An increase in the quality of the ligand docking can be gained by consideration of flexibility of the modeled system.
4. *Scoring and re-ranking*: In many of our projects (*see* **Note 4**), we used AMBER interaction energy to rescore and re-rank docking poses. We used the MM_PBSA/GBSA approach [35] to calculate the AMBER interaction energy [36]. This approach, while subject to the same limitations of all force field-based methods, was able to accurately predict relative binding affinities between the ligand and protein and was therefore selected as a reliable method to rescore and to rank docking poses [37].

5. *Flexible docking*—typical docking protocols keep the receptor (largely) rigid, and so cannot address the issue of receptor flexibility. As these protocols do not take into account the ligand-induced (or ligand-stabilized) conformation of the receptor, it makes it harder to rationalize the effects of ligand binding in terms of activation or deactivation (agonists and antagonists, respectively). Some docking approaches like induced fit docking (IFD) introduced in Autodock 4 [38], AutoDock Vina, and Schrödinger assign limited flexibility to the sidechains of key residues. However, this approach is slightly artificial and is an unsatisfactory solution to the general problem of receptor flexibility. The ensemble docking protocol, implemented in GOLD [33], performs docking into multiple states of the same receptor but it is highly governed by the availability of the structural information on the targeted receptor. The perfect scenario would be if the bioactive conformation of the docked ligand was known prior to the docking simulation.

6. *HGMP-C4XD* integrates HGMP with experimental NMR-based technology (C4XD) (Fig. 3). The C4XD [39] was developed by C4X Discovery Ltd. to explore how molecules behave in physiologically relevant solution. C4XD demonstrated that small molecules exist in relatively few conformations in the solution and that one of those conformations closely resembles

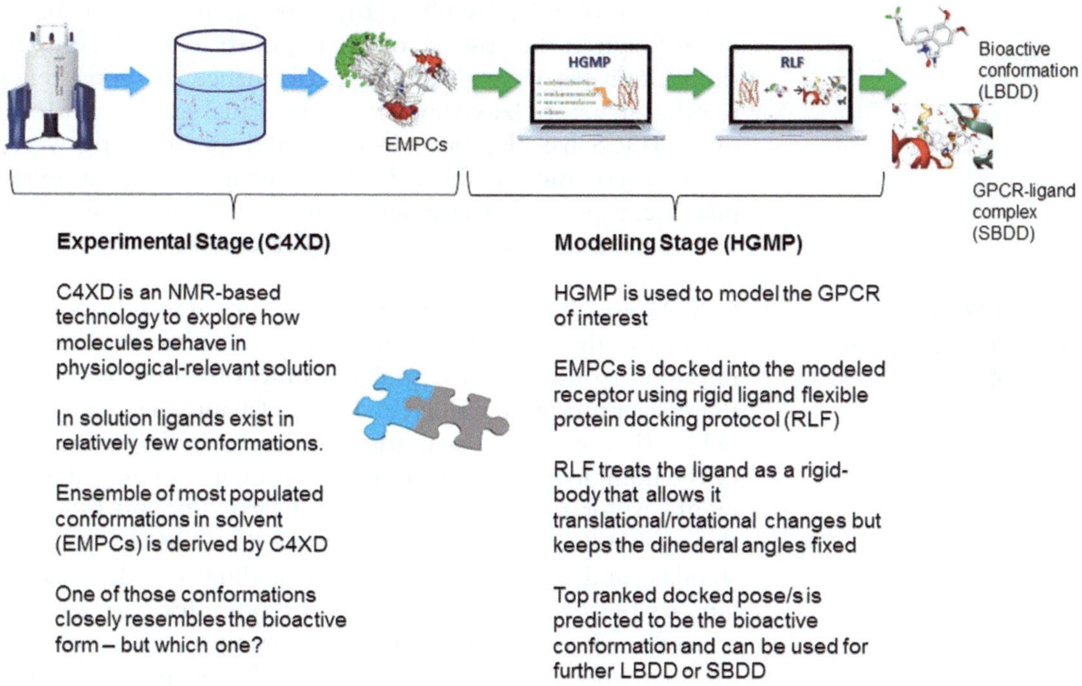

Fig. 3 HGMP-C4XD workflow

the bioactive form—but which one? Next, during the docking we limit the ligand conformation space only to the most populated conformations found by C4XD and assign the flexibility to the receptor. The combination of HGMP and the C4XD approaches allows the isolation of the bioactive conformation of the ligand and the identification of the key pharmacophoric features required for GPCR-ligand binding and selectivity. These structural insights are essential for the refinement of GPCR models, for addressing the ligand-induced receptor flexibility and for the rationalization of the ligand binding.

7. An additional way to place the ligand inside the receptor is to overlay it on the top of an already bound ligand (template) usually extracted from the crystal structure. The most common software to perform molecular overlays is ROCS, from OpenEye [40]. An additional minimization of the ligand within the active site is needed after in order to remove clashes with the receptor.
8. *ROCS protocol* [40] is the most common shape-based superposition method employed in the industry nowadays. ROCS performs shape-based overlays of conformers of candidate molecules to a query molecule (template) in one or more conformations. The overlays can be performed very quickly because the molecules can be described as atom-centered Gaussian functions. ROCS maximizes the rigid overlap of these Gaussian functions and thereby maximizes the shared volume between a template and a single conformation of a database molecule. ROCS is therefore used in ligand-based drug design in the absence of the target structure. Despite its simplicity it has shown a similar performance and consistency to other structure-based approaches in virtual screening. Moreover, ROCS has also been incorporated into docking workflows where the obtained ROCS overlay is used as initial placement/pose within the active site and has also been embedded in alignment-dependent 3D QSAR analyses.

2.3 Exploration of the Dynamic Nature of GPCRs

1. GPCRs are, by functional necessity, very dynamic entities. Molecular dynamic (MD) simulation therefore provides an important source of structural and functional information for these receptors (as described in detail in Chapter 6 of this book) [15]. MD can be used in a variety of ways including refinement of the homology model in a more realistic membrane environment, exploration of ligand-induced flexibility and function, the analysis of solvent, the effect of mutation on receptor stability, and exploration of ligand binding and dissociation kinetics [41, 42]. MD trajectories are often used to generate an ensemble of possible receptor substrates. The ProS and GLAS methods outlined in Subheading 2.1, **step 3** were developed to explore the structural data generated within MD simulations and to help distinguish between different GPCR substrates.

2. MD simulations also allow one to explore the possibility of allosteric and cryptic binding pockets. Cryptic binding pockets are not exposed to bulk solvent all of the time and so may be hidden in certain crystallographic structures. MD allows these sites to manifest themselves, enabling docking and similar protocols to be followed in the usual manner. Simulations are as well essential for the understanding of allosteric modulation [43, 44]. In some cases, however, full MD simulation may not be required, for example when just local refinement of a homology model is required. In these cases "low-mode" molecular dynamics (LowModeMD) simulation can provide a more rapid solution [45]. LowModeMD, as implement in MOE (Chemical Computing Group), is based on perturbing an existing conformation along a trajectory using initial atomic velocities with kinetic energy concentrated on the low-frequency vibrational modes, followed by energy minimization.
3. *Residence time and MD*—It has been recently demonstrated that GPCR modeling and MD simulation can be a promising tool for the exploration and structural rationalization of ligand-receptor residence time (RT) [15, 16, 18, 46, 47]. The definition of the RT is the length of time for which a small molecule stays bound to its receptor target [48]. The current challenge is the timescale: the millisecond timescales of conventional MD are incompatible with the typical RTs of drugs (up to hours) [15, 46]. To overcome this encounter new approaches to extending MD timescales have been developed. These include: (1) Markov State Models (MSM)—a very powerful method to describe dynamical processes between defined states in MD simulations [14] (2) Metadynamics-based approaches that employ MSM to calculate off-rates based on the transitions between the intermediate (calculated) and predefined end states, and (3) Scaled MD—another approximate approach to rank ligands by their off-rates [46, 47].

2.4 Exploring Receptor-Ligand Interactions

1. The understanding of binding interactions between a protein and a small molecule plays a key role in the rationalization of potency, selectivity and kinetics. However, even with the crystal structure in hand, visual inspection and force-field-based molecular mechanics calculations cannot always explain the full complexity of the molecular interactions that are so critical in both H2L and LO. Quantum mechanical methods have the potential to address this shortcoming, but the high computational cost has typically made the use of these calculations impractical.
2. *Fragment Molecular Orbital (FMO) method* [24] (Fig. 4a) is widely used by us for protein-ligand binding calculations and drug design because it offers substantial computational savings

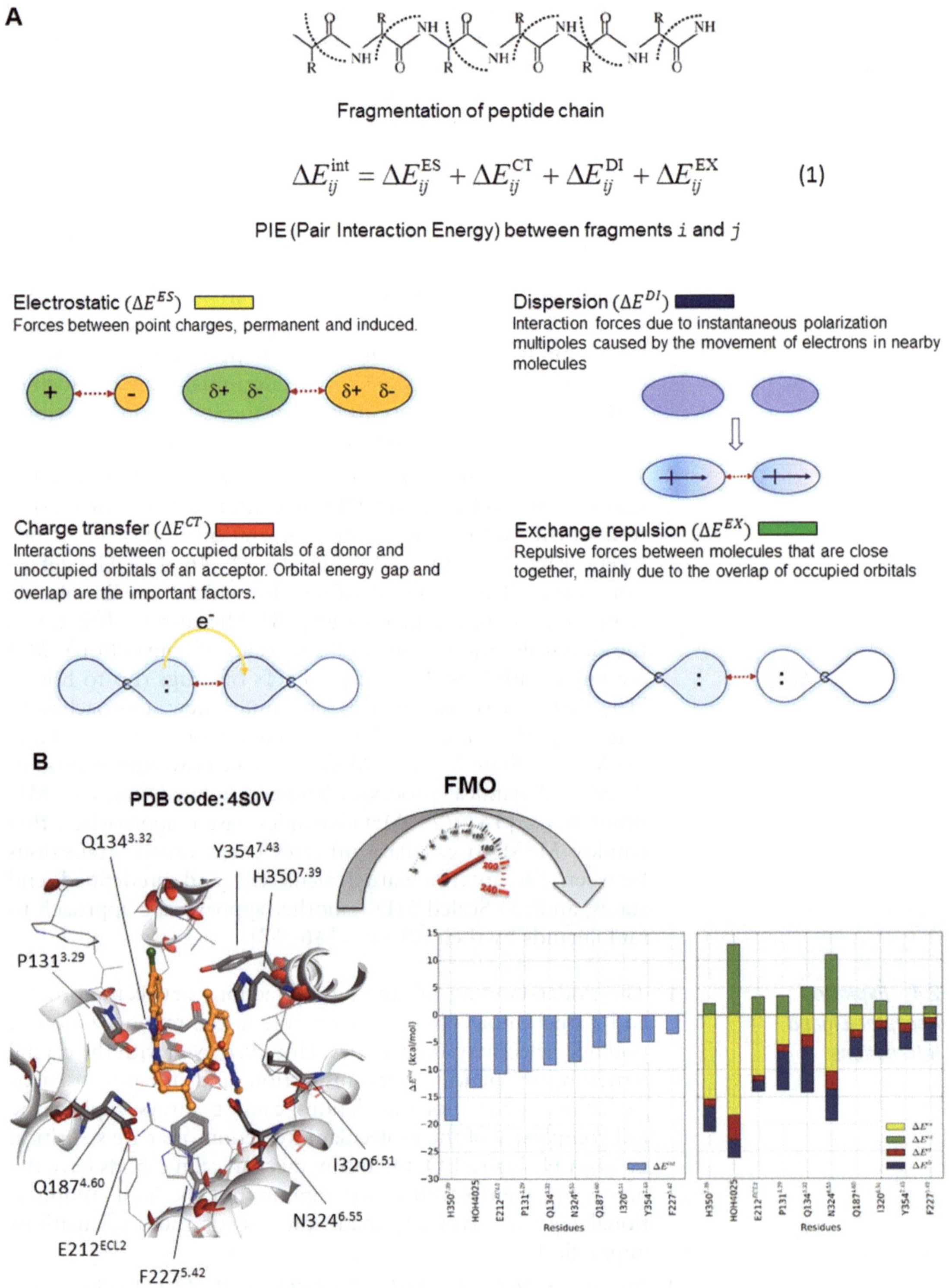

Fig. 4 Schematic summary of the FMO approach: (**a**) Workflow for PIEDA calculations and details on each of the PIE terms that are computed (**b**) FMO analysis of human adenosine OX_2 receptor in complex with Suvorexant (PDB ID 4SOV [47]). The carbon atoms of the ligand are shown in *light orange* and for the receptor

over traditional QM methods [24, 49]. By dividing the system, both the ligand and the receptor, into smaller pieces and performing QM calculations on these fragments, one can achieve high efficiency. A typical FMO calculation on a GPCR-ligand complex takes approximately 4 h on 36 CPU cores to complete, which is significantly faster than the equivalent classical QM calculations. Recently, we have demonstrated that FMO can be even faster (secs instead of hours) without compromising the accuracy by combining it with the density-functional tight-binding (DFTB) method [50].

3. Using FMO, one can take any protein-ligand complex and calculate a list of interactions and their chemical natures. Many of these interactions are difficult to detect or quantify with non-QM methods [49]. This information is very useful in guiding rational LO in terms of ligand modifications such as scaffold replacement and linking or the extension of chemical moieties to form stronger or new interactions with the protein [23].

2.5 Predicting Role of Water Molecules in Receptor-Ligand Binding

1. It is known that water-mediated interactions between ligands and receptor are extremely common and highly significant for binding and kinetics [17, 44]. Yet only high-resolution crystal structures are able to give any reliable indication as to the presence of water molecules. Displacement of these key water molecules can directly affect the ligand binding affinity and it is in the scope of SBDD programs to design compounds that can interact with or efficiently displace these water molecules. The prediction of water molecule networks and their perturbation is also critical in terms of its relationship to kinetics and residence time (*see* Chapter 9 of this book), as has been demonstrated for a series of adenosine A2A receptor antagonists [17].
2. Several methods (WaterMap [51], WaterFLAP [20], Water-Dock [52], AutoDock Vina and 3D-RISM [53]) enable a relatively rapid prediction of water molecule sites and estimation of the energy penalty for water displacement. They can help medicinal chemists to decide whether to interact with or displace a certain water molecule, if a particular subpocket of the receptor can be explored by hydrophobic moieties or if a displaced water has to be substituted by a group that mimics

Fig. 4 (continued) are *gray*. Nitrogen atoms are shown in *blue*, oxygen in *red*, and chlorine in *light green*. The fragmented bonds are marked as *red* discs. The *left-hand* bar plots describe the sorted PIE of the most significant residues, and the *right-hand* plots describe the pair interaction energy decomposition analysis (PIEDA) of these key interactions. PIE terms: electrostatics, dispersion, charge-transfer, and exchange-repulsion are color-coded in *yellow*, *blue*, *red*, and *green*, respectively. The figure is adapted from our previous publication

the hydrogen bond network. These methods are suitable for both H2L and LO.

3. Most of these methods are based on MD or Monte Carlo (MC) simulations and observing the peaks in water density can provide the location of water binding sites [54, 55]. However, these calculations can be time-consuming to run, especially with buried cavities, due to the long time it takes for water to permeate within the protein. Grand canonical MC methods [56] can significantly reduce the length of the simulation. This has led to a number of attempts to develop faster methods. JAWS, for example, is a grid-based MC method that estimates the free energy of displacing a water molecule into bulk. An integral theory approach (3D-RISM [53]) has also reported success in predicting solvation structure within ligand binding sites and protein cavities. Short molecular simulations can be used as the data for inhomogeneous fluid solvation theory (IFST). This method has the distinct advantage that the free energy can be broken down into enthalpic and entropic components. IFST also forms the framework for WaterMap [51].

2.6 Combining Individual Tools into Integrated Workflow Engines

1. GPCR modeling and SBDD is a multitask process comprised of sequential steps (Fig. 5). There is a desire to automate and standardize this process and make it more user-friendly so that less experience users can also work with it.

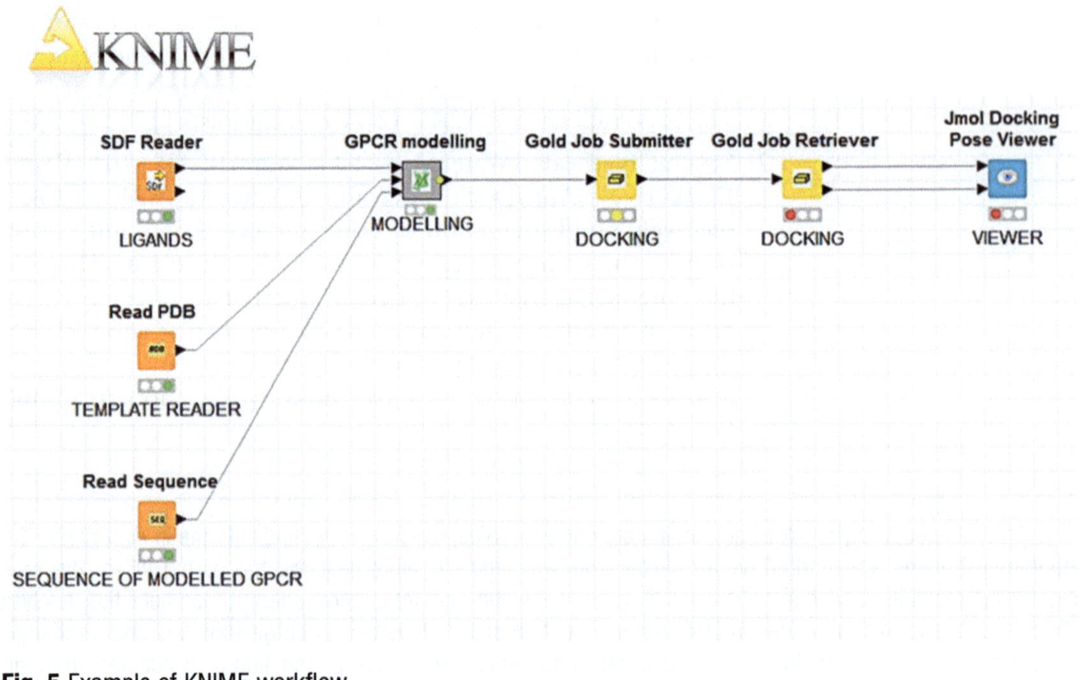

Fig. 5 Example of KNIME workflow

2. *Pipeline-Pilot and KNIME* [57, 58] are the most commonly used software packages (commercial and open source, respectively) that automate the modeling process and enable an easy concatenation of the individual tools (nodes) into an integrated workflow. Given the extensive interest in creating new therapeutics based on novel GPCR targets, modeling methodologies that are as streamlined, rapid, precise and accurate as possible are highly desirable and it is expected that an increasing number of workflows will become available in the future.

3 Notes

1. In the absence of the structural information of the receptor target, the design of new compounds in a medicinal chemistry program typically relies purely on SAR data. However, interpreting such data in isolation from specific knowledge of the protein can be challenging and even misleading [14]. Therefore, any additional means that can build confidence in the SAR interpretation and generate novel structure-based hypotheses is potentially very useful. As a result, GPCR modeling is used to bridge the gap and facilitate SBDD. The introduction of experimental data like SAR into a modeling process allows a refinement of the GPCR models to a degree that is not possible with homology modeling alone and provides a deeper rationalization of ligand binding and selectivity. In this way, modeling methods should be designed to accommodate experimental data in their algorithms and be flexible enough to deal with the wide variety of challenges that drug discovery programs face.
2. HGMP can take advantage of the experimental data that can be fed into the modeling process to add extra accuracy and confidence in the modeling outcomes. The use of the HGMP in "real" drug discovery projects is demonstrated below.
3. *Fighting obesity with a sugar-based library* [59]—Obesity is an increasingly common condition. Antagonism of the melanin-concentrating hormone-1 receptor (MCH-1R) has been widely reported as a promising therapeutic avenue for obesity treatment. However, discovery and optimization of new compounds targeting MCH-1R has been hindered by a lack of structural information about the MCH-1R and low high-throughput screening (HTS) success rates. In this H2L project, we combined HGMP (*see* Subheading 2.1, **step 3**) with the screening of a diverse library of sugar-based compounds from the VAST technology (Versatile Assembly on Stable Templates [59]). The GPCR-VAST method provides a good example of how ligand SAR data, when combined with modeling, can provide a useful source of structural information on GPCR binding sites and for SBDD.

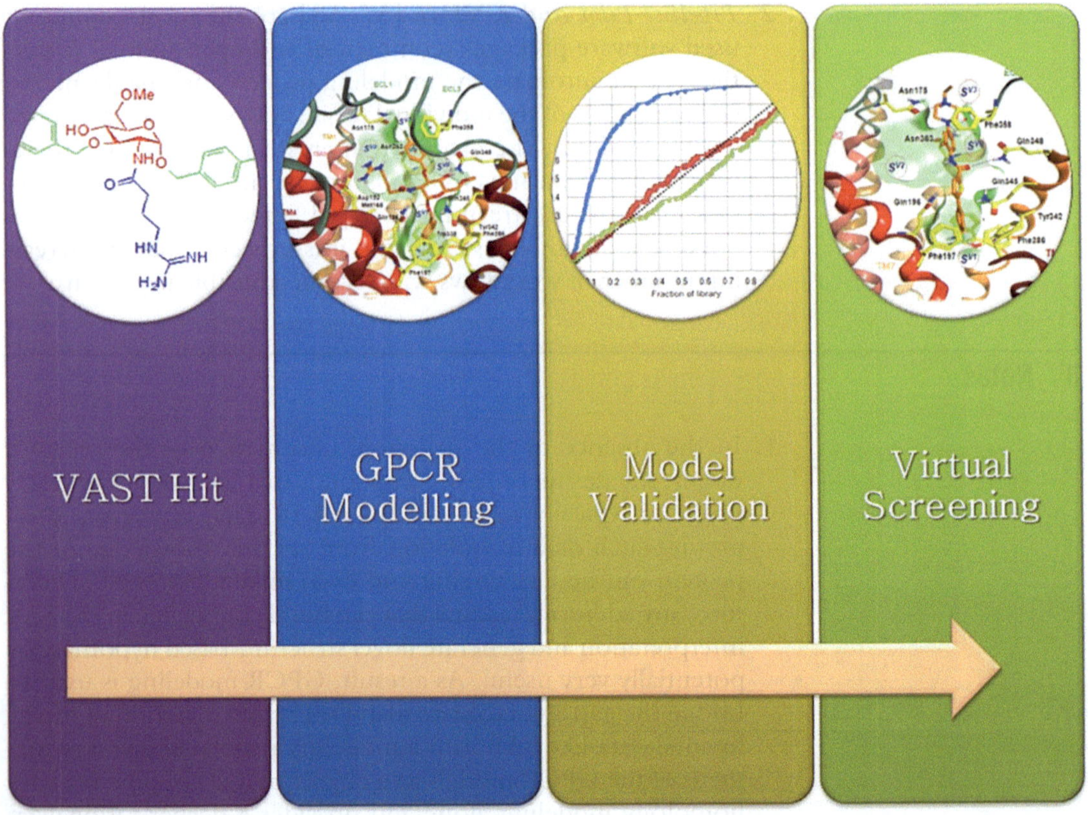

Fig. 6 Summary schematic of the VAST-GPCR modeling workflow that led to the discovery of new MCH-1R antagonists

(a) The 490 VAST compounds obtained from this library were screened against MCH-1R, resulting in the discovery of a moderately potent MCH-1R antagonist, ACL21823 ($IC_{50} = 306$ nM, *see* Fig. 6). The discovery of ACL21823 was utilized in the construction of a MCH-1R model and in the refinement of its binding site. We used HGMP (*see* Subheading 2.1, **step 3**) to model the MCH-1R and the flexible docking protocol of GOLD (*see* Subheading 2.2, **step 5**) to dock the VAST hits into the MCH-1R receptor model. The scoring and re-ranking was performed with AMBER interaction energy (*see* Subheading 2.2, **step 4**). The usefulness of this method in H2L was demonstrated by a structure-based VS, which achieved a hit rate of 14% and yielded 10 new chemotypes of MCH-1R antagonists including EOAI3367472 (IC_{50} = 131 nM) and EOAI3367474 ($IC_{50} = 213$ nM).

4. *Discovery of selective 5-HT$_{2C}$ agonists for the treatment of metabolic disorders* [60]—In this LO project, which was performed prior to the publication of the 5-HT$_{2B}$ and 5-HT$_{1B}$ crystal

structures, the challenge was to optimize 5-HT_{2c} binders and convert them into strong agonists that were unable to activate 5-HT_{2A} and 5-HT_{2B} receptors. It is known that for effective antagonism, it is sufficient for ligands just to occupy a relevant receptor site in order to inhibit the binding of endogenous ligands. However, agonist discovery has the additional complication and requirement that the ligand must not only be able to both occupy the receptor site but also be able to activate the receptor. Agonist binding should elicit conformational changes in the receptor that result in the activation of intracellular G-proteins and/or β-arrestins which, in turn, can modulate the activity of downstream effectors within the cell. The mechanism and structural changes associated with the activation of GPCRs remain unclear, making agonist design quite challenging.

(a) To explore 5-HT_{2C} activation mechanism and to design compounds that would promote receptor activation, HGMP was applied (*see* Subheading 2.1, **step 3**) to model both the active and inactive receptor conformations, referred to as 5-HT_{2C}^{active} and 5-$HT_{2C}^{inactive}$, respectively. Ensemble docking with GOLD (*see* Subheading 2.2, **step 5**) was used to predict the binding modes of lead compounds in 5-HT_{2A}, 5-HT_{2B} and 5-HT_{2C}. It was proposed that agonists enter deeply into 5-HT_{2C} binding site and interact simultaneously with both TM3 and TM6, thus increasing the overall stability of 5-HT_{2C}^{active} and promoting activation. In parallel, we modeled off-targets 5-HT_{2A} and 5-HT_{2B} to filter out compounds from the 5-HT_{2C}^{active} screen that might also bind to these two receptors. We also employed our hERG modeling [61] to take into account the hERG liability of our lead compounds. The final outcome was the discovery of a novel compound **10** ($EC_{50} = 8.4/762/73$ nM for 5-$HT_{2C/2A/2B}$ and hERG inhibition of 11% at 10 μM) [60].

5. *Case study 3: Discovery of potent and selective OX_2 receptor antagonists* [62]—The orexin receptors (OX_1 and OX_2) are linked to a range of different physiological functions including the control of feeding, energy metabolism, modulation of neuro-endocrine function, and regulation of the sleep-wake cycle. The key challenges of this project were to increase the OX_2 activity and selectivity of lead compounds over OX_1. This was particularly difficult as OX_1 and OX_2 receptors share over 80% sequence identity at the amino acid level. This project was completed before the crystal structures of OX_1 and OX_2 were released.

(a) HGMP was applied (*see* Subheading 2.1, **step 3**) to model both OX_1 and OX_2 receptors. We used MD simulation

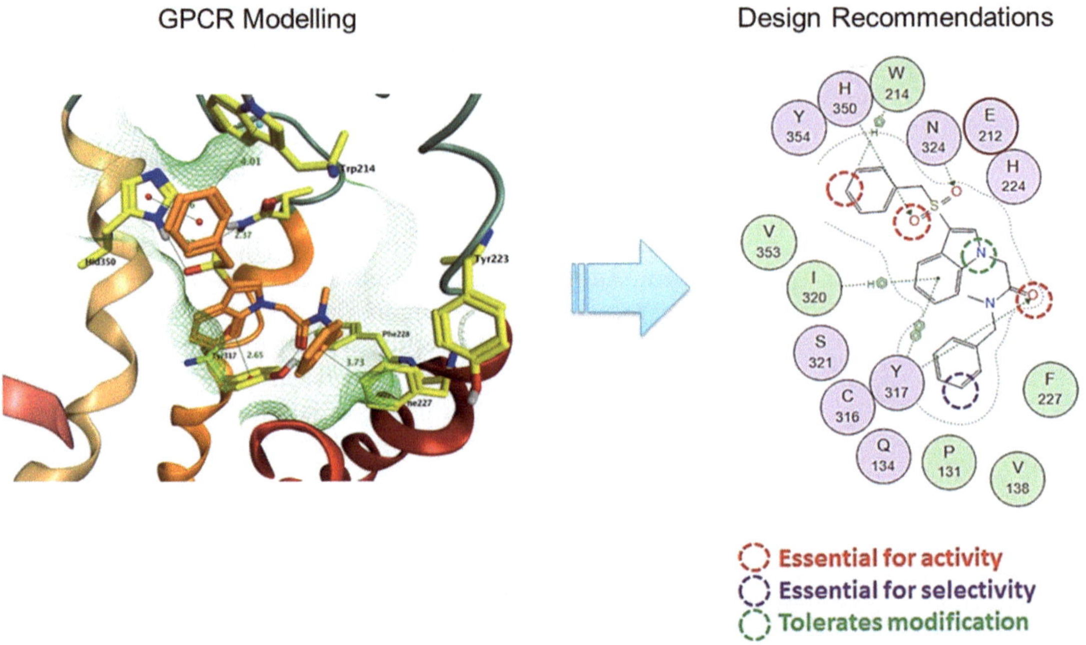

Fig. 7 Schematic summarizing how interaction maps derived from GPCR model for potent and selective OX_2 receptor antagonist

(*see* Subheading 2.3, **step 1**) as implemented in GROMACS [21, 63] to explore OX_1/OX_2 selectivity. MD suggested that differences in intra-helical interactions resulted in differences in TM conformation and in the topology of the binding pocket. The differences identified were small but sufficient to design molecules with OX_2 selectivity. This rational design significantly decreased the amount of synthesis required by focusing effort on the relevant portion of the ligand structure, as outlined in Fig. 7. The final compound, EP-009-0513, had K_i values of 4363 and 5.7 nM for OX_1 and OX_2, respectively.

6. *Conclusion*—Modern GPCR modeling protocols [64], such as the HGMP, have gone beyond the use of static models to allow for the type of detailed exploration of GPCR-ligand structures required to drive H2L and LO. These methods permit the prediction of GPCR substates in a way that is not possible with static homology modeling alone. The practicality and efficiency of GPCR modeling integrated with other modeling tools is enhanced by experimental data and by the availability of structural information on the targeted GPCR, satisfying the immediate need of the drug discovery process for the information needed to drive SBDD effectively.
7. *Future challenges*—Despite a huge effort by the pharmaceutical industry to design novel drugs for GPCR targets, there is

tremendous attrition along R&D pipelines [48]. Many promising drug candidates eventually fail in clinical trials due to a demonstrated lack of efficacy. A retrospective analysis of those that have successfully made it to the market has revealed that their beneficial effects in patients may be attributed to their long drug-target residence times (RTs)—the length of time for which a drug (ligand) stays bound to its receptor target [48]. There is substantial evidence that ~70% of long RT therapeutics displayed higher efficacy than comparable faster-dissociating drugs, supporting a growing recognition that drug-target RT may be of even greater importance than affinity, therapeutically [65].

8. Recently, several notable reviews [48, 65, 66] have emphasized the crucial role of RT optimization in the early phases of drug discovery, suggesting that detailed structure-based studies of RT should be introduced in the early phases of drug discovery to prevent "fail late, fail expensive" scenarios. Efforts to include RT in the drug development process have focused on the adoption of either experimental or computational approaches (*see* Subheading 2.3, **step 3**). Although each approach is very promising they only provide half of the whole picture. Experimental methods can measure the RT but cannot rationalize why certain compounds have longer RTs than the others or suggest ways to modify the structure of the ligand to improve its RT profile. On the other hand, computational methods are only able to provide this essential information if robust experimental data are available. Combining experimental and computational tools, as described in Chapter 15 of this book, is a highly encouraging step toward addressing the RT in early stages of H2L and LO.
9. Experience has shown that significant progress in technology R&D and "know-how" for GPCR SBDD can only be achieved when there are good interdisciplinary collaborations between the experimental and theoretical groups [1].

Acknowledgment

A.H. and A.T.-N. would like to acknowledge the support of EU H2020 CompBioMed project (http://www.compbiomed.eu/) and the BBSRC Flexible Interchanger Programme project (BB/P004245/1).

References

1. Heifetz A, Schertler GF, Seifert R, Tate CG, Sexton PM, Gurevich VV, Fourmy D, Cherezov V, Marshall FH, Storer RI, Moraes I, Tikhonova IG, Tautermann CS, Hunt P, Ceska T, Hodgson S, Bodkin MJ, Singh S, Law RJ, Biggin PC (2015) GPCR structure, function, drug discovery and crystallography: report from academia-industry

international conference (UK Royal Society) Chicheley hall, 1-2 September 2014. Naunyn Schmiedeberg's Arch Pharmacol 388:883–903

2. Shonberg J, Kling RC, Gmeiner P, Lober S (2015) GPCR crystal structures: medicinal chemistry in the pocket. Bioorg Med Chem 23:3880–3906
3. Wise A, Gearing K, Rees S (2002) Target validation of G-protein coupled receptors. Drug Discov Today 7:235–246
4. Rask-Andersen M, Masuram S, Schioth HB (2014) The druggable genome: evaluation of drug targets in clinical trials suggests major shifts in molecular class and indication. Annu Rev Pharmacol Toxicol 54:9–26
5. Dohlman HG (2015) Thematic minireview series: new directions in G protein-coupled receptor pharmacology. J Biol Chem 290:19469–19470
6. Jazayeri A, Andrews SP, Marshall FH (2017) Structurally enabled discovery of adenosine A2A receptor antagonists. Chem Rev 117:21–37
7. Jazayeri A, Dias JM, Marshall FH (2015) From G protein-coupled receptor structure resolution to rational drug design. J Biol Chem 290:19489–19495
8. Cooke RM, Brown AJ, Marshall FH, Mason JS (2015) Structures of G protein-coupled receptors reveal new opportunities for drug discovery. Drug Discov Today 20:1355–1364
9. Congreve M, Dias JM, Marshall FH (2014) Structure-based drug design for G protein-coupled receptors. Prog Med Chem 53:1–63
10. Topiol S, Sabio M (2009) X-ray structure breakthroughs in the GPCR transmembrane region. Biochem Pharmacol 78:11–20
11. Topiol S (2013) X-ray structural information of GPCRs in drug design: what are the limitations and where do we go? Expert Opin Drug Discov 8:607–620
12. Topiol S, Sabio M (2015) The role of experimental and computational structural approaches in 7TM drug discovery. Expert Opin Drug Discov 10:1071–1084
13. Tautermann CS, Gloriam DE (2016) Editorial overview: new technologies: GPCR drug design and function-exploiting the current (of) structures. Curr Opin Pharmacol 30:8–10
14. Biggin PC, Aldeghi M, Bodkin MJ, Heifetz A (2016) Beyond membrane protein structure: drug discovery, dynamics and difficulties. Adv Exp Med Biol 922:161–181
15. Tautermann CS, Seeliger D, Kriegl JM (2015) What can we learn from molecular dynamics simulations for GPCR drug design? Comput Struct Biotechnol J 13:111–121
16. Latorraca NR, Venkatakrishnan AJ, Dror RO (2017) GPCR dynamics: structures in motion. Chem Rev 117:139–155
17. Guo D, Pan AC, Dror RO, Mocking T, Liu R, Heitman LH, Shaw DE, IJ AP (2016) Molecular basis of ligand dissociation from the adenosine A2A receptor. Mol Pharmacol 89:485–491
18. Pan AC, Borhani DW, Dror RO, Shaw DE (2013) Molecular determinants of drug-receptor binding kinetics. Drug Discov Today 18:667–673
19. Dror RO, Arlow DH, Maragakis P, Mildorf TJ, Pan AC, Xu H, Borhani DW, Shaw DE (2011) Activation mechanism of the beta2-adrenergic receptor. Proc Natl Acad Sci U S A 108:18684–18689
20. Mason JS, Bortolato A, Weiss DR, Deflorian F, Tehan B, Marshall FH (2013) High end GPCR design: crafted ligand design and druggability analysis using protein structure, lipophilic hotspots and explicit water networks. In Silico Pharmacol 1:23
21. Heifetz A, James T, Morao I, Bodkin MJ, Biggin PC (2016) Guiding lead optimization with GPCR structure modeling and molecular dynamics. Curr Opin Pharmacol 30:14–21
22. Deprez-Poulain R, Deprez B (2004) Facts, figures and trends in lead generation. Curr Top Med Chem 4:569–580
23. Heifetz A, Aldeghi M, Chudyk E, Fedorov DG, Bodkin M, Biggin PC (2016) Using the fragment molecular orbital method to investigate agonist-orexin 2 receptor interactions. Biochem Soc Trans 44(2):574–581
24. Heifetz A, Chudyk EI, Gleave L, Aldeghi M, Cherezov V, Fedorov DG, Biggin PC, Bodkin MJ (2016) The fragment molecular orbital method reveals new insight into the chemical nature of GPCR-ligand interactions. J Chem Inf Model 56:159–172
25. Heifetz A, Storer RI, McMurray G, James T, Morao I, Aldeghi M, Bodkin MJ, Biggin PC (2016) Application of an integrated GPCR SAR-modeling platform to explain the activation selectivity of human 5-HT over 5-HT. ACS Chem Biol 11(5):1372–1382
26. Storer RI, Brennan PE, Brown AD, Bungay PJ, Conlon KM, Corbett MS, DePianta RP, Fish PV, Heifetz A, Ho DK, Jessiman AS, McMurray G, de Oliveira CA, Roberts LR, Root JA, Shanmugasundaram V, Shapiro MJ, Skerten M, Westbrook D, Wheeler S, Whitlock GA, Wright J (2014) Multiparameter optimization in CNS drug discovery: design of pyrimido[4,5-d]azepines as potent 5-hydroxytryptamine 2C (5-HT(2)C) receptor

agonists with exquisite functional selectivity over 5-HT(2)A and 5-HT(2)B receptors. J Med Chem 57:5258–5269

27. Tautermann CS (2014) GPCR structures in drug design, emerging opportunities with new structures. Bioorg Med Chem Lett 24:4073–4079
28. Bartuzi D, Kaczor AA, Targowska-Duda KM, Matosiuk D (2017) Recent advances and applications of molecular docking to G protein-coupled receptors. Molecules 22(2):E340
29. Kitchen DB, Decornez H, Furr JR, Bajorath J (2004) Docking and scoring in virtual screening for drug discovery: methods and applications. Nat Rev Drug Discov 3:935–949
30. Morris GM, Goodsell DS, Halliday RS, Huey R, Hart WE, Belew RK, Olson AJ (1998) Automated docking using a Lamarckian genetic algorithm and an empirical binding free energy function. J Comput Chem 19:1639–1662
31. Trott O, Olson AJ (2010) AutoDock Vina: improving the speed and accuracy of docking with a new scoring function, efficient optimization, and multithreading. J Comput Chem 31:455–461
32. Rarey M, Kramer B, Lengauer T, Klebe G (1996) A fast flexible docking method using an incremental construction algorithm. J Mol Biol 261:470–489
33. Verdonk ML, Cole JC, Hartshorn MJ, Murray CW, Taylor RD (2003) Improved protein-ligand docking using GOLD. Proteins 52:609–623
34. Friesner RA, Banks JL, Murphy RB, Halgren TA, Klicic JJ, Mainz DT, Repasky MP, Knoll EH, Shelley M, Perry JK, Shaw DE, Francis P, Shenkin PS (2004) Glide: a new approach for rapid, accurate docking and scoring. 1. Method and assessment of docking accuracy. J Med Chem 47:1739–1749
35. Kollman PA, Massova I, Reyes C, Kuhn B, Huo S, Chong L, Lee M, Lee T, Duan Y, Wang W, Donini O, Cieplak P, Srinivasan J, Case DA, Cheatham TE 3rd (2000) Calculating structures and free energies of complex molecules: combining molecular mechanics and continuum models. Acc Chem Res 33:889–897
36. Liu S, Wu Y, Lin T, Abel R, Redmann JP, Summa CM, Jaber VR, Lim NM, Mobley DL (2013) Lead optimization mapper: automating free energy calculations for lead optimization. J Comput Aided Mol Des 27(9). https://doi.org/10.1007/s10822-10013-19678-y
37. Sotriffer CA, Flader W, Winger RH, Rode BM, Liedl KR, Varga JM (2000) Automated docking of ligands to antibodies: methods and applications. Methods 20:280–291
38. Morris GM, Huey R, Lindstrom W, Sanner MF, Belew RK, Goodsell DS, Olson AJ (2009) AutoDock4 and AutoDockTools4: automated docking with selective receptor flexibility. J Comput Chem 30:2785–2791
39. Blundell CD, Packer MJ, Almond A (2013) Quantification of free ligand conformational preferences by NMR and their relationship to the bioactive conformation. Bioorg Med Chem 21:4976–4987
40. Hawkins PC, Skillman AG, Nicholls A (2007) Comparison of shape-matching and docking as virtual screening tools. J Med Chem 50:74–82
41. Marino KA, Shang Y, Filizola M (2017) Insights into the function of opioid receptors from molecular dynamics simulations of available crystal structures. Br J Pharmacol. https://doi.org/10.1111/bph.13774
42. Schneider S, Provasi D, Filizola M (2015) The dynamic process of drug-GPCR binding at either orthosteric or allosteric sites evaluated by metadynamics. Methods Mol Biol 1335:277–294
43. Kaczor AA, Rutkowska E, Bartuzi D, Targowska-Duda KM, Matosiuk D, Selent J (2016) Computational methods for studying G protein-coupled receptors (GPCRs). Methods Cell Biol 132:359–399
44. Bartuzi D, Kaczor AA, Matosiuk D (2015) Activation and allosteric modulation of human mu opioid receptor in molecular dynamics. J Chem Inf Model 55:2421–2434
45. Labute P (2010) LowModeMD--implicit low-mode velocity filtering applied to conformational search of macrocycles and protein loops. J Chem Inf Model 50:792–800
46. De Vivo M, Masetti M, Bottegoni G, Cavalli A (2016) Role of molecular dynamics and related methods in drug discovery. J Med Chem 59:4035–4061
47. Mollica L, Theret I, Antoine M, Perron-Sierra-F, Charton Y, Fourquez J-M, Wierzbicki M, Boutin JA, Ferry G, Decherchi S, Bottegoni G, Ducrot P, Cavalli A (2016) Molecular dynamics simulations and kinetic measurements to estimate and predict protein–ligand residence times. J Med Chem 59:7167–7176
48. Copeland RA (2016) The drug-target residence time model: a 10-year retrospective. Nat Rev Drug Discov 15:87–95
49. Heifetz A, Trani G, Aldeghi M, MacKinnon CH, McEwan PA, Brookfield FA, Chudyk E, Bodkin M, Pei Z, Burch JD, Ortwine DF (2016) Fragment molecular orbital method

applied to lead optimization of novel interleukin-2 inducible T-Cell Kinase (ITK) inhibitors. J Med Chem 59(9):4352–4363

50. Morao I, Fedorov DG, Robinson R, Southey M, Townsend-Nicholson A, Bodkin MJ, Heifetz A (2017) Rapid and accurate assessment of GPCR-ligand interactions using the fragment molecular orbital-based density-functional tight-binding method. J Comput Chem 38(23):1987–1990
51. Abel R, Young T, Farid R, Berne BJ, Friesner RA (2008) Role of the active-site solvent in the thermodynamics of factor Xa ligand binding. J Am Chem Soc 130:2817–2831
52. Ross GA, Morris GM, Biggin PC (2012) Rapid and accurate prediction and scoring of water molecules in protein binding sites. PLoS One 7:e32036
53. Truchon JF, Pettitt BM, Labute P (2014) A cavity corrected 3D-RISM functional for accurate solvation free energies. J Chem Theory Comput 10:934–941
54. Gerogiokas G, Southey MW, Mazanetz MP, Heifetz A, Bodkin M, Law RJ, Henchman RH, Michel J (2016) Assessment of hydration thermodynamics at protein interfaces with grid cell theory. J Phys Chem B 120: 10442–10452
55. Gerogiokas G, Southey MW, Mazanetz MP, Heifetz A, Bodkin M, Law RJ, Michel J (2015) Evaluation of water displacement energetics in protein binding sites with grid cell theory. Phys Chem Chem Phys 17:8416–8426
56. Vajda S, Guarnieri F (2006) Characterization of protein-ligand interaction sites using experimental and computational methods. Curr Opin Drug Discov Devel 9:354–362
57. Goldmann D, Zdrazil B, Digles D, Ecker GF (2016) Empowering pharmacoinformatics by linked life science data. J Comput Aided Mol Des 31(3):319–328
58. Mazanetz MP, Marmon RJ, Reisser CB, Morao I (2012) Drug discovery applications for KNIME: an open source data mining platform. Curr Top Med Chem 12:1965–1979
59. Heifetz A, Barker O, Verquin G, Wimmer N, Meutermans W, Pal S, Law RJ, Whittaker M (2013) Fighting obesity with a sugar-based library: discovery of novel MCH-1R antagonists by a new computational-VAST approach for exploration of GPCR binding sites. J Chem Inf Model 53:1084–1099
60. Tye H, Mueller SG, Prestle J, Scheuerer S, Schindler M, Nosse B, Prevost N, Brown CJ, Heifetz A, Moeller C, Pedret-Dunn A, Whittaker M (2011) Novel 6,7,8,9-tetrahydro-5H-1,4,7,10a-tetraaza-cyclohepta[f]indene analogues as potent and selective 5-HT(2C) agonists for the treatment of metabolic disorders. Bioorg Med Chem Lett 21:34–37
61. Davenport AJ, Moller C, Heifetz A, Mazanetz MP, Law RJ, Ebneth A, Gemkow MJ (2010) Using electrophysiology and in silico three-dimensional modeling to reduce human Ether-a-go-go related gene K(+) channel inhibition in a histamine H3 receptor antagonist program. Assay Drug Dev Technol 8:781–789
62. Heifetz A, Morris GB, Biggin PC, Barker O, Fryatt T, Bentley J, Hallett D, Manikowski D, Pal S, Reifegerste R, Slack M, Law R (2012) Study of human Orexin-1 and -2 G-protein-coupled receptors with novel and published antagonists by modeling, molecular dynamics simulations, and site-directed mutagenesis. Biochemistry 51:3178–3197
63. Barnoud J, Monticelli L (2015) Coarse-grained force fields for molecular simulations. Methods Mol Biol 1215:125–149
64. Gutierrez-de-Teran H, Keranen H, Azuaje J, Rodriguez D, Aqvist J, Sotelo E (2015) Computer-aided design of GPCR ligands. Methods Mol Biol 1272:271–291
65. Tummino PJ, Copeland RA (2008) Residence time of receptor-ligand complexes and its effect on biological function. Biochemistry 47:5481–5492
66. Guo D, Hillger JM, IJzerman AP, Heitman LH (2014) Drug-target residence time—a case for G protein-coupled receptors. Med Res Rev 34:856–892

Chapter 20

Cheminformatics in the Service of GPCR Drug Discovery

Tim James

Abstract

Cheminformatics is a broad discipline covering a wide range of computational approaches, including the characterization of molecular similarity, pattern recognition, and predictive modeling. The unifying theme that these apparently disparate methods have in common is the aim of extracting useable information from the increasing amounts of data that are associated with contemporary drug discovery projects. Both proprietary and publically available data can be exploited to help inform and improve the process of developing novel therapeutic molecules targeting the GPCR family of proteins.

Key words Cheminformatics, G protein-coupled receptor, Library design, Reaction mining, QSAR, Drug-likeness, Multi-parameter optimization

1 Introduction

A commonly quoted definition of cheminformatics is taken from a 1998 article in Annual Reports in Medicinal Chemistry by Frank Brown [1]:

"Chemoinformatics is the mixing of those information resources to transform data into information and information into knowledge for the intended purpose of making better decisions faster in the area of drug lead identification and optimization."

This definition, although descriptive, is incredibly broad in scope and encompasses almost any use of information technology in improving the drug discovery process. In this chapter I have therefore focussed on a number of areas where, in my experience, cheminformatics approaches are most frequently applied and have the greatest impact. These are listed in Fig. 1 against a schematic preclinical small-molecule drug discovery pipeline, to indicate where each approach might commonly be considered. However, as the nominally distinct stages of the process are largely arbitrarily defined the different cheminformatics techniques can be and are applied across the pipeline.

Alexander Heifetz (ed.), *Computational Methods for GPCR Drug Discovery*, Methods in Molecular Biology, vol. 1705,
https://doi.org/10.1007/978-1-4939-7465-8_20,

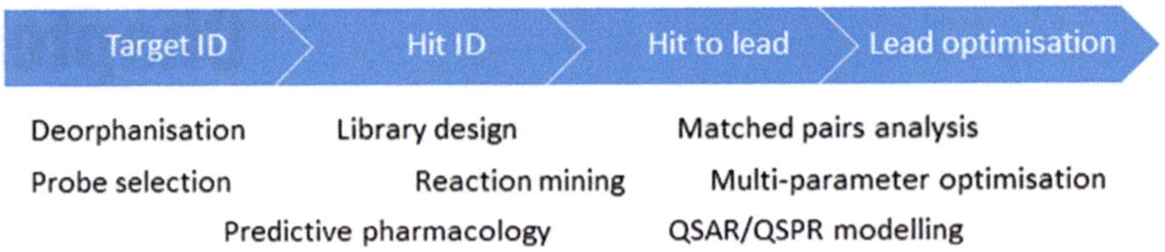

Fig. 1 Common cheminformatics approaches and an illustration of where they are applied in the preclinical drug discovery pipeline

Other therapeutic classes including peptides, antibodies, engineered proteins, and RNAi agents are increasingly important in pharmaceutical R&D, but will not be considered here. Some authors also include approaches such as virtual screening as part of the cheminformatics toolkit, but specific chapters exist to cover these in more detail and so I will instead focus here on complementary techniques.

2 Data Sources for Cheminformatics

All chemoinformatics endeavors start with one or more data sources. The most commonly used type of data is that related to biologically relevant compound activities, typically against individual proteins or protein complexes. Many pharmaceutical companies have the advantage of having access to considerable amounts of proprietary information of this type, accumulated over years of drug discovery research. Commercial compendia also exist, of which Integrity [2], Liceptor [3] and Reaxys Medicinal Chemistry [4] are well-known examples. However, there are an increasing number of public domain databases in this area. Perhaps the best known and most widely used of these is ChEMBL [5], which in its most recent release contains approximately 14 million activity values drawn from a variety of sources. Other examples of general-purpose bioactivity databases include PubChem BioAssay [6], BindingDB [7], the IUPHAR/BPS Guide to Pharmacology [8] and the Psychoactive Drug Screening Program's K_i database [9]. There are also a number of systems focused more specifically on GPCR research [10], including GLIDA [11] and GPCRdb [12].

The scientific literature and publically funded screening centers are comparatively well served by open source bioactivity databases. However, the patent literature remains more challenging. Specifically claimed compounds are generally available and captured by systems such as SureChEMBL [13], although Markush patterns can cover vast numbers of hypothetical compounds that are never likely to be synthesized. However, activity data in patents is often deliberately as vague as possible, and it can be difficult to associate such data that is included with the corresponding molecular structures using automated approaches. This is one area in which commercial databases currently offer an advantage [14].

Beyond bioactivities, a number of other types of data are commonly mined to drive cheminformatics applications. Publically accessible three-dimensional structural information relevant to drug discovery is consolidated to a large extent, both for small molecules (the Cambridge Structural Database [15]) and macromolecules (the Protein Databank [16]). By contrast, structured information related to chemical synthesis is largely confined to company electronic lab notebooks (ELNs) and commercial databases. Reaction informatics is a relatively immature area of research, but one that has shown a number of developments in recent years.

The proliferation of data sources offers new opportunities to the cheminformatician, but also presents a number of challenges. Assessing the overlap between different sources is an ongoing task [14], which is complicated by the fact that data abstraction from the primary sources does not appear to be performed in a consistent manner [17]. Even where information from a journal article or other experimental account is accurately transcribed, the underlying data quality should always be treated with a certain amount of healthy scepticism. For example, despite a number of well-known publications [18] describing compounds that are prone to showing artifactual behavior in biochemical assays, the counter-assays required to clarify the activity of such compounds are often not reported. There have been recent efforts to address these issues in the literature [19], but universal adoption of such standards appears to still be some way off. Beyond establishing the veracity of any individual data point, a further issue remains the challenge of combining information from different experiments performed in different labs by different scientists. Reproducibility in the biological sciences is never as high as one would like it to be [20], and analyses performed on inconsistent data sets are likely to suffer from decreased signal-to-noise ratios. The greatest advantage that proprietary corporate datasets offer over public compendia is therefore probably not the amount or diversity of information that is available but rather its consistency.

3 Target Identification

Although the extreme reductionist philosophy of drug discovery ("one target one disease") has never truly received universal subscription [21], the recent resurgence in the popularity of phenotypic screening suggests that a rebalancing between target-based and systems approaches is currently underway [22]. Nonetheless, many projects are still initiated with the hypothesis that modulating a specific target will offer therapeutic value in a particular disease setting. A number of GPCRs that, in principle, could be useful drug

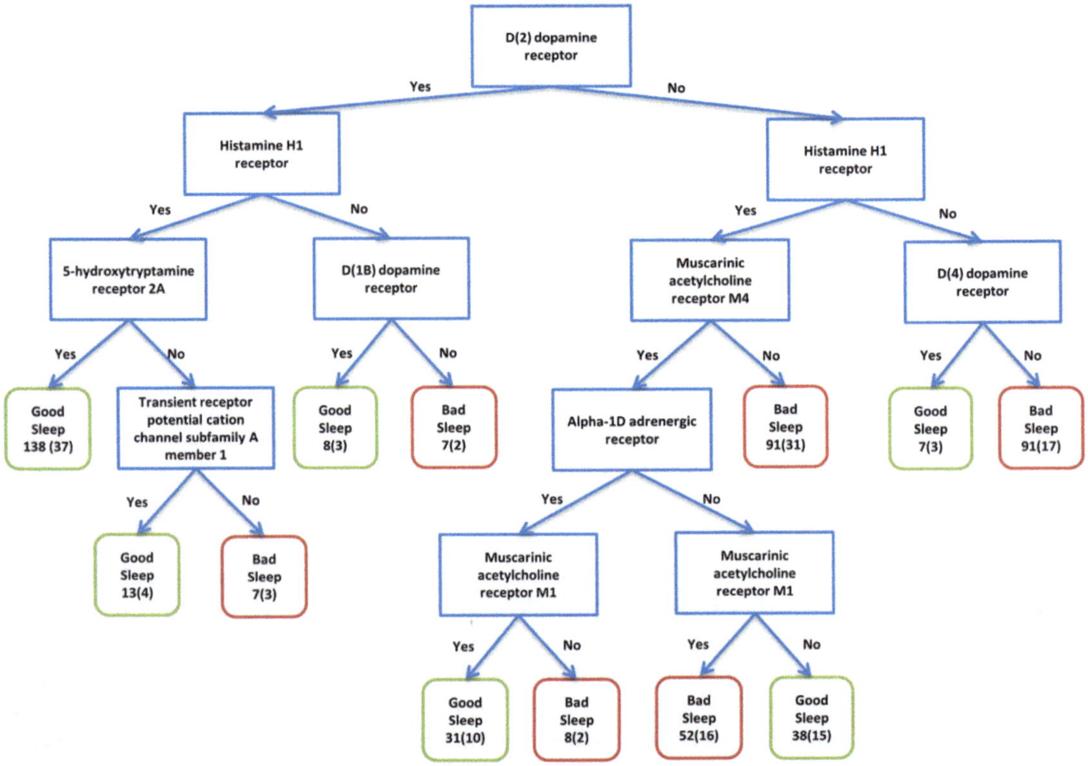

Fig. 2 A decision tree to classify compounds as either promoting a good or bad sleep pattern in rats using predicted protein-level activities. Reproduced with permission from [24]

targets remain uncharacterized both in terms of function and endogenous ligands. Cheminformatics techniques can be used to supplement bioinformatics analysis and assist in receptor de-orphanization. For example, van der Horst et al. [23] created a phylogenetic GPCR classification scheme using only known ligands and showed generally good agreement between this and sequence-based classification. In a virtual experiment, these authors showed that a combined chemogenomics strategy was able to successfully de-orphanize a variety of pseudo-orphan GPCRs by using the ligands of related receptors.

The efficacy of GPCR-targeting drugs seems likely to be related to their polypharmacology across a number of receptors within the family, at least in some therapeutic applications. It would therefore be useful to be able to determine what the important elements of these efficacious bioactivity profiles are. In a recently published work, Drakakis et al. performed such an analysis to investigate the sedative-hypnotic effects of compounds in rats [24]. The final decision tree from this analysis is shown in Fig. 2. Using predicted protein-level activities as inputs, this model achieved an overall accuracy of 68% in retrospective cross-validation on a training set of 491 molecules. Perhaps more impressively, 5 out of 7 of the compounds selected for prospective testing also showed promotion of good sleep patterns. Importantly, these authors found that models built on single targets alone did not perform as well as those that incorporated activities against multiple proteins.

Once an initial target (or profile) hypothesis has been generated, achieving an acceptable (preclinical) level of validation for that hypothesis typically involves a multi-faceted approach using both small molecules and genetic tools such as knockouts, CRISPR, or RNAi. The requirements for a compound to give a useful level of target-specific information—a chemical probe—have been the subject of recent discussion [25]. The commonly cited characteristics include a certain level of on-target activity and selectivity over relevant off-targets, activity in cells, and an absence of general toxicity and other features that are likely to interfere with assay readouts. Currently, a number of compounds employed as probes likely fail to fulfil some of these criteria. As an example, in a recent evaluation of 64 probes nominated by the NIH Molecular Libraries and Imaging initiative, 25% were qualitatively assessed to be of low confidence [26]. Attempts to introduce greater objectivity into the assessment process remain much debated [27, 28], but it seems clear that the increasing amounts of publically available bioactivity data will have a positive impact in this area.

4 Library Design

Tailoring the composition of screening libraries for specific applications often makes use of cheminformatics techniques. Libraries can be designed with a broad focus such as CNS or antibacterial screening, more particularly on the pathways associated with a certain disease or biological process, or on individual targets or target classes. In the context of GPCR drug discovery, most library design strategies are based around the fundamental hypothesis that similar compounds have similar properties [29]. The idea is that if a library is biased toward areas of chemical space that have already shown activity against GPCR targets, that library is likely to show higher hit rates when screened against additional proteins from the same class. One way to quantify chemical similarity is through the use of physicochemical properties. For example, Balakin et al. [30] explored the use of eight simple properties such as molecular weight and number of hydrogen-bond donors in combination with a neural network algorithm. These authors demonstrated that they were able to distinguish between known GPCR actives and non-GPCR ligands in this way, before applying the same model to the design of a GPCR-focused library of 30,000 molecules.

Structural similarity is perhaps the most frequently adopted approach to tailoring library properties. In this context, much attention has been paid to the idea of privileged substructures.

Originally defined as those molecular features that give rise to activity at more than one receptor [31], the concept has since been extended to encompass target-family-privileged substructures [32]. This more recent idea implies the existence of common pharmacophoric elements across the binding sites of a protein family, to which the putative privileged substructures are assumed to bind. Probably the archetypal example of this concept is the hinge binding region in kinase domains, where a group of very highly conserved hydrogen-bonding interactions with the protein backbone is frequently exploited in inhibitor design. Whether this approach can usefully be applied to GPCRs is perhaps less clear. For example, Bondensgaard and coworkers examined three pairs of GPCR ligands with common substructures that bind to multiple receptors with relatively distant homology [33]. By docking these ligands into receptor models they concluded that the conserved regions of the binding sites responsible for interacting with the privileged substructures are typically buried deep in the pocket and possessed of a largely hydrophobic and aromatic character. However, they also reported that some interactions are formed with non-conserved parts of the binding sites, implying that each privileged substructure would likely only be useful for a subset of receptors. Other studies have also reported substructures that are enriched in ligands for particular subgroups of GPCRs such as the histamine or adrenergic receptors [34], and the rationale for identifying common binding elements appears much clearer where the proteins in question are related by their endogenous ligands.

It is clear that one limitation of similarity-based library design is the degree of extrapolation that can reasonably be expected. For example, a collection designed around class A GPCR ligands—by far the most well explored in terms of chemical matter—seems unlikely to show much enrichment for class B or C GPCRs. Likewise, a library based on orthosteric binding site properties seems unlikely to yield many allosteric modulators. A competing priority that sometimes receives less attention in focussed library design is that of novelty. It can be argued that the chemical space occupied by ligands for a particular target or group of targets is more reflective of historical trends in synthesis methods than it is of true receptor binding preferences. Thus, while it may be true that choosing further molecules from this same space will yield an increased hit rate in a statistical sense, there is also the danger that one simply rediscovers that which is already known. One way to circumvent this issue is to define molecular similarity in a more abstract way than using substructures, for example by searching for common pharmacophoric features or field-based descriptors [35]. These approaches are considerably more computationally demanding than 2D property-based searches, requiring as they do the calculation of one or more 3D conformations for each molecule. This may explain why they appear to be rather less frequently employed.

5 Reaction Mining, Virtual Libraries, and De Novo Design

One of the allures of computer-aided drug discovery has always been the hypothetical ability to navigate the vastness of chemical space without the need to synthesize every compound. However, a significant issue with virtual molecules is that they are not always synthetically tractable, at least not with contemporary reaction schemes. Broadly speaking, the two approaches used to mitigate this issue are either to use synthetic considerations to guide the construction of the molecules in the first place, or to apply an automated synthetic accessibility scoring algorithm post-construction.

A comparatively small number of reaction types account for a substantial proportion of the synthetic steps that are carried out in medicinal chemistry programs [36]. This is, in large part, due to the time and other resource constraints that are placed on such projects, which mandate a strong preference for robust and versatile chemistries. Manual abstraction and mapping of these reaction types to an appropriate transformation language such as Daylight SMARTS, MDL RXN, or reaction vectors is therefore a feasible approach. One frequently referenced contemporary example of this is the work by Hartenfeller et al. [37]. Alternatively, the information can be mined from an appropriate source such as an ELN [38] or the patent corpus [39]. Automated mining not only requires a suitable parser to locate and extract the relevant information from each document, but also the ability to algorithmically clean that data and classify the various roles within each synthetic step as well as the overall reaction type.

One application of in silico chemical synthesis is the generation of virtual libraries. An example of this is the synthetically accessible virtual inventory (SAVI) database, which uses reversed retrosynthetic transforms from the LHASA program to generate a collection of approximately 610,000 products in its current form [40]. Due to the combinatorial nature of the process, even with modest numbers of reactions and reagents significantly larger virtual libraries than that represented by SAVI have been generated, at least up to 10^{16} compounds [41]. Efficient searching of these virtual spaces then becomes a significant issue, as brute force approaches and post-filtering do not scale appropriately. A more elegant solution is to adapt the search algorithm to operate in the reagent space, which eliminates the need to fully enumerate all of the virtual products [42, 43].

De novo design approaches also avoid the need to fully enumerate virtual chemistry space by applying some kind of objective function to construct the molecules in an iterative fashion. Objective functions can include the predicted strength of interaction with a particular target (or targets) using either 2D or 3D information,

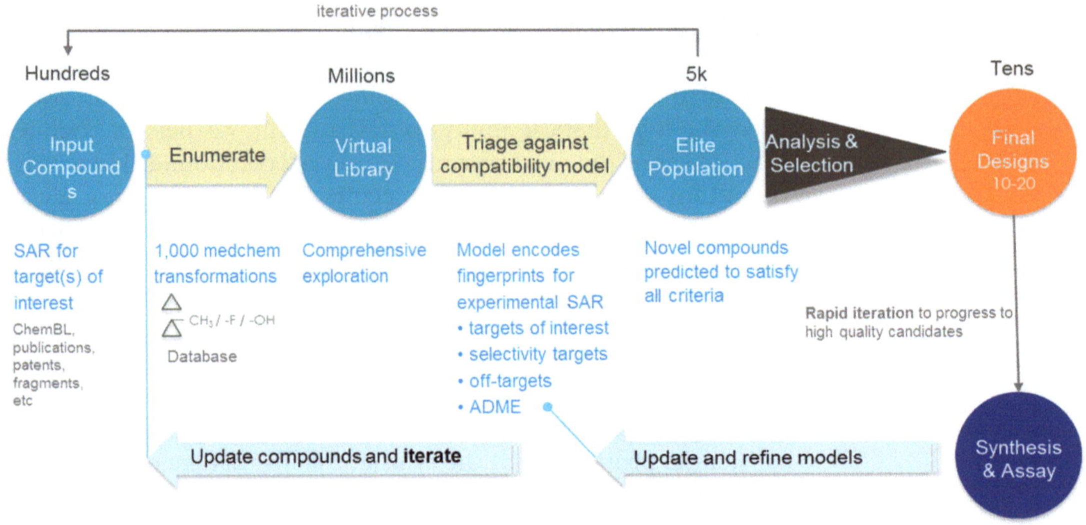

Fig. 3 An example of an evolutionary de novo design workflow of the type exemplified by the work of Besnard et al. At each iteration, virtual enumeration according to a set of medicinal chemistry transforms is followed by scoring and filtering to generate an elite population. After a number of generations have evolved compounds are selected for synthesis and experimental testing, and the results are fed back in to update both the models and the molecules for the next cycle

as well as other physicochemical properties in the typical case of a multi-parameter optimization. For example, Besnard et al. used an evolutionary algorithm similar in concept to that illustrated in Fig. 3 to design compounds with various profiles against aminergic GPCRs [44]. The objective function in this case included Bayesian models to predict a variety of on- and off-target activities, together with scores representing the likelihood of achieving CNS penetration as well as other ADME properties. The chemical transformation rules were derived from an analysis of analogue series in ChEMBL, and were therefore more akin to the kind of medicinal chemistry transformations exemplified by the Drug Guru approach [45] than true synthetic reactions. A synthetic accessibility filter was therefore applied to each generation to penalize unusual or overly complex molecules.

One question that seems unlikely to be resolved in the immediate future is that of the appropriate level of detail that should be included when describing a chemical reaction electronically. A description that specifies the exact reagents and conditions that have been experimentally validated will, in principle, be reproducible but will not usefully generalize. On the other hand, a description that only includes the direct reaction center without considering its environment is likely to be inappropriate nonspecific, giving rise to a significant proportion of virtual products that would not, in reality, be formed in any useful yield. The correct balance between these two extremes depends on the particular application and on the skills and experience of the medicinal chemistry team. However, it should always be expected that the type of approach described above will provide an enrichment in terms of synthetic tractability, rather than guaranteeing that every molecule will be readily accessible.

6 QSAR/QSPR Modeling

Quantitative structure-activity or property relationship (QSAR or QSPR, respectively) modeling is a substantial area of research that is applied across the discovery pipeline. Hence, only the briefest summary will be attempted here, and the interested reader is referred to the large number of recent [46] and not so recent [47] reviews in this field. Fundamentally, QSAR involves the construction of a mathematical model that relates a compound's structure to an experimentally measureable property of interest. This can range from superficially straightforward properties such as melting point or aqueous solubility to more complex biological endpoints such as human toxicity. A wide range of modeling or machine learning approaches have been and continue to be developed, and some contemporary algorithms are discussed below. First, however, some more general considerations for this type of approach are put forward.

Perhaps the most important aspect of QSAR modeling is choosing how to characterize the molecules; which known properties (or descriptors) to attempt to relate to the unknown property of interest. Examples of common solutions to this problem are illustrated in Fig. 4. The simplest approach is to use the molecular graphs themselves as descriptors. This is the method used in Free-Wilson analysis [48], which relates the structural elements of a molecule to a biological activity or other properties using multiple linear regression. On the assumption that the contributions from

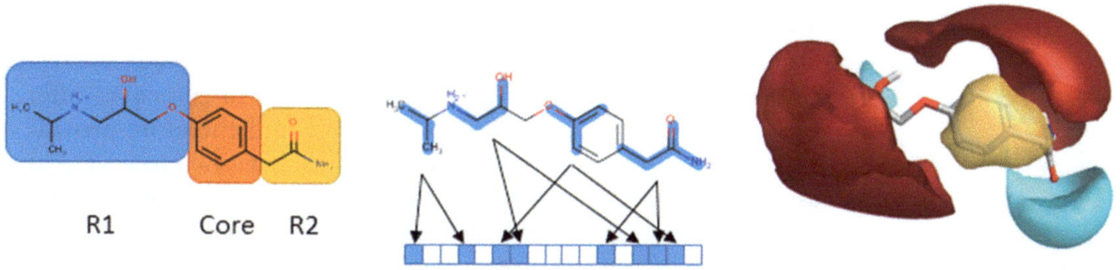

Fig. 4 Examples of different ways of characterizing a molecule during QSAR modeling, illustrated using the β-adrenoceptor antagonist Atenolol. (**a**) Division of the molecule into R groups and a core "scaffold", as would be performed during Free-Wilson or matched-pairs analysis. (**b**) Construction of a binary fingerprint from the molecular graph; in this case, a path-based fingerprint where each pattern sets two bits. (**c**) A 3D description of the molecule using electrostatic (*red*—positive, *blue*—negative) and shape (*yellow*) fields

different parts of a molecule are approximately additive, this type of approach can be useful if one is operating in a relatively well-characterized chemical space. It is therefore still applied, for example, in lead optimization programs to search for favorable combinations of previously exemplified R-groups. Free-Wilson is highly related to matched-molecular pairs analysis [49], another comparatively mature analysis method that has recently been extended to encompass matched-triplicate design [50] as well as matched molecular series [51]. Although they can offer good predictive performance, a significant drawback to these approaches is that they require a significant amount of experimental data to draw robust conclusions. Furthermore, they are entirely interpolative, and cannot be used to make predictions for compounds with novel substructures.

In order to reduce data requirements and build models capable of extrapolation, a more abstract level of compound description is required. A simple modification of the molecular composition approach is to use chemical fingerprints, either with predefined fragments such as MACCS keys or more general graph-based features such as atom paths or circular environments [52]. Predefined fragments offer easy interpretation but limited extrapolation, whereas hashed fingerprints are not limited in the same way but are harder to interpret. A large number of other graph theoretical indices can also be calculated from a molecular structure, although the practical utility of these descriptors is generally limited by their lack of interpretability. Alternatively, compounds can be described using physicochemical descriptors such as molecular weight, atom/ring counts, hydrogen-bond donor or acceptor counts, and so forth. Despite issues surrounding propagation of errors, it is relatively common practice to attempt to model one property such as a biological activity using other predicted properties. For example, descriptors that quantify molecular lipophilicity, such as the predicted octanol-water partition coefficient (logP or logD), are often found to correlate with other experimental properties of interest.

Moving beyond 2D descriptions, 3D QSAR approaches have also been developed. Perhaps the best known of these use the concept of fields to describe the shape and electrostatic character of a molecule [53]. A 3D approach should, in principle, offer a more complete description of a molecule and therefore provide a better foundation for predicting its behavior. However, this assumes that an accurate and representative conformation (or conformational ensemble) can reliably be generated, which is unfortunately not the case with present technology for many molecules. This additional complexity, computational cost and uncertainty probably contribute to the fact that 3D QSAR approaches are currently rather less frequently applied than their 2D counterparts.

Algorithms to model the relationship between a set of descriptors and a property (or properties) of interest are many and varied.

They range from comparatively simple approaches such as k nearest neighbors and recursive partitioning trees to more complicated ones such as random forests, support vector machines, and neural networks. Algorithms are sometimes divided into those used to model continuous variables and those used for classification. This is pertinent in drug discovery because a number of properties of interest such as aqueous solubility, permeability, and metabolic stability are often experimentally characterized as high, medium, or low, even where quantitative data is available. Most algorithms have a number of adjustable parameters that specifically tune their behavior, and indeed choosing the optimal parameters for a given approach can be a significant challenge. However, regardless of the exact method employed, it is key to try to capture the relationship between the descriptors and the property of interest in a generalizable way, rather than simply modeling the exact data set to hand. It is common practice to partition the available information and use cross-validation to check for overfitting, although the use of temporal datasets is probably more representative of the real world scenario in which QSAR models are employed [54]. Establishing the domain of applicability—the area of chemical or, more generally, descriptor space in which predictions above a certain level of accuracy are more likely—is also important. Studies have shown that even models built on large amounts of corporate data tend to show a deterioration in performance over time [55], and part of the reason for this is likely to be the introduction of new chemotypes that are distinct from those present in the training set.

As both the size of data sets and the available computing power increase, machine learning approaches offer ever-increasing levels of sophistication [56] and much attention is paid to which algorithm performs best for a particular data set. Pattern recognition alone is sometimes useful in drug discovery, for example in the analysis of high-throughput screening results where one is interested in identifying potentially erroneous data points based on their inconsistency with neighboring results [57]. However, a degree of understanding is usually required in addition to statistical correlation. In such situations the exact algorithm is, in many ways, the least important component of the QSAR process, and it is how the molecules are described that is critical. Some approaches allow for the use of mathematical transformations of the input descriptors and therefore, in principle, are able to recapitulate physical laws involving nonlinear relationships [58]. Unfortunately, this is only of use if all of the factors relevant to determining the property of interest are known and calculable. This may be the case, for example, if one is attempting to model the motion of a pendulum based on its physical characteristics and the local gravity field, but in essentially all drug discovery applications the important descriptors are either partly unknown, or cannot themselves be calculated accurately.

7 Drug-Likeness and Multi-parameter Optimization

Drug discovery is inherently a multi-parameter optimization (MPO) problem. Typically, it is necessary to consider a range of properties for a candidate therapeutic, including activity against the desired target or targets, activities against undesired "off-targets" and pharmacokinetic and ADMET properties such as solubility, oral bioavailability, and metabolic stability. The specific ranges that should ultimately be achieved for each property are assembled into an ideal profile, sometimes referred to as a target product profile. This is normally specified at the start of a project, although it may be necessary to adjust some of the requirements based on experience.

Unfortunately, it is nearly always the case that improvements in certain properties are accompanied by degradation of others. A balance between the various profile elements therefore needs to be achieved. Computationally, the factors can be kept separate and analyzed using an approach such as Pareto optimization, or they can be combined into an aggregated objective function or desirability metric. Combined metrics are appealing because they reduce multiple factors to a single number, and have a long history of application in medicinal chemistry programs [59]. They can range from simple equations such as those defining ligand efficiency and ligand-lipophilic efficiency up to more complicated functions involving multiple measured or calculated properties. The nature of the MPO function employed typically varies throughout a project's lifetime, with the complexity increasing to reflect the breadth of data that is available and the additional criteria that need to be met. For example, in the analysis of HTS results the experimental information is often limited to single concentration responses and so a function such as ligand efficiency might be appropriate. By contrast, during lead optimization the focus is on identifying candidate compounds for in vivo studies and so a more sophisticated MPO scheme involving multiple ADMET properties is likely to be used.

Aside from theoretical concerns about specific formulations [60], two general issues regarding the use of combined metrics are the subjective nature of the functions employed and the propagation of uncertainties. The real goal of any drug discovery program is the development of therapeutics that are both safe and efficacious in humans and, despite our best efforts, currently available preclinical assays are generally not terribly predictive of this. Therefore, constructing an optimization function using these preclinical readouts relies, in large part, on the expertise and experience of the project team. Subjectivity inevitably arises in a number of areas including the choice of properties, the functional forms of the individual metrics (as illustrated in Fig. 5), and the relative weight assigned to each during the aggregation. Furthermore, all measurements (and predictions) have an associated level of uncertainty, and

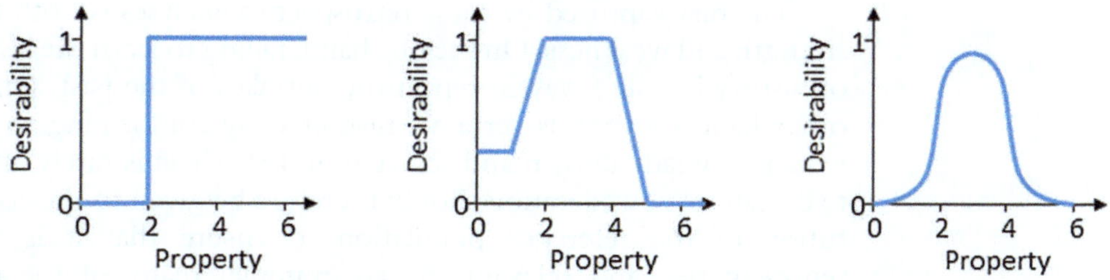

Fig. 5 Examples of individual desirability functions that might be combined into an overall MPO objective. (**a**) A hard cutoff at property values below two, exemplifying early binary classification schemes such as Lipinski's rule of five. (**b**) A more complicated function representing a preferred property range from 2 to 4, with asymmetric plateaux at low and high values and a graduated penalisation between the two. (**c**) A desirability function based on a pre-existing distribution of property values, as employed in the QED approach

the greater the number of properties that are considered, the greater the overall uncertainty will be in the combined outcome [61]. Probabilistic approaches to MPO are therefore much more representative of the true underlying data than hard cutoffs or filters, although they are more complicated to implement. The functional form of the optimization metric and the influence of uncertainties can both be challenged using sensitivity analysis [62], which is important when making compound prioritization decisions in order to avoid inappropriate exclusion of potentially useful series.

One specific application of multi-parameter thinking that is perhaps worthy of separate discussion is the characterization of drug-likeness. Work in this area typically seeks to characterize an area of chemical space based on the known compounds that already display some property of interest, such as being an approved drug. A measure of how close novel molecules are to this space is then derived, on the assumption that revisiting historically precedented space is more likely to yield future success. In an early and now much imitated study, Lipinski and coworkers analyzed the physicochemical properties of known drugs and clinical candidates to identify factors that were associated with passive membrane permeability and oral absorption [63]. This analysis was encapsulated as a series of individual property thresholds that became known as the "rule of 5" because all of the thresholds are multiples of that number. Although the philosophy remains the same, hard cutoffs have subsequently been superseded by continuous metrics familiar to other contemporary MPO applications, for example in the quantitative estimate of drug-likeness (QED) parameter [64] and the Pfizer CNS MPO score [65]. Such general metrics tend not to be monitored continuously during a drug discovery project, where a more tailored optimization function of the type discussed above would likely be developed. However, they may be of use in pre-project and early stage activities where little project-specific information is available, such as compound library design or HTS analysis.

The bias captured by these retrospective analyses is both their strength and weakness. On the one hand, failing to learn the lessons of history is a sure way to repeat the mistakes of the past. On the other hand, novelty is certainly restricted by conforming to that which is already known and reinventing the wheel is rarely attractive. Careful consideration should therefore be given to the constitution of the reference population to ensure that it is both representative and relevant. As an example, many of the early protein-protein interaction inhibitors were comparatively large molecules, reflecting attempts to mimic elements of protein secondary structure. An analysis of the chemical space at that stage would therefore likely have yielded a drug-likeness metric that favored these types of molecules. However, more recent research has shown that it is clearly possible to identify PPI modulators with more similar properties to drugs from other target classes, at least for some types of interaction. Immunomodulatory drugs such as lenalidomide, which alter the ability of the cereblon-containing E3 ubiquitin ligase to target various proteins for proteasomal degradation, are one such example [66]. Antibacterial compounds are another example where earlier analyses of physicochemical property requirements have recently been brought into question [67]. Historical data sets are generally reasonably good at indicating previous success, but distinguishing regions with previous high failure rates from those that are merely poorly explored usually receives rather less attention.

8 Personal Perspective

Even a cursory glance at the popular literature will tell you that we are currently entering an era of "big data", and drug discovery is certainly no exception to this. We are awash with data, and this trend only looks to continue for the foreseeable future. The task of the cheminformatician—turning this data into information, and the information into knowledge—therefore becomes ever more crucial. Simple increases in data volume are relatively easily accommodated by improvements in processing power, but issues surrounding data quality, consistency, and interpretation do not scale so straightforwardly. Cheminformatics is principally concerned with summarizing and presenting the relevant information to inform and expedite decision making in drug discovery, and this is a valuable activity. Computer algorithms are, comparatively speaking, excellent at pattern recognition, and therefore ideally suited to identifying statistical correlations in our sea of numbers. However, algorithms have almost nothing to say about the meaning of these correlations and, despite the enthusiasm around somewhat misleading monikers like "machine learning" and "artificial intelligence", this situation seems unlikely to change soon.

Improving our understanding of human biology and how to manipulate it are ultimately the keys to creating more effective therapeutics, and no amount of processing power can substitute for thinking in this endeavor.

References

1. Brown FK (1998) Chemoinformatics: what is it and how does it impact drug discovery? Annu Rep Med Chem 33:375–384
2. Clarivate Analytics, Integrity, https://clarivate.com/products/integrity
3. Evolvus, Liceptor Database, http://www.evolvus.com/products/databases/liceptordatabase.html
4. Elsevier, Reaxys Medicinal Chemistry, https://www.elsevier.com/solutions/reaxys/reaxys-medicinal-chemistry
5. Bento AP, Gaulton A, Hersey A, Bellis LJ, Chambers J, Davies M, Kruger FA, Light Y, Mak L, Overington JP (2014) The ChEMBL bioactivity database: an update. Nucleic Acids Res 42:1083–1090
6. Wang Y, Xiao J, Suzek TO, Zhang J, Wang J, Zhou Z, Han L, Karapetyan K, Bryant SH (2012) PubChem's bioassay database. Nucleic Acids Res 40:D400–D412
7. Gilson MK, Baitaluk M, Nicola G, Hwang L, Chong J (2016) BindingDB in 2015: a public database for medicinal chemistry, computational chemistry and systems pharmacology. Nucleic Acids Res 44:D1045–D1063
8. Southan C, Sharman JL, Benson HE, Faccenda E, Pawson AJ, Alexander SP, Buneman OP, Davenport AP, Davies JA (2016) The IUPHAR/BPS guide to PHARMACOLOGY in 2016: towards curated quantitative interactions between 1300 protein targets and 6000 ligands. Nucleic Acids Res 44:D1054–D1068
9. Roth BL, Kroeze WK, Patel S, Lopez E (2000) The multiplicity of serotonin receptors: uselessly diverse molecules or an embarrasment of riches? Neuroscientist 6:252–262
10. Southan C (2016) Retrieving GPCR data from public databases. Curr Opin Pharmacol 30:38–43
11. Okuno Y, Tamon A, Yabuuchi H, Niijima S, Minowa Y, Tonomura K, Kunimoto R, Feng C (2008) GLIDA: GPCR—ligand database for chemical genomics drug discovery—database and tools update. Nucleic Acids Res 36: D907–D912
12. Isberg V, Mordalski S, Munk C, Rataj K, Harpsøe K, Hauser AS, Vroling B, Bojarski AJ, Vriend G, Gloriam DE (2016) GPCRdb: an information system for G protein-coupled receptors. Nucleic Acids Res 44:D356–D364
13. Papadatos G, Davies M, Dedman N, Chambers J, Gaulton A, Siddle J, Koks R, Irvine SA, Pettersson J, Goncharoff N, Hersey A, Overington JP (2016) SureChEMBL: a large-scale, chemically annotated patent document database. Nucleic Acids Res 44:D1220–D1228
14. Southan C, Várkonyi P, Muresan S (2009) Quantitative assessment of the expanding complementarity between public and commercial databases of bioactive compounds. J Cheminform 1:10
15. Groom CR, Bruno IJ, Lightfoot MP, Ward SC (2016) The Cambridge structural database. Acta Crystallogr B72:171–179
16. Berman HM, Westbrook J, Feng Z, Gilliland G, Bhat TN, Weissig H, Shindyalov IN, Bourne PE (2000) The protein data bank. Nucleic Acids Res 28:235–242
17. Tiikkainen P, Franke L (2012) Analysis of commercial and public bioactivity databases. J Chem Inf Model 52:319–326
18. Baell J, Walters MA (2014) Chemical con artists foil drug discovery. Nature 513:481–483
19. Aldrich C, Bertozzi C, Georg GI, Kiessling L, Lindsley C, Liotta D, Merz KM, Schepartz A, Wang S (2017) The ecstasy and agony of assay interference compounds. ACS Central Sci 3:143–147
20. Baker M (2016) Is there a reproducibility crisis? Nature 533:452–454
21. Wermuth CG (2004) Multitargeted drugs: the end of the "one-target-one-disease" philosophy? Drug Discov Today 1:826–827
22. Swinney DC, Anthony J (2011) How were new medicines discovered? Nat Rev Drug Discov 10:507–519
23. van der Horst E, Peironcely JE, Ijzerman AP, Beukers MW, Lane JR, van Vlijmen HW, Emmerich MT, Okuno Y, Bender A (2010) A novel chemogenomics analysis of G protein-coupled receptors (GPCRs) and their ligands: a potential strategy for receptor de-orphanization. BMC Bioinformatics 11:316

24. Drakakis G, Wafford KA, Brewerton SC, Bodkin MJ, Evans DA, Bender A (2017) Polypharmacological in silico bioactivity profiling and experimental validation uncovers sedative-hypnotic effects of approved and experimental drugs in rat. ACS Chem Biol 12:1593–1602
25. Arrowsmith CH, Audia JE, Austin C, Baell J, Bennet J, Blagg J, Bountra C, Brennan PE, Howe T (2015) The promise and peril of chemical probes. Nat Chem Biol 11:536–541
26. Oprea TI, Bologa CG, Boyer S, Curpan RF, Glen RC, Hopkins AL, Sklar LA (2009) A crowdsourcing evaluation of the NIH chemical probes. Nat Chem Biol 5:441–447
27. Workman P, Collins I (2010) Probing the probes: fitness factors for small molecule tools. Chem Biol 17:561–577
28. Frye SV (2010) The art of the chemical probe. Nat Chem Biol 6:159–161
29. Johnson AM, Maggiora GM (1990) Concepts and applications of molecular similarity. John Willey & Sons, New York
30. Balakin KV, Tkachenko SE, Lang SA, Okun I, Ivashchenko AA, Savchuk NP (2002) Property-based design of GPCR-targeted library. J Chem Inf Comput Sci 42:1332–1342
31. Evans BE, Rittle KE, Bock MG, DiPardo RM, Freidinger RM, Whitter WL, Lundell GF, Veber DF, Anderson PS, Hirshfield J (1988) Methods for drug discovery: development of potent, selective, orally effective cholecystokinin antagonists. J Med Chem 31:2235–2246
32. Schnur DM, Hermsmeier MA, Tebben AJ (2006) Are target-family-privileged substructures truly privileged? J Med Chem 49:2000–2009
33. Bondensgaard K, Ankersen M, Thøgersen H, Hansen BS, Wulff BS, Bywater RP (2004) Recognition of privileged structures by G-protein coupled receptors. J Med Chem 47:888–899
34. van der Horst E, Okuno Y, Bender A, Ijzerman A (2009) Substructure mining of GPCR ligands reveals activity-class specific functional groups in an unbiased manner. J Chem Inf Model 49:348–360
35. Mason JS, Cheney DL (2000) Library design and virtual screening using multiple 4-point pharmacophore fingerprints. Pac Symp Biocomput 5:573–584
36. Roughley SD, Jordan AM (2011) The medicinal chemist's toolbox: an analysis of reactions used in the pursuit of drug candidates. J Med Chem 54:3451–3479
37. Hartenfeller M, Eberle M, Meier P, Nieto-Oberhuber C, Altmann K, Schneider G, Jacoby E, Renner S (2011) A collection of robust organic synthesis reactions for in silico molecule design. J Chem Inf Model 51:3093–3098
38. Patel H, Bodkin MJ, Chen B, Gillet VJ (2009) Knowledge-based approach to de novo design using reaction vectors. J Chem Inf Model 49:1163–1184
39. Schneider N, Lowe DM, Sayle RA, Tarselli MA, Landrum GA (2016) Big data from pharmaceutical patents: a computational analysis of medicinal chemists' bread and butter. J Med Chem 59:4385–4402
40. National Cancer Institute, Synthetically Accessible Virtual Inventory (SAVI) Database, https://cactus.nci.nih.gov/download/savi_download/
41. Klinger F, Gastreich M, Mazanetz MP, Dawson G, Bodkin M 2016, KNIME-ing through the EVOSpace of FTrees, CCG UGM
42. Boehm M, Wu T, Claussen H, Lemmen C (2008) Similarity searching and scaffold hopping in synthetically accessible combinatorial chemistry spaces. J Med Chem 51:2468–2480
43. Lessel U, Wellenzohn B, Lilienthal M, Claussen H (2009) Searching fragment spaces with feature trees. J Chem Inf Model 49:270–279
44. Besnard J, Ruda GF, Setola V, Abecassis K, Rodriguiz RM, Huang X, Norval S, Sassano MF, Shin AI, Webster LA, Simeons FRC, Stojanovski L, Prat A, Seidah NG, Constam DB, Bickerton GR, Read KD, Wetsel WC, Gilbert IH, Roth BL, Hopkin AL (2012) Automated design of ligands to polypharmacological profiles. Nature 492:215–222
45. Stewart KD, Shiroda M, James CA (2006) Drug guru: a computer software program for drug design. Bioorg Med Chem 14:7011–7022
46. Lavecchia A (2015) Machine-learning approaches in drug discovery: methods and applications. Drug Discov Today 20:318–331
47. Hansch C (1980) Use of quantitative structure-activity relationships (QSAR) in drug design. Pharm Chem J 14:678–691
48. Free SM, Wilson JW (1964) A mathematical contribution to structure-activity studies. J Med Chem 7:395–399
49. Griffen E, Leach AG, Robb GR, Warner DJ (2011) Matched molecular pairs as a medicinal chemistry tool. J Med Chem 54:7739–7750
50. Waring MJ, Bennett SNL, Boyd S, Campbell L, Davies RDM, Gerhardt S, Hargreaves D, Martin NG, Robb GR, Wilkinson G (2013) Matched triplicate design sets in the optimisation of glucokinase activators – maximising medicinal chemistry information content. Med Chem Commun 4:657–662

51. O'Boyle NM, Boström J, Sayle RA, Gill A (2014) Using matched molecular series as a predictive tool to optimize biological activity. J Med Chem 57:2704–2713
52. Cereto-Massagué A, Ojeda MJ, Valls C, Mulero M, Garcia-Vallvé S, Pujadas G (2015) Molecular fingerprint similarity search in virtual screening. Methods 71:58–63
53. Cramer RD, Patterson DE, Bunce JD (1988) Comparative molecular field analysis (CoMFA). 1. Effect of shape on binding of steroids to carrier proteins. J Am Chem Soc 110:5959–5967
54. Gavaghan CL, Arnby CH, Blomberg N, Strandlund G, Boyer S (2007) Development, interpretation and temporal evaluation of a global QSAR of hERG electrophysiology screening data. J Comput Aided Mol Des 21:189–206
55. Rodgers SL, Davis AM, van de Waterbeemd H (2007) Time-series QSAR analysis of human plasma protein binding data. QSAR Comb Sci 26:511–521
56. Ramsundar B, Kearnes S, Riley P, Webster D, Konerding D, Pande V (2015) Massively multitask networks for drug discovery. arXiv 1502:02072
57. Posner BA, Xi H, Mills JEJ (2009) Enhanced HTS hit selection via a local hit rate analysis. J Chem Inf Model 49:2201–2210
58. Schmidt M, Lipson H (2009) Distilling free-form natural laws from experimental data. Science 324:81–85
59. Hopkins AL, Keserü GM, Leeson PD, Rees DC, Reynolds CH (2014) The role of ligand efficiency metrics in drug discovery. Nat Rev Drug Discov 13:105–121
60. Shultz MD (2013) Setting expectations in molecular optimizations: strengths and limitations of commonly used composite parameters. Bioorg Med Chem Lett 23:5980–5991
61. Segall MD, Champness EJ (2015) The challenges of making decisions using uncertain data. J Comput Aided Mol Des 29:809–816
62. Segall MD, Yusof I, Champness EJ (2016) Avoiding missed opportunities by analysing the sensitivity of our decisions. J Med Chem 59:4267–4277
63. Lipinski CA, Lombardo F, Dominy BW, Feeney PJ (1997) Experimental and computational approaches to estimate solubility and permeability in drug discovery and development settings. Adv Drug Deliv Rev 23:3–25
64. Bickerton GR, Paolini GV, Besnard J, Muresan S, Hopkin AL (2012) Quantifying the chemical beauty of drugs. Nat Chem 4:90–98
65. Wager TT, Hou X, Verhoest PR, Villalobos A (2010) Moving beyond rules: the development of a central nervous system multiparameter optimization (CNS MPO) approach to enable alignment of druglike properties. ACS Chem Neurosci 1:435–449
66. Lu G, Middleton RE, Sun H, Naniong M, Ott CJ, Mitsiades CS, Wong K, Bradner JE, Kaelin WG Jr (2014) The myeloma drug lenalidomide promotes the cereblon-dependent destruction of ikaros proteins. Science 343:305–309
67. Ebejer J, Charlton MH, Finn PW (2016) Are the physicochemical properties of antibacterial compounds really different from other drugs? J Cheminform 8:30

Chapter 21

Modeling and Deorphanization of Orphan GPCRs

Constantino Diaz, Patricia Angelloz-Nicoud, and Emilie Pihan

Abstract

Despite tremendous efforts, approximately 120 GPCRs remain orphan. Their physiological functions and their potential roles in diseases are poorly understood. Orphan GPCRs are extremely important because they may provide novel therapeutic targets for unmet medical needs. As a complement to experimental approaches, molecular modeling and virtual screening are efficient techniques to discover synthetic surrogate ligands which can help to elucidate the role of oGPCRs. Constitutively activated mutants and recently published active structures of GPCRs provide stimulating opportunities for building active molecular models for oGPCRs and identifying activators using virtual screening of compound libraries. We describe the molecular modeling and virtual screening process we have applied in the discovery of surrogate ligands, and provide examples for CCKA, a simulated oGPCR, and for two oGPCRs, GPR52 and GPR34.

Key words GPCR, Orphan GPCR, Molecular model, Homology modeling, Molecular dynamics, Structure, Virtual screening, Surrogate ligand, CCKA, GPR34, GPR52

1 Introduction

G protein-coupled receptors (GPCRs) are cell surface sensors which are activated by extracellular transmitters with a wide diversity of sizes, shapes, and chemical properties, including photons, ions, biogenic amines, nucleotides, amino acids, peptides, and lipids [1, 2]. GPCRs play a key role in cell-to-cell communication and they regulate a wide array of physiological and pathophysiological processes. These membrane proteins form the largest human receptors superfamily with more than 800 members [3]. Approximately 360 non-olfactory receptors are potentially druggable [4, 5], of which about 60 have been exploited as pharmaceutical targets and 120 remain orphan: their endogenous ligands and functions are unknown [6, 7].

Orphan GPCRs (oGPCRs) represent an unexplored set of targets for novel drug discovery [8, 9]. Two important steps in the deorphanization of oGPCRs are the identification of their highly selective natural ligand [10], and the understanding of their physiological function and possible role in diseases. The

Alexander Heifetz (ed.), *Computational Methods for GPCR Drug Discovery*, Methods in Molecular Biology, vol. 1705,
https://doi.org/10.1007/978-1-4939-7465-8_21,

most common strategy for ligand identification is high-throughput screening (HTS) of focused libraries of small molecules, lipids, peptides, and tissues extracts, using eukaryotic cells overexpressing the oGPCR of interest, and a second messenger readout [11–13]. In the absence of ligand, phenotypical characterization of organisms overexpressing or silencing oGPCRs may help in deciphering their function [14]. Identification of synthetic surrogate ligands is an alternative strategy to classical deorphanization approaches. Finding activators opens the door for studying the function of the receptor and provides matter for medicinal chemistry for optimizing compounds potency, selectivity, and biological profile [15–17].

Molecular modeling of oGPCRs and virtual screening (VS) contribute to identifying surrogate ligands with potential for therapeutic development [18, 19]. Here, we present the protocols we have used for identifying modulators for oGPCRs. We provide results for CCKA, a simulated oGPCR, and for two oGPCRs, GPR52 and GPR34.

2 Materials

2.1 Molecular Dynamics

The CHARMm force field in InsightII [http://accelrys.com/] was used. The production phase was during 1 ns at 300 K, with a distance-dependent dielectric term, a 14 Å non-bonded cut-off distance, and 0.5 fs steps. No restraint was used.

2.2 Homology Modeling

Homology models for CCKA, GPR52, and GPR34 were built with MOE 2015.10 [https://www.chemcomp.com/], based on NTS1 structures. The models were full-length, including extracellular and intracellular loops.

2.3 Compounds Libraries

2.3.1 CCKA

A CCKA library with 3375 compounds, containing 117 human CCKA agonists, 195 antagonists, and 3063 decoys, was built for the evaluations. Agonists and antagonists were retrieved from the ChEMBL v21 database [20]. The included actives had activating or inhibiting profile, values EC50, IC50, or Ki below 4 μM, and molecular weights between 300 and 800 g/mol. Decoys were collected from the DUD-E database [21]. They had size and physicochemical properties similar to the actives (e.g., molecular weight, LogP) but dissimilar topology. We randomly selected decoys among those proposed for building a testset with a ratio 4% CCKA agonists and 96% decoys.

2.3.2 GPR52

A GPR52 library with 10179 compounds containing 15 human GPR52 agonists and 10164 decoys was built for VS evaluations with a true oGPCR. Agonists were retrieved from the ChEMBL v21 database. The included agonists had EC50 values below 1 μM, and molecular weights between 414 and 486 g/mol. Decoys were randomly collected from the ZINC database [22]. They had molecular weights between 300 and 600 g/mol.

2.3.3 GPR34

A GPR34 library with 10239 compounds containing 78 human GPR34 antagonists and 10161 decoys was built. Antagonists were retrieved from the patent EP1849465A1 [23] with SureChEMBL [24]. The included antagonists exhibited IC50 below 1 μM and molecular weights between 459 and 589 g/mol. Decoys were randomly collected from the ZINC database. They had molecular weights between 300 and 600 g/mol.

2.4 Preparation of the Chemical Libraries

The libraries were prepared through a Knime workflow containing the successive ChemAxon nodes [https://www.chemaxon.com/]:

1. Major microspecies, to keep the major tautomers, at pH = 7.4,
2. Stereoisomers, to consider undefined stereocenters and generate stereoisomers, and,
3. Conformers, to build the lowest energy conformer.

2.5 Virtual Screening

2.5.1 Docking

Docking sites in the GPCR models were defined using Site Finder in MOE.

The compound libraries were docked in the GPCR models using Gold 2016 [25] and four different scoring functions PLP, ASP, ChemScore, and GoldScore. For each compound, ten docking poses were generated for each stereoisomer, and the best scoring value among all poses and all stereoisomers was considered for the compound. All docking parameters were those by default. No post-docking process was done: no rescoring, and no visualization or validation of the poses.

2.5.2 Analysis

VS efficiency was evaluated considering:

1. The enrichment curve, where the percentage of found actives is plotted against the ranked library.
2. The enrichment factor (EF) of a hit list composed by the top 1% of the ranked library.

$$EF = (Actives_{hit\ list}/N_{hit\ list})/(Actives_{total}/N_{total})$$

   in which $Actives_{hit\ list}$ and $N_{hit\ list}$ are respectively the number of actives and the number of compounds in the hit list, $Actives_{total}$ and N_{total} are respectively the number of actives and the number of compounds in the full library.

3. The hit rate (HR) of a hit list composed by the top 1% of the ranked library.

$$HR = 100 \times (Actives_{hit\ list})/(N_{hit\ list})$$

3 Methods

Residues in trans-membrane domains are designated by their position in the sequence followed by their universal numbering proposed by Ballesteros and Weinstein [26].

3.1 Templates for Building Active Molecular Models for oGPCRs

Two alternative approaches were used:

- Building of an active molecular model for a GPCR, using a constitutively activated mutant (CAM) and molecular dynamics.
- Identification of GPCR structures, in the PDB database [27], which are in an active conformation.

3.1.1 Active Molecular Model Built Using a CAM

Activation of GPCRs

GPCRs share a common architecture: (1) an extracellular region containing the N-terminus and three extracellular loops EL1-3, (2) a transmembrane domain comprising seven α-helices H1-7, and (3) an intracellular region with three intracellular loops IL1-3 and the C-tail. The intracellular and extracellular regions show a high variability in size and sequence across GPCRs, while the transmembrane domain reveals a higher sequence conservation. For a wild-type GPCR, binding of an agonist to parts of the extracellular and transmembrane domains of the receptor modifies its conformation and interaction with cytosolic effectors such as G-proteins and β-arrestins, thus activating the downstream signaling [28].

GPCR with constitutive activating mutations show spontaneous activity in an agonist-independent manner. The first CAM was reported for the α1-adrenergic receptor [29], rapidly followed by many other GPCRs [30]. CAMs are of considerable interest because their study shed light on structural differences between active and inactive GPCR conformations. It is worth noting that naturally occurring CAMs are associated with human diseases [31, 32].

The NTS1-V308E CAM

For the NTS1 receptor, a CAM was produced by a single $V308E^{6.40}$ mutation. The spontaneous activity of the $V308E^{6.40}$ mutant was eightfold higher than wild-type receptor when expressed in COS-3 cells, and assessed for basal activity of Inositol Phosphate production [33].

Active Molecular Model for the NTS1-V308E CAM

First, models for NTS1 and NTS1-V308E were built by homology modeling using the inactive structure of rhodopsin 1F88 [34], as follows. Bovine rhodopsin, human NTS1, and other class-A GPCR sequences were aligned using ClustalW. The positions of motifs in class-A GPCRs for each TM helix were considered [35] and manual modifications in the alignment were made where needed, producing the multi-alignment for the 7 TM helices shown in Fig. 1. Residues in the 7 TM helices of the rhodopsin structure that were

```
bRhod    34 PWQFSMLAAYMFLLIMLGFPINFLTLYVTVQ
hNTS1    60 IYSKVLVTAVYLALFVVGTVGNTVTAFTLAR
hCCKA    38 EWQPAVQILLYSLIFLLSVLGNTLVITVLIR
hGPR34   51 KLLSTVLTTSYSVIFIVGLVGNIIALYVFLG
hGPR52   37 VDVCIFETVVIVLLTFLIIAGNLTVIFVFHC

bRhod    71 PLNYILLNLAVADLFMVFGGFTTTLYTSLH
hNTS1   100 TVHYHLGSLALSDLLTLLLAMPVELYNFIW
hCCKA    75 VTNIFLLSLAVSDLMLCLFCMPFNLIPNLL
hGPR34   88 SIQIYLLNVAIADLLLIFCLPFRIMYHINQ
hGPR52   75 TTSYFIQTMAYADLFVGVSCLVPTLSLLHY

bRhod   107 PTGCNLEGFFATLGGEIALWSLVVLAIERYVVV
hNTS1   138 DAGCRGYYFLRDACTYATALNVASLSVERYLAI
hCCKA   111 SAVCKTTTYFMGTSVSVSTFNLVAISLERYGAI
hGPR34  124 VILCKVVGTLFYMNMYISIILLGFISLDRYIKI
hGPR52  111 SLTCQVFGYIISVLKSVSMACLACISVDRYLAI

bRhod   150 ENHAIMGVAFTWVMALACAAPPLVG
hNTS1   182 RSRTKKFISAIWLASALLTVPMLFT
hCCKA   155 KSHALKVIAATWCLSFTIMTPYPIY
hGPR34  168 TKQSIYVCCIVWMLALGGFLTMIIL
hGPR52  155 PCRLRICIILIWIYSCLIFLPSFFG

bRhod   200 NESFVIYMFVVHFIIPLIVIFFCYGQLVF
hNTS1   233 VKVVIQVNTFMSFIFPMVVISVLNTIIAN
hCCKA   206 QQSWHTFLLLILFLIPGIVMMVAYGLISL
hGPR34  215 GEAIFNFILVVMFWLIFLLIILSYIKIGK
hGPR52  200 SAYFTGFIVCLLYAPAAFVVCFTYFHIFK

bRhod   250 VTRMVIIMVIAFLICWLPYAGVAFYIFT
hNTS1   301 GVRVLRAVVIAFVVCWLPYHVRRLMFCY
hCCKA   311 VIRMLIVIVVLFFLCWMPIFSANAWRAY
hGPR34  264 TARNSFIVLIIFTICFVPYHAFRFIYIS
hGPR52  263 YAMVLFRITSVFYMLWLPYIIYFLLESS

bRhod   286 IFMTIPAFFAKTSAVYNPVIYIMMN
hNTS1   344 YFYMVTNALFYVSSTINPILYNLVS
hCCKA   350 TPISFILLLSYTSSCVNPIIYCFMN
hGPR34  307 KTNEIMLVLSSFNSCLDPVMYFLMS
hGPR52  297 TLSFLTTWLAISNSFCNCVIYSLSN
```

Fig. 1 Sequence alignment for the 7TM helices of bovine rhodopsin (bRhod), human NTS1 (hNTS1), human CCKA (hCCKA), human GPR52 (hGPR52), and human GPR34 (hGPR34) receptors. The conserved residues for class-A GPCRs are shown in *bold*. The Val308 residue in hNTS1 is in *bold* and *underlined*

different to those of NTS1 were changed for the corresponding NTS1 residues, with InsightII. Residues outside the 7 TM helices were removed, generating a bundle with 7 independent TM helices. Hydrogen atoms were added to relevant positions. Steric clashes between side chains were checked, and manual modifications of side chain rotamers were performed where needed, using the rotamers library of InsightII. The NTS1 molecular model was submitted to the V308E substitution for producing a NTS1-V308E molecular model. Both models were minimized, generating the NTS1 and NTS1-V308E homology models.

Then, the two models were submitted to molecular dynamics (*see* Subheading 3). The simulations resulted in NTS1 and NTS1-V308E molecular dynamics models.

In the NTS1 homology model, Val308$^{6.40}$ is situated in helix H6 and is oriented toward the center of the TM helices bundle. It participates in a central hydrophobic cluster composed by Leu105$^{2.43}$, Leu108$^{2.46}$, Val159$^{3.43}$, and Leu162$^{3.46}$ (Fig. 2a, b).

In the 1 ns NTS1 molecular dynamics model, the packing of the 7 TM helices bundle and the hydrophobic cluster remain essentially unchanged (Fig. 2c).

In the 1 ns NTS1-V308E molecular dynamics model, the substitution V308E$^{6.40}$ in helix H6 introduced a strong polar group perturbing the hydrophobic core and leading to structural reorganization of the helices bundle. The negatively charged side chain of Glu308$^{6.40}$ formed a salt bridge with the positively charged part of the side chain of Asn158$^{3.42}$ in helix H3, producing a concerted anticlockwise rotation of H6 and a small clockwise rotation of H3, in an extracellular view. The side chains of the

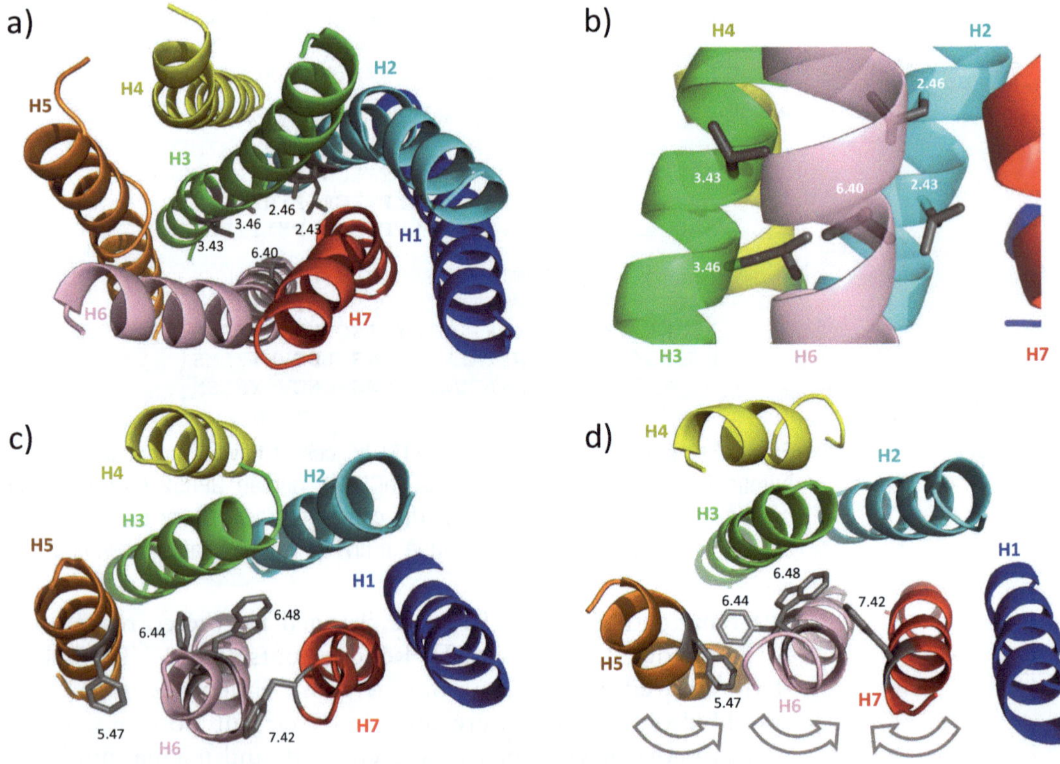

Fig. 2 (**a**) Extracellular and (**b**) membrane views of the NTS1 homology model built using the inactive structure of rhodopsin 1F88. Residues participating in the hydrophobic core are shown. (**c**) Extracellular view of the NTS1 molecular dynamics model. (**d**) Extracellular view of the NTS1-V308E molecular dynamics model. Main changes compared with the NTS1 molecular dynamics model are shown by *arrows*

conserved Phe312$^{6.44}$ and Trp316$^{6.48}$ moved in the direction of H5. An anticlockwise rotation of H5 and a clockwise rotation of H7 also occurred, as well as changes in tilts for H4 and H5, and kink for H6, resulting in the activation scheme shown in Fig. 2d.

CAMs with a Mutation at $X^{6.40}$

The residue $X^{6.40}$ (where X is a hydrophobic residue Val, Ile, Leu, or Met) is a hot spot, as mutations decreasing its hydrophobicity generated CAMs for diverse GPCRs including rhodopsin, muscarinic M5, histamine H1, angiotensin AT1A, and opioid receptors [30]. The hydrophobic core that includes in particular the conserved Leu$^{3.43}$ in helix H3 and $X^{6.40}$ in helix H6 was proposed to hold H3 and H6 in place in the inactive state for class-A GPCRs. Rearrangement of the interactions between the residues involved in this hydrophobic core enables the movement of H6 in the activation process, with co-occurring movements of H5 and H7.

3.1.2 Active Structures of GPCRs

The first reported structure of a GPCR was in 2000, for rhodopsin, a special GPCR with a covalently bound ligand, the retinal [34]. Seven years were necessary to develop successful receptor stabilization and crystallization techniques for GPCRs with diffusible ligands. In 2007, a structure was published for the β2-adrenergic receptor [36], followed in 2008 by structures of β1-adrenergic and adenosine A2A receptors [37, 38]. Since then, there has been an almost exponential growth in the number of published GPCR structures (Fig. 3). In early 2016, 146 structures were available in the PDB repository for 32 different GPCRs in class-A, -B, -C, and -F [39]. Structural coverage of the GPCR phylogenic tree is underway, providing atomic details for drug discovery and drug design using computational methods [8].

Most GPCR structures were crystallized in complex with antagonists or inverse agonists, and are therefore in an inactive state. GPCR structures in a semi-active state cocrystallized with an agonist, or in a fully active state stabilized in the extracellular and intracellular sides by, respectively, an agonist and a G protein or a G protein surrogate, are available for rhodopsin, β2 and β1 adrenergic, adenosine A2A, muscarinic M2, serotonin 5-HT1B and 5-HT2B, purinergic P2Y12, FFAR1, SMO, and neurotensin NTS1 receptors [40, 41]. These structures provide invaluable templates for building increasingly accurate active models for GPCRs and oGPCRs.

3.2 Homology Modeling of oGPCRs

The modeling process involves the following steps.

3.2.1 Listing of Potential Templates

For searching activators, active GPCR structures or validated molecular models of CAMs are needed. The GPCRdb database [39] provides an updated list of GPCR structures.

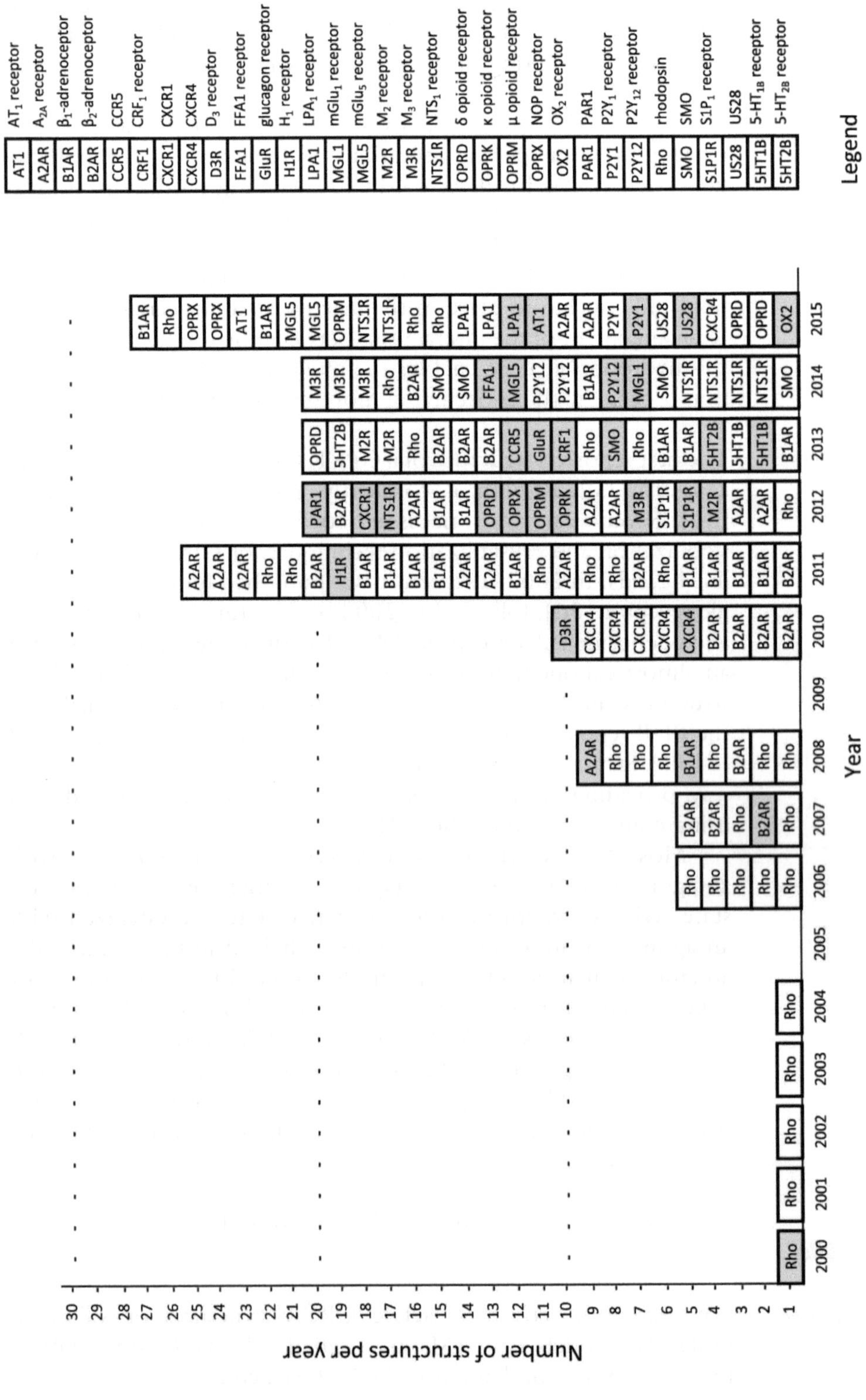

Fig. 3 *Bar chart* showing the number of new GPCR structures deposited in the PDB database over time. Structures for new receptors are shown in *gray*

3.2.2 Sequence Alignment Between the Target and the Templates

The sequences are aligned with multiple-sequences alignment tools like ClustalW [42] and TCoffee [43]. Errors in the initial alignment are common, and appropriate corrections are generally required, based on the conserved motifs in the 7 TM helices [35], and the conserved cysteine in EL2 loop.

3.2.3 Selection of a Template

The selection is based on the highest target-template sequences similarity. Three types of rankings can be used:

1. Considering the whole sequences,
2. Based on the full 7 TM sequences,
3. Restricting the sequences to 30 [44], 40 [45], or 44 [46] residues pointing into the generic binding pocket of GPCRs.

Other factors may also be considered for selecting the template: the identical position in the target-template sequences alignment of key amino acids like prolines, the presence of the conserved cysteine in the EL2 loop, and similar target-template lengths for EL2a and EL2b loop. The EL2 loop must be considered with attention, as it shows variable lengths and secondary structures for diverse GPCRs [45, 47].

3.2.4 Building of a Crude Homology Model

Various software packages perform the building of a homology model, given a 3D template and target-template sequences alignment. We have used MOE. In the model generation, a large number of models are built, and the best model in terms of energy is further refined. Visual inspection terminates the building of the crude homology model.

3.3 Structure-Based Virtual Screening

Molecular docking is a powerful technique for the discovery of new ligands, and it has been successfully applied to various GPCRs [19]. The main steps are the definition of the docking site, and docking and scoring of the compounds in the docking site.

The docking of a compound consists of geometrically fitting a flexible compound into the docking site which is mostly assumed to be rigid. The scoring assesses the interactions between the docked compound and the receptor using force-field, empirical or knowledge-based functions. The widely used VS programs include Gold [25], Dock [48], Glide [49], and LigandFit [50]. The docking and scoring performances vary depending on the target protein. Benchmarks considering sets of protein structures cocrystallized with ligands provide objective evaluations [51].

4 Notes

4.1 A Simulated oGPCR: CCKA

The cholecystokinin receptor (CCKA) was selected as a simulated orphan receptor for evaluating diverse approaches for the identification of surrogate ligands. No CCKA structure is available, and for the present study, previous published work on CCKA was not considered.

4.1.1 VS with an Active CCKA Model Built Using the MD Template

A crude active CCKA molecular model (CCKA-MD) was built by homology modeling from the NTS1-V308E MD model, according to the alignment in Fig. 1. The CCKA-MD model had only the 7 TM helices, the conserved Cys196$^{45.50}$ of the extracellular loop EL2, and the conserved disulfide bridge Cys114$^{3.25}$-Cys196$^{45.50}$ with the same geometry as the corresponding cysteine pair in bovine rhodopsin [34]. The docking site was between the TM helices and under Cys196$^{45.50}$.

The enrichment curve resulting from the docking of the CCKA library into the CCKA-MD model with Gold and the PLP scoring function is shown in Fig. 4. Enrichment factor (EF) and hit rate

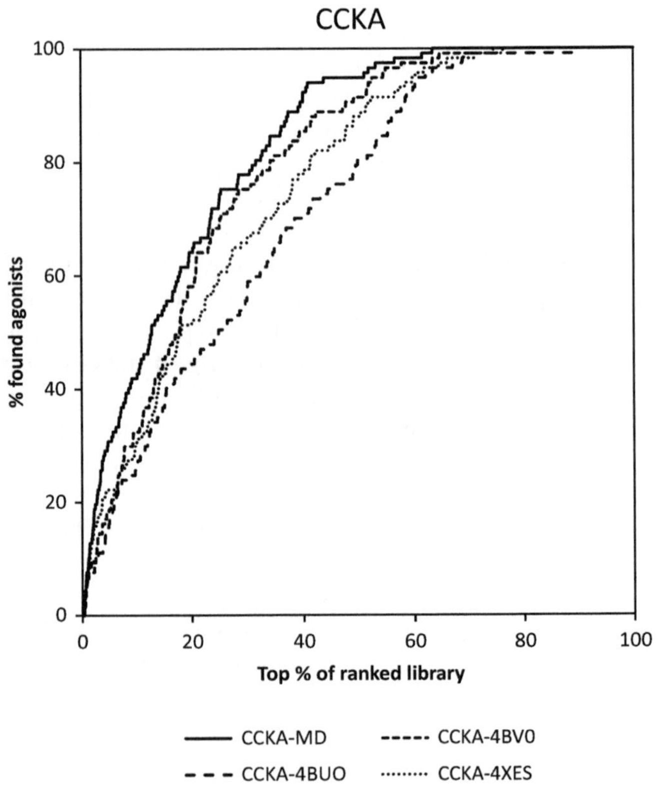

Fig. 4 Enrichment curves obtained by docking the CCKA library into the four molecular models CCKA-MD, CCKA-4BUO, CCKA-4BVO, CCKA-4XES with Gold and the PLP scoring function

Table 1
Enrichment factors (EF) and hit rates (HR) for CCKA agonists resulting from the docking of the CCKA library into the four molecular models CCKA-MD, CCKA-4BUO, CCKA-4BV0, CCKA-4XES with Gold and the PLP, ASP, ChemScore, and GoldScore scoring functions. EF and HR are given for the top 1% of the ranked library

	PLP		ASP		ChemScore		GoldScore	
Models	**EF**	***HR (%)***	**EF**	***HR (%)***	**EF**	***HR (%)***	**EF**	***HR (%)***
CCKA-MD	7.9	*27*	2.6	*9*	1.7	*6*	0.9	*3*
CCKA-4BUO	7.9	*27*	0.9	*3*	0.9	*3*	0.0	*0*
CCKA-4BV0	7.0	*24*	6.1	*21*	0.9	*3*	4.4	*15*
CCK1-4XES	6.1	*21*	1.7	*6*	1.7	*6*	1.7	*6*

(HR), considering the top-scored 1% of the ranked library, are shown in Table 1. Excellent results were obtained with PLP: EF = 7.9 and HR = 27%.

4.1.2 VS with Active CCKA Models Built Using PDB Structures

To be consistent with the use of a NTS1 molecular dynamics model as a template for building a CCKA model for VS, here we explored the use of NTS1 structures for building CCKA models for VS and for comparing the results. The three rat NTS1 active structures with PDB Id 4BUO, 4BV0 [52], and 4XES [53] were used for building crude active CCKA models by homology modeling, named respectively CCKA-4BUO, CCKA-4BV0, and CCKA-4XES. Extracellular loops were modeled based on those of the NTS1 structures. Importantly, the lengths of loops EL2a and EL2b, respectively before and after the conserved cysteine in EL2, are similar for rat NTS1 (respectively 18 and 8 amino acids) and human CCKA (17 and 9 amino acids).

Enrichment curves for the VS of the CCKA library with the three models and the PLP scoring function are shown in Fig. 4. Enrichment factors and hit rates with the four scoring functions, considering the top-scored 1% of the ranked library, are shown in Table 1. With the PLP scoring function, similar results were obtained for the model built using the NTS1-V308E MD template, and the three models built using the three NTS1 active structures. For the four models, the PLP scoring function generated the best performance. As the goal was to identify agonists, only the 117 CCKA agonists were considered actives. The 195 CCKA antagonists in the library were considered to be negative compounds like the 3063 decoys.

4.2 GPR52

GPR52 is an orphan receptor [www.guidetopharmacology.org] for which 15 agonists were recently described [54]. The NTS1 active structure 4BV0 was used for building a crude active GPR52 molecular model by homology modeling: GPR52-4BV0.

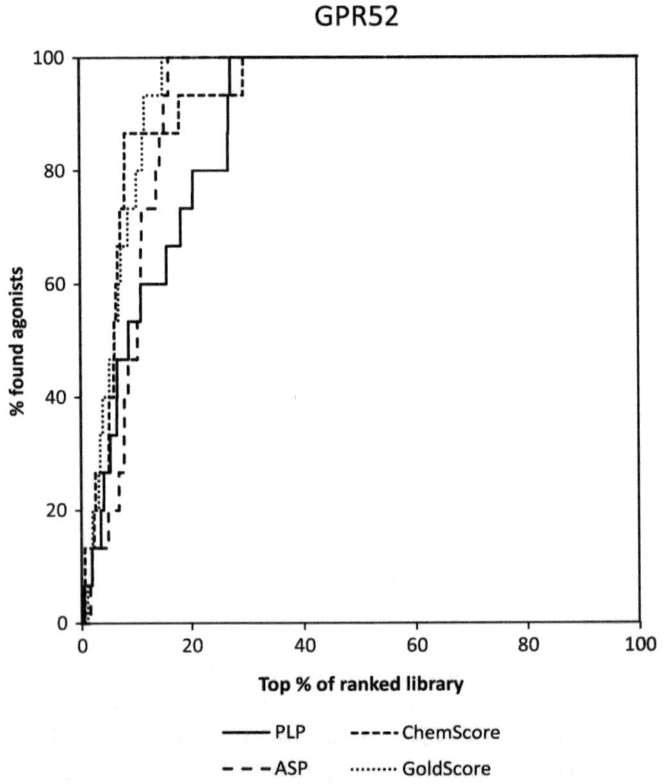

Fig. 5 Enrichment curves obtained by docking the GPR52 library into the GPR52-4BV0 molecular model with Gold and PLP, ASP, ChemScore, and GoldScore scoring functions

Table 2
Enrichment factors (EF) and hit rates (HR) for GPR52 agonists resulting from the docking of the GPR52 library into the GPR52-4BV0 molecular model with Gold and the PLP, ASP, ChemScore, and GoldScore scoring functions. EF and HR are given for the top 1% of the ranked library

	PLP		ASP		ChemScore		GoldScore	
Model	EF	*HR (%)*	EF	*HR (%)*	EF	*HR (%)*	EF	*HR (%)*
GPR52-4BV0	6.7	*1*	0.0	*0*	13.4	*2*	0.0	*0*

Enrichment curves for the VS of the GPR52 library with the four scoring functions are shown in Fig. 5. Enrichment factors and hit rates, for the top-scored 1% of the ranked library, are shown in Table 2. Best results were obtained with PLP and ChemScore. With PLP, the enrichment factor for GPR52 is excellent and similar to that of CCKA.

4.3 GPR34

GPR34 is an orphan receptor [www.guidetopharmacology.org]. Based on its sequence, GPR34 is closely related to P2Y12, P2Y13, and P2Y14 [17]. It is activated by lysophosphatidylserine (LPS), but LPS has no or very weak agonist activity at most vertebrate GPR34 orthologues [55], suggesting that the search for the endogenous agonist should continue. Recent studies suggest constitutive activity [56]. Small compound antagonists, but no small compound agonist, were published for GPR34 [23]. In this third example, we assess if an active molecular model can be used for searching antagonists for an oGPCR.

The NTS1 active structure 4BV0 was used again for building a crude active GPR34 model by homology modeling.

Enrichment curves for the VS of the GPR34 library with the four scoring functions are shown in Fig. 6. Enrichment factors and hit rates, considering the top-scored 1% of the ranked library, are shown in Table 3. Excellent results were obtained with the four scoring functions, the best results being generated by PLP with EF = 57.9 and HR = 44.1%.

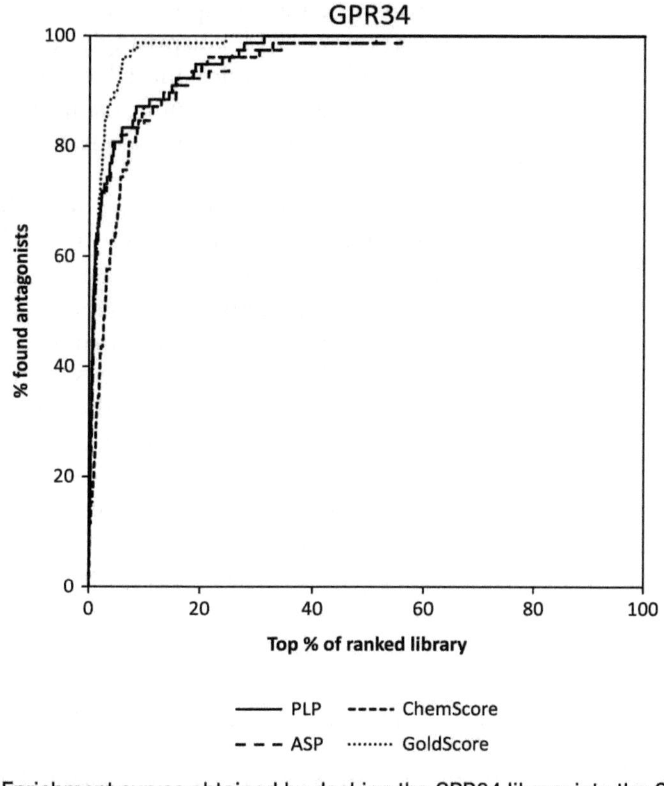

Fig. 6 Enrichment curves obtained by docking the GPR34 library into the GPR34-4BV0 molecular model with Gold and PLP, ASP, ChemScore, and GoldScore scoring functions

Table 3
Enrichment factors (EF) and hit rates (HR) for GPR34 antagonists resulting from the docking of the GPR34 library into the GPR34-4BV0 molecular model with Gold and the PLP, ASP, ChemScore, and GoldScore scoring functions. EF and HR are given for the top 1% of the ranked library

	PLP		ASP		ChemScore		GoldScore	
Model	**EF**	***HR (%)***	**EF**	***HR (%)***	**EF**	***HR (%)***	**EF**	***HR (%)***
GPR34-4BV0	57.9	*44*	50.2	*38*	23.2	*17*	51.5	*39*

4.4 Perspectives

The three retrospective studies show that, using the molecular modeling and virtual screening techniques described, active compounds were found in VS-lists representing 1% of the screened libraries. These excellent results were obtained without requiring prior knowledge of ligands for building the molecular models of the receptors. Interestingly, among the four tested scoring functions PLP, ASP, ChemScore, and GoldScore, the best results were, on average, obtained with the PLP scoring function. Recently, a comparison of 20 scoring functions over a set with 195 diverse protein-ligand complexes also found Gold/PLP among the top performers [51]. Surprisingly, the CCKA model built using the molecular dynamics model of the NTS1-V308E CAM and the three CCKA models built using NTS1 active structures generated similar VS results when considering the VS-lists with 1% of the screened library.

In a previous prospective study [57], we built a GPR34 molecular model by homology to the NTS1-V308E MD model described in the methods paragraph. The GPR34 model was refined using active compounds found in an experimental screening, and it was used to perform the virtual screening of a corporate library. Three inverse agonists with new chemical structures were discovered, thus validating the use of molecular models for searching modulators for orphan GPCRs.

In the last years, the almost exponential growth in the number of published GPCR structures opens up new perspectives for building increasingly accurate molecular models for oGPCRs, and searching surrogate ligands using VS.

References

1. Civelli O, Saito Y, Wang Z, Nothacker HP, Reinscheid RK (2006) Orphan GPCRs and their ligands. Pharmacol Ther 110 (3):525–532. https://doi.org/10.1016/j.pharmthera.2005.10.001
2. Chung S, Funakoshi T, Civelli O (2008) Orphan GPCR research. Br J Pharmacol 153 (Suppl 1):S339–S346. https://doi.org/10.1038/sj.bjp.0707606
3. Fredriksson R, Lagerström MC, Lundin LG, Schiöth HB (2003) The G-protein-coupled receptors in the human genome form five main families. Phylogenetic analysis, paralogon groups, and fingerprints. Mol Pharmacol 63

(6):1256–1272. https://doi.org/10.1124/mol.63.6.1256

4. Vassilatis DK, Hohmann JG, Zeng H, Li F, Ranchalis JE, Mortrud MT, Brown A, Rodriguez SS, Weller JR, Wright AC, Bergmann JE, Gaitanaris GA (2003) The G protein-coupled receptor repertoires of human and mouse. Proc Natl Acad Sci U S A 100(8):4903–4908. https://doi.org/10.1073/pnas.0230374100
5. Tautermann CS (2014) GPCR structures in drug design, emerging opportunities with new structures. Bioorg Med Chem Lett 24 (17):4073–4079. https://doi.org/10.1016/j.bmcl.2014.07.009
6. Stockert JA, Devi LA (2015) Advancements in therapeutically targeting orphan GPCRs. Front Pharmacol 8(6):100. https://doi.org/10.3389/fphar.2015.00100
7. Jacobson KA (2015) New paradigms in GPCR drug discovery. Biochem Pharmacol 98:541–555. https://doi.org/10.1016/j.bcp.2015.08.085
8. Kumari P, Ghosh E, Shukla AK (2015) Emerging approaches to GPCR ligand screening for drug discovery. Trends Mol Med 21 (11):687–701. https://doi.org/10.1016/j.molmed.2015.09.002
9. Shore DM, Reggio PH (2015) The therapeutic potential of orphan GPCRs, GPR35 and GPR55. Front Pharmacol 6:69. https://doi.org/10.3389/fphar.2015.00069
10. Tang XL, Wang Y, Li DL, Luo J, Liu MY (2012) Orphan G protein-coupled receptors (GPCRs): biological functions and potential drug targets. Acta Pharmacol Sin 33 (3):363–371. https://doi.org/10.1038/aps.2011.210
11. Kotarsky K, Nilsson NE (2004) Reverse pharmacology and the de-orphanization of 7TM receptors. Drug Discov Today Technol 1 (2):99–104. https://doi.org/10.1016/j.ddtec.2004.07.003
12. Jimonet P, Jäger R (2004) Strategies for designing GPCR-focused libraries and screening sets. Curr Opin Drug Discov Devel 7 (3):325–333
13. Bradshaw HB, Lee SH, McHugh D (2009) Orphan endogenous lipids and orphan GPCRs: a good match. Prostaglandins Other Lipid Mediat 89(3-4):131–134. https://doi.org/10.1016/j.prostaglandins.2009.04.006
14. Ahmad R, Wojciech S, Jockers R (2015) Hunting for the function of orphan GPCRs – beyond the search for the endogenous ligand. Br J Pharmacol 172(13):3212–3228. https://doi.org/10.1111/bph.12942
15. Thomsen W, Leonard J, Behan DP (2004) Orphan GPCR target validation. Curr Opin Mol Ther 6(6):640–656
16. Wise A, Jupe SC, Rees S (2004) The identification of ligands at orphan G-protein coupled receptors. Annu Rev Pharmacol Toxicol 44:43–66. https://doi.org/10.1146/annurev.pharmtox.44.101802.121419
17. Lecca D, Abbracchio MP (2008) Deorphanisation of G protein-coupled receptors: a tool to provide new insights in nervous system pathophysiology and new targets for psycho-active drugs. Neurochem Int 52(3):339–351. https://doi.org/10.1016/j.neuint.2007.08.002
18. Eberini I, Daniele S, Parravicini C, Sensi C, Trincavelli ML, Martini C, Abbracchio MP (2011) In silico identification of new ligands for GPR17: a promising therapeutic target for neurodegenerative diseases. J Comput Aided Mol Des 25(8):743–752. https://doi.org/10.1007/s10822-011-9455-8
19. Bermudez M, Wolber G (2015) Structure versus function – The impact of computational methods on the discovery of specific GPCR–ligands. Bioorg Med Chem 23(14):3907–3912. https://doi.org/10.1016/j.bmc.2015.03.026
20. Bento AP, Gaulton A, Hersey A, Bellis LJ, Chambers J, Davies M, Krüger FA, Light Y, Mak L, McGlinchey S, Nowotka M, Papadatos G, Santos R, Overington JP (2014) The ChEMBL bioactivity database: an update. Nucleic Acids Res 42:D1083–D1090. https://doi.org/10.1093/nar/gkt1031
21. Mysinger MM, Carchia M, Irwin JJ, Shoichet BK (2012) Directory of useful decoys, enhanced (DUD-E): better ligands and decoys for better benchmarking. J Med Chem 55 (14):6582–6594. https://doi.org/10.1021/jm300687e
22. Irwin JJ, Sterling T, Mysinger MM, Bolstad ES, Coleman RG (2012) ZINC: a free tool to discover chemistry for biology. J Chem Inf Model 52(7):1757–1768. https://doi.org/10.1021/ci3001277
23. Kohara Y, Aramaki Y, Mori M, Kimura E, Imai T, Ito F, Ogi K, Sugo T, Kobayashi H, Hayase Y (2007) Agent for controlling function of GPR34 receptor. European Patent Application EP1849465A1, 31 Oct 2007
24. Papadatos G, Davies M, Dedman N, Chambers J, Gaulton A, Siddle J, Koks R, Irvine SA, Pettersson J, Goncharoff N, Hersey A, Overington JP (2016) SureChEMBL: a large-scale, chemically annotated patent document database. Nucleic Acids Res 44(D1):D1220–D1228. https://doi.org/10.1093/nar/gkv1253

25. Verdonk ML, Cole JC, Hartshorn MJ, Murray CW, Taylor RD (2003) Improved protein-ligand docking using GOLD. Proteins 52 (4):609–623. https://doi.org/10.1002/prot.10465
26. Ballesteros JA, Weinstein H (1995) Integrated methods for the construction of three-dimensional models and computational probing of structure-function relations in G protein-coupled receptors. Methods Neurosci 25:366–428
27. Berman HM, Kleywegt GJ, Nakamura H, Markley JL (2014) The protein data bank archive as an open data resource. J Comput Aided Mol Des 28(10):1009–1014. https://doi.org/10.1007/s10822-014-9770-y
28. Venkatakrishnan AJ, Deupi X, Lebon G, Heydenreich FM, Flock T, Miljus T, Balaji S, Bouvier M, Veprintsev DB, Tate CG, Schertler GF, Babu MM (2016) Diverse activation pathways in class A GPCRs converge near the G-protein-coupling region. Nature 536 (7617):484–487. https://doi.org/10.1038/nature19107
29. Cotecchia S, Exum S, Caron MG, Lefkowitz RJ (1990) Regions of the alpha 1-adrenergic receptor involved in coupling to phosphatidylinositol hydrolysis and enhanced sensitivity of biological function. Proc Natl Acad Sci U S A 87(8):2896–2900
30. Tehan BG, Bortolato A, Blaney FE, Weir MP, Mason JS (2014) Unifying family A GPCR theories of activation. Pharmacol Ther 143 (1):51–60. https://doi.org/10.1016/j.pharmthera.2014.02.004
31. Schöneberg T, Schulz A, Biebermann H, Hermsdorf T, Römpler H, Sangkuhl K (2004) Mutant G-protein-coupled receptors as a cause of human diseases. Pharmacol Ther 104(3):173–206. https://doi.org/10.1016/j.pharmthera.2004.08.008
32. Tao YX (2008) Constitutive activation of G protein-coupled receptors and diseases: insights into mechanisms of activation and therapeutics. Pharmacol Ther 120 (2):129–148. https://doi.org/10.1016/j.pharmthera.2008.07.005
33. Diaz C, Leplatois P, Angelloz-Nicoud P, Lecomte M, Josse A, Delpech M, Pecceu F, Loison G, Shire D, Pascal M, Ferrara P, Ferran E (2011) Differential virtual screening (DVS) with active and inactive molecular models for finding and profiling GPCR modulators: case of the CCK1 receptor. Mol Inf 30 (4):345–358. https://doi.org/10.1002/minf.201000180
34. Palczewski K, Kumasaka T, Hori T, Behnke CA, Motoshima H, Fox BA, Le Trong I, Teller DC, Okada T, Stenkamp RE, Yamamoto M, Miyano M (2000) Crystal structure of rhodopsin: a G protein-coupled receptor. Science 289 (5480):739–745
35. Baldwin JM, Schertler GF, Unger VM (1997) An alpha-carbon template for the transmembrane helices in the rhodopsin family of G-protein-coupled receptors. J Mol Biol 272 (1):144–164
36. Cherezov V, Rosenbaum DM, Hanson MA, Rasmussen SG, Thian FS, Kobilka TS, Choi HJ, Kuhn P, Weis WI, Kobilka BK, Stevens RC (2007) High-resolution crystal structure of an engineered human beta2-adrenergic G protein-coupled receptor. Science 318 (5854):1258–1265. https://doi.org/10.1126/science.1150577
37. Warne T, Serrano-Vega MJ, Baker JG, Moukhametzianov R, Edwards PC, Henderson R, Leslie AG, Tate CG, Schertler GF (2008) Structure of a beta1-adrenergic G-protein-coupled receptor. Nature 454 (7203):486–491. https://doi.org/10.1038/nature07101
38. Jaakola VP, Griffith MT, Hanson MA, Cherezov V, Chien EY, Lane JR, Ijzerman AP, Stevens RC (2008) The 2.6 angstrom crystal structure of a human A2A adenosine receptor bound to an antagonist. Science 322 (5905):1211–1217. https://doi.org/10.1126/science.1164772
39. Isberg V, Mordalski S, Munk C, Rataj K, Harpsøe K, Hauser AS, Vroling B, Bojarski AJ, Vriend G, Gloriam DE (2016) GPCRdb: an information system for G protein-coupled receptors. Nucleic Acids Res 44(D1): D356–D364. https://doi.org/10.1093/nar/gkv1178
40. Cooke RM, Brown AJ, Marshall FH, Mason JS (2015) Structures of G protein-coupled receptors reveal new opportunities for drug discovery. Drug Discov Today 20(11):1355–1364. https://doi.org/10.1016/j.drudis.2015.08.003
41. Shonberg J, Kling RC, Gmeiner P, Löber S (2015) GPCR crystal structures: medicinal chemistry in the pocket. Bioorg Med Chem 23(14):3880–3906. https://doi.org/10.1016/j.bmc.2014.12.034
42. Larkin MA, Blackshields G, Brown NP, Chenna R, McGettigan PA, McWilliam H, Valentin F, Wallace IM, Wilm A, Lopez R, Thompson JD, Gibson TJ, Higgins DG (2007) Clustal W and clustal X version 2.0. Bioinformatics 23(21):2947–2948. https://doi.org/10.1093/bioinformatics/btm404
43. Notredame C, Higgins DG, Heringa J (2000) T-coffee: a novel method for fast and accurate

multiple sequence alignment. J Mol Biol 302 (1):205–217. https://doi.org/10.1006/jmbi.2000.4042

44. Surgand JS, Rodrigo J, Kellenberger E, Rognan D (2006) A chemogenomic analysis of the transmembrane binding cavity of human G-protein-coupled receptors. Proteins 62 (2):509–538. https://doi.org/10.1002/prot20768
45. Venkatakrishnan AJ, Deupi X, Lebon G, Tate CG, Schertler GF, Babu MM (2013) Molecular signatures of G-protein-coupled receptors. Nature 494(7436):185–194. https://doi.org/10.1038/nature11896
46. Gloriam DE, Foord SM, Blaney FE, Garland SL (2009) Definition of the G protein-coupled receptor transmembrane bundle binding pocket and calculation of receptor similarities for drug design. J Med Chem 52 (14):4429–4442. https://doi.org/10.1021/jm900319e
47. Costanzi S (2013) Modeling G protein-coupled receptors and their interactions with ligands. Curr Opin Struct Biol 23(2):185–190. https://doi.org/10.1016/j.sbi.2013.01.008
48. Ewing TJ, Makino S, Skillman AG, Kuntz ID (2001) DOCK 4.0: search strategies for automated molecular docking of flexible molecule databases. J Comput Aided Mol Des 15 (5):411–428
49. Friesner RA, Murphy RB, Repasky MP, Frye LL, Greenwood JR, Halgren TA, Sanschagrin PC, Mainz DT (2006) Extra precision glide: docking and scoring incorporating a model of hydrophobic enclosure for protein-ligand complexes. J Med Chem 49(21):6177–6196. https://doi.org/10.1021/jm051256o
50. Venkatachalam CM, Jiang X, Oldfield T, Waldman M (2003) LigandFit: a novel method for shape-directed rapid docking of ligands to protein active sites. J Mol Graph Model 21 (4):289–307
51. Li Y, Han L, Liu Z, Wang R (2014) Comparative assessment of scoring functions on an updated benchmark: 2. Evaluation methods and general results. J Chem Inf Model 54 (6):1717–1736. https://doi.org/10.1021/ci500081m
52. Egloff P, Hillenbrand M, Klenk C, Batyuk A, Heine P, Balada S, Schlinkmann KM, Scott DJ, Schütz M, Plückthun A (2014) Structure of signaling-competent neurotensin receptor 1 obtained by directed evolution in Escherichia coli. Proc Natl Acad Sci U S A 111(6): E655–E662. https://doi.org/10.1073/pnas.1317903111
53. Krumm BE, White JF, Shah P, Grisshammer R (2015) Structural prerequisites for G-protein activation by the neurotensin receptor. Nat Commun 6:7895. https://doi.org/10.1038/ncomms8895
54. Setoh M, Ishii N, Kono M, Miyanohana Y, Shiraishi E, Harasawa T, Ota H, Odani T, Kanzaki N, Aoyama K, Hamada T, Kori M (2014) Discovery of the first potent and orally available agonist of the orphan G-protein-coupled receptor 52. J Med Chem 57 (12):5226–5237. https://doi.org/10.1021/jm5002919
55. Ritscher L, Engemaier E, Stäubert C, Liebscher I, Schmidt P, Hermsdorf T, Römpler H, Schulz A, Schöneberg T (2012) The ligand specificity of the G-protein-coupled receptor GPR34. Biochem J 443(3):841–850. https://doi.org/10.1042/BJ20112090
56. Xiao SH, Reagan JD, Lee PH, Fu A, Schwandner R, Zhao X, Knop J, Beckmann H, Young SW (2008) High throughput screening for orphan and liganded GPCRs. Comb Chem High Throughput Screen 11(3):195–215
57. Diaz C, Labit-Le Bouteiller C, Yvon S, Cambon-Kernëis A, Roasio A, Jamme MF, Aries A, Feuillerat C, Perret E, Guette F, Dieu P, Miloux B, Albène D, Hasel N, Kaghad M, Ferran E, Lupker J, Ferrara P (2013) A strategy combining differential low-throughput screening and virtual screening (DLS-VS) accelerating the discovery of new modulators for the orphan GPR34 receptor. Mol Inform 32 (2):213–229. https://doi.org/10.1002/minf.201200047

INDEX

Alexander Heifetz (ed.), *Computational Methods for GPCR Drug Discovery*, Methods in Molecular Biology, vol. 1705, https://doi.org/10.1007/978-1-4939-7465-8, © Springer Science+Business Media LLC 2018

D

E

F

G

H

I

K

L

M

N

O

P

U

V

W

X

Z

MIX
Papier aus verantwortungsvollen Quellen
Paper from responsible sources
FSC® C105338

If you have any concerns about our products,
you can contact us on
ProductSafety@springernature.com

In case Publisher is established outside the EU,
the EU authorized representative is:
Springer Nature Customer Service Center GmbH
Europaplatz 3, 69115 Heidelberg, Germany

Printed by Libri Plureos GmbH
in Hamburg, Germany